NEUTRON STARS, BLACK HOLES AND BINARY X-RAY SOURCES

ASTROPHYSICS AND SPACE SCIENCE LIBRARY

A SERIES OF BOOKS ON THE RECENT DEVELOPMENTS
OF SPACE SCIENCE AND OF GENERAL GEOPHYSICS AND ASTROPHYSICS
PUBLISHED IN CONNECTION WITH THE JOURNAL
SPACE SCIENCE REVIEWS

VOLUME 48

NEUTRON STARS, BLACK HOLES AND BINARY X-RAY SOURCES

Edited by

HERBERT GURSKY

Center for Astrophysics,
Smithsonian Astrophysical Observatory/Harvard College Observatory,
Cambridge, Mass., U.S.A.

and

REMO RUFFINI

Joseph Henry Laboratories, Princeton University, Princeton, N.J., U.S.A.

D. REIDEL PUBLISHING COMPANY

DORDRECHT-HOLLAND / BOSTON-U.S.A.

Library of Congress Cataloging in Publication Data

Main entry under title:

Neutron stars, black holes, and binary X-ray sources.

(Astrophysics and space science library ; v. 48)
'Articles based on a session of the annual meeting
of the American Association for the Advancement of Science,
held in San Francisco in February 1974.'
Includes bibliographical references.
1. Neutron stars—Congresses. 2. Black holes
(Astronomy)—Congresses. 3. X-ray astronomy—Congresses.
I. Gursky, Herbert. II. Ruffini, Remo. III. American
Association for the Advancement of Science. IV. Series.
QB843.N4N48   523   75–15716
ISBN 90–277–0541–0
ISBN 90–277–0542–9 pbk.

Published by D. Reidel Publishing Company,
P.O. Box 17, Dordrecht, Holland

Sold and distributed in the U.S.A., Canada and Mexico
by D. Reidel Publishing Company, Inc.
306 Dartmouth Street, Boston,
Mass. 02116, U.S.A.

APPENDIX I / CLASSIC PAPERS

APPENDIX II / CONTEMPORARY PAPERS LEADING TO THE DISCOVERY OF GRAVITATIONALLY COLLAPSED STARS

PREFACE

This book contains a set of articles based on a session of the annual meeting of the American Association for the Advancement of Science held in San Francisco in February, 1974. The reason for the meeting arose from the need to communicate to the largest possible scientific community the dramatic advances which have been made in recent years in the understanding of collapsed objects: neutron stars and black holes. Thanks to an unprecedented resonance between X-ray, γ-ray, radio and optical astronomy and important new theoretical developments in relativistic astrophysics, a new deep understanding has been acquired of the physical processes occurring in the late stages of evolution of stars. This knowledge may be one of the greatest conquests of man's understanding of nature in this century.

This book aims to give an essential and up-to-date view in this field. The analysis of the physics and astrophysics of neutron stars and black holes is here attacked from both theoretical and experimental points of view. In the experimental field we range from the reviews and catalogues of galactic X-ray sources (H. Gursky and E. Schreier) and pulsars (E. Groth) to the observations of the optical counterpart of X-ray sources (P. Boynton) to finally the recently discovered gamma-ray bursts (I. Strong) and pulse astronomy R. B. Partridge). In the theoretical field, we range from a fresh overview of the theoretical progress made in the understanding of supernova processes (S. Colgate) to the evolution of binary X-ray sources (R. Kraft) to the analysis of the physics of neutron stars and black holes (R. Ruffini). Each article has been written with the aim of giving an up-to-date and simple presentation of the most important new results together with references to current researches. Extensive bibliographies are given in each article. This should answer to the two different requirements of beginners and advanced researchers in the field. A set of classic papers of the theoretical works on gravitational collapse has been reproduced at the end of the book. Also reproduced is a set of basic articles related to the discovery and the interpretation of neutron stars and black holes. The knowledge of these papers is essential in understanding the few key ideas which have most influenced this entire field. With the kind permission of the Reidel Publishing Company, we have been able to add a comment regarding the discovery of a remarkable radio pulsar in a binary system, along with a reprinting of the discovery paper.

To cover the theoretical background of this field of research, the readers are referred to the books – Ya. B. Zel'dovich and I. Novikov, *Relativistic Astrophysics*, Vol. I (Univ. Chicago Press, Chicago, 1971); M. Rees, R. Ruffini and J. A. Wheeler, *Black Holes, Gravitational Waves and Cosmology* (Gordon and Breach, New York, 1974), C. DeWitt and B. DeWitt, *Black Holes* (Gordon and Breach, New York, 1973), and for the aspects related to the classical relativistic theories to the following text –

L. Landau and I. Lifshitz, *The Classical Theory of Fields* (Addison-Wesley Press, Cambridge, Massachusetts, 1962); S. Weinberg, *Gravitation and Cosmology* (Wiley, New York, 1972); C. Misner, K. Thorne, and J. A. Wheeler, *Gravitation* (W. H. Freeman, San Francisco, 1973).

It is a pleasure to thank Dr Harold Greyber, Director of Meetings of the AAAS for his assistance in arranging the Colloquium on which this book is based; we also wish to thank Ms Diane Jarmac, whose editorial asistance has been invaluable in the preparation of the manuscript.

ACKNOWLEDGEMENTS

S. Chandrasekhar: 'The Highly Collapsed Configuration of a Stellar Mass' *Monthly Notices Roy. Astron. Soc.* **91**, 456 (1931). Reprinted with permission from the *Monthly Notices Roy. Astron. Soc.* **91**, Copyright, Blackwell Scientific Publications, Ltd., 1931.

L. Landau: 'On the Theory of Stars', *Phys. Z. Soviet* **1**, 285 (1932). Reprinted with permission from *Phys. Z. Soviet* **1**, Copyright, 1932.

W. Baade and F. Zwicky: 'Supernovae and Cosmic Rays' (Abstract), *Phys. Rev.* **45**, 138 (1934). Reprinted with permission from *Phys. Rev.* **45**, Copyright, American Physical Society, 1934.

J. R. Oppenheimer and G. M. Volkoff: 'On Massive Neutron Cores', *Phys. Rev.* **55**, 374 (1939). Reprinted with permission from *Phys. Rev.* **55**, Copyright, American Physical Society, 1939.

J. R. Oppenheimer and H. Snyder: 'On Continued Gravitational Collapse', *Phys. Rev.* **56**, 455 (1939). Reprinted with permission from *Phys. Rev.* **56**, Copyright, American Physical Society, 1939.

S. Chandrasekhar: 'Some Remarks on the State of Matter in the Interior of Stars', *Z. Astrophys.* **5**, 321 (1932) Reprinted from *Z. Astrophys.* **5**.

J. Weber: 'Detection and Generation of Gravitational Waves', *Phys. Rev.* **117**, 306 (1960). Reprinted with permission from *Phys. Rev.* **117**, Copyright, American Physical Society, 1960.

R. Giacconi, H. Gursky, F. R. Paolini, and B. B. Rossi: 'Evidence for X-Rays from Sources Outside the Solar System', *Phys. Rev. Letters* **9**, 439 (1962). Reprinted with permission from *Phys. Rev. Letters* **9**, Copyright, American Institute of Physics, Inc., 1962.

Ya. B. Zel'dovich: 'The Fate of a Star and the Evolution of Gravitational Energy Upon Accretion', *Soviet Phys.–Dokl.* **9**, 195 (1964). Reprinted with permission from *Soviet Phys.–Dokl.* **9**, Copyright, American Institute of Physics, Inc., 1964.

I. S. Shklovskii: 'The Nature of the X-Ray Source Sco X-1', *Soviet Astron.–AJ* **11**, 749 (1968). Reprinted with permission from *Soviet Astron.–AJ* **11**, Copyright, American Institute of Physics, Inc., 1968.

A. Hewish, S. J. Bell, J. D. Pilkington, P. F. Scott, and R. A. Collins: 'Observation of a Rapidly Pulsating Radio Source', *Nature* **217**, 709 (1968). Reprinted with permission from *Nature* **217**, Copyright, Macmillan Journals, Ltd., 1968.

T. Gold: 'Rotating Neutron Stars as the Origin of the Pulsating Radio Sources', *Nature* **218**, 731 (1968). Reprinted with permission from *Nature* **218**, Copyright, Macmillan Journals, Ltd.,

R. Penrose: 'Gravitational Collapse: The Role of General Relativity', *Riv. Nuovo*

Cim., Numero Speciale **1**, 252 (1969). Reprinted with permission from *Riv. Nuovo Cim., Numero Speciale* **1**, Copyright, Societa Italiana di Fisica, 1969.

R. Ruffini and J. A. Wheeler: 'Introducing the Black Hole', *Phys. Today* **24**, 30 (1971). Reprinted with permission from *Phys. Today* **24**, Copyright, American Institute of Physics, Inc., 1971.

V. F. Shvartsman: 'Halos around Black Holes', *Soviet Astron.–AJ* **15** 377 (1971). Reprinted with permission from *Soviet Astron.–AJ* **15**, Copyright, American Institute of Physics, Inc., 1971.

N. I. Shakura: 'Disk Model of Gas Accretion on a Relativistic Star in a Close Binary System', *Soviet Astron.–AJ* **16**, 756 (1973). Reprinted with permission from *Soviet Astron.–AJ* **16**, Copyright, American Institute of Physics, Inc., 1973.

E. Schreier, R. Levinson, H. Gursky, E. Kellogg, H. Tananbaum, and R. Giacconi: 'Evidence for the Binary Nature of Centaurus X-3 from UHURU X-ray Observations', *Astrophys. J.* **172**, L79 (1972). Reprinted with permission from *Astrophys. J.* **172**, Copyright, The University of Chicago Press, 1972.

C. E. Rhoades and R. Ruffini: 'Maximum Mass of a Neutron Star', *Phys. Rev. Letters* **32**, 324 (1974). Reprinted with permission from *Phys. Rev. Letters*, **32**, Copyright, American Institute of Physics, Inc., 1974.

R. A. Hulse and J. H. Taylor: 'Discovery of a Pulsar in a Binary System', *Astrophys. J.* **195**, L51 (1975). Reprinted with permission from *Astrophys. J.* **195**. Copyright, The American Astronomical Society, 1975.

INTRODUCTION

HERBERT GURSKY and REMO RUFFINI

In this century, a revolution swept through physics and astronomy. Quantum mechanics, elementary particle physics, special and general relativity have drastically modified our traditional concepts of matter, motion and space time structures. Similar advances have been made in astronomy and astrophysics. For the first time in human history, we have solid information on the nature of stars and the structure of the Universe. This revolution is continuing and we present here one of its latest chapters: the formation, appearance and evolution of neutron stars and black holes as we see them in our Galaxy.

These objects have a dual role in the natural sciences. For one, they represent stars in their most condensed state, formed by the process of gravitational collapse and in which the laws of quantum mechanics and relativity dominate over all classical descriptions. Neutron stars and black holes can then be considered an ideal testing ground to study quantum mechanics and general relativity on a large scale in much the same way as high energy elementary particles have been used to probe physical laws on a small scale.

In their second role, these objects are stars in the classical sense of the word, in that they are massive bodies held together by their own gravitational fields and have evolved to their present condition from other, more 'normal' stars. They release enormous amounts of energy, of the order of 10% and perhaps as high as 40% of their total mass energy. This is to be compared to the less than 1% liberated by nuclear burning during their entire previous existence. Unlike the thermal radiation from a stellar surface, the energy emission from neutron stars and black holes is complex and varied. Some of this energy originates in a single, cataclysmic event, the supernova explosion which signals their birth. Some of it emerges in the form of high energy photons – principally X-rays – and much of it may appear in the form of cosmic rays.

Progress in this field has occurred because of the interplay of observations over the whole range of modern astronomy. X- and γ-ray astronomy, radio and infrared astronomy, cosmic ray investigations, and ordinary ground-based optical observations have all contributed to the knowledge we have of these objects. However all of this would have been a simple catalogue of facts without the solid foundation and understanding of the fundamental theoretical physics involved in these objects. In contrast to the key observations which emerged in the past decade, the various theoretical ideas were distinctly enunciated beginning almost half a century ago. In this Introduction, we attempt to trace the origin of the most fundamental concepts in the field and also the primary observations.

* The references in this Introduction which appear with an asterisk (e.g. Chandrasekhar, 1931*) are reprinted in the Appendix.

1. Early History

At the beginning of this century, with the aid of large telescopes, spectrometers and photography, astronomers were able to study the physical characteristics of individual stars and to construct models of their internal structure. Stars were found to be large gaseous bodies in which gravitational forces held together matter against the ordinary pressure of gas and radiation. The material, very hot and highly ionized, was described by classical concepts of thermodynamics and radiation physics. This view has proven to be essentially correct, at least for the early portion of a star's life, and with the addition of energy generation by nuclear reactions is the basis for present theories of stellar structure and evolution.

However, even the earliest observations indicated that conditions existed which could not be simply accounted for by this classical approach. One of the first versions of the Hertzsprung-Russell diagram as prepared by Russell (1914) in which the main sequence is clearly outlined, also revealed that the single star, 40 Eridanus B, had an unaccountably low luminosity for its color.

Stars of this kind came to be called white dwarfs, 'white' because of their color temperature, and 'dwarf' because they were substantially smaller than other known stars. The astounding characteristic of these objects was their density, greater by many order of magnitude than that of others stars or of any known material, at least $10^5 \mathrm{~g~cm}^{-3}$.

The observational data were irrefutable. For a few of these objects, such as the white dwarf companion of Sirius, the distance was accurately measured from its parallax. Thus, one knows its absolute luminosity which then relates to ist radius and surface temperature T simply as

$$L = 4\pi R^2 \sigma T^4,$$

where T is found from the spectrum of the stellar radiation. The mass of the white dwarf may be found from an analysis of the orbital elements of the binary system, and the density follows directly.

The nature of these objects was inexplicable until more progress was made in quantum mechanics. Classically such high densities implied the existence of such high internal temperatures that these stars should have been much brighter than actually observed. During the early 1920's, Bose, Einstein, Fermi and Dirac worked out the fundamental, statistical properties of matter based on the new quantum mechanics and the application of the Pauli Exclusion principle. In the same year that Fermi (1926) developed his quantum statistics, R. N. Fowler (1926) proposed that these new concepts should be applied in treating the high density regime found in white dwarfs. Fowler's ideas were applied and developed further by Milne, Stoner and Chandrasekhar, and by about 1930, the theory of the fundamental properties of white dwarfs had been developed. Thus, within a decade, astronomers and physicists had discovered and explained a new state of matter that could only be understood in terms of twentieth century concepts. In dwarfs, the electrons, already decoupled from the nuclei, are as densely packed as permitted by the Pauli exclusion principle, a condition known as degeneracy. Under these conditions the electron gas is in its lowest energy state, and can not radiate photons; the material is effectively very cold in spite of its high internal energy. With the discovery of the neutron, it became apparent that one could also have a star built of a degenerate neutron gas. However, the density of such a system would be orders of magnitude greater than is the case for a white dwarf. Such a configuration was explicitly described by Gamow in 1936.

At about the same time, a most remarkable theoretical discovery was made. Chandrasekhar (1930, 1931*, 1932*) and Landau (1932)* found that as a consequence of a special relativistic effect the rate of change of pressure with density would decrease at very high densities and a maximum stable mass existed for stars which have exhausted their internal sources of energy. (For an historical recollection of the steps which led to this fundamental discovery, see Chandrasekhar, 1969, 1972.) Landau, in particular, stated that the addition of matter over this critical value would lead to the collapse of the star without limit because of the ever increasing gravitational forces.

2. The Basic Theoretical Works on Gravitational Collapse

We can here summarize the main points of the reasoning that led to a mass limit for cold stars and to a prediction of their eventual collapse. We shall use a description of a star which follows closely the lines of the Thomas-Fermi treatment of an atom. In this model of the atom the electrons are described by a degenerate Fermi gas constrained to a finite volume by the Coulomb attraction of the nucleus.

At the endpoint of thermonuclear evolution, a star is expected to be composed either of nuclei embedded in a gas of electrons (all the orbitals having been destroyed by the great compression of the material) as in the case of a white dwarf, $\varrho \sim 10^5-10^8$ gm cm^{-3}, or of a gas of neutrons, protons and electrons in beta equilibrium, a neutron star with $\varrho \sim 10^{14}-10^{16}$ gm cm^{-3} (matter having undergone neutronization under the effect of pressure). In both cases, the material of the star may be described as a degenerate Fermi gas constrained to a finite volume by the self gravitation of the system. The pressure required to prevent the system from collapsing is provided, in both cases, by the quantum pressure due to the Pauli exclusion principle. In the case of the white dwarfs, the main contribution to the pressure comes from the gas of degenerate electrons while the nuclei mainly contribute to the density distribution; in the case of neutron stars, both pressure and density are generated by the gas of neutrons.

The main formulae describing these states can be obtained from the theory of Fermi-Dirac statistics (cf. Ruffini, 1973).

For a degenerate Fermi gas the density of particles is simply given by

$$n = \frac{N}{V} = \frac{8\pi}{h^3} \int_0^{p_F} p^2 \, \mathrm{d}p$$

and the Fermi momentum

$$p_F = (3\pi^2 \hbar^3 N/V)^{1/3}.$$

The kinetic energy of a particle is given by

$$\varepsilon = p^2/2m.$$

The total energy is simply

$$E = \int_0^{p_F} V \frac{4\pi}{\hbar^3} \frac{p^4}{m} \, \mathrm{d}p$$

and the pressure

$$P = -\partial E/\partial V = (3\pi^2)^{2/3} (\hbar^2/5m) (N/V)^{5/3}. \tag{2}$$

If, however, the kinetic energy at the top of the Fermi sea becomes comparable to the rest energy of the particle, then the particle energy is no longer given by Equation (1) but by the usual relativistic relation

$$\varepsilon = c(p^2 + m^2 c^2)^{1/2}$$

or in the extreme relativistic regime by

$$\varepsilon = cp. \tag{3}$$

The pressure is then in this limit given by

$$P = -\partial E/\partial V = (3\pi^2)^{1/3} (\hbar c/4) (N/V)^{4/3}. \tag{4}$$

The increase of pressure corresponding to an increase in density is clearly much smaller in Equation (4) than in Equation (2). To have clearly understood and expressed this relativistic softening of the equation of state and to have uncovered the essential role the transition to a relativistic regime plays in the physics of a degenerate star has been the basic contribution of S. Chandrasekhar and L. Landau. In turn their work led to the revolutionary concept of a critical mass against gravitational collapse.

There exists therefore a fundamental difference between the theoretical description of an atom and that of a degenerate star. Never do the electrons in an atom reach an energy sufficiently relativistic as to be in the regime described by Equation (4). Their pressure, given by Equation (2), reacts to a density change in such a way that it always balances the attractive effects of the electromagnetic field of the nucleus. On the contrary, in a star at the end-point of thermonuclear evolution, the conditions under which Equation (4) applies are always reached if the star is massive enough. This is a direct consequence of the non-screening and long range character of gravitational interactions. The more massive the star is, the more it contracts, and the more the spatial volume of the phase space occupied by the system is reduced and correspondingly the volume in the momentum space expanded. Hence, the Fermi momentum of the system can always reach relativistic regimes. In the transition from Equation (2) to Equation (4) the pressure dependence on the density softens considerably; the system is not able to support itself up in equilibrium and thus gravitationally collapses. As Landau said, the star should shrink down to a point practically in free fall unless some new physical phenomena intervened.

Following these pioneering works the systematic analysis and detailed computations of the configurations of equilibrium for white dwarfs were presented by Chandrasekhar (1935) and those for neutron stars by Oppenheimer and Volkoff (1938)*. The value of the critical mass for white dwarfs was found to be $M_{crit} \sim 1.39\ M_\odot$, the one for neutron stars $M_{crit} \sim 0.7\ M_\odot$.

Then Oppenheimer and Snyder (1939)* focused attention on the process of gravitational collapse itself. For the first time, it became evident that this phenomenon is of basic importance for our understanding of the nature of space and time. They showed that the phenomenon of gravitational collapse with its process of time dilatation, light deflection, and gravitational redshifts, is a unique place where a fully relativistic theory of gravitation could be seen at work and the validity of general relativity may be confronted with experimental evidence. The work of Oppenheimer and Snyder appears today as one of the cornerstones of relativistic astrophysics and one of the most profound works ever written in gravitation physics. The authors not only give the basis for the entire field of what will be later known as 'black hole' physics, but they also give a fully analytic solution of the collapse of a cloud of dust (zero pressure or free fall, which is a very good approximation to a realistic gravitational collapse) with a detailed analysis of 'self-closure' as seen by a distant observer as a consequence of the general relativistic effects.

Still more time was needed to realize that the process of gravitational collapse is the ideal place to generate a theoretically detectable amount of gravitational radiation. in The first days of general relativity, Einstein himself had shown that gravitational fields, like any other relativistic mass zero field, must propagate with the speed of light and that energy can be carried by gravitational waves. It was not conceived, however, even by a gedanken process how gravitational radiation could be detected. The key to this problem, a fundamental contribution, was made by Weber (1959, 1960*) and by Bondi (1959). Weber (1961) presented not only the main principles on which the detectability of gravitational radiation should be based, but also the technical details on the ground of which a detector of gravitational radiation could be built. The relevance of Weber's work, quite apart from its impact on experimental techniques, has been also very large in generating much thinking on astrophysical processes to produce detectable amounts of gravitational radiation.

3. The Discovery of Neutron Stars and Black Holes

As with the theoretical concepts in this field, the important observations also started early in this century with the discovery of inexplicable effects – the supernovae. They had been observed in ancient times but it was only with the establishing of the stellar and galactic distance scales that their true enormity was realized; namely, the release of $\sim 10^{50}$ erg within a matter of days. It was Baade and Zwicky (1934)* who first suggested that the supernova process was the result of the transition from a normal star to a neutron star; the essential point (Zwicky, 1939) being that the energy release in such a process is comparable to the change in gravitational potential energy of a star which collapses from its 'normal' size of 10^6 km down to the size of a neutron star of the order of 10 km.

In the 1950's, work on stellar nucleosynthesis (cf. Burbidge *et al.*, 1957) led to physically realistic models of stars prior to supernova explosions. The supernova process was seen as the result of a catastrophic change of state occurring in the core of a highly evolved star; one possibility being the transformation of an iron core to a helium core. Cameron (1958) instead suggested that this degenerate iron core would collapse to a neutron core through inverse beta decay. In this model, the resulting implosion would blow off the outer envelope of the star and leave behind the core as a neutron star. Thus, the 20 year old idea of Baade and Zwicky received at least a general qualitative explanation. Colgate's contribution in this volume describes our present state of knowledge regarding the late stages of a star's life, its collapse and the ensuing explosion.

Meanwhile, the conditions of matter under extremely high pressures and the conditions following collapse were reexamined by Ambarzumian, Cameron, Wheeler and their students and collaborators (cf. Ambarzumian and Saakyan, 1961; Cameron, 1959; Harrison *et al.*, 1958). The major emphasis in their work was a critical examination of the assumptions made in the classical works by Oppenheimer and his students. Alternative models of neutron stars were advanced, the emphasis being directed at times toward a better treatment of the surface layers, and at other times toward the understanding of the role played by nuclear interactions in the treatment of the core. Different models lead to different ranges of masses, radii and density distributions for the equilibrium configurations of neutron stars. This work was accelerated when the first X-ray sources were discovered by Giacconi *et al.* (1962)*. It was believed that neutron stars could radiate large fluxes of X-rays, possibly as a consequence of very high surface temperature. However, subsequent observational effort yielded no positive evidence for the existence of neutron stars.

Many believed that it might be impossible to actually detect neutron stars or black holes. Since no internal energy sources existed in these objects, it was believed that they would be invisible for all practical purposes. One possible idea pointing to their observation was that they might be powerful radiation sources because of accretion of matter – either from the interstellar medium or from a neighboring star. This idea was first advanced to account for the powerful radiogalaxies and quasars, and both Salpeter (1964) and Zel'dovich (1964)* clearly stated how the energy release could come about. Another idea that was pursued was to exploit their invisibility and to search for their presence as an unseen, massive companion in a binary system (Zel'dovich and Guseynov, 1965).

Great impetus was given to the field when Scorpius X-1, the brightest X-ray source in the sky, was identified optically as a faint, blue, star-like object (Sandage *et al.*,

(1966). The similarity of this to the old novae lead to conjectures that the X-rays were being produced by accretion onto a white dwarf or a collapsed object in a binary system. (For a review of the many discussions on the binary nature of X-ray sources that emerged at the time of the optical discovery of Sco X-1, see Burbidge, 1972). Ginzburg (1967) emphasized the relevance of gravitational energy for the explanation of the energy source of Sco X-1 ("We have such a large amount of gravitational energy in such a binary system; we must use it! of course!"). Zel'dovich and Novikov (1966) pointed out the possibility of producing X-rays by accretion into a collapsed object "The baryon or hardened star (later called 'Black Holes') can be a component of a double star.... In a double system the accretion on the collapsed star is fed by the stellar wind from the normal star. It shall lead to strong X-ray and γ-ray emission. In the case of accretion the falling gas is heated by a stationary shock wave and radiates like an optically thin hot layer (bremsstrahlung) instead of the black body radiation". Shklovsky (1968*) gave a detailed picture of Scorpius X-1 as a close binary system in which one of the members is a neutron star and the other member a cool dwarf star. "A stream of gas glowing out of the second component is permanently incident on the neutron star," this accreting material, enormously compressed, should then emit X-rays; "The optical object accompanying the X-ray source might be a cool dwarf star with half of its surface heated by a strong flux of hard X-rays from the source". As we shall see, ideas presented in these articles became a few years later the key for the understanding of the binary X-ray sources.

Suddenly, however, the observational situation changed in an entirely different direction. In 1967, a new type of radio telescope intended for the study of interplanetary scintillation came into operation in Cambridge, England under the direction of A. Hewish. For this purpose, the telescope was designed with two features not particularly common to such installations – operation at low frequency and with high time resolution. Within a month, a radio source was detected whose emission was not steady but rather pulsed with a precise period of 1.33 s. This object, known as CP1919, was the first pulsar to be discovered. In their initial paper, Hewish *et al.* (1967)* suggested that this phenomena was associated with a compact star, possibly the radial pulsations of a white dwarf or a neutron star.

The idea that these objects where rotating magnetized neutron stars was advanced among others, by Gold (1968)* and by Pacini (1968), and within a short time, observational evidence was found to confirm this view. Brief, sporadic pulses were detected from the direction of the Crab Nebula by Staelin and Reifenstein (1968) and a pulsar with a period of ~ 33 ms was identified as their source by Comella *et al.* (1969). The lengthening of the pulse that had been suggested by Gold was also soon discovered, with a fractional change of $\sim 1:240$ per year (Richards and Comella, 1969). Ironically, the pulsar was the same star studied intensively by Baade and Minkowski who believed it to be the remnant of the SN explosion.

As shown by Gold (1969) these observations provided very compelling evidence that one was dealing with a rotating neutron star. In the first place, only a neutron star among all known stellar configuration could be stable when rotating (or vibrating) with such a high frequency. Secondly, the pulse period was observed only to decrease (slow down) as expected from a rotating object which was losing rotational energy. A vibrating object would tend to increase its frequency. Finally, the slow down rate was consistent with a loss of energy of $\sim 10^{38}$ erg s^{-1} if one assumed the object to be a neutron star of ~ 1 solar mass. This amount of energy is comparable to the radiated energy from the entire Crab Nebula.

Thus, with this single object, a variety of astronomical problems were clarified. First, it provided strong evidence for the existence of neutron stars; second, it provided a simple mechanism for the observed energy generation in the Crab Nebula; and third, it provided confirmation of the ideas of nucleosynthesis that tied together the formation of neutron stars with supernova explosions. The whole subject of pulsars is reviewed in this volume by Groth.

One of the important by-products of the discovery of pulsars has been that it dramatically brought back to the attention of a large section of theoretical and experimental astrophysicists the basic issues of the physics of gravitationally collapsed objects. The clear evidence for the existence of neutron stars, together with the existence of a critical mass against gravitational collapse (cf. Oppenheimer and Snyder, 1938; Wheeler, 1963) made the existence of 'black holes' inside our Galaxy inescapable. On the other side the theoretical picture to be expected at this ultimate endpoint of gravitational collapse appeared more and more complex and presented fundamental issues on the physical processes occurring in these ultrarelativistic regimes (cf. Penrose, 1969*; Ruffini and Wheeler, 1971*).

From all this naturally developed the need for launching a realistic effort in order to detect and differentiate the two families of collapsed objects: neutron stars and black holes. Ruffini and Wheeler (1971)* emphasized the importance of capitalizing on the Shklovsky accretion picture not only in the case when a neutron star is a member of a close binary system, but also in the case where a black hole attracts and pulls material from the companion star. The short time variability of the X-ray source and the statistical analysis involved would then be a most powerful tool for the determination of the 'form factor' of a black hole. Meanwhile, Zel'dovich and Shakura (1969) had shown that spherical accretion onto a neutron star would result in a hot, X-ray emitting plasma in the vicinity of the star, and Shakura (1973)* showed that accretion onto a rotating neutron star would lead to the formation of a disk from which one could expect copious X-ray emission. Shakura and Sunyaev (1973) and Pringle and Rees (1972) developed this mode, especially from the point of view of predicting some detailed observational features of the X-ray sources. These works already considered classic papers have provided the theoretical basis now being used to unravel many of the X-ray and optical observations. Shvartsman (1971)* developed the picture of spherical accretion onto a black hole. Novikov and Thorne (1973) have later analyzed a disclike structure around a black hole. An important theoretical point that developed during this time was the realization that significant energy release could take place in the vicinity of a black hole, contrary to the earlier notion of a black hole as an inert object from which no radiation could emerge. In particular, Shvartsman pointed out that at least $\sim 10\%$ of the rest mass of infalling material could reemerge as radiation simply as a result of falling through the gravitational potential up to the Schwarzchild radius. Even more striking was the result that rotational and electromagnetic energy could comprise up to 50% of the total mass energy of a black hole which could be augmented or depleted at will (Christodoulou and Ruffini, 1971).

In late 1970, the observational situation in X-ray astronomy improved considerably with the launch of NASA's satellite, UHURU, a project which was conceived and directed by R. Giacconi, while at American Science and Engineering in Cambridge, Massachusetts. Whereas almost all previous results in X-ray astronomy were derived from very brief sounding rocket or balloon flights, UHURU provided for continuous observations of X-ray sources with good sensitivity, time resolution and angular resolution.

One of the first objects to be studied by UHURU was Cygnus X-1. This source apparently was seen during the discovery observations in 1962 (Gursky *et al.*, 1963) and was the first source found to be variable (Bowyer *et. al*, 1965). In 1966, following the discovery of the optical counterpart of Sco X-1, the AS&E group surveyed the Cygnus region with the specific intent of locating sources with sufficient precision to search for optical and radio candidates. The search was apparently successful in the case of Cyg X-2, but not successful in the case of Cyg X-1 (Giacconi *et al.*, 1967).

The UHURU data on Cyg X-1 revealed that its X-ray emission was highly variable on a very short time scale. In many instances, the emission appeared to be organized in single pulsations or bursts with a suggestion of regularity (Oda *et al.*, 1971). The conclusion, particularly stressed by Giacconi, was that possibly one was dealing with a periodicity comparable to or less than the UHURU time resolution of 0.1 s in turn requiring an object smaller than a white dwarf. Since there was no evidence of a supernova explosion, it was argued that this was not a neutron star, but rather a black hole. These arguments are now known to be naive; nevertheless, the following chain of reasoning and subsequent observations lead to the conclusion that Cyg X-1 is a black hole.

(1) The object is a binary system with a massive B-star companion.

(2) The energetics of the source is simply explained by the release of gravitational energy into kinetic and thermal energy during the process of mass accretion.

(3) Analysis of the binary elements leads to a lower limit of $\sim 6\ M_\odot$ for Cyg X-1, well above the absolute upper limit for the mass of a neutron star (Rhoades and Ruffini, 1974)*.

(4) The short time scale variability for the X-ray emission requires a compact source.

There are several other X-ray sources with characteristics similar to those of Cyg X-1 although with less stringent limits on the masses. Also, the X-ray emission of two X-ray binaries, Hercules X-1 and Centaurus X-3, are found to be regularly pulsing with periods of 1.2 s and 4.8 s, respectively, which is obviously reminiscent of the radio pulsars. Centaurus X-3 was the first X-ray source shown definitely to be a binary system based on the observation of the Doppler shift of the 4.8 s pulse period and of regular eclipses (Schreier *et al.*, 1972)*. The contributions of Gursky and Schreier and of Boynton in this volume present our present knowledge of these and other X-ray sources.

The feedback of these various observations in the theoretical field has been very impressive. Essential to the understanding of these binary systems is a knowledge of the process of gravitational collapse and the means by which these systems are formed (see Ruffini, this volume). To understand the process of X-ray emission, a fully relativistic magnetohydro-dynamics has to be developed, together with more classical analysis, of disc-like structures around the collapsed object in equilibrium under the two opposite effects of gravitational attraction and centrifugal forces. The physics of binary star systems is essential in reaching an understanding of the evolution of these binary systems. For the massive binary X-ray sources, such as Cygnus X-1 and Cen X-3, a consistent evolutionary picture seems available (Van den Heuvel and Heise, 1972). Our present knowledge of the evolution of binary stars leading to X-ray sources is presented in this volume by Kraft. An accurate analysis of the gravitational and electromagnetic structure of the collapsed objects is essential for understanding the 'core' of these X-ray sources. Finally, criteria to differentiate between the two classes of collapsed objects are much needed.

The great progress currently being made in all these theoretical fields together with the possibility of having satellites with X- and γ-ray instruments with larger collecting area and better time resolution, as well as large X-ray telescopes, possibly as part of orbiting national facililities similar in scale to radio and optical national facilities, make one hope that in the next five or ten years, a complete understanding of the final configuration of gravitationally collapsed stars will be gained.

If we turn from this program of research to be achieved in the next few years to a much larger program on longer time scales, two different directions of research naturally present themselves. The first is clearly to try to observe directly the transition from a normal star into a collapsed object by observing 'the moment of gravitational collapse'. This program points to the need for developing an entire new branch of theoretical and experimental astrophysics, what we may call 'pulse astronomy' as discussed by Partridge in this volume. In this sense the recent results on γ-bursts (described by Strong in this volume), the preliminary work of K. Lande and colleagues at the University of Torino (Lande *et al.*, 1974) and the large effort being made in building a new more sensitive set of gravitational wave detectors at the universities of Louisiana, Rome and Stanford (Fairbanks, 1974) can only be viewed with the greatest expectation.

The second direction points toward even a larger program of capitalizing on what we are learning of the physics of collapsed objects to gain a deeper knowledge of the fundamental field equations governing physical processes. It is conceivable that the analysis of these fully relativistic processes will be of great help in understanding of other apparently disconnected fields of physics, like e.g., elementary particle physics.

References

Ambarzumian, V. A. and Saakyan, G. S.: 1971, *Soviet Astron.* **6**, 601.
Baade, W. and Zwicky, F.: 1934, *Phys. Rev.* **45**, 138.
Bondi, H.: 1959, Royamount Conference, Royamount, France.
Bowyer, C. S., Byram, E. T., Chubb, T. A., and Friedman, H.: 1965, *Science* **147**, 394.
Burbidge, E. M., Burbidge, G., Fowler, W. A., and Hoyle, F.: 1957, *Rev. Mod. Phys.* **29**, 547.
Cameron, A. G. W.: 1958, *Mem. Soc. Roy. Sci. Liège, 5th Ser.* **3**, 163.
Cameron, A. G. W.: 1959, *Astrophys. J.* **130**, 884.
Chandrasekhar, S.: 1930, *Astrophys. J.* **74**, 81.
Chandrasekhar, S.: 1931, *Monthly Notices Roy. Astron. Soc.* **91**, 456.
Chandrasekhar, S.: 1932, *Z. Astrophys.* **5**, 321.
Chandrasekhar, S.: 1935, *Monthly Notices Roy. Astron. Soc.* **95**, 207.
Chandrasekhar, S.: 1969, *Am. J. Phys.* **37**, 577.
Chandrasekhar, S.: 1972, *Observatory* **92**, 160.
Christodoulou, D. and Ruffini, R.: 1971, *Phys. Rev.* **4**, 3442.
Comella, J. W., Craft, H. D., Jr., Lovelace, R. V. E., Sutton, J. M., and Tyler, G. L.: 1969, in *Pulsating Stars* **2**, Plenum Press, New York, p. 2.
Fairbanks, W.: 1974, in B. Bertotti (ed.), *Proc. Varenna Summer School*, Academic Press, New York.
Fermi, E.: 1926, *Z. Phys.* **36**, 902.
Fowler, R. N.: 1926, *Monthly Notices Roy. Astron. Soc.* **87**, 117.
Gamow, G.: 1936, *Structure of Atomic Nuclei and Nuclear Transformations*, Oxford, Univ. Press.
Giacconi, R., Gursky, H., Paolini, F. R., and Rossi, B. R.: 1962, *Phys. Rev. Letters* **9**, 439.
Giacconi, R., Gorenstein, P., Gursky, H., Usher, P. D., Waters, J. R., Sandage, A., Osmer, P., and Peach, J.: 1967, *Astrophys. J.* **148**, L119.
Ginzburg, V.: 1967, in H. van Woerden (ed.), 'Radio Astronomy and the Galactic System', *IAU Symp.* **31**, 411.
Gold, T.: 1968, *Nature* **218**, 731.
Gold, T.: 1969, *Nature* **221**, 27.
Gursky, H., Paolini, F. R., Giacconi, R., and Rossi, B. R., 1963, *Phys. Rev. Letters* **11**, 524.
Harrison, B. K., Wakano, M., and Wheeler, J. A.: 1958, *La Structure et l'évolution de l'universe*, Stoops, Brussels; also, Harrison, B. K., Wakano, M., Wheeler, J. A., and Thorne, K., 1964.
Hewish, A., Bell, S. J., Pilkington, J. P. H., Scott, P. F., and Collins, R. A.: 1967, *Nature* **217**, 709.
Landau, L. D.: 1932, *Phys. Z. Sow.* **1**, 285.
Lande, K., Bozoki, G., Frati, W., Lee, C. K., Fenyves, E., and Saavedra, O.: 1974, *Nature Phys. Sci.* **251**, 485.
Novikov, I. and Thorne, K.: 1973, in B. DeWitt and C. DeWitt (eds.), *Black Holes*, Gordon and Breach, New York, p. 343.
Oda, M., Gorenstein, P., Gursky, H., Kellogg, E., Schreier, E., Tananbaum, H., and Giacconi, R.: 1971, *Astrophys. J. Letters* **166**, L1.
Oppenheimer, R. J. and Volkoff, G. M.: 1938, *Phys. Rev.* **55**, 374.
Oppenheimer, R. J. and Snyder, R.: 1939, *Phys. Rev.* **56**, 455.
Pacini, F.: 1968, *Nature* **219**, 145.
Penrose, R.: 1969, *Riv. Nuovo Cimento*, Num. Spec. I.
Pringle, J. E. and Rees, M.: 1972, *Astron. Astrophys.* **21**, 1.
Rhoades, C. E. and Ruffini, R.: 1974, *Phys. Rev. Letters* **32**, 324; also Rhoades, C. E.: 1971, Ph.D. Dissertation, Physics Department, Princeton, University, University Microfilms, Michigan.
Richards, D. W. and Comella, J. W.: 1969, *Nature* **222**, 551.
Ruffini, R.: 1973, in B. DeWitt and C. DeWitt (eds.), *Black Holes, Les Houches*, 1972, Gordon and Breach, New York, p. 451.
Ruffini, R. and Wheeler, J. A.: 1971, in H. V. Hardy (ed.), *Cosmology from Space Platforms*, ESRO SP52, Paris.
Russell, H. N.: 1914, *Pop. Astron.* **22**, 285.
Salpeter, E.: 1964, *Astrophys. J.* **140**, 796.
Sandage, A. R., Osmer, P., Giacconi, R., Gorenstein, P., Gursky, H., Waters, J. R., Bradt, H., Garmire, G., Sreekantan, B. V., Oda, M., Osawa, K., and Jukago, J.: 1966, *Astrophys. J.* **146**, 316.

Schreier, E., Levinson, P., Gursky, H., Kellogg, E., Tananbaum, H., and Giacconi, R.: 1972, *Astrophys. J.* **172**, L79.
Shakura, N. I.: 1973, *Soviet Astron.* **16**, 756.
Shakura, N. I. and Sunyaev, R.: 1973, in H. Bradt and R. Giacconi (eds.), X-and Gamma-Ray Astronomy', *IAU Symp.* **55**, 143.
Shvartsman, V. F.: 1971, *Soviet Astron.* **15**, 377.
Shklovskii, I. S.: 1968, *Soviet Astron.* **11**, 749.
Staelin, D. H. and Reifenstein, E. C.: 1968, *Science* **162**, 1481.
Van den Heuvel, E. and Heise, J.: 1972, *Nature Phys. Sci.* **239**, 67.
Weber, J.: 1959, Royamount Conference, Royamount, France.
Weber, J.: 1960, *Phys. Rev.* **117**, 306.
Weber, J.: 1961, *General Relativity and Gravitational Waves*, Interscience Publ. Co., New York.
Wheeler, J. A.: 1963, in H. Y. Chiu and W. F. Hoffman (eds.), *Gravitation and Relativity*, W. A. Benjamin, Inc., New York, pp. 195ff.
Zel'dovich, Ya. B.: 1964, *Soviet Phys. Dokl.* **9**, 246.
Zel'dovich, Ya. B.: 1965, *Astrophys. J.*
Zel'dovich, Ya. B.: 1966, *Supp. Nuovo Cimento* **IV**, 810.
Zel'dovich, Ya. B. and Shakura, N. I.: 1969, *Soviet Astron.* **13**, 175.
Zwicky, F.: 1939, *Phys. Rev.* **55**, 726.

SUPERNOVAE

STIRLING A. COLGATE

New Mexico Institute of Mining and Technology, Socorro, N.M. 87801, U.S.A.

1. Introduction

Supernovae are the explosion of a star. This awesome and exciting realization came about at the turn of the century from the following sequence of observations. The sudden flaring up of a star had been recognized many times by astronomers and had been called a nova, but then the pioneering work of Hubble indicated that galaxies or associations of stars of hundreds of billions at a time, called galaxies, populated our Universe. When a star brightens up in a distant galaxy to a luminosity that is comparable to all the stars in the galaxy, then this phenomenon came to be called a supernova. The color temperature of these objects and the fact that the light curve rose to maximum in something like a week are sufficient alone to deduce that something like a solar mass of matter must have expanded at a velocity corresponding to energies of several MeV nucleon^{-1} and therefore represented a kinetic energy greater than any binding energy of stars then known. Since the violence of the explosion is so great, it was natural to consider exotic mechanisms for its origin. Fritz Zwicky in his pioneering work on supernovae even made the wild suggestion that the formation of a neutron star could be the source of the energy and indeed I believe he was correct because now we have seen neutron stars (pulsars) at the center of the remnants of supernovae – the Crab and the Vela. In the past, astrophysicists had to grasp almost at science-fiction to find an explanation for supernovae, but now it appears feasible to transport the energy released in the binding of a neutron star to the outer layers of the imploding star by an extraordinarily large neutrino flux, thereby pushing and ejecting the outer layers. Conversely, it is equally plausible that most stars evolving to a carbon-oxygen core will thermonuclearly detonate and destroy themselves. The neutrino transport theory of supernovae has recently been bolstered by the new theoretical considerations concerning the weak interaction force and how this relates to resonant neutrino scattering off larger nuclei. The thermonuclear origin of supernovae has resisted all attempts at finding some way to avoid detonation before implosion to a neutron star. The theoretical picture of the origin of supernovae now has two equal alternatives to choose from. Much of the heat input to galaxies, to cosmic rays, the gas motions, partially ionized state of matter, in large part, are due to supernovae. The chemical elements that have made us have been ejected from supernovae. Some of us even attempt to explain quasistellar sources as multiple supernovae in galactic nuclei. The range of dynamical variables that describe the phenomena of supernovae covers a wider range – more orders of magnitude – than any other single astrophysical event short of the universe itself. It is small wonder then that so many astrophysicists try to explain and so many astronomers try to observe this awesome phenomenon.

H. Gursky and R. Ruffini (eds.), Neutron Stars, Black Holes and Binary X-Ray Sources, 13–27. All Rights Reserved.
Copyright © 1975 by D. Reidel Publishing Company, Dordrecht-Holland.

2. Energetics of Supernovae

The phenomenon that we call a supernova is believed to be the explosion of a star at the end of its life. The supernova is by all odds the most dramatic single event that occurs in the 'normal life of a galaxy' unless perchance the quasistellar phenomenon is part of such a normal life, and even a quasar may be a succession of supernovae. It is difficult to recognize when supernovae were first identified as the awesome explosive disruption of an entire star, but for this to be even plausible requires the fundamental recognition of the existence of other galaxies of hundreds of billions of stars in the pioneering work of Hubble. Once the size and distance of these vast assemblages of stars was recognized in the early 1900's, then the phenomenon of single stars flaring up to a luminosity of 10^{10} suns could be recognized (Figure 1). The first to make this

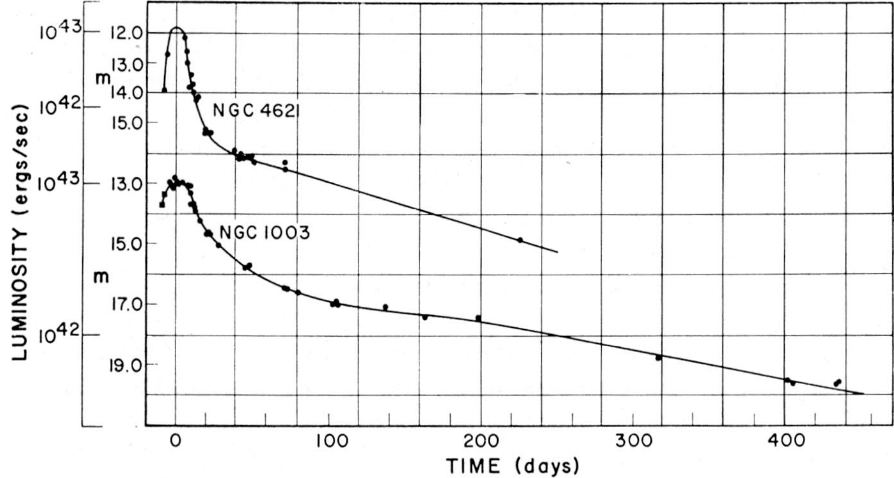

Fig. 1. Two typical supernova light-curves are shown giving a peak luminosity of 10^{43} erg s^{-1} at a time $\geqslant 5$ days. (From Minkowski, Reference 5.) Published by the University of Chicago Press; © 1939, University of Chicago Press.

association was Lundmark [1]. Once having been made, the implication was immediate; a search of thousands of galaxies could possibly produce several supernovae a year for astronomers to study and understand. Zwicky [2, 3, 4] was the first to organize a search on a large and effective scale, and with Baade this search produced dramatic and immediate results; some dozen supernovae were found in three years of work so that a mean rate of stellar explosions per galaxy of one per four hundred years could be estimated. Once it was understood that the stellar explosions occurred in all types of galaxies, one was led to the obvious conclusion that the phenomenon of an exploding star must have been occurring in our own Galaxy, and indeed, supernovae were associated with Tycho's star, Kepler's star and the famous Crab nebula in the constellation of Taurus. This latter dramatic event was associated with the observations of the Chinese in 1054 A.D. and it became the historical supernovae of our own Galaxy. The human mind is challenged by the question, 'Why a star explodes?' and after half a century of work by literally hundreds of scientists many still doubt whether we understand.

As I have already stated, supernovae are associated with the observation that a star in some distant galaxy or our own, increases in luminosity over a period of a week such that it is shining brighter than 10^{10} stars like our Sun; and in a matter of months, it slowly decays back to a luminosity that is characteristic of an average star.

Finally, in the early part of Zwicky's work, and even earlier in a crude fashion and finally in far greater detail by Minkowski [5], the spectra of supernovae have been observed. Although a distinction could be made into two and perhaps more classes, the important features are that the spectrum is by and large that of a black body, and that the black body temperature as observed by the color distribution is not vastly different from that of ordinary stars; namely, the surface temperature of supernovae appear to be in the range of 5000 to perhaps 10000 deg. The spectral lines that are observed are usually quite broad and only stand out modestly above a continuum, and it is this continuum or black body that is the essence of the most important aspect of supernovae; namely, the conclusion that an entire star is exploding. The way in which one comes to this conclusion is to compare the luminosity of 10^{10} suns of the distant supernova with the calculated luminosity per unit area of a black body at the temperature determined by the spectroscopy. One then arrives at a surface area of the supernovae as a function of time. Since one knows the minimum mass of matter that is thick enough in $\mathrm{gm\ cm}^{-2}$ at these temperatures to act like a black body, it is a simple matter to multiply area times mass per unit area and arrive at a minimum mass for the object. This mass turns out to be a minimum of a tenth of a solar mass. If the origin of the radiation is heat diffusing from the interior, then one arrives at a relatively simple argument based upon Compton opacity and time, that at least one solar mass of matter is radiating in most supernovae.

Finally, the radius that one derives for the emitting surface is at least 10000 times greater than the Sun. It is plausible to assume that the expelled matter reaches this radius in a time characteristic of the rise time of a supernova light curve, which results in a velocity of this matter nearly a tenth that of light. The total kinetic energy of this rapidly expanding solar mass is an extraordinary number. It is of the order of a hundred times the total energy in light that is emitted and it corresponds to an energy per nucleon that could only be available in a nuclear explosion or very possibly, and this is the exciting point of the last decade, from the binding energy of a neutron star.

Let us now make more detailed calculations of the total mass and kinetic energy of the ejecta of the most energetic supernova. The key question to many astrophysicists is, 'What is the source of the energy of the explosion?', and of course, the key point in such an explanation is the amount of energy that must be produced, usually measured in MeV per nucleon, in order to eject matter at the required velocities.

As stated earlier, the brightest supernovae are approximately 25 mag. brighter at maximum than the Sun. This corresponds to a luminosity 10^{10} times that of the Sun, or about 4×10^{43} erg s^{-1}. The light curve will typically have a half-width of about one week, or approximately 10^6 s. Consequently, the optical energy radiated during the light curve maximum is about 4×10^{49} erg. If the supernova surface were at the same temperature as the Sun, then the luminosity would be proportional to R^2, so that the radius of the supernova at maximum would have to be 10^5 times the solar radius, or 7×10^{15} cm. On the other hand, if the supernova temperature were 30000 K or 5 times greater than the Sun, the optical output would increase only by a factor of 5 or 6, compared to a solar temperature of 6000 K (This is a consequence of the shape of the Planck spectrum, and the energy radiated in the invisible ultraviolet would be very large (10^{52} erg).) Therefore, it is more reasonable to assume a temperature of about 10000 K, the temperature which is inferred from the specta. For this temprature, the radius will be about 3×10^{15} cm; if we assume that the supernova reaches maximum in 10^6 s by the radial expansion of the ejected matter, it follows that the velocity of the surface is 1.5 to 2×10^9 cm s^{-1}, which corresponds to about 1 to 2 MeV nucleon^{-1}.

We can make a rough calculation based on diffusion theory arguments of the total mass involved. The density of the expanding material is such that bound-bound transitions are negligible and the only opacity is provided by Compton scattering. If we assume that less heat energy is being generated at the time of maximum than earlier, that is, that the energy source has been turned off somewhat early, then diffusion theory predicts that the maximum luminosity will occur at approximately the time when a diffusion wave has penetrated to about one third the radius of the sphere. Therefore, the diffusion velocity should be one third of the expansion velocity of the surface:

$$v_{\text{diff}} = \tfrac{1}{3} u_{\text{s}}. \tag{1}$$

From diffusion theory, the skin depth ∂ is given by $\partial^2 = Dt$, where D is the diffusivity. We substitute $D = \lambda c/3$, where $\lambda =$ the photon mean free path and c the velocity of light so that

$$\frac{\partial}{\lambda} = \frac{t}{\partial}\frac{c}{3}. \tag{2}$$

If we consider ∂/t to be the diffusion velocity then from Equations (1) and (2) we obtain $\partial/\lambda = c/u_s = 10$ (since we previously calculated u_s to be 3×10^9 cm s^{-1}). We interpret this result to mean that the medium is 10 mean free paths thick. This then requires that $\varrho r/3 = 10/K$, where K is the Compton opacity; taking $1/K = 5$ g cm^{-2} gives $\varrho r = 150$ g cm^{-2}. A computer calculation [6] using a more complicated diffusion theory, including expansion and a variable velocity distribution gives a value of $\varrho r = 50$ g cm^{-2}. The total mass will then be $4\pi r^2 (\varrho r)/3$; using the more conservative value of $\varrho r = 50$ gives a mass of 2×10^{33} gms, which is one solar mass. Therefore, from the very simplest observations; namely, that a supernova occurs in distant assemblages of stars and has a brightness corresponding to 10^{10} suns and rises to this maximum in a week and has a black body spectrum corresponding to a temperature of 10000 deg, we can infer that something like a solar mass of matter is expanding at the extraordinary energy of 1 to 2 MeV nucleon^{-1}.

It should be noted that the inferred kinetic energy is at least 100 times the energy radiated in the optical. By way of substantiating this picture, one can observe in some supernova remnants and in some early line shifts evidence for these high velocities. Minkowski [7] infers this velocity for the expansion of Tycho's nebula which is consistent with these estimates, and for the cases where the He I and He II spectra have been identified late in time for type II supernovae, the velocities from the line shifts are again consistent with these estimates. However, the average supernova remnant and the line shifts observed for the hydrogen lines of type II supernovae, have considerably lower velocities, a factor of 2 or more, either indicative of a slower supernova or that the matter which is observed is a smaller, inner mass fraction moving at lower velocities.

3. The Origin of Supernovae

We have referred in the previous paragraph to a type I and type II supernova and, indeed, there may be a more complicated morphology as suggested by Zwicky. In general, most investigators would agree that there is major distinction in the light curves and the spectra for two kinds of supernovae, type I supernovae have a more rapidly rising light curve from our previous discussion, and therefore higher specific energy, and occur in old population stars. The spectra of these supernovae have not been uniquely understood despite the work of many investigators (Minkowski [7], Kirshner *et al.* [8], but there is general agreement that very little hydrogen, less than 10%, if any, contributes to this unexplained spectrum. This is very odd matter indeed; these supernovae have less than 10% hydrogen whereas the normal matter of the Universe is at least 90% hydrogen, and so this observation alone is indicative of a star that is probably high evolved and has consumed most of its original hydrogen prior to the time of explosion.

The second type of supernovae, type II, on the other hand, rises more slowly and in particular has a spectrum that appears as if it is composed of normal matter with dominantly hydrogen lines appearing and, furthermore, with a velocity shift and spread of these lines that is consistent with a slower optical rise time and therefore indicative of slower expansion velocities. These supernovae occur in spiral galaxies and therefore in younger population stars and even with a high concentration at the star formation edge of each spiral arm as noted by Elliott Moore [9].

The conceptual origin of supernovae fits into a very broad scheme of stellar evolution. The structure of a star is presumed to be a balance between two forces: (1) the attraction of gravity, and (2) a repulsion from the gradient of pressure which in turn is derived from the gradient of temperature and density of the star. These two forces are in equilibrium and normally a very stable equilibrium at that. A very slow evolution takes place because of radiation, namely, the heat loss from the star (although in the final stages neutrino losses may dominate). The stable equilibrium occurs because if the system is perturbed, that is, either imagine a giant hand squeezing the star slightly or pulling it apart slightly, the energy generation by nuclear reactions or from adiabatic compression or expansion of the matter is just such as to cause the pressure gradients to react in such a way as to return the star to a stable condition. During the bulk of the history of the star, the nuclear energy generation and the adiabatic compressibility are strongly stabilizing factors. As the star evolves by burning the nuclear fuel at its center, both these stabilizing factors at some point in the evolution inevitably reverse in sign. This inevitable evolution toward instability was first recognized by Chandrasekhar [10] as the limit to the maximum mass of a cold stable star. However, the implications for supernovae were first developed at length in the now famous article by Burbidge *et al.* [12] on the origin of elements. An integral part of the theory of the origin of elements was the rapid nucleosynthesis that would take place under the extreme conditions of a supernovae explosion. They further recognized that at a critical stage of stellar evolution a dynamical collapse of a large star was inevitable and that this collapse would lead to the conditions necessary for rapid nucleosynthesis. The presumed structure at the point of dynamical collapse was a star whose inner core was composed mostly of iron. It may seem a strange element to postulate as the endpoint of nucleosynthesis in a star, but it just so happens that the assemblage of nucleons that has the greatest stability, that is, the most binding energy per nucleon (in other words, the 'ashes' of nuclear burning) is an assemblage of 26 protons and 30 neutrons which is iron. This also happens to be why we have so much of it around. To a nuclear physicist iron means not so much steel and red dirt, but more the fact that it is the minimum in the packing fraction curve of nuclear binding. We are all familiar with the fact that fusion of light nuclei can lead to the release of energy and equally well that the fission of heavy nuclei like uranium leads to the release of energy. It just turns out that iron is the minimum in between, so that the inner core of a star that has evolved by nucleosynthesis, called burning, to the endpoint of its evolution inevitably must end up as elements that have nuclei close to iron. Thus these elements are literally the ashes of nuclear burning.

When one assembles more and more mass of iron so that the gravitational attraction becomes greater and greater, there becomes a point where the electron pressure of the cold matter is less than the force of attraction of the gravitational field. This mass of iron (called the Chandrasekhar limit) of 1.3 $M_\odot$ is the limit above which unstable collapse can occur.

Stars whose mass is smaller than 1.3 $M_\odot$ will never reach this instability but instead quietly cool off and become stable by virtue of the electron pressure, which is the case for most stars because they are born less massive than the critical 1.3 $M_\odot$. In addition, mass loss occurs that reduces the mass during evolution.

The observational astronomer calls these cold dense stars 'white dwarfs'. It was the triumph of the original theoretical calculations that this critical mass could be predicted on elementary physical grounds, and that the observations strongly support it. It also leads to the conclusion that only some stars whose mass is greater than 1.3 $M_\odot$ can possibly evolve to supernovae, but we are confronted with a larger dichotomy that the frequency of supernovae, one per tens to hundreds of years per galaxy, is still small enough such that only a fraction of those stars born and evolving with initial mass greater than 1.3 $M_\odot$ can possibly become supernovae. Currently, the discrepency is explained by the mass loss predicted and observed to take place during what is called the 'red giant' stage. A few very massive stars avoid the red giant stage and enough mass accumulates in the core to cause collapse. Their number may be sufficient to explain the occurrence of type II supernova that occur almost exclusively in young Population I stars. The other type I supernovae occur in other stars, and therefore require some form of mass accretion to take place after ejecting their envelope in the red giant phase. It has been suggested by Whelan and Iben [11] that the mass accretion can take place by the exchange of red giant envelopes in a binary pair of stars. Such a mass accretion might even manifest itself as an X-ray source as suggested by Cameron [11a].

4. The Mechanisms of Supernova Explosions

Burbidge *et al.* [12] suggested that a dynamical collapse led inevitably to the triggering of a thermonuclear explosion in the less dense outer layers of carbon and oxygen. It was recognized that the matter, once collapsed as iron, is very difficult to eject from a star because of the ever increasing gravitational binding energy. However, the matter that is still only partially evolved to iron, say, in the carbon-oxygen stage of nucleosynthesis is potentially explosive and releases 0.5 MeV nucleon^{-1} of thermonuclear energy, and could very possibly blow itself off the star. The remainder would collapse to a neutron star. The reason that hydrogen or helium, both of which have a greater potential energy available in nucleosynthesis, were not considered as potential thermonuclear fuels is because their reaction rates to synthesize something heavier and release the potential energy of binding are far too small to contribute significantly within the dynamic time scales of such an explosion. A thermonuclear explosion is therefore limited to the energy available in synthesizing or burning carbon and oxygen up to the most stable element – iron. This energy of complete synthesis is only 0.5 MeV nucleon^{-1} and is inadequate to explain some type I supernovae explosions as interpreted in the preceding elementary basis. On the other hand, factors of 2 or 4 are small indeed in the range of astrophysical variables, and it has always been considered possible that this elementary interpretation of the velocity of expansion may have some significant error which can allow thermonuclear burning to explain the origin of supernovae.

5. The Explosion Process and Neutron Star Formation

It was some years after this major work of Burbidge *et al.* that Richard White and myself [13] started evaluating such an explosion using the hydrodynamic numerical computer calculations that were then available at Lawrence Laboratory in Livermore for nuclear weapons research. Our initial calculations explored the possibility that dynamical collapse alone would cause the matter to bounce on a hypothetical 'hard core', possibly a neutron star. Even when no heat loss due to neutrino emission was included in these initial calculations it was evident that a dynamical collapse of an already tightly bound star would lead to only a minuscule ejection of matter, and this minuscule fraction of ejected matter depended upon an extraordinarily sensitive transfer of energy by shock waves in the many, many orders of magnitude variation of density of a star. The phenomenon was governed by the generation of a reflected shock wave of the in-falling matter on the presumed neutron star core, and this reflected shock wave had to propagate out through in-falling matter through 10 to 12 orders of magnitude (10^{10} to 10^{12}) change in density to the outer layers before reaching a strength sufficient to eject only the very outermost surface mass fraction of a star. Under these conditions, the ejected matter was some 10^{-3} to 10^{-5} of the mass of the star and its velocity at least an order of magnitude less than what could explain supernova light curves. At this point, we decided that the logical conclusion was that the thermonuclear energy had to be released in the outer layers triggered by this relatively weak reflected shock wave from neutron star core. We did not fully understand thermonuclear detonation waves in carbon and oxygen and there were still many questions of cross sections involved. We took the extreme view and placed in the envelope of a 10 $M_\odot$ star at the optimum position, three solar masses of carbon and oxygen. We then released 0.5 MeV nucleon^{-1} energy, expecting to see the ejection of several solar masses of matter.

Contrary to this native and even optimistic expectation, the detonated matter with a 0.5 MeV nucleon^{-1} internal energy was swallowed into the imploding neutron star just as if it had not been heated at all! The reason for this became more evident in retrospect. Without the pressure of a 'center' to push on, the inner boundary of the exploded carbon and oxygen saw the reduced pressure created by imploding iron into the neutron star. Without an inner boundary to push against, the exploded matter simply expanded inward toward the neutron star and was swallowed in a transformation from normal matter to neutron matter. There seemed to be no way out of this dilemna because the gravitational binding energy of the carbon and oxygen at the point where it would be synthesized in such a star was always greater than the potential thermonuclear energy that could be obtained by burning it to the endpoint – iron. Therefore, it appeared to be almost impossible to use the gravitational collapse to a neutron star core as a means of triggering a thermonuclear explosion that would blow apart a massive star.

6. Binding Energies

I would like to digress briefly into some comments about the binding energies of matter in various forms. If we considered a zero point to be separated protons and electrons in the Universe, (ionized hydrogen) then we would say that such matter could evolve to a lower energy state; namely, the hydrogen atom with 13.6 electron volts of binding energy, and this energy would be radiated in photons corresponding to the various transitions of the electrons of the quantum shells. We can envisage a further decrease in the energy state for an increase in the binding energy of hydrogen if the hydrogen atoms combine into molecules and gain an additional electron volt of energy. If we proceed to the interior stars where the densities and temperatures are high enough to allow the hydrogen nucleus to transform and combine to make helium, then some 8 MeV nucleon^{-1} becomes available as radiation. This is the radiation from the Sun.

Finally, as helium nuclei are combined to carbon, oxygen and ultimately, iron, additional energy is available, and one reaches the state of maximum binding energy of a nucleon in the nucleus in iron of roughly 10 MeV nucleon^{-1}. This represents the maximum binding energy available in what we might describe as normal matter. But there is another state of matter that was predicted many years ago; first by Gamow [14], who recognized that if matter were squeezed hard enough by gravity, a transformation to a neutron state could take place, and that this matter would represent a very much larger binding energy than that occurring in the iron nucleus. This binding energy occurs only in a collective state of matter because it depends upon the interaction of the gravitational force acting over a very large assemblage of nucleons (10^{57}), and so cannot exist in the small unit of nuclei of normal matter. Oppenheimer and Volkoff [15] were the first to construct models of such neutron stars based upon the new theory of general relativity of Einstein and nuclear physics.

These models were considered almost akin to science finction in their day, although the physics that went into constructing them was believed to be the most likely basis for the foundation of the real world. Yet, I can remember personally when I was a child considering the extraordinary properties of neutron star matter where a match-box-full weighs a billion tons, and believing that this was sheer fantasy, and yet now, the observation of pulsars has made the existence of neutron stars an accepted certainty, and furthermore the discovery of a pulsar in the Crab nebula remnant makes their association with supernovae raised from the level of inevitable to dramatic certainty.

Two scientists who in the early days did not take the neutron star hypothesis as science-fiction were Zwicky and Baade [16]. They were the first to propose that neutron stars might be the origin and the result of supernovae. Yet White and I, in the first numerical calculations, used a perfectly rigid neutron star as a 'hard' core to the imploding supernovae and, one might comment, with depressingly modest results. At this stage we did not know how to deal with the binding energy of a neutron star and, as we pointed out above, this is the largest binding energy, the lowest negative energy state, that has been postulated for matter. The exact value of this binding energy has been open to relatively wide uncertainties due to the complicated interaction of the nuclear equation of state of such dense matter with the subtleties of the general theory of relativity, and as a consequence, the maximum binding energy of a neutron star has varied from 50 MeV nucleon^{-1} up to 200 MeV nucleon^{-1}, almost an order of magnitude. Yet, even the smallest of these limits is 100 times larger than the available energy from the thermonuclear detonation of oxygen and carbon, the most energetic explosive stellar fuel; so that the question was, 'How could we use this 100-fold greater energy source to blow up a star?' We knew that this energy source had to appear as heat in the newly formed neutron star, but the question was: 'How could this heat be used or conducted to the outer layers which reside at a lower gravitational potential and so lead to the internal pressure that could blow them off or eject them from the star?'

7. Neutrinos in Supernovae

It was at this point, that a chance discussion with Robert Christy at Caltech led me to understand how neutrinos, when they are in thermal equilibrium, acted like a lepton gas that had properties similar to a relativistic electron gas or even to a photon gas, and would exhibit all the properties of pressure and equation of state that were more familar to us for electrons and photons. If one could conceive of the heat of formation of this neutron star being in thermodynamic equilibrium with neutrinos as well as photons and electrons, then indeed, there was a possibility that this newly formed star would radiate a neutrino flux like a black body characteristic of the implied extraordinary temperatures of formation. One can relatively easily derive these temperatures because one knows the rate of energy release from the in-falling matter to the neutron star state. The total energy available is 0.1 to 0.2 $M_\odot c^2$ and it is released in the time necessary for the matter to free fall from the initial radius of the highly evolved core of the initial star. This radius ($\simeq 10^3$ km) is determined by the gravitational binding necessary to hold the matter together at the temperatures at which the final nucleosynthesis of the star takes place. The resulting free-fall time of 10 to 20 ms has been characteristic of all the calculations performed (Colgate and White [13], Arnett [17], Wilson [18], and Schwartz [19]) and determines the temperature of emission. The binding energy of the neutron star is then emitted in the free-fall time as a black-body neutrino flux from the surface of the neutron star. The equivalent radiation temperature is then determined by equating the Planck emission rate to the energy release or

$$\frac{c}{4} a T^4 (4\pi R^2) = 0.2 M_\odot c^2 / \tau . \tag{3}$$

The inferred temperature is then roughly 30 MeV. The characteristic energy of a relativistic particle at such a temperature – neutrinos, electrons, or photons – is $3kT$ so that particle reactions characteristic of 100 MeV energies must be considered. Most important of all, such a hot region will emit neutrinos as a black body flux and these neutrinos will have a mean energy of up to 100 MeV. The cross section of such neutrinos on the imploding matter is such that the imploding matter is roughly one to several neutrino scattering and absorption mean free paths thick, and, as a consequence, White and myself [13] recognized the possibility that this neutrino flux could represent the heating or energy transport mechanism that could carry the binding energy of the neutron star to the outer layers of the still imploding original star. The heat and energy deposited might be sufficient to result in the explosion and ejection of the remaining matter. The original hydrodynamic calculations (Figure 2) made the relatively crude assumption of black body emission from the neutron star surface and deposition of the heat of the neutrinos nearly proportional to the cross section in the outer layers. Subsequent calculations [17, 18, 19] included a hierarchy of neutrino transport theories and modifications to the cooling history of the imploding matter and although the first calculations showed a relatively straightforward mass

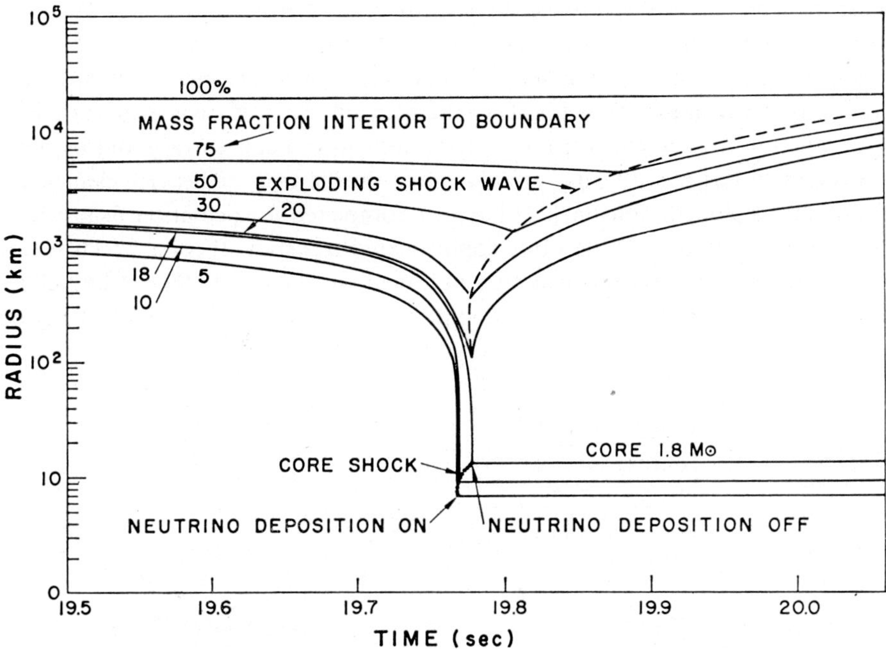

Fig. 2. Radius vs time for 10 $M_\odot$ supernova with neutrino deposition. During the initial collapse the neutrino energy is assumed lost from the star, but at the time of formation of a core shock wave (heavy dots) a fraction of the neutrino energy is deposited in the envelope. The deposition ceases when the explosion terminates the imploding shock wave on the core. (From Colgate and White, Reference 13.) Published by the University of Chicago Press; © 1966, University of Chicago Press.

ejection, later calculations by others found limitations to the circumstances in which enough mass would be ejected to prevent the catastrophe of collapse to a black hole. This latter question is, of course, at the heart of much of the speculation about supernovae. Phrased another way, 'Is there a mechanism for transporting the binding energy of a neutron star to the outer layers of the imploding star that is sufficiently effective to prevent further collapse? Can the neutron star always manage to eject sufficient mass such as to prevent its own mass from accumulating beyond the limit where it, too, will start imploding?' This mass limit is roughly 2 $M_\odot$, and this important question is not yet resolved.

At any rate, some ten years after the first calculations, there has been major concern that the neutrino transport mechanism may not be efficient enough to explain the implied energies of supernovae even though the available energy may be as much as 200 MeV nucleon^{-1}, and the energy to be explained or inferred from the observations is only 1 to 2 MeV nucleon^{-1}. It is indeed difficult to demonstrate by numerical calculations, the efficiency of a process that depends upon such a complicated hierarchy of phenomena.

Very recently, however, the new universal neutrino theory of Weinberg [20] has had a major impact upon supernova theory. As Freedman [21] has pointed out, a coherence phenomenon occurs in large nuclei that increases the cross section per nucleon for neutrino scattering and absorption proportional to the atomic weight; so that if the configuration at the start of implosion is composed of heavy nuclei, then the transformation to neutron matter as it falls to the neutron star surface is ideal for allowing the free emission of neutrinos from the forming neutron star surface and then a disproportionately larger scattering cross section on the iron or heavier elements that have not yet imploded. Wilson [22] has just completed a calculation using the new neutrino cross sections and the more sophisticated transport theory, which seems to imply that indeed the heat of formation of a neutron star is an efficient and effective mechanism for causing the ejection and explosion of stars (see Figure 3). The ejection

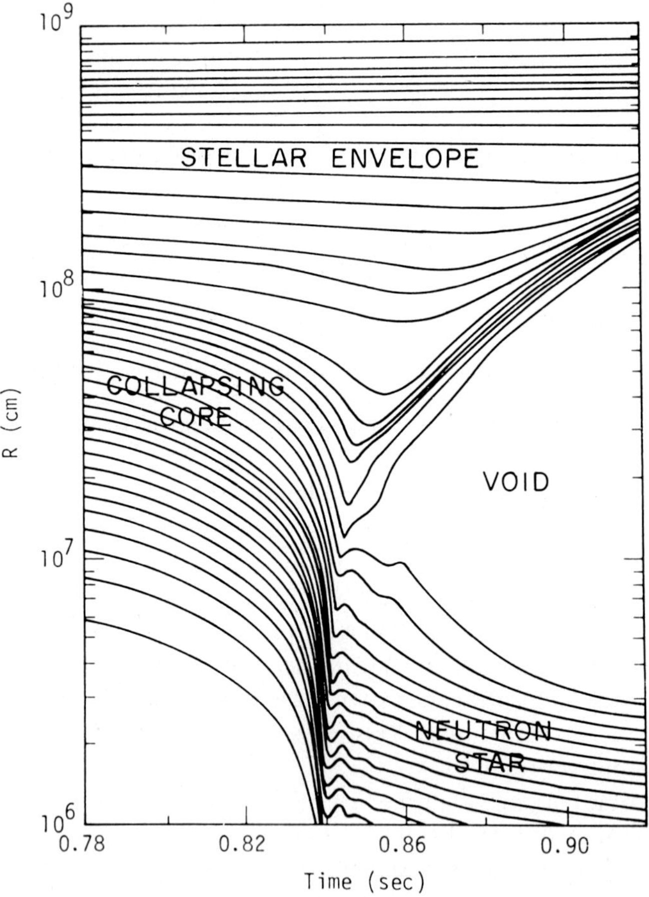

Fig. 3.   The radius vs time of mass points in a star whose iron core is collapsing. Each line is the trajectory of unique fluid element in the star. Only the inner $1.68\ M_\odot$ of a $12\ M_\odot$ star is considered and comprises a $1.49\ M_\odot$ iron core ($R < 10^8$ cm) and envelope. The neutron star itself is formed at ~ 0.84 s. At this point, the collapse is halted and an outward going shock wave is formed. The combined force of the shock wave and the radiation force of the emitted neutrinos is sufficient to blow off the envelope. (From Wilson, Reference 22.) Published by American Physical Society; © 1974, American Institute of Physics.

energies appear to be reasonably close to those inferred from the observations so that
there appears to be significant hope of associating the binding energies of a neutron
star with the explosion energy observed in supernovae.

On the other hand, a completely different view has been proposed and developed at
great length; first by Fowler and Hoyle [23] and then in greater detail by Arnett [24].
This is the view that a smaller star, less than 8 $M_\odot$, evolves to a carbon core or carbon
and oxygen core in such a fashion that its temperature is relatively low and supported
primarily by degeneracy pressure. Under these circumstances, carbon or carbon and
oxygen is a potentially explosive thermonuclear fuel.

Then if a detonation of the carbon is initiated, there is present the central support
pressure necessary to explode the star. This is opposite to the case where an iron core
implodes to a neutron star and an outer shell of carbon and oxygen is denotated. In
this latter case, there is nothing for the exploding matter to push against and so it is
swallowed into the neutron star. It turns out that during the continuing evolution of a
carbon-oxygen core, the matter reaches the state where a very small fluctuation in
temperature at an exceedingly high density (10^9 to 10^{10} gm cm^{-3}) initiates a runaway
thermonuclear reaction. Detailed calculations by Arnett [25] and Wheeler and Hanson
[26] have shown that the suggestions of Fowler and Hoyle [23] are indeed correct
that such a detonation will completely disintegrate the star (Figure 4). The ejection

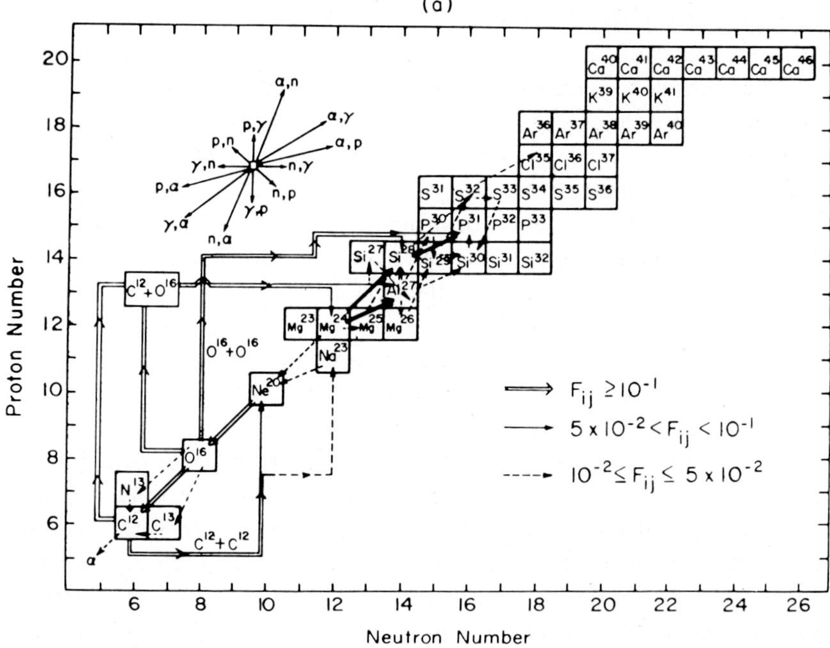

Fig. 4.   Primary flows during explosive oxygen burning. Principal flows of the strength and type
illustrated are shown at the 'standard' explosive-oxygen-burning run ($T_{9i} = 3.6$, $\varrho_i = 2.0 \times 10^5$). Figure 4 is a 'snapshot' at time 7.53×10^{-3} s at which point $X(^{24}Mg) = 0.18$ and $X(^{16}O) = 0.49$. (From
Woosley *et al.*, *Astrophys. J. Suppl. Ser.* **26**, No. 231 (1973).) Published by the University of Chicago
Press; © 1973, American Astronomical Society.

velocities are relatively modest corresponding to less than 0.5 MeV nucleon^{-1}, but again the interpretation of supernovae phenomena may agree with this. There has been a great deal of effort by Barkat *et al.* [27] and Bruenn [28] to attempt to find an evolutionary path for the center of such stars such that they can avoid or circumvent the detonation catastrophe. There is still a distinct possibility that Paczyński's [29] suggestion of a convective enhanced UCRA process might help in leading to such an exception. The reason for the effort to circumvent the thermonuclear state is that the present conclusion would be that all less massive stars evolving to a carbon-oxygen core would necessarily undergo a thermonuclear explosion that would disintegrate the star and leave to remnant neutron star. Yet we observe sufficient numbers of neutron stars and, in addition, a neutron star at the center of the Crab and the Vela supernovae remnants so that we are confronted with the following dichotomy. There are enough neutron stars or pulsars observed such that their number is consistent with the notion that supernovae produce neutron stars as remnants. In the one case where a definite association can be made, the Crab, apparently a neutron star is left after a type I supernova; namely, the smaller, hotter, older population supernovae. Yet it is just this particular class of supernovae that one would expect might evolve through the carbon-oxygen core thermonuclear detonation. It would be very neat if we could find a mechanism such that a carbon-oxygen core would evolve stably up to a density of 2×10^{10} gm cm^{-3}.

As Bruenn [30] has pointed out, the further implosion and denotation of a carbon-oxygen star results in sufficient neutrino energy loss at the time of the detonation such that the star continues to implode to a neutron star state, yet, at the same time the outer layers have been detonated in a carbon-oxygen thermonuclear reaction. Then, as the neutron star is formed by implosion of the inner layers, the Planck neutrino emission from the hot neutron star will be absorbed in the already detonated outer layers augmenting the initial energy form the thermonuclear detonation and ejecting the outer layers at the high energies inferred from the observation. Wheeler *et al.* [31] may have found just such a path when the delayed ignition of the initial collapse is considered.

The matter that would be observed later in time would be characteristic of the debris from such a thermonuclear detonation, in particular, a distribution of elements from carbon up to iron some of which would be characteristic of silicon burning.

This picture has the added attraction that the initial star would be highly compact and that the surface layers would correspond to a very small radius, approximately several thousand kilometers, so that the shock wave associated with the detonation and subsequent neutrino heating that would travel through the outer layers, would speed up and form the energy distribution that Johnson and I [32] have claimed is the origin of cosmic rays. The particular compact structure is ideal for this phenomenon, and the energy imparted to particles greater than a GeV in the shock wave agrees with what is necessary to form cosmic rays in our Galaxy.

Finally, we note that pulsars are observed because of a large magnetic field that is undoubtedly frozen into the neutron star at the time of its formation. This magnetic field means that a more modest but still very large (10^8 G) magnetic field must surround the initial star before the explosion. There is a significant probability of observing the electromagnetic pulse from the relativistic ejection of the outer layers occurring within such a magnetic field. Noerdlinger and I [33] have calculated that it may yet be possible to observe such pulses from red shifts as large as 2 or greater. It would be my longterm hope that we can learn to detect distant supernovae at the time of their formation by the electromagnetic pulse and observe them throughout their initial expansion to luminosity at maximum. If we can do this in large numbers, we have an ideal standard candle for measuring the distance scale of the Universe. We need this standard candle to confirm the interpretation of the red shift distance relationship for our universe and insure that the fantastic construction of observation and deduction which is our present picture of the Universe is substantiated in one more major observational way.

References

1. Lundmark, K.: *Svenska Vetenkapsakad Handlingar* **60**, No. 8 (1920).
2. Zwicky, F.: *Astrophys. J.* **88**, 529 (1938).
3. Zwicky, F.: *Astrophys. J.* **96**, 28 (1942).
4. Zwicky, F.: *Handbuch der Physik* **51**, 776 (1958).
5. Minkowski, R.: *Astrophys. J.* **89**, 156 (1939).
6. Colgate, S. A. and McKee, C.: *Astrophys. J.* **157**, 623 (1969).
7. Minkowski, R. L.: *Stars and Stellar Systems* 7, Chapter 11.
8. Kirshner, R. P., Oke, K. B., Penston, M. V., and Searle, L.: *Astrophys. J.* **185**, 303 (1973).
9. Moore, E. P.: *Astron. Soc. Pacific* **85**, 564 (1973).
10. Chandrasekhar, S.: *An Introduction to the Study of Stellar Structure*, University of Chicago Press, Chicago, 1939.
11. Whelan, J. and Iben, I.: *Astrophys. J.* **186**, 1007 (1973).
11a. Cameron, A. G. W.: 'Hot Vibrating White Dwarf Models of Pulsating X-ray Sources', Harvard College and Smithsonian Observatory. Submitted to *Astrophys. Space Sci.*
12. Burbidge, c. M., Burbidge, G. R., Fowler, W. A., and Hoyle, F.: *Rev. Mod. Phys.* **29**, 547 (1957).
13. Colgate, S. A. and White, R. H.: *Astrophys. J.* **143**, 626 (1966).
14. Gamow, G.: *Atomic Nuclei and Nuclear Transformations*, Oxford, 1936.
15. Oppenheimer, J. R. and Volkoff, G. M.: *Phys. Rev.* **55**, 374 (1939).
16. Baade, W. and Zwicky, F.: *Phys. Rev.* **45**, 138 (1934).
17. Arnett, W. D.: *Can. J. Phys.* **45**, 1621 (1967).
18. Wilson, J. R.: *Astrophys. J.* **163**, 209 (1971).
19. Schwartz, R. A.: *Ann. Phys.* **43**, 42 (1967).
20. Weinberg, S.: *Phys. Rev. Letters* **27**, 1688 (1971).
21. Freedman, D. Z.: National Accelerator Laboratory, Pub-73/76-THY, Batavia, Illinois, 1973.
22. Wilson, J. R.: *Phys. Rev. Letters* **32**, 849 (1974).
32. Hoyle, F. and Fowler, W. A.: *Astrophys. J.* **132**, 565 (1960).
24. Arnett, W. D.: *Nature* **219**, 1344 (1968).
25. Arnett, W. D.: *Astrophys. Space Sci.* **5**, 180 (1969).
26. Wheeler, J. C and Hansen, C J : *Astrophys Space Sci* **11**, 373 (1971)
27 Barkat, Z , Buchler, J R , and Wheeler, J C : *Astrophys J Letters* **6**, 117 (1970)
28 Bruenn, S W : *Astrophys J* **168**, 203 (1971)
29 Paczyński, B : *Acta Astron* **20**, 47 (1970)
30 Bruenn, S W : *Astrophys J* **177**, 459 (1972)
31. Wheeler, J. C., Buchler, J.-R., and Barkat, Z. K.: *Astrophys. J.* **184**, 897 (1973).
32. Colgate, S. A. and Johnson, M. H.: *Phys. Rev. Letters* **5**, 235 (1960).
33. Colgate, S. A. and Noerdlinger, P. D.: *Astrophys. J.* **165**, 509 (1971).

PULSE ASTRONOMY:
SHORT TIME SCALE PHENOMENA IN ELECTROMAGNETIC AND GRAVITATIONAL WAVE ASTRONOMY

R. B. PARTRIDGE

Haverford College, Haverford, Pa., U.S.A.

Patience is generally reckoned to be a virtue in astronomers. The reason is very simple: astronomical processes generally occur on extremely long time scales. The motions of planets in our solar system take place on a time scale of the order of years or decades; in double star systems, the orbital periods range from days to millennia. Likewise, changes in the properties of individual astronomical bodies generally take place on extremely long time scales. For instance, it is now well established that our Sun, like all other stars, is an evolving object. In the process of generating the energy which warms the Earth, its interior properties are slowly changing; but it is quite apparent that changes in the Sun have been negligibly small over historical time. Indeed we are fairly confident that the Sun has been essentially as it is now for the past five billion years, and will remain in essentially the same state for the next five billion years.

In most astronomical situations, therefore, changes – both motions and evolutionary processes – are slow compared to the pace of a human life. To quote the psalmist, 'Behold, Thou hast made my days as an handbreath; and mine age is as nothing before Thee.'

One can show in fact that there is a fundamental limit to the speed at which ordinary astronomical objects can change. Changes in the properties of an entire body cannot take place on a time scale shorter than $t \approx R/c$, where R is a characteristic dimension

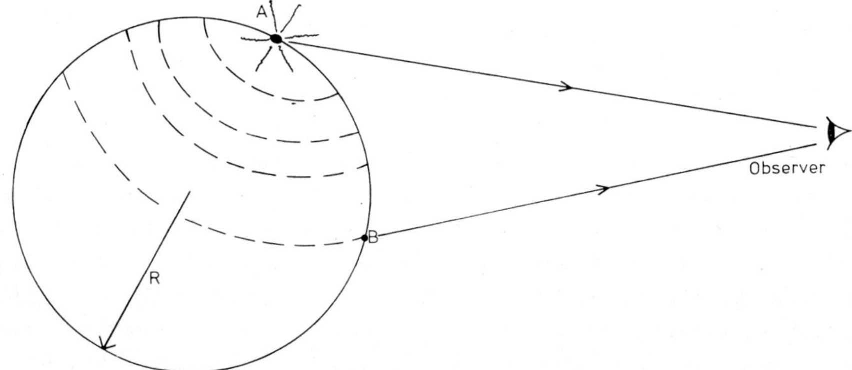

Fig. 1.   An extended luminous body, of radius R, cannot in general vary its luminosity appreciably in a time shorter than $\sim R/c$. Let us assume we want the entire surface to change in brightness. A disturbance initiated at A will take $\sim R/c$ to reach a general point like B. Even if the entire surface were to brighten simultaneously (which might happen if a disturbance initiated at the center), light from A would take $\sim R/c$ longer to reach the observer than light from B.

H. Gursky and R. Ruffini (eds.), Neutron Stars, Black Holes and Binary X-Ray Sources, 29–45. All Rights Reserved.
Copyright © 1975 by D. Reidel Publishing Company, Dordrecht-Holland.

of the body and c is the velocity of light. Suppose some change in the properties of a star (assumed spherical) were to be initiated at point A (see Figure 1). We know from Special Relativity that it would take a finite time for information about this change to propagate to a point such as B. Thus the *entire* star cannot change its properties in less than the light travel time from A to B, or roughly speaking $t \simeq R/c$. Even if every point on the star's surface changed simultaneously (as might happen if the change were initiated at the center), light would take $\sim R/c$ longer to reach the observer from point A than from point B. For stars whose radii are of the order of hundreds of thousands of miles, the characteristic limit on the time scale for changes is a few seconds.

To sum up, when they are engaged in ordinary stellar astronomy, astronomers have little choice but to wait.

We now recognize that this view is simply too restrictive. Over the past four to five years, a number of dramatic and exciting discoveries in astronomy have shown that there are phenomena in astronomy whose time scales lie well below the value, derived above, of a few seconds. Many of the physicists and astronomers who have contributed to this volume are in fact those who have helped make these exciting discoveries. I view my task as providing an introduction to some of the experimental and theoretical results which they will discuss. The main burden of my paper will be to present five separate reasons of why *pulse astronomy* has recently come to the fore – astronomy on a time scale a few seconds or less. Four of these topics will be treated in depth in other papers of this volume. The remaining topic – a search for pulsed electromagnetic and other emissions – will be examined here.

1. The Death of Stars

The possibility of rapid phenomena in astronomy was first raised in connection with the very late stages of stellar evolution. Over the past 15 years, research workers have come to recognize that stars will evolve rapidly, even violently, as they reach the end of their life. (For early work on the late stages of stellar evolution see Hoyle and Fowler, 1960; and Fowler and Hoyle, 1964. A more recent paper on supernova hydrodynamics is Colgate and White, 1966; see also Colgate's contribution to this volume. For general reviews, see Shklovskii, 1968; Brancasio and Cameron, 1969; Cosmovici, 1974.) Why it is that these later stages of stellar evolution proceed so much more rapidly than the earlier stages? How is it that our Sun, which stays essentially at the same luminosity for 10^{10} years, will eventually embark on a hectic second childhood? One reason for the more rapid evolution is that the central temperature of stars increases as they age. The nuclear processes which provide the energy for stars are highly temperature dependent. Consequently, as the central temperature increases, nuclear fusion proceeds more and more rapidly. Runaway thermonuclear processes can result: one example is the supernova detonation discussed by Colgate (note the time scales of his Figures 2 and 3). The violence of the explosion is clear from Figure 2, a photograph of the Crab nebula, a jumbled mass of gas resulting from a supernova explosion 920 years ago.

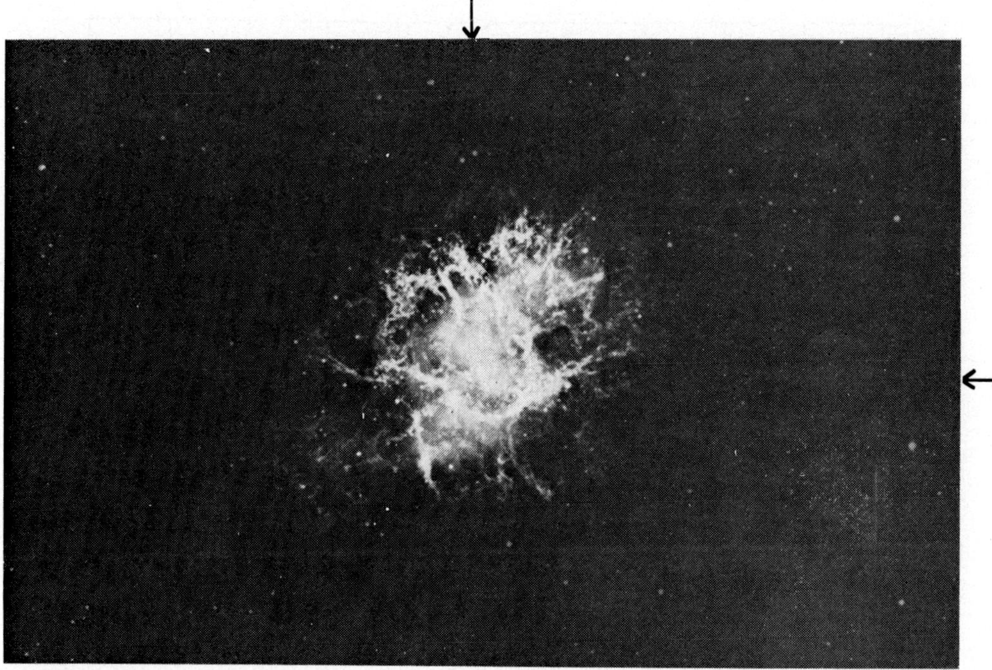

Fig. 2.   The Crab Nebula, remnant of the supernova of 1054 *A.D.* (Mt. Wilson and Palomar Observatories, Copyright by the California Institute of Technology and the Carnegie Institution of Washington.)

More germane to our study of pulse astronomy is the fact that the final stages of evolution of a star do not take place throughout its entire bulk. As Paczyński (1970), was the first to show, stars in the mass range which is thought to produce supernovae ($\sim$3–10 $M_\odot$) will form small but extremely dense cores. Rapid evolutionary changes, such as the supernova collapse process discussed by Colgate, occur in the condensed central core, which is characteristically the size of the Earth. The light travel time across such a core is of the order of a tenth of a second or less, rather than the several seconds which one would expect for ordinary stars. Amusingly enough, this small concentrated core develops in the interior of a star which itself has become very large: in the later stages of stellar evolution the interior regions of the star contract while the exterior is puffed up. The contraction of the core of evolved stars is illustrated schematically in Figure 3, which compares the mass distributions in a star on the main sequence and in the same star when it reaches the red giant branch (freely adapted from Iben, 1967).

By the late 1960's, the theory of supernovae had been worked out in some detail (for a recent review of the topic, see Wheeler, 1973). One result was the prediction that explosive effects would occur on time scales as short as a tenth of a second (again, see Figures 2 and 3 of Colgate's contribution). The only problem for astronomers is that supernova events occur very rarely in our Galaxy – perhaps one every thirty to one hundred years. We are thus dealing with a strange mixture of pulse astronomy and

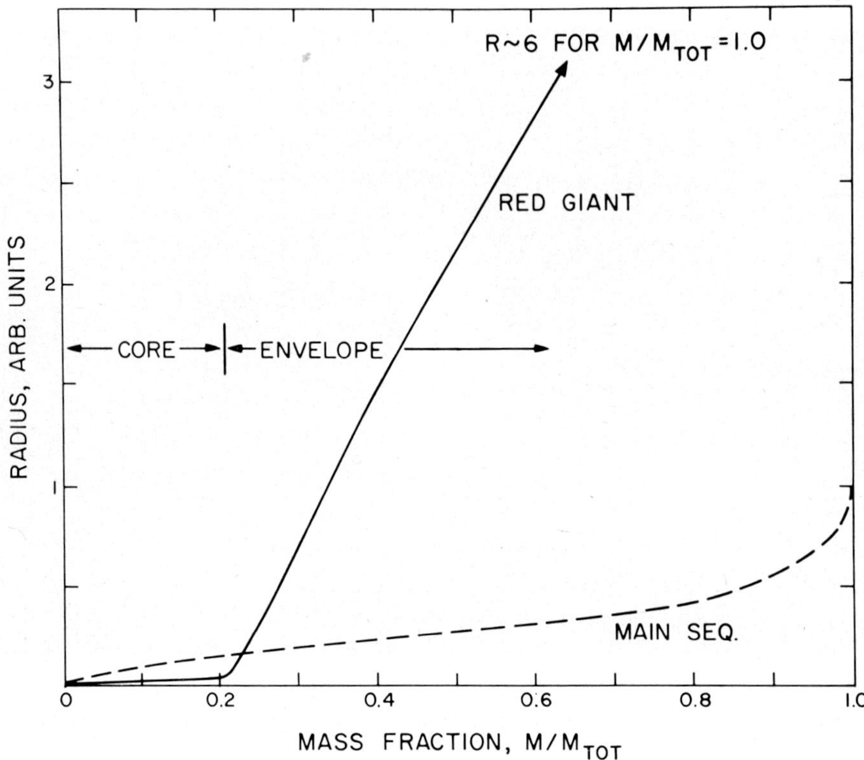

Fig. 3.   Radius as a function of mass-fraction for a star on the main sequence and the same star as a red giant. M/M_{TOT} represents the fraction of the total mass lying within the indicated radius (freely adapted from Iben, 1967).

the old astronomy of patience. At least, however, the possibility of short time scale astronomy had been opened.*

2. Pulses of Gravitational and Electromagnetic Radiation (and Neutrinos)

A tremendous impetus to the field of short time scale astronomy was provided by the important work of Jospeh Weber. In 1960, he proposed an experiment to search for pulses of *gravitational radiation* (see appendix). In this paper and a later one in *Physics Today* (Weber, 1968), he describes the properties of gravitational radiation and means of detecting it. It is convenient to introduce gravitational radiation by comparing it with electromagnetic radiation. It is well known that the acceleration of an electric charge will generate electromagnetic radiation. For instance, one can con-

---

* Long before the detailed dynamics of supernovae had been explored, Zwicky suggested and orga-nized an optical supernova 'patrol' which continues to this day. This project may represent the first systematic attempt to search for transient stellar events in astronomy (see Zwicky's contribution in the volume by Brancazio and Cameron cited above).

sider the forced motion of the electrons in a dipole transmitting antenna. Likewise, the motion, or more specifically the acceleration, of a mass can generate gravity waves. These waves also propagate at the speed of light. Unlike electromagnetic radiation, the lowest mode for the emission of gravitational radiation is the quadrupole mode (gravitational dipoles do not exist). Since gravitational waves can be generated by the motion of mass, they can also be *detected* by observing the motion of a suitably designed mass upon which the gravity waves impinge. An idealized detector of gravitational radiation can be represented by two masses attached by a spring (Figure 4).

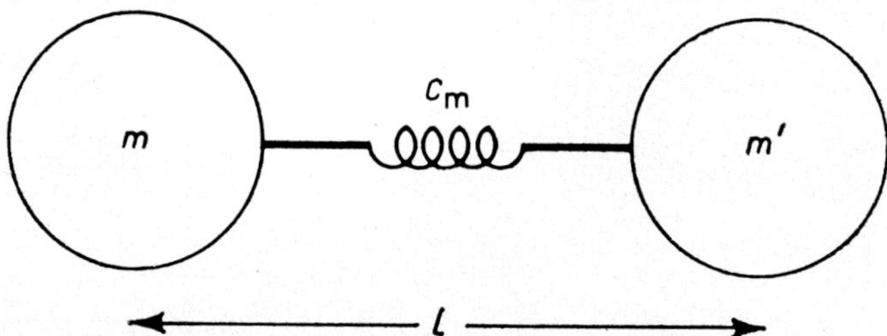

Fig. 4.   An idealized detector of gravitational radiation. A gravitational wave incident from any direction, save along the axis of symmetry, will cause the system to oscillate. (From J. Weber, *Radio and Electronic Engineer* **41**, 394, 1971).

If a gravitational wave is incident on this system, the two masses will oscillate, alternately stretching and contracting the spring (Weber, 1968).

While there is little conceptual difficulty in developing a gravitational radiation detector, there are very severe practical problems. These are all a consequence of the fact that gravitational waves interact extremely weakly with ordinary matter.* For instance, the Earth is virtually transparent to gravitational radiation, whereas it is obviously not transparent to electromagnetic radiation. It follows from this weak interaction that even an intense gravitational wave will produce only a very slight effect on a detector. One must therefore construct a detector which can be very highly isolated from other background disturbances, such as earthquakes or other vibrations, which might hide or mimic the effect of an incoming gravitational wave.

These problems were faced by Weber about 15 years ago when he constructed the first detector of gravitational radiation. His device is a cylinder of aluminum, one meter in diameter and two meters long, carefully suspended to isolate it from the surrounding environment (Figure 5). The cylinder of aluminum acts both as the pair of masses in the idealized detector, and also as the spring. A gravitational wave sweeping over the aluminum will cause the cylinder alternately to lengthen and contract. It is a measure of the weak coupling between gravitational waves and ordinary matter (and of the sensitivity of Weber's detector) that the change in length of the cylinder produced by a gravitational wave just strong enough to detect is only 10^{-14} cm. I hasten

---

* It is frequently pointed out that the ratio of the gravitational coupling constant to the electromagnetic coupling constant is of order 10^{-40}!

Fig. 5.   One of Weber's gravitational wave detectors. The aluminum cylinder is suspended by a wire sling from a shock absorbing mount. Further insulation from thermal and mechanical disturbances is provided by the heavy vacuum chamber. (Photograph courtesy of J. Weber.)

to add that Weber does not try to detect the change in length of the cylinder directly. He instead measures changes in length indirectly, by looking for the strain produced along the axis of the cylinder.

This, then, is Weber's detector. Beginning in 1969 he reported the detection of short bursts of what appeared to be gravitational radiation (Weber, 1969, 1970a, b, 1972). These bursts occur at a rate of one per day or more, and appear to be coming from the very general direction of the galactic center.* To help verify that the events are real, not random fluctuations in a detector, Weber uses two independent detectors spaced about 1000 km apart. Only those events which register in both detectors are counted.

To indulge in real understatement, his results astounded astronomers and physicists. Because of the weak coupling between gravitational waves and the detector, very large energy fluxes are needed to produce a detectable change in Weber's detectors, approx-

---

* Or the anticenter direction – recall that gravitation radiation is quadrupole in nature, so that his detector has a quadrupole 'beam' (with very low angular resolution, $\sim 45°$): see Tyson and Douglass, 1972.

imately 3×10^5 erg cm^{-2} s^{-1} Hz^{-1}. Let us suppose, for instance, that the source of the events Weber detects is indeed at the center of our Galaxy, some 30000 light years away. Then for *each* single pulse he detects a mass equivalent to at least that of Jupiter would have to be converted to pure gravitational radiation energy! Weber's results presented physicists and astronomers with a real challenge: to explain the nature of a source which for a second or so is 10^{16} times more powerful than the Sun.

There are known sources of gravity waves, as theorists were quick to point out: close double star systems and pulsars (see the contributions by Kraft and Groth, respectively). Both classes of astronomical objects are expected to emit gravitational radiation, but not in bursts, and not with anything like the intensity needed to trigger Weber's detectors (Sejnowski, 1974). Gravitational collapse of a star, a dramatic event to be discussed by Ruffini, could trigger the detectors, but the present view is that only a few stars (at most) die per year in our Galaxy, and that most of these are not massive enough to end their life in catastrophic gravitational collapse. The number of stellar collapses is clearly insufficient to explain Weber's reported results. If, indeed, Weber has been detecting pulses of gravitational radiation, we have no acceptable theory to explain their source.

Experimental physicists and astronomers also rose to the challenge raised by Weber results. A number of groups in the United States and abroad set out to repeat Weber's work with similar, or improved, detectors. Others of us devised observational programs to try to determine the nature of the sources emitting the pulses Weber detected.

Let us consider the latter class of observations first. The idea behind these efforts is simple: a dramatic event of the kind necessary to produce the gravitational wave flux claimed by Weber might very well produce some electromagnetic radiation as well. For instance, it is expected that the collapse of a star in the supernova process will generate an intense pulse of electromagnetic energy (see, for example, Colgate *et al.*, 1972). Starting early in 1970, a number of experimental groups set out to search for pulses of electromagnetic radiation that might accompany the pulses reported by Weber. Recall that both light and gravity waves will travel at the same velocity, so both kinds of signals should reach the earth simultaneously.

Now, as I have pointed out, the interaction of electromagnetic radiation with matter is far stronger than the interaction of gravitational radiation with matter. Therefore, with relatively simple apparatus, it is possible to search for electromagnetic pulses of far lower intensity than the gravitational wave pulses Weber claimed to detect. The first experimental searches for electromagnetic pulses were carried out with relatively unsophisticated equipment. An example is the small radio telescope I employed in such a search in 1970 (Figure 6). The results of this and other early searches were negative. That is, there was no statistically significant tendency for electromagnetic pulses to arrive at the times Weber detected apparent pulses of gravity waves (Charman *et al.*, 1970; Partridge, 1971).

In the past five years, more sophisticated and more sensitive searches for pulsed

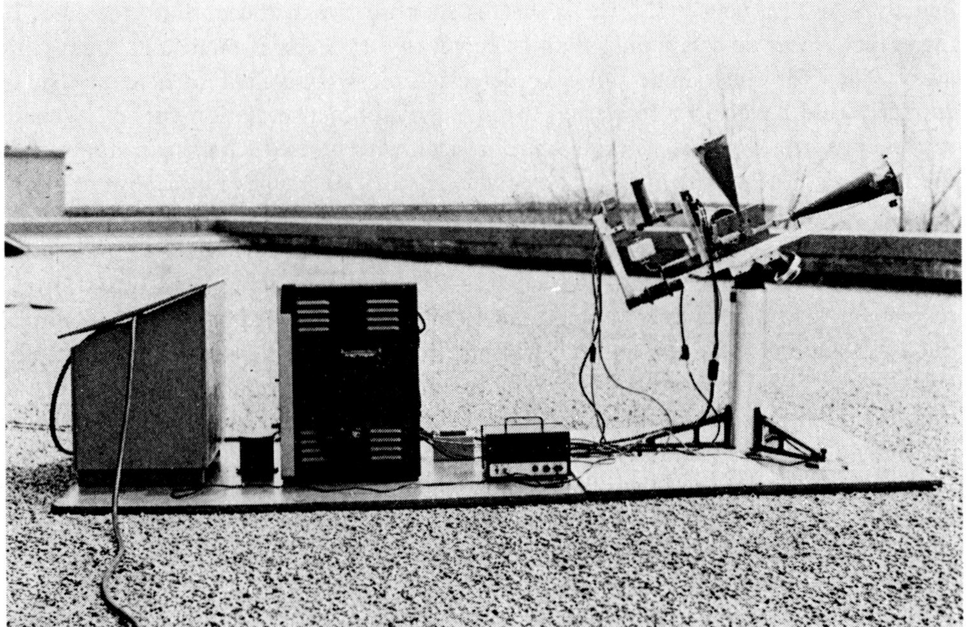

Fig. 6.   A small 19 GHz radiometer used to track the galactic center region in an early search for
radio pulses associated with Weber events.

electromagnetic radiation have been carried out (see, for instance, Baird and Pomerantz, 1972; Slusher and Tyson, 1972; Dube *et al.*, 1973; Partridge *et al.*, 1973). Since
these efforts represent the first systematic attempts to search for aperiodic rapid events
on a galactic scale, let us look at the results in a bit more detail.

Consider first searches for pulses at radio wavelengths. As any observational radio
astronomer well knows, it is all too easy to find pulses in the output of a radio telescope. The vast majority of the pulses arise from man-made, or at least terrestrial,
interference. Three techniques have been employed to distinguish between 'signal'
pulses from astronomical sources and 'noise' pulses from lightning, auto ignitions,
and so on. The first (employed by Hughes and Retallack, 1973) can be used only when
there is a known or suspected astronomical source of pulses. The observer allows the
putative source of radio pulses to drift through the beam of a radio telescope: if more
pulses are detected while the source is in the beam, he concludes the source is emitting
radio pulses. The drawback to this technique is obvious: it is not possible to identify
any single pulse as being astrophysical in origin rather than local. In addition, observing time is wasted, since the source is not always in the beam. The second technique is to employ two independent ratio telescopes some distance apart, both
tracking the assumed source. Only pulses observed simultaneously, or within a narrow time interval, at *both* sites are accepted (e.g., Partridge *et al.*, 1973*). Finally, at

* The two large radio telescopes used in this search are shown in Figures 7 and 8. This technique
closely resembles Weber's method.

longer radio wavelengths, $\lambda \gtrsim 10$ cm, one may make use of the fact that radio pulses traveling through an ionized interstellar medium will be dispersed so that a time delay, Δt, is introduced between the arrival of the high frequency components of a pulse and the low frequency components:

$$\Delta t = t_1 - t_2 = 4 \times 10^{15} \left(\frac{1}{v_1^2} - \frac{1}{v_2^2} \right) \int_0^L n_e \, dl \,,$$

where $v_2 > v_1$ are the two frequencies used for the observations and the integral is the dispersion measure of the source, usually given in units of cm^{-3} pc. The presence of dispersion provides the 'cleanest' test of the extraterrestrial origin of radio pulses. For this reason, it has been widely employed (see, for instance, Ó'Mongáin, 1973; Huguenin and Moore, 1974; or Edwards *et al.*, 1974). However, a glance at the equation above will show that Δt becomes small, and therefore difficult to measure accurately for high frequencies. Furthermore, if low frequencies are employed, pulses will also be dispersed across the finite bandwidth of the detectors. This has the effect of lengthening, and reducing the intensity of, the pulses.

What have the results been of all these searches for pulsed radio emission? First, where comparison with Weber's data was possible, there was no statistically significant correlation between gravitational and electromagnetic pulses: our work lent no support to Weber's findings. Second, many hours were spent observing the galactic center* and nearby regions at wavelengths ranging from 1.5 cm to 250 cm with only one report of a positive result (Hughes and Retallack, 1973). The parameters (and results) of the various observational programs at radio wavelengths are gathered together in Table I.

The emphasis in the measurements appearing in Table I was on high sensitivity over a limited solid angle. There is always the possibility that interesting sources of pulsed radio waves might lie in parts of the sky other than the galactic center.** Several groups (Charman *et al.*, 1970; Colgate *et al.*, 1972; Troitsky *et al.*, 1973; Huguenin and Moore, 1974) have therefore searched for pulses over a wide solid angle, with necessarily reduced sensitivity. Here too, the results were negative: in the most recent experiment, no genuine extraterrestrial radio pulses exceeding 10^{-22} W m^{-2} Hz^{-1} (10^4 fu) were found (see Huguenin and Moore, 1974).

In addition to the work at radio frequencies, searches were carried out in the infrared (Slusher and Tyson, 1972), optical (Baird *et al.*, 1972), and X-ray (Baird and Pomerantz, 1972) regions of the spectrum. Once again, the results were negative. Either no statistically significant pulses were seen, or there was no correlation between electromagnetic and gravitational pulses. In each case, the sensitivity of the electro-

---

* Two groups also observed the Crab Nebula (see Meikle *et al.*, 1972). It was selected because it is astrophysically interesting, containing a central source which is known to change rapidly, and because it lies in the general direction of the galactic anticenter, another possible location for the source of Weber's pulses.
** As is the case for the γ-ray pulses to be discussed below.

Fig. 7. The 20 foot horn antenna of the Bell Telephone Laboratories, Holmdel, N.J. Together with the radiotelescope shown in Figure 8, it was used in a sensitive search for coincident radio pulses on time scales down to 10^{-3} s. (Photograph courtesy of Bell Telephone Laboratories, Holmdel, New Jersey.)

Fig. 8. The 100 foot dish of the Instituto Argentino de Radioastronomia, La Plata, Argentina.

TABLE I

Frequency (MHz)	Mode of observation	Beam size	Time resolution (s)	Flux limit (fu)	Hours of observation	Reference
125	dispersion, 1 telescope	$\sim 14°$	see reference		~ 60	EHM
150	dispersion, several telescopes	$\sim 15°$	0.3–3.5	2.5×10^{-17} Jm^{-2} per event		C
430	dispersion, 1 telescope	$\sim 1°$	0.1	$\lesssim 30$	33	DGRW
858	dispersion, 1 telescope	$\sim 2°$	1.0	pulses detected above threshold of 84 fu	270	HR
1500	2 telescopes	28′ and 2°	0.5	$\lesssim 30$	50	PWPT
1500	2 telescopes	28′ and 2°	0.001	$\lesssim 4000$	50	PWPT
1700	dispersion, 1 telescope	29′	0.1	$\lesssim 63$	28	O
~ 17000	2 telescopes	12′ and 17°	0.5	$\lesssim 100$	90	PW

Results of recent radio searches for pulses from the region of the galactic center. More details on the entries in column 2 are given in the text. One flux unit (fu) $= 10^{-26}$ W m^{-2} Hz^{-1}. The references are: C (Charman et al., 1971); DGRW (Dube et al., 1973); EHM (Edwards et al., 1974); HR (Hughes and Retallack, 1973); O (Ó'Mongáin, 1973); PWPT (Partridge et al., 1973); PW (Partridge and Wrixon, 1972)

magnetic detector was orders of magnitude better than Weber's gravitational wave detectors.

The failure to detect electromagnetic pulses was disappointing. Could it be that the events giving rise to gravitational wave pulses are electromagnetically 'silent'?* The suggestion has been raised. There is, however, a further check available, a search for pulsed neutrino emission. I have already noted that the violent gravitational collapse of a massive object could produce a pulse of gravitational radiation. Such a process would also produce a sudden increase in the temperature and density of matter, leading in turn to copious neutrino emission (see Colgate's contribution for a discussion of neutrino emission in the supernova process). Even an electromagnetically 'silent' collapse might produce a neutrino pulse. Consequently, searches for neutrinos were made (Bahcall and Davis, 1971; Reines et al., 1971; Lande et al., 1974). Three sorts of particles were searched for: – electron neutrinos (v_e) in the approximate energy range 1 MeV $< E <$ 100 MeV; v_μ with $E > 10^3$ MeV; and $\bar{v}_e$ with $E \sim 50$ MeV.

The first search employed the 10^5 gallon solar neutrino 'telescope' of Davis (Davis et al., 1968). In outline, the search proceeded as follows. Neutrinos passing through the 'telescope' tank can convert ^{37}Cl atoms of the C$_2$Cl$_4$ in the tank to argon:

$$v_e + {}^{37}\text{Cl} \rightarrow e^- + {}^{37}\text{Ar}.$$

* In view of the high sensitivity of the searches, the 'silence' would have to be virtually complete: $E^e{}_m/E_{\text{grav}} \lesssim 10^{-20}$ at radio wavelengths.

The argon atoms are periodically removed by sweeping the tank with helium, and then counted. Clearly Davis's tank is an integrating detector: he cannot isolate individual astrophysical neutrino pulses. Instead, Bahcall and Davis (1971) used the low experimental limit on the integrated neutrino flux to set limits on the ratio of the integrated neutrino flux to the integrated energy flux of gravitational radiation: the ratio is less than unity for neutrinos in the 1–100 MeV energy range.

An underground cosmic ray muon neutrino detector has been used by Reines *et al.* (1971) to search for energetic v_μ bursts in coincidence with Weber events. No statistically significant coincidence was reported: consequently, the authors set an upper limit on the ratio E_v/E_g of $\sim 10^{-7}$–10^{-8}. It is not clear, however, that large v_μ fluxes would be expected from collapsing objects of stellar mass.

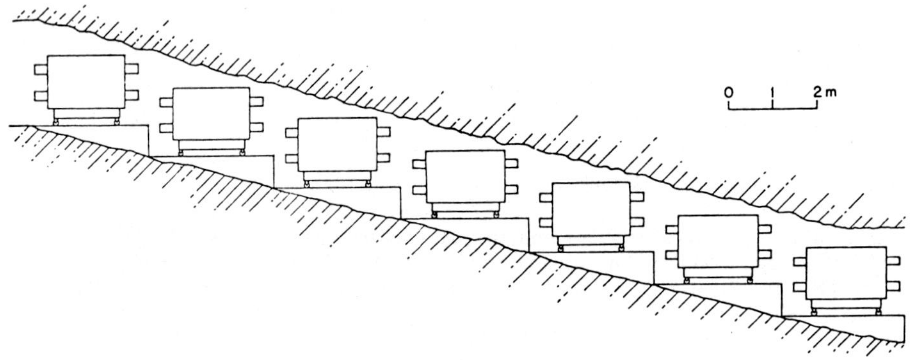

Fig. 9.   The disposition of the 7 Cerenkov counters used to search for astrophysical $\bar{v}_e$ pulses. They are located in a mine tunnel 1500 m underground (from Lande *et al.*, 1974).
© 1974, MacMillan Journals, Ltd.

In this respect, the most recent experimental result is of particular interest – the apparent detection of a $\bar{v}_e$ burst. Antineutrinos as well as neutrinos are expected from gravitational collapse to high density. In late 1973, Lande and his collaborators started a search for energetic $\bar{v}_e$ bursts using an array of water filled Cerenkov counters (the expected reaction is

$$\bar{v}_e + p^+ \rightarrow n + e^+$$

and the e^+ produces the Cerenkov radiation). Figure 9, taken from Lande *et al.* (1974), shows the experimental arrangement. On January 4, 1974, a multiple pulse of energetic $\bar{v}_e$ was observed, consisting of 4 (or more) one microsecond pulses, separated by approximately millisecond intervals (see Figure 10). The experimenters have considered and rejected several instrumental effects and other possible sources of spurious pulses. While the single event is intriguing, it is too early to speculate fruitfully on possible astrophysical events which might have produced it. If more such events occur, we will at least have a rate to compare to the expected rate of stellar collapses in our Galaxy.

No Weber event was observed at the time of the $\bar{v}_e$ event; nor has there been any

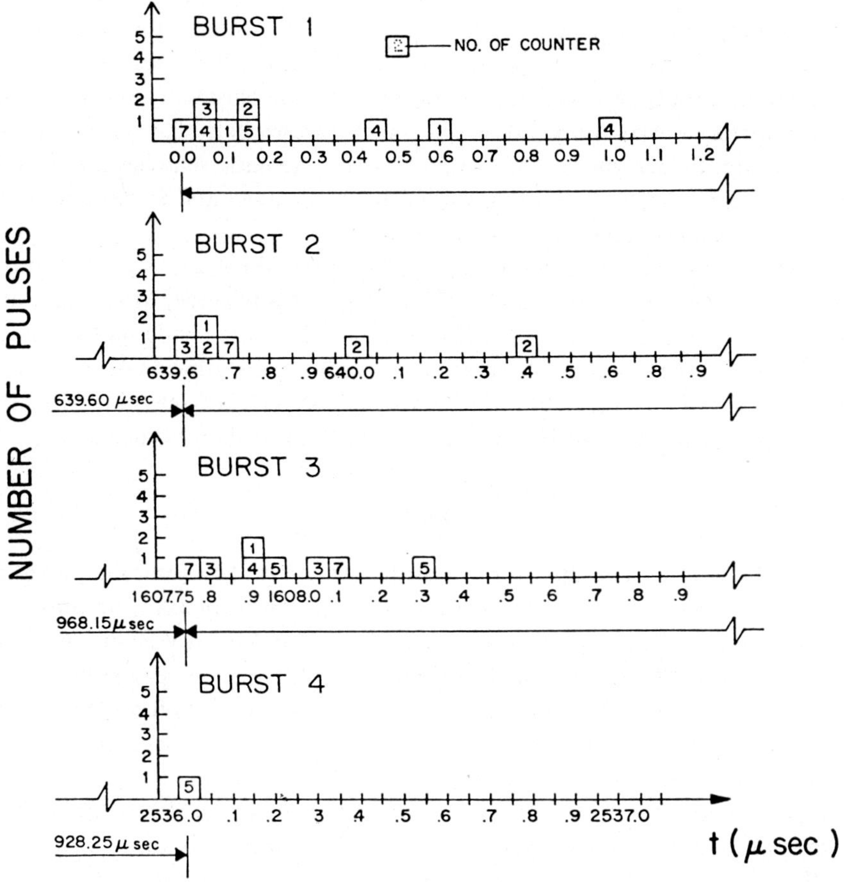

Fig. 10.   The time distribution of the multiple burst observed on January 4, 1974. Note that the interval between bursts is ~ 10⁻³ s (from Lande *et al.*, 1974).
© 1974, MacMillan Journals, Ltd.

report of an accompanying electromagnetic pulse (either in the radio region or the γ-ray region – see below).

In summation, we are left in the unhappy position of having few positive results to show for our searches for pulsed emission, and of having no evidence to help confirm Weber's findings. Indeed, the *absence* of pulsed emission raised early doubts about the straightforward interpretation of Weber's events as pulses of gravitational radiation from the galactic center.

In the past year, additional experimental evidence has come forward which casts far more severe doubt on Weber's work. Experimentalists who set out to duplicate Weber's work (in some cases with improved detectors) have not detected pulsed gravitational radiation (Braginskii *et al.*, 1972; Tyson, 1973; Drever *et al.*, 1973). At the moment (mid 1974) Weber's work is in considerable doubt. Nevertheless, there is no denying that his papers stimulated an active interest in short-lived, energetic, phenomena in astrophysics.

In this section, I have devoted considerable attention to searches for electromagnetic pulses, as well as gravity wave pulses, for the simple reason that these searches represent a substantial body of experimental evidence on the existence (or lack of it!) of pulse phenomena in astronomy. Although one must conclude that the initial results are discouraging, longer, more sensitive and more systematic searches are underway, both for gravitational pulses (Fairbanks and his associates) and for electromagnetic pulses (Jelley, Palumbo and their associates).

While we were laboring in the vineyard, however, three groups, at Cambridge University, at American Science and Engineering in Cambridge, Mass., and at Los Alamos, made separate and dramatic discoveries which have made milliseconds and seconds as much part of the observational astronomer's time scale as days and centuries. Each of these three discoveries was an example of scientific serendipity: each reinforces the wisdom of Heraclitus' advice, 'Expect the unexpected'.

3. Pulsars

Historically the first of these discoveries was the detection of periodically pulsed radio emission from objects now known to lie within our own Galaxy: the pulsars, discovered by Jocelyn Bell and Anthony Hewish of the Mullard Radio Astronomy Laboratory in Cambrdige, England (see Hewish, 1968). The properties of these remarkable objects will be fully described by Groth later in this volume. In this paper, I wish to stress only two rather straightforward aspects of pulsar astrophysics in connection with my topic, pulse astronomy.

The first concerns the range of pulsar periods, from about 4 s down to 0.033 s, which lie at or below the characteristic time limit derived early in this contribution. It follows that we are dealing with highly luminous astronomical bodies which are substantially *smaller* than ordinary main sequence stars. The most extreme case is the pulsar NP 0532 in the Crab Nebula (see Figure 2), with a period of only 0.033 s. Even more remarkable is the fact that the pulsar turns on in a time of order 10^{-3} s (Horowitz *et al.*, 1972; see Groth, this volume). By the argument given earlier, a turn-on time of 10^{-3} s sets an upper limit of ~ 300 km on the size of the emitting region (see Groth's contribution for a confirming calculation, based on the idea that pulsars are actually rotating bodies). Both white dwarfs and main sequence stars are one or more orders of magnitude larger than 300 km. We are driven to the conclusion, now widely accepted, that pulsars are neutron stars, the smallest stable configurations of matter in the range of stellar masses. Note that the short time scale involved in pulsar phenomena was the essential clue to the nature of these objects.

A second important point is that pulsars combine short time scales with a kind of permanence: they vary rapidly, but are not transient. Consequently, we can study them in far greater detail than a single burst. For instance, not just the period but the first and second time derivatives of the period of pulsars can be measured (see Boynton *et al.*, 1972, for an example). The precision of the measurements, especially for the Crab Nebula pulsar, has produced a wealth of information about the physics of neutron stars. For instance, a small but sharp increase in the frequency of the Crab pulsar has been interpreted as evidence for a 'starquake', a sudden cracking of the crystalline crust of the neutron star. The need for precise time measurements is shown by the size of the observed frequency change, $\sim 2 \times 10^{-7}$ Hz (Boynton *et al.*, 1972), corresponding to a change in the neutron star radius of a few microns. We can claim, with only a little overstatement, to be studying the physics of superdense matter ($\varrho \sim 10^{14}$ gm cm^{-3}) in a laboratory 5000 light years away, using a telescope and a precise clock as our apparatus.

4. The Uhuru Satellite and Pulsating X-Ray Sources

Luck and alertness led to the discovery of pulsars (Hewish, 1968). The discovery of pulsating and rapidly varying X-ray sources, on the other hand, required a remarkable improvement in technology as well. I refer to the Uhuru X-ray satellite, launched in December, 1970, by a group including Giacconi and Gursky (Giacconi *et al.*, 1971). Until Uhuru began to operate, our view of the X-ray sky had been limited to brief rocket flights. In its first day of operation, Uhuru more than doubled our observing time at X-ray wavelengths. New X-ray sources were discovered; the positions of known sources were improved.

Of greater interest to us is the discovery of rapid fluctuations in the X-ray intensity of many galactic sources (see, for instance, Oda *et al.*, 1971; Tananbaum *et al.*, 1972). X-ray variability of some sources, generally of the 'now you see it, now you don't sort, had been suggested tentatively before Uhuru. But before consistent measurements with the same instrument became available, it was not clear whether intensity variations were real or whether they arose from improper comparison of data from different rocket flights. Within a year, Uhuru established that there are essentially two types of variable sources: X-ray pulsars with periods of a few seconds (for example Her X-1; see Tananbaum *et al.*, 1972) and irregular variables like Cyg X-1 (Rothschild *et al.*, 1974). The former will be discussed by Boynton, Gursky and Kraft below. The latter class, including Cyg X-1 and Cir X-1, may be the most exciting discovery of this remarkable decade in astronomy. It would be unfair to steal the thunder of either Ruffini or Gursky, who will discuss Cyg X-1 below. Let me emphasize, however, that the rapid fluctuations reported for Cyg X-1 again imply a very small source for the X-rays, far too small to be an ordinary star. How very extraordinary the 'star' is will emerge in Ruffini's paper.

5. Gamma Ray Bursts

To complete our list, the fifth and most recent discovery which has shaken the paradigm of patience in astronomy is the detection of γ-ray bursts (Klebesadel *et al.*, 1973). These intense – and unexpected* – events will be discussed by Strong in his contribu-

---

* Amusingly enough, Field *et al.* (1969) did suggest searching for γ-ray pulses, which they argued might be emitted by the source or sources of Weber's gravitational radiation pulses.

tion. The short time scales suggest sources of stellar size or smaller: the high energy flux (and the very high energy of the individual photons involved) suggest a very exotic source. At the moment, we have no idea what the source of γ-ray pulses might be. We do not even know whether the sources are local (nearby stars) or extragalactic (see Strong *et al.*, 1974). Attempts to find correlations between γ-ray pulses and other pulsed electromagnetic emission have not so far been successful, except in the X-ray region. Clearly the puzzles presented by this branch of pulse astronomy will be with us for some time.

6. Conclusions

Virtually every issue of *Nature* or of the *Astrophysical Journal* now contains an article dealing with astronomical phenomena occurring on a time scale of seconds or less. These phenomena are being studied not just for their own sake, but also for the light they can shed on a wide range of astrophysical problems, from stellar evolution to the properties of collapsed matter. In the next decade, we may even see the popular image of astronomy widen to include fast and accurate clocks as well as telescopes, photographic plates, patience, and plenty of hot coffee.

Acknowledgements

I acknowledge the support of the Alfred P. Sloan Foundation in the preparation of this report. Convivial, informative, discussions with many of the other contributors to this volume have added much to what I have written.

References

The references are grouped by topics, in the same order as they are treated in this paper. No claim is made that the references are complete (or even very representative), except those on the search for electromagnetic pulses (2 below), which I believe to be complete as of March 1, 1974.

1. *The Death of Stars*

Brancasio, P. J. and Cameron, A. G. W.: 1969, *Supernovae and Their Remnants*, Gordon and Breach, N. Y.
Colgate, S. A. and White, R. H.: 1966, *Astrophys. J.* **143**, 626.
Cosmovici, C. B. (ed.): 1974, *Supernovae and Supernovae Remnants*, D. Reidel, Dordrecht.
Fowler, W. A. and Hoyle, F.: 1964, *Astrophys. J. Suppl.* **9**, 201.
Hoyle, F. and Fowler, W. A.: 1960 *Astrophys. J.* **132**, 565.
Iben, I.: 1967, *Astrophys. J.* **147**, 624.
Paczyński, B. E.: 1970, *Acta Astron.* **20**, 47.
Shklovskii, I. S.: 1968, *Supernovae*, John Wiley Interscience, N. Y.
Wheeler, J. C.: 1973, *American Scientist* **61**, 42. See also Colgate: this volume, p. 13.

2. *Pulses of Gravitational (and Electromagnetic) Radiation*

Bahcall, J. N. and Davis, R.: 1971, *Phys. Rev. Letters* **26**, 662.
Baird, G. A., Delaney, T. J., and Lawless, B. G.: 1972, *Observatory* **92**, 233.
Baird, G. A. and Pomerantz, M. A.: 1972, *Phys. Rev. Letters* **28**, 1337.
Braginskii, B. B., Manukin, A. B., Popov, E. I., Rudenko, V. N., and Khorev, A. A.: 1972, *Zh.E.T.F. Pis'ma Red.* **16**, 157. In English in *J.E.T.P. Letters* **16**, 108.

Charman, W. N., Jelley, J. V., Fruin, J. H., Hodgson, E. R., Scott, P. F., Shakeshaft, J. R., Baird, G. A., Delaney, T. J., Lawless, B. G., Drever, R. W. P., and Meikle, W. P. S.: 1970, *Nature* **228**, 346.
Charman, W. N., Jelley, J. V., Fruin, J. H., Hodgson, E. R., Scott, P. F., Shakeshaft, J. R., Baird, G. A., Delaney, T. J., Lawless, B. G., Drever, R. W. P., Meikle W. P. S., Porter, R. A., and Spencer, R. E.: 1971, *Nature* **232**, 177.
Colgate, S. A., McKee, C. R., and Blevins, B.: 1972, *Astrophys. J. Letters* **173**, L87.
Davis, R., Harmer, D. S., and Hoffman, K. C.: 1968, *Phys. Rev. Letters* **20**. 1205.
Drever, R. W. P., Hough, J., Bland, R., and Lessnoff, G. W.: 1973, *Nature* **246**, 340.
Dube, R., Groth, E. J., Rudnick, L., and Wilkinson, D. T.: 1973, *Nature Phys. Sci.* **245**, 17.
Edwards, P. J., Hurst, R. B., and McQueen, M. P. C.: 1974, *Nature* **247**, 444.
Field, G. B., Rees, M. J., and Sciama, D. W.: 1969, *Comments Astrophys. Space Sci.* **1**, 187.
Hughes, V. A. and Retallack, D. S.: 1973, *Nature* **242**, 105.
Huguenin, G. R. and Moore, E. L.: 1974, *Astrophys. J. Letters* **187**, L57.
Lande, K., Bozoki, G., Frati, W., Lee, C. K., Fenyves, E., and Saavedra, O.: 1974, *Nature* **251**, 485.
Meikle. W. P. S., Drever, R. W. P., Haynes, R. F., Shakeshaft, J. R., Charman, W. N., and Jelley, J. V. 1972, *Monthly Notices Roy. Astron. Soc.* **160**, 5pp.
Ó'Mongáin, E.: 1973, *Nature Phys. Sci.* **242**, 136
Partridge, R. B.: 1971, *Phys. Rev. Letters* **26**, 912.
Partridge, R. B. and Wrixon, G. T.: 1972, *Astrophys. J. Letters* **173**, L75.
Partridge, R. B., Wrixon, G. T., Pena, H., and Turner, K. C.: 1973, *Nature Phys. Sci.* **245**, 53.
Reines, F., *et al.*: 1971, *Phys. Rev. Letters* **26**, 1451.
Sejnowski, T. J.: 1974, *Physics Today* **27**, 40.
Slusher, R. E. and Tyson, J. A.: 1973, *Nature* **243**, 25.
Troitsky, V. S., *et al.*: 1973, *Radiophysika* **16**, 323.
Tyson, J. A.: 1973, *Phys. Rev. Letters* **31**, 326.
Tyson, J. A. and Douglass, D. H.: 1972, *Phys. Rev. Letters* **28**, 991.
Weber, J.: 1960, *Phys. Rev.* **117**, 306.
Weber, J.: 1968, *Physics Today* **21**, 34.
Weber, J.: 1969, *Phys. Rev. Letters* **22**, 1320.
Weber, J.: 1970a, *Phys. Rev. Letters* **24**, 276.
Weber, J.: 1970b, *Phys. Rev. Letters* **25**, 180.
Weber, J.: 1971, *Nuovo Cimento* **4B**, 197.
Weber, J.: 1972, *Nature* **240**, 28.

3. *Pulsars*

Boynton, P. E., Groth, E. J., Hutchinson, D. P. Nanos, G. P., Partridge, R. B., and Wilkinson, D. T.: 1972, *Astrophys. J.* **175**, 217.
Hewish, A.: 1968 in *Pulsating Stars* (p. vii and 5), Plenum Press, N. Y.
Horowitz, P., Papaliolios, C., and Carleton, N. P.: 1972, *Astrophys. J. Letters* **172**, L51.
See also Groth: this volume, p. 121.

4. *Pulsating X-Ray Sources*

Giacconi, R., Kellogg, E., Gorenstein, P., Gursky, H., and Tananbaum, H.: 1971, *Astrophys. J. Letters* **165**, L27.
Oda, M., Gorenstein, P., Gursky, H., Kellogg, E., Schreier, E., Tananbaum, H., and Giacconi, R.: *Astrophys. J. Letters* **166**, L1.
Rothschild, R. E., Boldt, E. A., Holt, S. S., and Serlemitsos, P. J.: 1974, *Astrophys. J. Letters* **184**, L13.
Tananbaum, H., Gursky, H., Kellogg, E. M., Levinson, R., Schreier, E., and Giacconi, R.: 1972, *Astrophys. J. Letters* **174**, L143. See also Boynton, Gursky, Kraft and Ruffini: this volume, resp. pp. 221, 175, 235 and 59.

5. *Gamma-Ray Bursts*

Field, G. B., Rees, M. J., and Sciama, D. W.: 1969, *Comments Astrophys. Space Sci.* **1**, 187.
Klebesadel, R. W., Strong, I. B., and Olson, R. A.: 1973, *Astrophys. J. Letters* **182**, L85.
Strong, I. B., Klebesadel, R. W., and Olson, R. A.: 1974, *Astrophys. J. Letters* **188**, L1.
See also Strong: this volume, p. 47.

COSMIC GAMMA-RAY BURSTS

IAN BALFOUR STRONG

University of California, Los Alamos Scientific Laboratory, Los Alamos, N.M., 87544, U.S.A.

1. Introduction

We describe the recent discovery by the Vela satellite system of short, extremely intense, bursts of gamma rays originating outside the solar system, and the subsequent confirmation by a number of other independent observers. The known properties of these so-called gamma-ray bursts will be given together with outlines of some of the more promising explanations produced so far. Essentially all these explanations invoke compact stellar objects – that is, dwarf stars, neutron stars or black holes.

This chapter on cosmic gamma-ray bursts will be divided into three parts. I will begin with an 'historical' introduction, since even so recently discovered a phenomenon has a history. I will describe the satellites and instruments used, how the bursts were discovered, and how the original findings were corroborated and even extended by other workers.

Next, the properties of these gamma-ray bursts will be described, to the extent we presently know them. This will include their durations and fluctuations, their spectra, source directions, and the energies involved for different possible cosmic distances. We then make some plausible deductions from this data about the likely distances to the known sources. Finally, I will outline some of the imaginative, and at first sight perhaps implausible models that have been suggested to account for this phenomenon.

2. The Discovery of Cosmic Gamma-Ray Bursts

New scientific discoveries are made in many different ways, but most can probably be classified into three categories. The popular image is sudden (or slow) inspiration, usually the culmination of a long period of study. Then there are findings which result primarily from the development of new instruments or techniques, where one starts with a certain objective in mind. Finally, there is the accidental discovery – serendipity – looking for one thing and finding something else, quite unexpected. This must be distinguished from what might be called 'dumb luck', since it involves a certain receptiveness involving both curiosity and background knowledge and preparation.

I think this story involves all three of these categories. The discovery required the latter two – development of new techniques, together with a fair amount of lucky accident, properly followed up. The interpretations to be described later are certainly inspired.

The original discovery was made possible by the development of the Vela satellite system to monitor the Limited Test Ban Treaty of 1963. Over the last decade six pairs of Vela satellites have been launched into large circular orbits about the Earth – about

H. Gursky and R. Ruffini (eds.), Neutron Stars, Black Holes and Binary X-Ray Sources, 47–58. All Rights Reserved.
Copyright © 1975 by D. Reidel Publishing Company, Dordrecht-Holland.

Fig. 1. Artists conception of the Vela Satellites at the time of separation in orbit. (Courtesy, TWR Systems, Redondo Beach, California.)

one hundred and fifty thousand miles in diameter. Each satellite can view almost every direction in space, unshielded by the Earth, and can simultaneously monitor half the Earth at a time. Since the satellites are always launched in pairs, shown at their moment of deployment in Figure 1, with one on each side of the Earth, the whole of space, and the whole of the Earth's surface is constantly under surveillance.

The satellites have always included X-ray, neutron, and gamma-ray detectors, the latter to detect gamma rays emitted as a nuclear weapon is detonated. But some people worried that a bomb might be secretly tested behind the Moon which would shield it from the Vela satellites. To counter this possibility, new detectors were added, designed to respond to the gamma radiation emitted by the radioactive debris as it expanded from behind the limb of the Moon. Each satellite was equipped with six of these small detectors, scattered around the inside, so that radiation from any direction could be seen. Such detectors also respond to the fluxes of charged particles that populate space, in particular to the well known cosmic ray particles. This results in a fairly steady counting rate even with no gamma-rays present, and is used to indicate the instruments are working.

If, however, the rate rises suddenly, the electronic logic systems aboard go into action, storing the next fifteen minutes of data into a memory for later transmission to Earth, together with the exact time at which the rate increase occurred.

Unfortunately, several other occurrences can trigger the satellite this way, for example blasts of charged particles from the Sun, which occur fairly frequently. Because of this, such data were usually ignored unless we had other reasons to be suspicious.

Now for the serendipity part. Bursts of charged particles are purely local events or at least considerable time for them to travel from one satellite to another would be required, so that only one satellite system would be triggered at a time. Blasts of gamma rays, on the other hand, would never need more than 0.8 s to trigger a detector system and then cross the orbit to trigger another satellite (since the orbit is 0.8 light second in diameter).

My colleague at Los Alamos, Ray Klebesadel, was checking the accidental triggering rate and concluded that at nearly the same time two satellites were triggering at a rate several hundred times greater than would be expected on a purely accidental basis.

It took several months to eliminate all other possibilities but we finally convinced ourselves that we were indeed seeing blasts of gamma radiation in space. Furthermore, by checking the time of passage at more than one satellite it is possible to find out something about the source direction. Clearly, any radiation travelling along the line joining two satellites would take the full 0.8 s to do so. Radiation coming broadside to this line would arrive simultaneously at the two satellites. In-between directions would require times between zero and 0.8 s. By developing this technique we were able to show that the gamma rays were not coming from any known object in the solar system, including the Earth and Sun.

These results were published last June in the *Astrophysical Journal* [1] and six of the events were promptly confirmed by Cline and Desai of [2] NASA's Goddard Space

Flight Center, using the IMP-6 satellite data. They also made spectral measurements from 100 keV to 1 MeV and showed that the spectrum was 'hard', meaning a lot of high energy photons compared to the number of low-energy photons. In fact, most of the energy was around 150 keV. Next, Peterson's X-ray astronomy group at the University of California in San Diego were able to find one of those events in their old records, and they both measured the X-ray part and provided a location, because their detector only scans a small area of the sky at time a [3]. The direction agreed almost perfectly with one of two possibilities that we had proposed.

Since the original publication, one or more events in the Vela List have been seen by more than a dozen different groups of investigators, not only American but also Italian, German and, recently, a Russian group under E. P. Mazets of the Ioffe Institute in Leningrad [4].

3. The Characteristics of Gamma-Ray Bursts

Let us now take a look at some examples of gamma-ray bursts as seen by the Vela satellites, starting with one of the briefest in Figure 2, The figure shows four seconds of

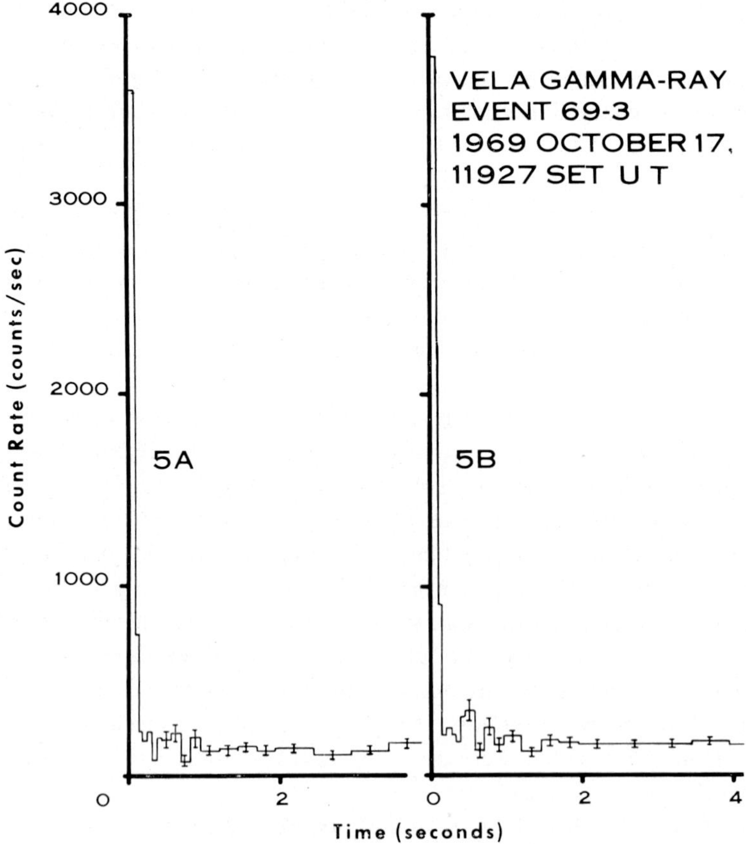

Fig. 2.   Event 69-3, on 1969 October 17, beginning 11927 s UT.

data for both the Vela 5A and the Vela 5B satellites, placed side by side for comparison. For almost the whole four seconds the detectors are counting at the background rate of about 150 counts s^{-1}. (Even this is not due to gamma rays. The steady low flux of gamma rays is thousands of times weaker than this. Cosmic-ray particles contribute almost all the counts here.) On the very left edge of each graph is what we have recorded of the gamma-ray burst. To the left of 'zero seconds' we do not know exactly what happened, but the indications are that it was like background until almost the point marked zero seconds and then rose abruptly to the peak we see at about 3500 counts s^{-1}.

In Figure 3 we have a less extreme case of a gamma-ray burst. A number of features are noticeable here. The first peak is wider than in the previous example, about half a second long altogether. This peak also has 'structure', in the shape of two sharp spikes, a very common phenomenon. There is also an 'extra' peak at about 2.5 s from the beginning. This kind of trailing pulse occurs in about two-thirds of the gamma-ray bursts we have analyzed. Notice that the count rate is back at the background rate between the two peaks – it stops completely and starts again. One final feature of this

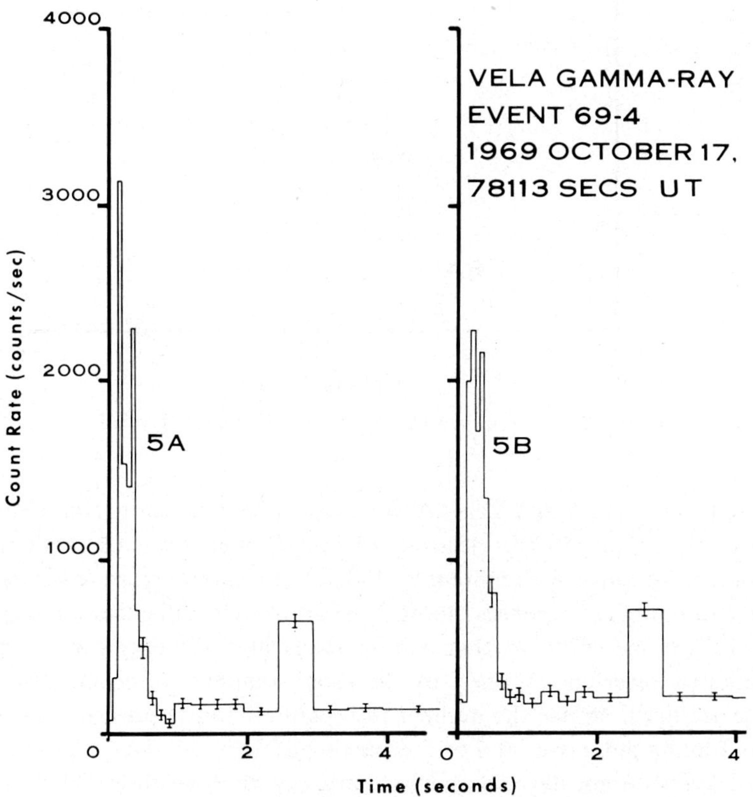

Fig. 3.   Event 69-4, on 1969 October 17, beginning 78113 s UT.

event. At zero seconds the burst does not appear to have already begun as it did in
Figure 2. In fact, it can be seen to climb in the Vela 5A data. This is one of those misleading effects we must watch for. The satellites would not even have recorded this
data unless they had detected an increased rate, as I described above. We conclude
that the burst had one or more unrecorded spikes to the left of zero seconds, and that
the resulting counts triggered the systems. We thus see that not *all* of the burst is recorded here.

Figure 4 is a more typical gamma-ray burst, in that it is not ultra-brief. Here we have

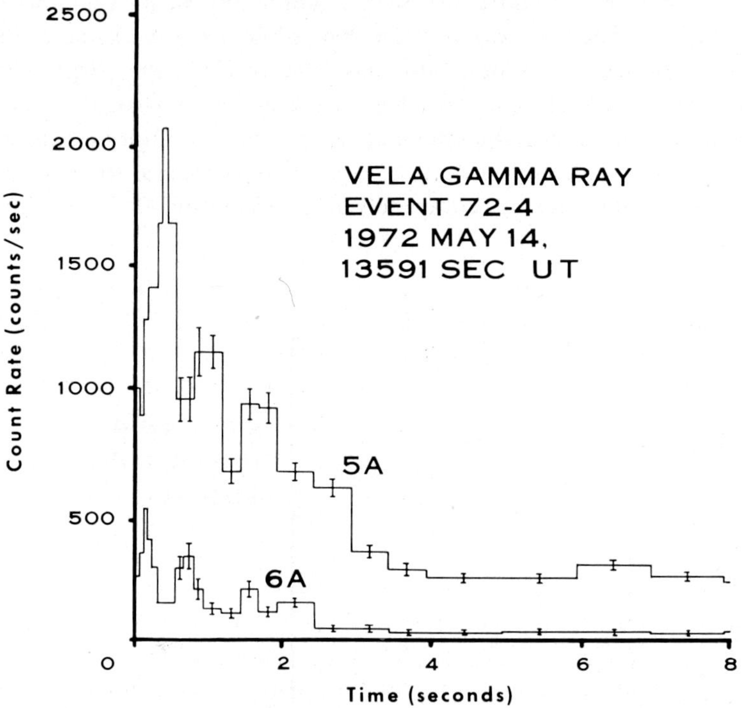

Fig. 4.   Event 72-4, on 1972 May 14, beginning 13591 s UT.

the response from Vela 5A and Vela 6A. Since the Vela 5 detectors count gamma-ray
photons with more than 150 kilo electron-volts (keV) of energy each, while the Vela 6
detectors only count those with more than 300 keV, the latter record fewer counts. We
are also able to put both responses on the same graph. The differences in response tell
us how 'hard' the spectrum is – that is, how many higher energy gamma rays there
are compared to lower energy ones. By the usual standards of cosmic-ray measurements these are hard. Notice the number of separate bumps appearing in the graph,
and a weak trailing pulse over at 6 or 7 s, barely visible in 6A. This is also a common
effect. The trailers do not have so many gamma-ray photons above 300 keV.

Our final example from the Vela data, Figure 5 is in another sense the first. It is the

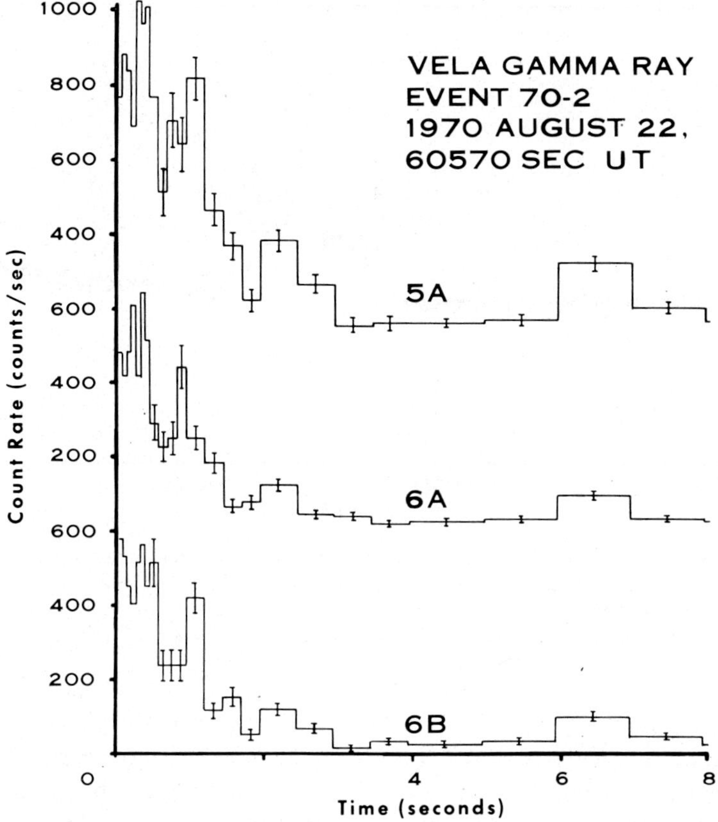

Fig. 5.   Event 70-2, on 1970 August 22, beginning 60571 s UT.

one used in the original paper in the Astrophysical Journal by Klebesadel *et al.* [1] when reporting the discovery, although here plotted differently. It behaves in more or less the same fashion as the one in Figure 4 down to the trailing pulse at between 6 and 7 s.

Partly to show that not everybody's data looks like the Vela data when plotted, we show in Figure 6 the beautiful recording of event number 72-1 (meaning the first one seen in the 1972 data) by the Kosmos 461 satellite, using E. P. Mazets' figure (translated!) [4]. The Kosmos 461 satellite was quite close to the Earth and has a large background rate compared to Vela. Also it does not trigger, but records everything all the time. This way we see clearly how outstanding these gamma-ray bursts can appear in the data.

Let us take a look at the directions from which these bursts are coming. First we take note that we can find directions for only about a dozen; still worse, even for these the calculations give us two possible answers. The reason, in non-mathematical language, is more or less that the individual detectors cannot give directions, only a combination of at least three detectors, which are all in the same plane (the orbit of the

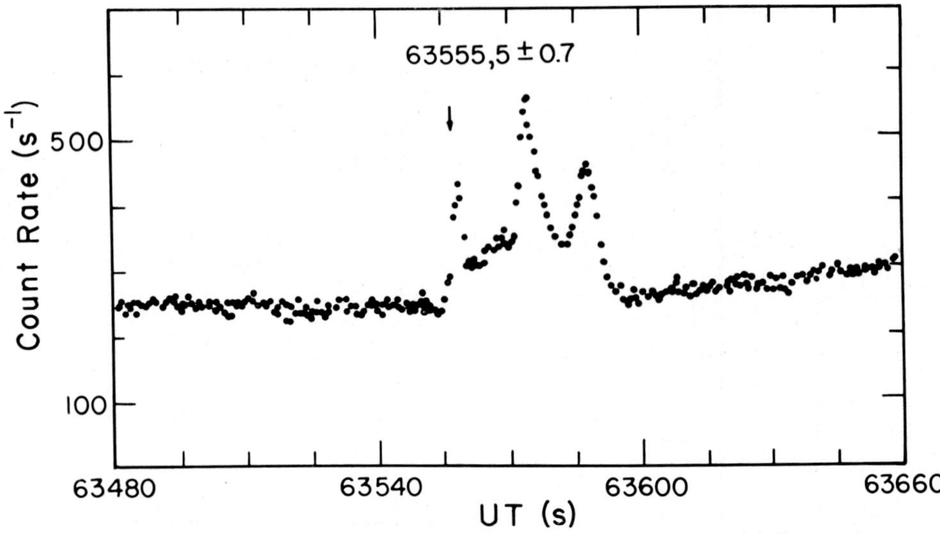

Fig. 6.   Intense burst of gamma-radiation in the range 0.05–0.3 MeV. Kosmos 461 data on event 72–1. All three prominent peaks appear in the Vela data together with a fourth peak at ∼ 63565 s UT. In the Vela data the count rate for the two later peaks is reduced relative to the Kosmos 461 data, indicating a softening of the spectrum (after Mazets *et al.* [4]).

satellites). The trouble is that this gives directions away from the plane, but the set of detectors cannot tell one side of the plane from the other – hence we get two answers, one right and one wrong. In a number of cases we have been able to eliminate the wrong one. Sometimes it would not be visible from a satellite which recorded the burst because the Earth was in the way. Sometimes another detector which responds like a telescope, only when pointing at the source, happened to see it.

Figure 7 is a map of the sky, where the Milky Way, if shown, would run right to left across the middle. The black discs are sources we are sure of. The ambiguous ones are shown as pairs of open circles joined by a dotted line.

We can learn much even from this little information. If they originate in our Galaxy, then we are not seeing all of them (the detectors are not big enough, or sensitive enough) since if we did, they would be mostly around the center of the map where most of the stars in the Galaxy are. We are not even seeing, say, a tenth of them, since even then they would string across the middle of the map. We conclude that they are either fairly close, within a thousand light years, or *very* far away, outside our Galaxy at a million light years or more.

Let us at this point make a summary of the data presented above, together with some other data, and deductions therefrom. First, the events are brief – typically a few seconds suffice for most of the radiation to arrive, although a few rumble on slowly for a minute or more. About half a dozen of the Vela-recorded events last as little as a tenth of a second. Next, they are very intense. During those few seconds, the gamma radiation from these sources exceeds by many orders of magnitude all the steady

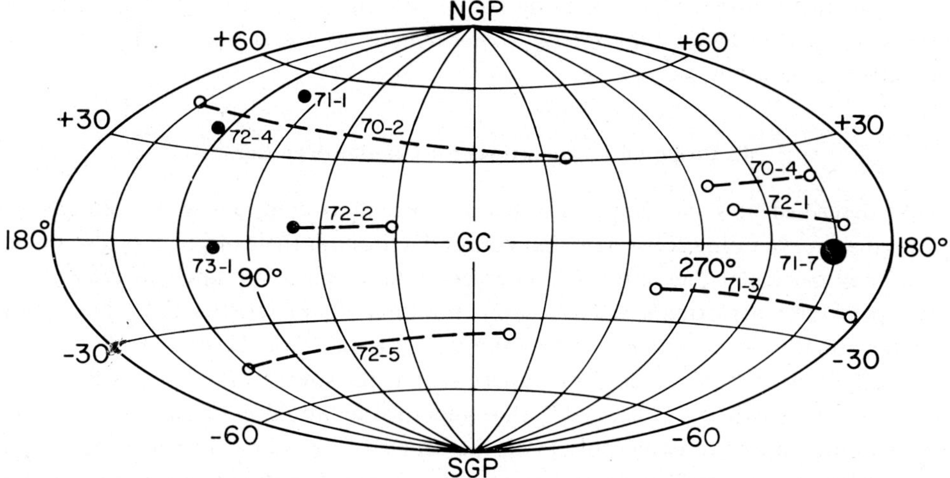

Fig. 7. Distribution of gamma-ray burst sources in galactic coordinates. Where alternative positions exist these are shown with open circles, connected by a dashed line. Unique directions are shown by a black disk. Event number 71-7 has a greater uncertainty in position than the others, and has a larger disk. Event 72-2 has one circle in black. It appears likely that one position for event 71-2 (not shown) is very close to this point and may represent a repeat burst from the same object (see Strong *et al.* [12]).

gamma-ray radiation from all regions of the cosmos. The bursts contain great power in energetic photons in the gamma-ray region and less in X-radiation, unlike, for example, solar flares. They have fast fluctuations, changing significantly within a hundredth of a second, thus showing us that the sources are small, $\sim$a few hundredths of a light-second in diameter, at the very most, corresponding to a few thousands miles. They are not frequent, at least at the level to which we are sensitive. We have discovered just under thirty in four years of data. They sometimes have a complex structure, with several independent peaks. Their source directions, where known, do not correspond to any obvious objects in the sky – such as supernovae, novae, nor to the X-ray sources. They are scattered apparently randomly over the sky. This tells us that either they are relatively close and in our Galaxy, no more than a few hundred light years away, or, alternatively they are extra-galactic. If they were in our Galaxy but at *more than* a few hundred light years, they would appear clustered along the Milky Way. Assuming that at the source the energy is radiated uniformly in all directions, they put out roughly $\sim 10^{39}$ erg if in our Galaxy or $\sim 10^{48}$ erg if they lie in other galaxies. At even the minimum likely distance the energy release is very large.

Numbers like 10^{39} really do not mean anything, even to scientists, unless they have something about that size to compare against. Let us try to do that.

The Sun puts out 4×10^{33} erg of energy each second, almost all of it optical, infra-red and ultra violet. It takes about three days to radiate 10^{39} erg. Of this energy, the Earth only recovers four parts in every ten billion. It therefore would take about twenty million years for the whole Earth to get 10^{39} erg of sunlight. Why choose the Earth –

is this not just a little unfair – making something big look even bigger? Not really, because the place where all this energy *comes* from is no bigger than the Earth, and it all comes out in a few seconds!

4. Proposed Explanations for Gamma-Ray Bursts

I have described how the gamma-ray bursts were discovered and what their known properties are. In doing so we saw that certain facts about their origins are also determined – in particular they are small, energetic and, as far as we know at present, do not repeat their performance. To account for these characteristics there have been thus far about a dozen articles written suggesting models for these events.

Initially, speculation centered on supernovae as the sources. Predictions had been made earlier by Stirling Colgate [5] that supernovae would emit gamma radiation which would appear to come only from the center of the star's surface as seen by a given observer. Edward Teller was also of this opinion. The merits of this idea are the ease in explaining the enormous energies even for extra-galactic sources, and the short time scale of the events. Difficulties include the measured spectrum and multiplicity of peaks.

A second class of models assumes a dwarf star to be involved. Such small (Earth-sized) hot, dense stars have been known for half a century or more. Floyd Stecker and Kenneth Frost [6] of NASA propose that some of these stars, which have much stronger magnetic fields than the sun, also emit flares as does our Sun. By extrapolating from known solar flare properties, it might be possible to imagine flares billions of times more productive of X-rays or gamma rays than the Sun. Philip Morrison and Kenneth Brecher [7] of MIT in a somewhat different variation assume the flare emits the radiation in a concentrated beam, thus cutting the energy requirements by several orders of magnitude, but this requires the events to be much more frequent. Both these theories have advantages, such as accounting for the time variatons, but suffer disadvantages in that they have to extrapolate so far from what is presently known about flares on the Sun or on so-called 'flare stars'. Don Clayton of Rice University and Fred Hoyle [8] recently suggested that the source may be a class of binary stars related to the novae, in which a very small dwarf star constantly attracts material from a giant companion and occasionally reaches a state of instability. The star then may erupt as a nova, or in some cases erupt by explosive emission of material and radiation, seen as gamma-ray bursts but not as an optical nova.

A third class of models utilizes a neutron star as the source. If we drop material onto an ordinary dwarf star, each nucleon (e.g. proton, or hydrogen atom) would have up to 1 MeV in energy of motion on contact. This is not enough to give directly the gamma-ray energies per photon that have been measured. Neutron stars, however, can give up to 100 MeV per nucleon, more than enough. Donald and Fred Lamb, and David Pines [9] of Illinois have proposed an ingenious theory in which a flare star emits material which flows away and then 'crashes' onto a companion neutron star; and then through complex interactions with material in orbit around the neutron star, bursts of gamma radiation are emitted. Martin Harwit and E. Salpeter of Cornell [10] have proposed similar models except that the flare star is replaced as a source of material by a comet. The comet approaches so close to the neutron star that it is disrupted, some of the material falling onto the star and the rest going off into a modified orbit. Note that in none of these neutron star theories does the matter strike the neutron star *directly* – putting it crudely, it is just too small to hit! The most radical model has been proposed by Grindlay and Fazio [11] of the Center for Astrophysics. They envisage a neutron star emitting small, mostly iron particles, during one of their occasional rearrangements. These particles, travelling at relativistic velocities would, in some cases, arrive in the neighborhood of another star, such as our Sun. At distances well outside the orbit of Pluto these 'dust grains' would scatter back the sunlight photons which would appear to us, on return, as gamma rays. A further class of sources which has been suggested by Reuven Ramaty of NASA and Remo Ruffini of Princeton envisages emission of the radiation as a dying dwarf star collapses.

Very recently, the author, with his colleague Ray Klebesadel [12], have proposed in a paper that of the two possible ranges of source distance (near or far – see earlier in this chapter) the weight of the evidence is on the sources being nearby – about a thousand to several thousand lightyears away. While analyzing the data on directions for that paper the author also came across information suggesting that the famous Cygnus X-1, the peculiar X-ray star suspected of containing a 'black hole', was a possible source for two gamma-ray bursts happening a year apart. We have now formally pointed out, in a paper submitted to the *Astrophysical Journal*, the evidence for suspecting this association*. It is far from being firm, and Cyg X-1 itself is only *suspected* of being a black hole, but it is very tantalizing, since if gamma-ray bursts originate in black holes (strictly speaking, in the associated 'accretion disc' of material, orbiting and spiralling into the black hole rather like Saturn's rings) then a few of them may be companions of stars visible to the naked eye, well away from the obsuring material of the Milky Way!

Perhaps the most fascinating thing about gamma-ray bursts is that today, a year after the announcement of their discovery, we are still so unsure as to their origin. Perhaps Philip Morrison is right when he suggested that of all the proposed models (including his own, and perhaps ours) probably none is yet correct.

Acknowledgements

The author is greatly indebted to his colleague Ray Klebesadel, also of the Los Alamos Scientific Laboratory of the University of California, who has performed most of this data analysis. The author acknowledges the support of the U.S. Atomic Energy Commission which, jointly with the U.S. Department of Defense, sponsors the Vela Satellite Program.

References

1. Klebesadel, R. W., Strong, I. B., and Olson, R. A.: *Astrophys. J. Letters* **182**, L85 (1973).

Strong, I. B., and Klebesadel, R. W.: *Nature* **251**, (1974).

Strong, I. B., Klebesadel, R. W., and Olson, R. A.: *Astrophys. J. Letters* **188**, L1 (1974).

2. Cline, T. L., Desai, U. D., Klebesadel, R. W., and Strong, I. B.: *Astrophys. J. Letters* **185**, L1 (1973).

3. Wheaton, W. A., Ulmer, M. P., Baity, W. A., Datlowe, D. W., Elcan, M. H., Peterson, L. E., Klebesadel, R. W., Strong, I. B., Cline, T. L., and Desai, U. D.: *Astrophys. J. Letters* **185**, L57 (1973).

4. Mazets, E. P., Golenetskii, S. V., and Il'Inskii, V. N.: *Zh.E.T.F. Pis'ma* **19**, 126 (1974).

5. Colgate, S. A.: *Astrophys. J.* **187**, 333 (1974).

6. Stecker, F. W. and Frost, K. J.: *Nature Phys. Sci.* **245**, 70 (1973).

Strong, I. B. (ed.): 'Transient Cosmic Gamma-and X-ray Sources', LA Report 5505-C (1974).

7. Brecher, K. and Morrison, P.: *Astrophys. J. Letters* **187**, L97 (1974).

8. Clayton, D. D. and Hoyle, F.: *Astrophys. J. Letters* **187**, L101 (1974).

9. Lamb, D. Q., Lamb, F. K., and Pines, D.: *Nature Phys. Sci.* **246**, 52 (1973).

10. Harwit, M. and Salpeter, E. E.: *Astrophys. J. Letters* **186**, L37 (1973).

11. Grindlay, J. E. and Fazio, G. G.: *Astrophys. J. Letters* **187**, L93 (1974).

12. Strong, I. B., Klebesadel, R. W., and Olson, R. A.: *Astrophys. J. Letters*, submitted to Editor, May 1974.

THE PHYSICS OF GRAVITATIONALLY COLLAPSED OBJECTS

REMO RUFFINI

Joseph Henry Physical Laboratories, Princeton, N.J. 08540, U.S.A.

1. Introduction

For a long time it has been recognized that Einstein's theory of general relativity is very likely the most elegant theoretical framework in modern physics. However, all conceivable effects predicted by this theory and observable inside our own solar system are largely negligible and could be taken into account by a simple 'tabulation' of the correction factors from the traditional Newtonian physics. If these observable effects were found to be so negligible in the entire Universe, the relevance of general relativity, despite its mathematical elegance, would certainly have been very limited. However, it has become more and more clear since the pioneering work of Landau [1], Chandrasekhar [2], Baade and Zwicky [3] that to properly describe the processes occurring at the late stages of evolution of a star after all the sources of its thermonuclear energy have been exhausted, a fully relativistic theory of gravity is needed and very large deviations from a Newtonian approach are to be expected. Then the process of gravitational collapse appears to be the natural testing ground where one may probe some of the most novel and unique predictions of Einstein's theory. From an astrophysical point of view, this process is also of the greatest relevance since it represents energetically, by far the most important part of the life of a star. (See Section 7).

We have today credible evidence for the existence, and in some cases direct observation, of a large number of collapsed objects (neutron stars or black holes) inside our own Galaxy. Their number is most likely larger than 10^8 (See Section 7).

Nothing did more for the development of this entire field of research in relativistic astrophysics than the clear identification in 1968 of a pulsar (NP0532) at the center of the Crab Nebula [4]. That the Crab Nebula, still expanding today at a velocity of ~ 800 km s^{-1}, was the remnant of the supernova explosion recorded with great scientific accuracy by the Chinese and Japanese astronomers in 1054 has been known for a long time [5]. But it was not until very recently that a large number of astrophysicists took seriously the idea presented by Baade and Zwicky [3] that a neutron star should be expected to be found at the center of this expanding envelope remnant of the supernova explosion. The unquestionable evidence for the identification of the star of 16.6th mag. at the center of the Crab Nebula with the pulsar NP0532 and its further identification with a neutron star came from the simultaneous observations in the radio [4] and in the optical wavelengths [6] of a sharply-defined pulsational period in the electromagnetic radiation emitted by this star (see Figure 1). The arguments forcing the identification with a neutron star [7] are most convincing, thanks to their simplicity:

(1) The total electromagnetic energy emitted by the pulsar NP0532 over the entire electromagnetic spectrum from radio waves to X-rays, is of the order of 5×10^{35} erg s^{-1}

H. Gursky and R. Ruffini (eds.), Neutron Stars, Black Holes and Binary X-Ray Sources, 59–118. All Rights Reserved.
Copyright © 1975 by D. Reidel Publishing Company, Dordrecht-Holland.

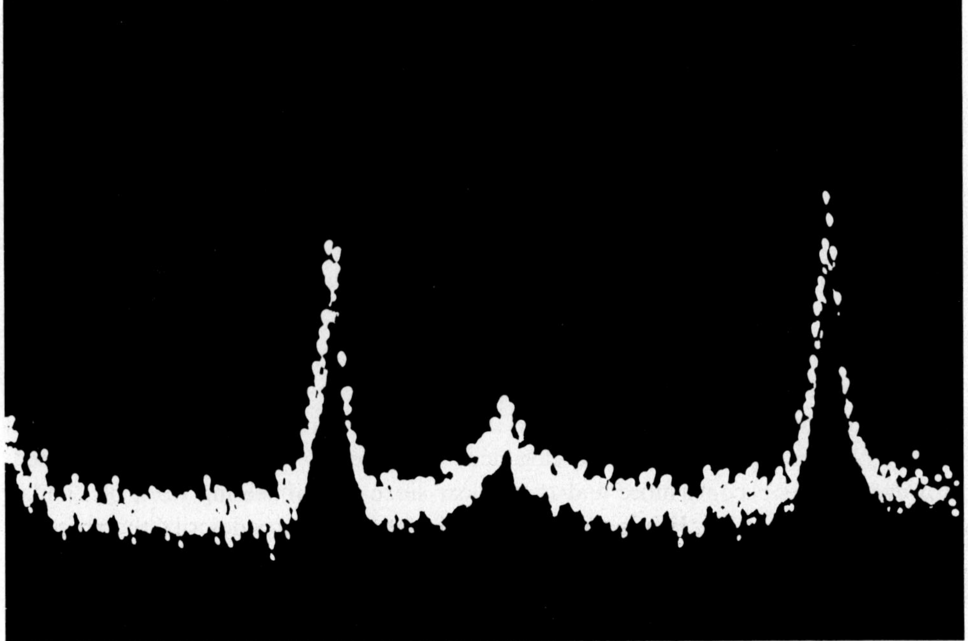

Fig. 1.   Pulse shape of the pulsar NP0532 in the Crab Nebula. The two successive main peaks are
~ 33 ms apart. An interpulse follows 12 msec after the main pulse. (Picture taken with the 36 in. optical telescope at Princeton University and reproduced by courtesy of E. J. Groth III.)

$(\sim 10^2 \, L_\odot)$ [8]. This object has therefore to be very massive $(\sim M_\odot)$ in order to account
for this very large steady emission of radiation.

(2) The radiation is emitted in sharp pulses with a period $P \sim 33$ ms. The object
has therefore to be very compact in order to modulate that very large amount of
energy over such a short period of time.

(3) The period of pulsation increases monotonically with time [9] with

$$\frac{\mathrm{d}P}{\mathrm{d}t} \simeq 13.5 \; \mu\mathrm{s} \; \mathrm{yr}^{-1} \, .$$

These three points and the entire energetics of the source and of the remnant were
explained at once by the assumption that the pulsar was indeed a neutron star. The
period of pulsation and its monotonic increase could then be interpreted as given by
the rotational period of the neutron star and by a loss of rotational energy due to the
braking processes occurring in its magnetosphere [10]. In turn if a value of the moment
of inertia predicted by the theoretical computations of neutron star equilibrium configuration was adopted for the pulsar, then this loss of rotational energy came out to be
of the same order of magnitude as the observed radiation flux from the pulsar and the
nebula [11]

$$\left(\frac{\mathrm{d}E}{\mathrm{d}t}\right)_{\mathrm{rot}} = I_{\mathrm{theor}} \omega_{\mathrm{obs}} \left(\frac{d\omega}{\mathrm{d}t}\right)_{\mathrm{obs}} \sim \left(\frac{\mathrm{d}E}{\mathrm{d}t}\right)_{\mathrm{obs}}^{\mathrm{emitted}} \, .$$

This explanation of the nature of the pulsar NP0532 gave rise to a deeper understanding of the Crab Nebula; but, even more important, it promoted a renewed and profound interest in the theoretical and experimental analysis of the two possible outcomes of gravitational collapse; neutron stars and black holes. Astrophysicists with a new spirit dictated by the experimental evidence of the existence in our own Galaxy of these most extreme regimes of pressure, density and gravitational fields, returned with much attention to a detailed examination of the fully relativistic analyses presented in the works of Oppenheimer and his students [12, 13, 14] with the aim of observing and verifying some of the most novel and revolutionary predictions of general relativity.

2. Neutron Stars

Oppenheimer and Volkoff [13] in 1939 were the first to give a detailed treatment of the equilibrium configuration of neutron stars. Their assumptions were very clear:

(1) Describe the microphysical structure of a neutron star by an equation of state obtained from the quantum mechanical treatment of a degenerate ($T=0$) relativistic gas of neutrons fulfilling Fermi statistics.

(2) Describe the macroscopic structure of a neutron star (mass, radius, density distribution) by the use of Einstein equations as applied to a perfect fluid distribution of matter.

The two major conclusions of the article were of comparable clarity:

(1) Stable equilibrium configurations of neutron stars can only exist in a finite range of masses and densities

$$0.1 \, M_\odot \lesssim M \lesssim 0.7 \, M_\odot$$
$$1.0 \times 10^{14} \lesssim \varrho \lesssim 3.6 \times 10^{15} \, \text{g cm}^{-3}.$$

(2) There is a critical value of the mass of a neutron star over which no equilibrium configuration can possibly exist. If the initial mass of a star is large enough, unless fission due to rotation, or ejection of mass reduces the star to a mass smaller than this critical value, then, after the exhaustion of thermonuclear sources of energy, the star will gravitationally collapse and contract indefinitely, never reaching true equilibrium.

Since the initial work of Oppenheimer and Volkoff, much work has been done in this field and the major assumptions adopted in their work have been critically reanalized. The major criticisms have ranged from the validity of any use of the concept of an equation of state for matter in these most extreme gravitational fields [15], to the use of the Einstein theory of gravitation for the computation of the equilibrium configuration [15], and the use of a degenerate non-interacting gas of Fermions (free neutrons) for the description of the neutron star material [16]. This last point particularly has lately generated much theoretical research [17]. In one direction there have been attempts to generalize to neutron star matter the two-body potentials obtained from laboratory experiments on the collision of two nucleons [18]. In a different direction attempts have been made to generalize to neutron star matter the statistical treatment [19] first introduced by Fermi in the analysis of high energy collisions

[20]. A summary of some of the different equations of state which have been purported is given in Figure 2 where a direct comparison is made with the free nucleon equations of state used by Oppenheimer and Volkoff.

It is remarkable that after more than thirty years of theoretical research, not many of the conclusions reached by Oppenheimer and Volkoff have been changed. Recent analyses have shown that the effects of nuclear interactions in computing the masses of the equilibrium configurations of neutron stars are indeed far from negligible (see Figure 3). However, neutron stars can still reach stable equilibrium configurations only in a finite range of masses:

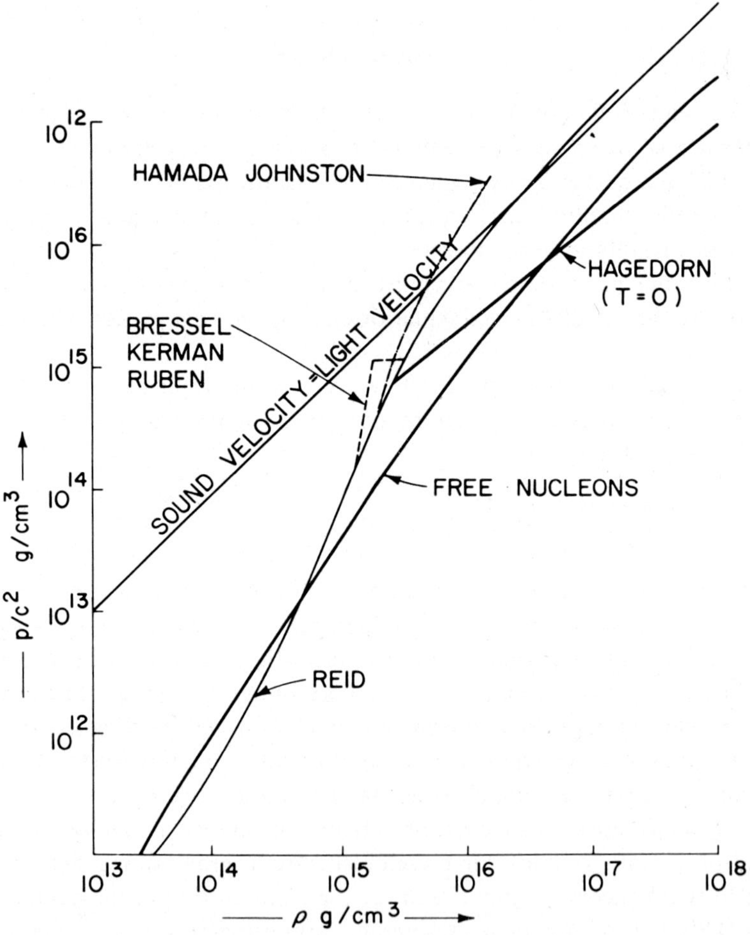

Fig. 2.   Pressure vs density for selected equations of state describing neutron star material. Both the equations of state with a Reid and an Hamada-Johnston two body nuclear potential violate causality at supra-nuclear densities. The free-nucleons approximation for densities smaller (larger) than 4.6×10^{15} g cm^{-3} gives values of the pressure systematically larger (smaller) than the one given by an equation of state taking into account nuclear interactions. The masses of neutron stars with central densities $\varrho \lesssim 4.6 \times 10^{15}$ g cm^{-3} ($\varrho \gtrsim 4.6 \times 10^{15}$ g cm^{-3}) computed with the free neutrons equation of state will always be larger (smaller) than the one computed out of a realistic equation of state. (Details in Reference 17, see also Figure 3.) © 1973, Gordon and Breach.

$$0.1\ M_\odot \lesssim M \lesssim 1.45\ M_\odot.$$

Apart from this mere change in the numerical value of the critical mass in no way has it been possible to overcome the necessity for the existence of the process of gravitational collapse.

Much work has also been done in the analysis of the composition of the crust of a neutron star. The following regimes are expected to be encountered as one goes down in depth from the 'atmosphere' of a neutron star: magnetic fields of the order of 10^{12} G are expected to exist on the surface. In the outermost layers of matter, as suggested by Ruderman [21] matter might exist in the form of very dense (average density on the order of 10^4 g cm^{-3}) one dimensional 'hairs' parallel to the lines of force of the magnetic field. The next shell of material in the density range between 10^4 g cm$^{-3} \lesssim \varrho \lesssim$ $\lesssim 10^7$ g cm^{-3} is expected to behave as a lattice of nuclei embedded in a degenerate gas

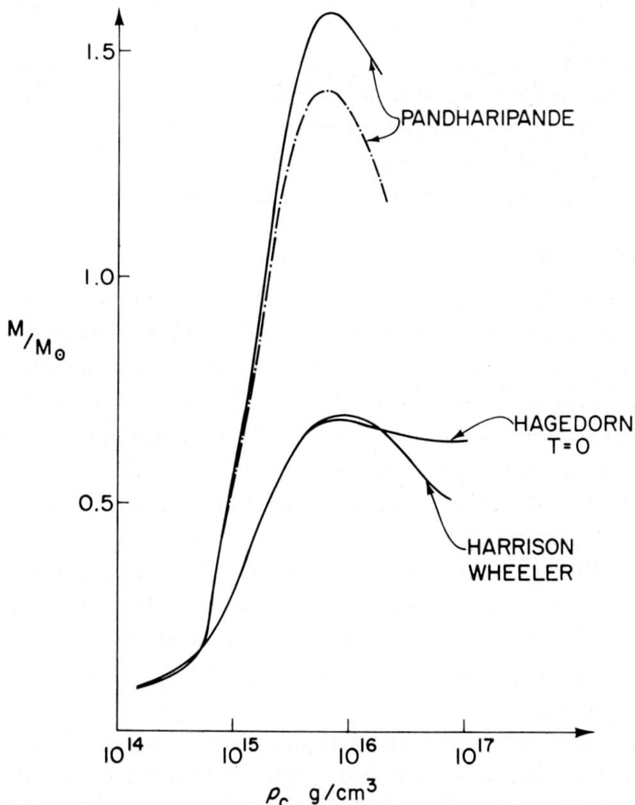

Fig. 3.   Masses of the equilibrium configuration of neutron stars plotted as a function of the central density for selected equations of state. The Pandharipande equation of state takes into account the strong interactions between nucleons. The Harrison-Wheeler equation of state neglects all the nuclear interactions and uses substantially a 'free particle' approximation. The Hagedorn equation of state is based on a thermodynamic approach derived from the theoretical analysis of high energy collisions between elementary particles and applies only asymptotically for $\varrho \gtrsim 5 \times 10^{15}$ g cm^{-3}. No matter what the different assumptions in the equation of state, the value of the critical mass is contained in a finite range $0.69 \lesssim m_{\mathrm{crit}} \lesssim 1.45\ M_\odot$. (Details in Reference 17.) © 1973, Gordon and Breach.

of relativistic electrons (white dwarf material), the reason being that the Fermi energy of the electrons is very much higher than the ionization energy of the atoms. Since the material is expected to be at the complete endpoint of nuclear evolution, the nuclei are thought to be mainly iron nuclei. However, as pointed out by Dyson [22], in this material some incompleteness of combustion could occur (as, for example, H – burned to helium, but not burned to iron) and the formation of some compounds could still be possible. The details of the thermonuclear reactions taking place in the collapse of a white dwarf material are not sufficiently well known to state what nuclear material and in what amount it should be expected in the upper layers of a neutron star. The important point is just to give a conceivable example of the incompleteness of combustion implying the presence in any given layer of more than one nuclear species. The simpler example considered by Dyson clearly shows that a particularly stable configuration is given by a lattice with NaCl structure and with Fe–He composition (iron-helide).

At densities between 10^7 g cm$^{-3} \lesssim \varrho \lesssim 10^{11}$ g cm^{-3} relativistic electrons transform bound protons into neutrons. Under normal circumstances, in fact, the total packing of a nucleus, under the two conflicting effects of nuclear and electrostatic forces is minimized for a value of $Z = 28$ and $A = 56$. A relativistic electron transmutes a nucleus of charge Z and atomic number A by inverse beta decay

$$e + (Z, A) \rightarrow (Z - 1, A) + v.$$

The nuclei become neutron rich compared to nuclei unpressured by electrons. For these neutron rich nuclei the mass number $A = 56$ no longer represents the point of maximum stability. Stability shifts to higher A values. The details of this shifting process are far from being well understood. For any electron pressure there corresponds a nucleus with a fixed value of Z and A which is in beta equilibrium with the electrons and has the most favorable packing fraction. At still higher densities in the range 10^{11} g cm$^{-3} \lesssim \varrho \lesssim 5 \times 10^{12}$ g cm^{-3} nuclei become so heavy ($A \sim 122$) and so neutron rich ($N/Z \sim 83/39$) that neutron 'drip' occurs. An 'atmosphere' of unbound free neutrons is formed. With a further increase in density, the Fermi energy of the electrons increases and the nuclei become even more neutron rich. The number of free electrons decreases further. Three different components characterize this range of densities: (a) an ultrarelativistic degenerate gas of electrons, (b) a system of heavy nuclei, (c) a degenerate neutron gas.

The contribution of the nuclei to the pressure is always negligible while the contribution of neutrons becomes more and more important with the increase of density. At densities already of the order of $\sim 5 \times 10^{12}$ g cm^{-3} the pressure of the degenerate gas of neutrons, extremely large by comparison to the nuclei pressure, is comparable to the pressure of the ultrarelativistic degenerate electron gas. To a further small increase in the density, there corresponds the disappearance of nuclei as such. The material of the star is uniquely formed of electrons, neutrons and protons in equilibrium against beta decay.

The properties of the material of the crust of a neutron star have been analyzed in depth in recent years largely using notions of solid state physics. Major contributions in this analysis have been made by Pines [23] and collaborators, Ruderman [24], as well as Dyson [22], Smoluchowsky [25], and Rhoades [26]. The major directions of research have been toward the determination of the composition, strength and conductivity of the material contained in the crust. The strength of this material is so small when compared with gravitational forces existing at the surface of the neutron star, that only 'mountains' of a few centimeters or less could be supported on the surface. This entire analysis could, indeed, prove to be of importance for the explanation of the tiny 'spin up' observed in the period of Pulsars [27, 28] ($\Delta P/P \sim 2 \times 10^{-6}$ in the Vela Pulsar and $\Delta P/P \sim 10^{-9}$ in the case of the Crab Nebula Pulsar) as well as in the understanding of the electrodynamic processes taking place near the surface of a neutron star. The detailed treatment of the crust of a neutron star is also of relevance for the determination of the value of the moment of inertia of a low mass neutron star. It can be, however, totally neglected in the computation of the value of the critical mass against gravitational collapse.

The reason is simply stated: The crust of a neutron star which extends a few tenths of kilometers in the case of a configuration of equilibrium corresponding to a central density $\varrho_c \sim 10^{14}$ g cm^{-3} becomes extremely thin for configurations of equilibrium with larger values of the central density. For a neutron star with a central density $\varrho_c \sim 5 \times 10^{15}$ g cm^{-3} the entire configuration of equilibrium has shrunk to a radius of ~ 10 km, the crust is only a few hundred meters thick and only a few percent of the total mass of the star is contained at a density $\varrho \lesssim 10^{13}$ g cm^{-3} [17].

If we focus, therefore, on the fundamental issue of the unavoidability of a neutron star reaching a critical mass against gravitational collapse, our attention is mainly directed to the physical processes occurring at nuclear and supranuclear densities. However, in no way from our knowledge of laboratory nuclear physics can we hope to infer a realistic equation of state for these regimes of densities and for a system of 10^{57} nucleons.

Despite these complications, recently, on the ground of a completely general variational principle, it has been shown by Rhoades and Ruffini [29] that quite independently from any detail of the equation of state at nuclear and supranuclear densities, an absolute maximum mass to the neutron star equilibrium configuration can beestablished.

This variational principle applies in complete generality to any distribution of a perfect fluid in general relativity and simply establishes that the maximum mass of an equilibrium configuration for a fixed central density is obtained for an equation of state which maximizes at every density the velocity of sound of the material. Therefore, an absolute upper limit to the neutron star mass can be immediately obtained under the following conditions:

(1) At densities lower than 4.6×10^{14} g cm^{-3} we choose the equation of state of a degenerate ($T=0$) non interacting neutron gas since this equation maximizes the speed of sound of the neutron star material by comparison to any realistic equation of state taking into account nuclear interactions substantially attractive in this range of densities.

(2) At densities larger than 4.6×10^{14} g cm^{-3} nothing is known for certain on the

66   REMO RUFFINI

equation of state of neutron star material, we then choose that most extreme equation
of state uniquely consistent with the conservation of causality with a velocity of sound
equal to the speed of light.

On the ground of these two assumptions it is then possible to establish an absolute
upper limit to a neutron star mass: $M < 3.2\ M_\odot$. It is hopeless to try to establish the
effective value of the critical mass or, for that matter, of the radius, density distribution, or moment of inertia of a neutron star by direct theoretical arguments. Progress
at this moment can be made only through collection of experimental data on neutron
stars and through a direct comparison with existing theoretical predictions.

It is also clear that data cannot be collected from detailed analysis of known pulsars
for at least two different reasons:

(1) Out of the 120 pulsars observed [30] none is in a binary system and in no way
can we then obtain a direct measurement of the mass of the neutron star.

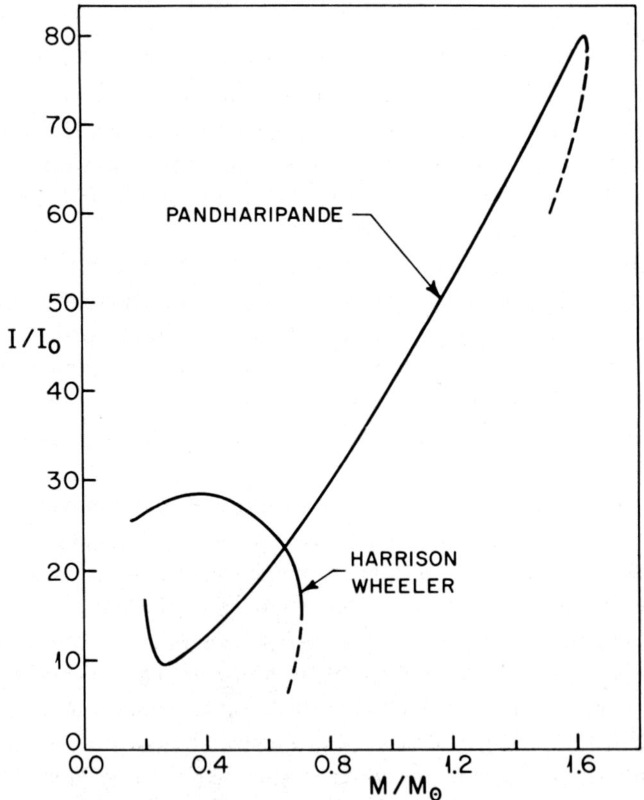

Fig. 4.   Moment of inertia plotted as a function of the neutron star mass for selected equations of
state. The dotted lines correspond to unstable equilibrium configurations. The major difference
between the Pandharipande and the Harrison-Wheeler equations of state comes from the treatment
of nuclear forces between nucleons at nuclear and supranuclear densities (details in Reference 17). It
is important to realize that in a process of accretion, the mass of a neutron star increases and possible
effects due to the change of the moment of inertia could be detected through the change of rotational
period of the neutron star (see Section 5). I_0 is here given by $M_\odot$ km^2. (Details in Reference 89.)

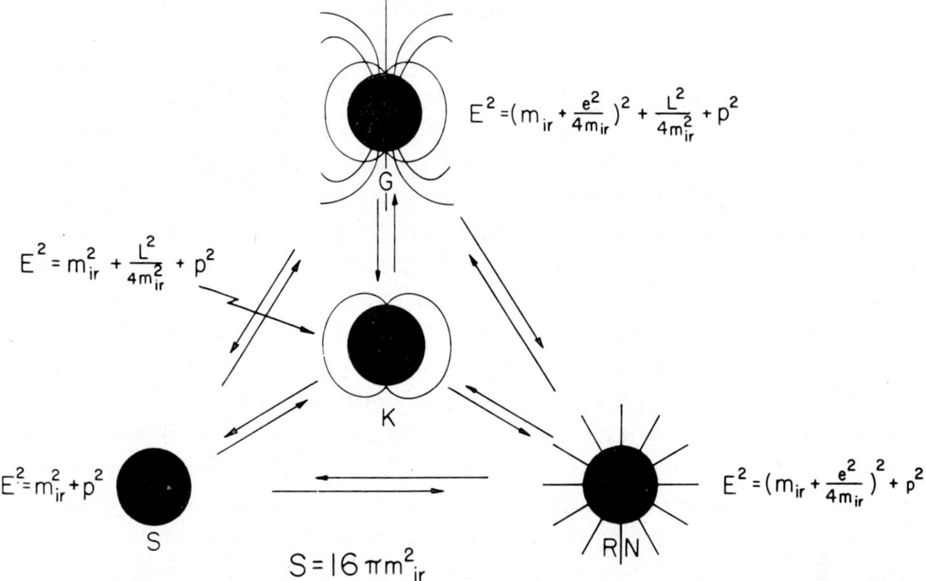

Fig. 5. The entire set of regular black holes is here summarized with the general formulae governing their total mass energy as a function of their characteristic parameters [33]. Mass m, charge e, and angular momentum L are expressed in geometrical units. Reversible and irreversible transformations can lead from one kind of black hole to another through loss or augmentation of charge and angular momentum. The only black hole deprived of an ergosphere (region around a black hole from which it is possible to extract a finite amount of the total mass energy of the collapsed object) is the Schwarzschild black hole. The Reissner-Nordstrøm (R.N.) black hole is endowed with charge (e) and mass (m), the Kerr (K) with mass and angular momentum (L), the Kerr-Newmann (G for general) with mass, charge and angular momentum. Extraction of energy is possible if the black hole is endowed with angular momentum [38, 34], charge [35] or both these parameters [36]. The effective ergosphere extends from the horizon r_+ to r_{erg} [35, 36]: $m + (m^2 - a^2 - e^2)^{1/2} = r_+ \leqslant r \leqslant r_{\mathrm{erg}} = m + [m^2 - e^2 \times (1 - q^2/\mu^2)]^{1/2}$ where q/μ is the charge to mass ratio of the test particle which reduces the mass energy of the black hole and $a = L/M$. Up to $29\%(50\%)$ of the total mass energy can be extracted in the transition from a Kerr (Reissner-Nordstrøm) to a Schwarzschild black hole. In all these cases, the surface area of a black hole is most simply expressed by $S = 16\pi m^2{}_{\mathrm{ir}}$. The result that the irreducible mass of a black hole can never decrease was independently obtained through a different derivation by Hawking [40] (details of this entire diagram in Reference 17).

(2) Apart from the glitches and microglitches [27, 28] the emission from pulsar is extremely steady and not giving any information on the internal structure of the neutron star.

In sharp contrast, all the binary X-ray sources observed (see Sections 5 and 6) are in binary systems and the processes of accretion occurring in the collapsed object gives us a very large amount of information both on their electromagnetic structure and internal composition.

For the sake of an example it is interesting to stress how the knowledge of macroscopic parameters of a neutron star like the moment of inertia as a function of its mass can be used as a probe into the equation of state of neutron star material (see Figure 4).

3. Black Holes

If we turn now to the analysis of black holes, once again the fundamental work in this field goes back to Oppenheimer. In 1938, Oppenheimer and Snyder [14], in one of the most beautiful papers ever written in general relativity were able to describe with a few essential formulae all the major features of a star undergoing gravitational collapse: "The radius of the star approaches asymptotically its gravitational radius, light from the surface of the star is progressively reddened, and can escape over a progressively narrower range of angles". "The total time of collapse for an observer co-moving with the stellar matter is finite... an external observer sees the star asymptotically shrinking to its gravitational radius" [14]. Much has been learned since 1938 about the physics of these totally collapsed objects [31] but, again, none of the conclusions reached by Oppenheimer has been modified or disproved.

We understand today that quite apart from the Schwarzschild black holes originally investigated by Oppenheimer and Snyder and uniquely characterized by their mass there exists an entire class of collapsed objects characterized by three different parameters m, mass, e, charge and L, angular momentum (see Figure 5).

One major direction of research has been aimed toward a deeper understanding of analogies and differences between these different collapsed objects. One of the most powerful tool to advance in this analysis has been the study of gedanken processes of capture of test particles by the black hole and how transitions can occur from one kind of black hole to another by accretion of selected particles and by gain or depletion of charge, mass and angular momentum. Examples of these processes are shown in Figures 6 and 7. The most striking result in the analysis of these transformations has been the possibility of differentiating in the accretion process between two radically different kind of transformations: reversible and irreversible [32, 33].

By capture of charged test particles endowed with angular momentum, we can always modify the mass (m), charge (e) and angular momentum (L) of a black hole:

$$m' = m + \delta m$$
$$e' = e + \delta e$$
$$L' = L + \delta L.$$

By further capture of a test charge of opposite sign and opposite angular momentum, a black hole can reacquire its initial value of charge and angular momentum:

$$m'' = m' + \delta m'$$
$$e'' = e' - \delta e = e$$
$$L'' = L' - \delta L = L.$$

Usually $m'' > m$ since two particles have been captured by the black hole. However, between all the possible transformations there exists a subset of transformations, the reversible ones, for which $m'' = m$.

As a direct consequence of these transformations, it has been possible to give a very simple formula governing the energetics of black hole physics. Christodoulou and Ruffini [30] have shown that the total mass energy of a black hole can be simply split

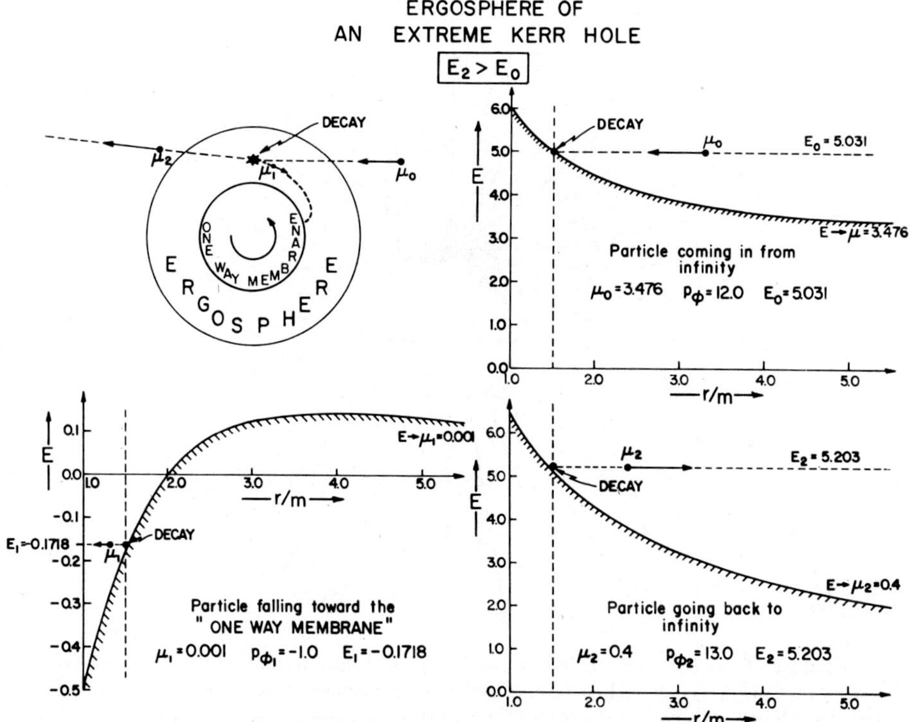

Fig. 6. An example of how to slow down a black hole *and* reduce its total mass energy by the extraction of rotational energy. A particle of mass μ_0 coming from infinity with total energy E_0 and a positive value of the angular momentum p_ϕ, can penetrate the ergosphere of an extreme Kerr hole and there decay into two particles [38, 39]. One particle of mass μ_1, negative value of the angular momentum p_ϕ and a negative value of the total energy E_1, falls towards and penetrates the horizon. The second particle of mass μ_2, positive value of the angular momentum $p_{\phi2}$ and a positive value of total energy E_2, goes back to infinity. The remarkable feature in this process is that the energy E_2 of the particle coming back to infinity is larger than the energy E_0 of the particle coming in. In the detailed computations of the process here presented [39] we have assumed beside the conservation of the energy also the conservation of the total momentum of the system during the process of fragmentation of the particle μ_0 into the two particles μ_1 and μ_2. As is clear from the example here presented this energy extraction process can be made at the expenses of a very large reduction in the rest mass of the particle. On the upper left side a qualitative diagram shows the main feature of the decay process in the equatorial plane of the ergosphere of a Kerr hole. In the upper right side is the effective potential (energy required to reach r as a turning point) for the incoming particle. The effective potential is plotted at the lower left and the lower right side for the particle falling toward the horizon and for the particle going back to infinity. This 'energy gain process' critically depends on the existence and on the size of the ergosphere which in turn depends upon the value a/m of the hole. In the case of a Kerr black hole considered here the ergosphere extends between the horizon and the infinite red-shift surface

$$m + (m^2 - a^2)^{1/2} \leqslant r \leqslant 2m$$

and when $a/m = 0$, (Schwarzschild black hole) the horizon expands and coincides with the infinite red shift surface, wiping out the ergosphere. The particle falling towards the one way membrane will in general alter and reduce the ratio a/m of the black hole. Details in References 38, 39, and 17. The corresponding processes for extraction of electromagnetic energy has been given in References 35 and 36. In the case of the existence of an electromagnetic field in the field of the black hole. The ergosphere will again start at the horizon of the black hole and will extend out to a surface $r_{\mathrm{erg}} = m + [m^2 - e^2$. $\times \times (1 - q^2/\mu^2)]^{1/2}$ which is usually a function of the charge to mass ratio q/μ of the test particle falling into the black hole and reducing the total energy of the black hole. Christodoulou and Ruffini [37], to stress the difference between this more general case and the case of a Kerr solution, have called the region in which energy can be extracted the 'effective ergosphere'. © 1973, Gordon and Breach.

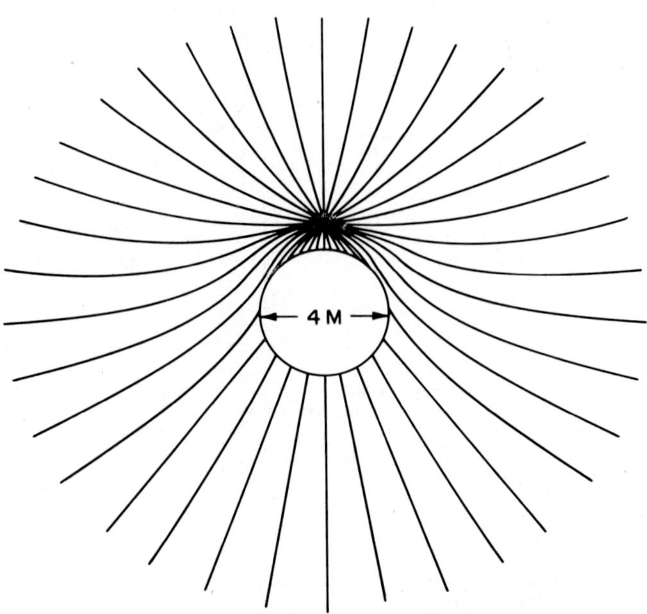

Fig. 7.   Capture of a particle by a Schwarzschild black hole and transition to a Reissner-Nordstrøm black hole. Since the early days of black hole physics, it has been well known that it takes an *infinite time* (as seen by a far-away observer) for a test particle to reach the horizon of a black hole. It is, therefore, natural to ask how the transition can occur from a Schwarzschild to a Reissner-Nordstrøm solution, since in this last case, the electric field should appear to a far away observer completely radial *as if* the charge was concentrated at the center of the black hole although the particle can never even cross the horizon as seen by that observer. This paradox was solved by the work presented in Reference 41. The test particles 'induces' charges on the surface of the black hole. The closer it approaches the Schwarzschild horizon, the larger the magnitude of the induced charge is, the 'transfer' of the charge to the black hole occurs, therefore as a polarization effect. The electromagnetic field appears very distorted to an observer near the Schwarzschild radius and radially directed to a far away observer as if the test charge was indeed at the center of the black hole. (From S. Hanni ana R. Ruffini, *Phys. Rev.* **8**, 1973.)
Published by American Physical Society; © 1973, American Institute of Physics.

into three contributions: the rest energy (the irreducible mass), the Coulomb energy and the rotational energy.

$$m^2 = (m_{ir} + e^2/4m_{ir})^2 + L^2/4m_{ir}^2.$$

Here we have used geometrical units with $(G = c = 1)$. To obtain the conventional units we have

$$m = Gm_{conv}/c^2, \qquad e = G^{1/2}e_{conv}/c^2, \qquad L = GL_{conv}/c^3.$$

The area of the black hole is given simply by [33]

$$S = 16\pi \, m_{ir}^2.$$

The irreducible mass is left constant in all reversible transformations and is monotonically increased by irreversible transformations [32, 33]. In contrast, the Coulomb and the rotational energy can be added and subtracted at will from the black hole under the following limitation [34]:

$$L^2/m^2 + e^2 = a^2 + e^2 \leqslant m^2.$$

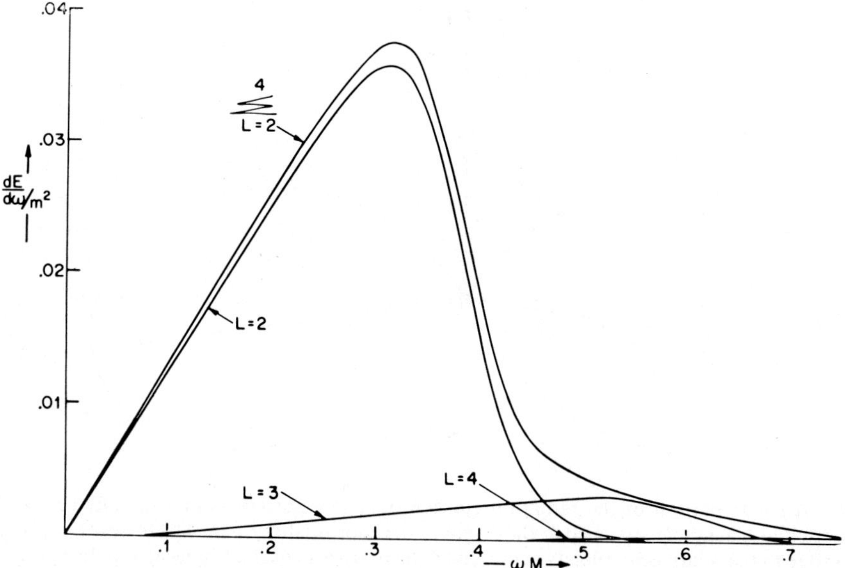

Fig. 8.   Spectrum of gravitational radiation emitted by a test particle of mass m falling radially into a black hole of mass M (geometrical units $G = c = 1$). This work based on the Regge-Wheeler-Zerilli [42, 43] formalism is the fully relativistic generalization of the semi-relativistic treatment presented by Ruffini and Wheeler [39]. In both treatments, the spectrum is broad with a sharp decrease both in the low and high frequency limits. The major amount of radiation is emitted in the quadrupole ($l = 2$) mode and the peak of the radiation occurs at the $\omega \sim 0.4\ c^3/GM$. The total amount of radiation emitted, integrated over frequency and multipole distribution is:

$$\Delta E \sim 0.0104\ mc^2\,(m/M).$$

It is immediately clear that the highest efficiency can be reached in the limit $m \sim M$, however in that approximation, the validity of the approximation used in References 42 and 43 ceases to be valid. In turn, it is also clear that the radiation of a one solar mass 'test particle' into a very large black hole of $10^8\ M_\odot$ or larger is negligible (details in References 44 and 45). (From M. Davis *et al.*, *Phys. Rev. Letters* **27**, 1971.) Published by American Physical Society; © 1971, American Institute of Physics.

The processes of extraction of energy from a black hole are mediated in the effective ergosphere of a black hole [33, 35, 36, 37]. It is interesting to notice that for an extreme black hole ($a^2 + e^2 = m^2$) up to 50% of its mass-energy can be stored in rotational and electromagnetic energy and is therefore extractable.

A second important line of research has been directed toward a detailed analysis of the radiation processes occurring in the field of black holes. The most powerful tool to carry out this program has certainly been the perturbation analysis introduced in 1957 by Regge and Wheeler [42] and further developed since 1971 by Zerilli [43]. The main idea of this approach can be easily summarized: to analyze the processes occurring in the background field of a collapsed object we can analyze small perturbations away from a given background metric $g^{(0)}{}_{\mu\nu}$

$$g_{\mu\nu} = g_{\mu\nu}{}^{(0)} + h_{\mu\nu} \tag{2}$$

the Einstein equations for the metric $g_{\mu\nu}$ then give the following set of equations for the perturbation field $h_{\mu\nu}$

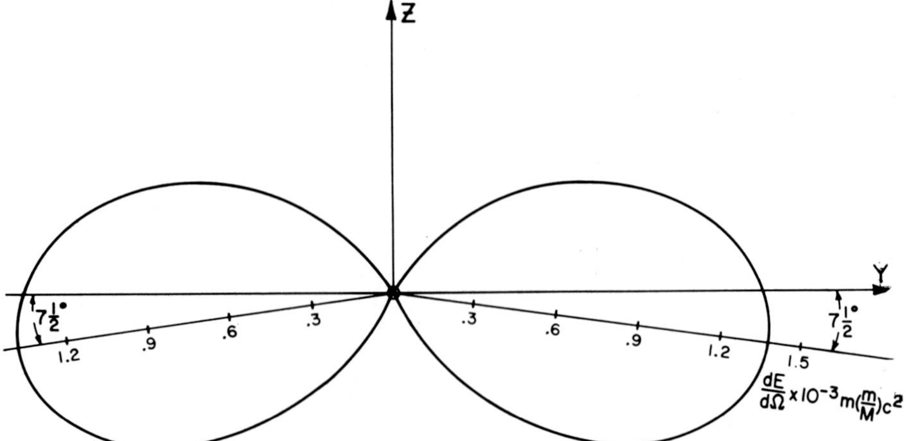

Fig. 9. Angular pattern of the radiation emitted by a test particle falling radially (Z axis) into a Schwarzschild black hole, assumed at the origin of the coordinate system. The forward beaming of the lobes is due to the relativistic velocity acquired by the particle in its final approach to the Schwarzschild horizon. Details on the intensity of the radiation ingoing into the black hole and of the polarization of the radiation as well as the details of this analysis are given in Reference 46. (From M. Davis, *Phys. Rev.* **5**, 1972.) Published by American Physical Society; © 1972, American Institute of Physics.

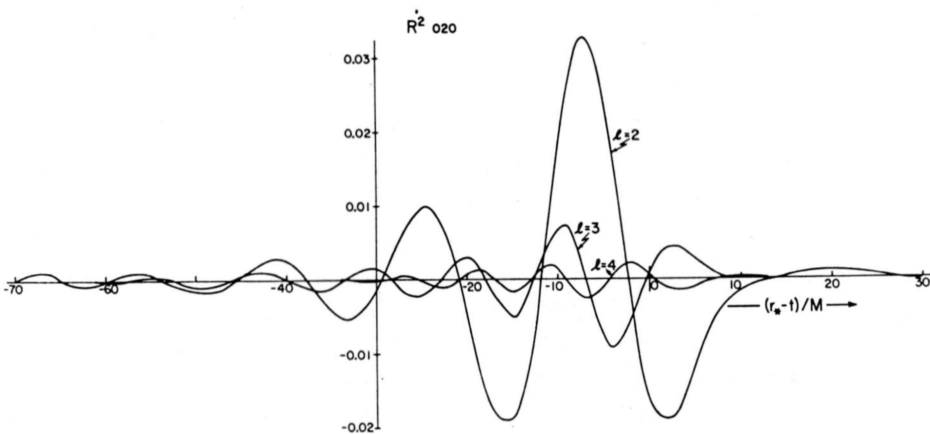

Fig. 10. A burst of gravitational radiation emitted by a particle of mass m falling into a black hole of mass M. The component of the Riemann tensor relevant for the detection of the burst of gravitational radiation (see also Figure 13) as well as the intensity of the radiation are here plotted as a function of the retarded-time coordinate. A detector of gravitational radiation located far away from the source would first receive a weak signal emitted by the particle approaching the black hole (the 'precursor' $5 \lesssim (r^* - t)/M \lesssim 30$) and then receive a very sharp pulse emitted during the last approach of the particle to the throat of the collapsed object ($-2 \lesssim (r^* - t)/M \lesssim 5$) where the particle reaches extreme relativistic velocities. Finally a sequence of impulses of decaying intensity is emitted during the amalgamation of the falling particle to the black hole. A new larger black hole has been formed by the capture of the test particle m (details in Reference 46). Published by American Physical Society; © 1972, American Institute of Physics.

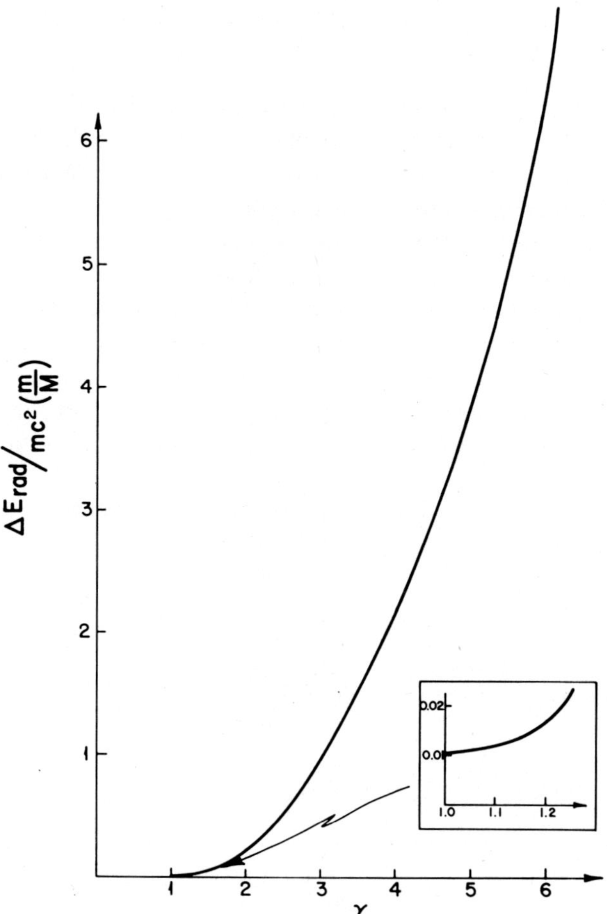

Fig. 11. Two major limitations govern the spectrum and the total energy emitted in the fall of a particle into a Schwarzschild black hole:

(a) As shown in Figure 8, the total amount of energy is

$$\Delta E \sim 0.0104 \, m \, c^2 \, (m/M) \, .$$

(b) The spectrum is peaked at a frequency

$$\omega \sim 0.4 \, c^3/GM \, .$$

An extensive analysis has been made to see if under suitable conditions both constraints (a) and (b) can be relaxed, namely if we could have both a larger energy emission and at higher frequencies. The fulfillment of one or both these conditions would greatly facilitate any detection of gravitational wave bursts (see also Figure 13). Of all the conceivable methods, one has proved to be very effective in largely enhancing the energy emission but only slightly shifting the peak of the radiation toward higher frequency: the projection of a particle into a black hole. Here the total energy emitted by a test particle of mass m projected into a Schwarzschild black hole of mass M is plotted as a function of the value $\gamma = 1/(1 - v^2/c^2)^{1/2}$, v being the velocity of the particle at infinity before starting its implosion. Interesting as it is for a better understanding of the radiation processes this process cannot be considered of astrophysical interest. (Details in Reference 44). © 1973, Gordon and Breach.

$$h_{\mu\nu;\alpha}{}^{;\alpha} - (f_{\mu;\nu} + f_{\nu;\mu}) + 2\,R^{\varrho}{}_{\mu}{}^{\alpha}{}_{\nu}\,h_{\varrho\alpha} + h^{\alpha}{}_{\alpha;\mu;\nu} +$$
$$+ g_{\mu\nu}(f_{\alpha}{}^{;\alpha} - h^{\alpha}{}_{\alpha;\lambda}{}^{;\lambda}) = -16\pi\Delta T_{\mu\nu}, \tag{3}$$

where $f_{\mu} = h_{\mu}{}^{\alpha}{}_{;\alpha}$ and $R^{\alpha}{}_{\beta}{}^{\gamma}{}_{\delta}$ is as usual the Riemann tensor of the background metric and $\Delta T_{\mu\nu}$ is the physical perturbation driving the field $h_{\mu\nu}$. In this set of equations all the non-linear terms, quadratic in $h_{\mu\nu}$ have been neglected. When the background metric is given by the Schwarzschild metric, the set of Equations (3) can be greatly simplified expanding the tensor perturbation $h_{\mu\nu}$ into a complete set of tensor spherical harmonics. Regge and Wheeler [42] and Zerilli [43] have then shown how to reduce the system of ten partial differential Equations (3) to a system of two Schrödinger-like equations governing the most general perturbations of a Schwarzschild background.

This mathematical formalism has allowed for the first time a fully relativistic analysis of gravitational and electromagnetic radiation processes in the field of collapsed objects. The spectrum (see Figure 8), the directional properties (see Figure 9),

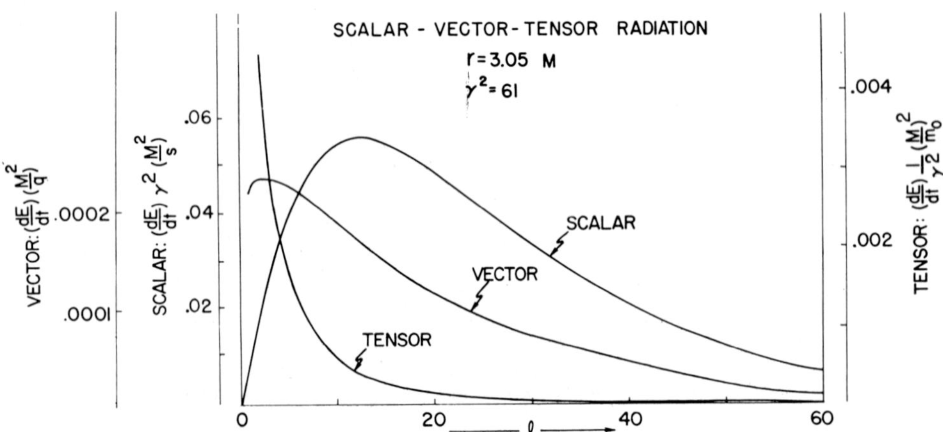

Fig. 12. Following a suggestion and a precise example given by Misner [48] and Misner *et al.* [49] that beaming effects could be found in the emission processes of gravitational radiation emitted by particles in relativistic orbit around a black hole, a systematic search and analysis of the spectral distribution of both electromagnetic and gravitational radiation emitted by test particles in relativistic circular orbits around black holes was initiated and completed [50, 51, 52]. The interest in the possibility of beaming effects of gravitational radiation can be simply summarized: If it happens by symmetry reasons that the Earth and our detection system is in the sharp beam of a source of gravitational rotation, then the required energy emitted by the source of the radiation will be smaller the stronger the beaming processes are. The fully relativistic analysis of electromagnetic and gravitational radiation has shown that any beaming process strongly depend upon the spin of the field under consideration: the higher the spin, the smaller are the beaming processes. Better than any further comment, this result is most clearly expressed in the present figure where the spectrum of scalar electromagnetic and gravitational radiation are compared and contrasted for the motion of a test particle in a relativistic orbit around a Schwarzschild black hole. The major conclusion is that in no realistic astrophysical conditions can we possibly find relativistic beaming processes of any relevance. Moreover in *no* way can a spin two field have beaming processes of the kind observed in electromagnetic radiation in flat space. The physical reasons behind these very interesting defocusing effects which forbid the beaming processes have been given in Reference 51 and 52. In this figure, m is the mass of the particle q the electric charge, s the scalar charge of the particle and l the multipole of the radiation under consideration ($l = 1$ dipole, $l = 2$ quadrupole, etc.). (From M. Davis, *Phys. Rev. Letters* **28**, 1972.)
Published by American Physical Society; © 1972, American Institute of Physics.

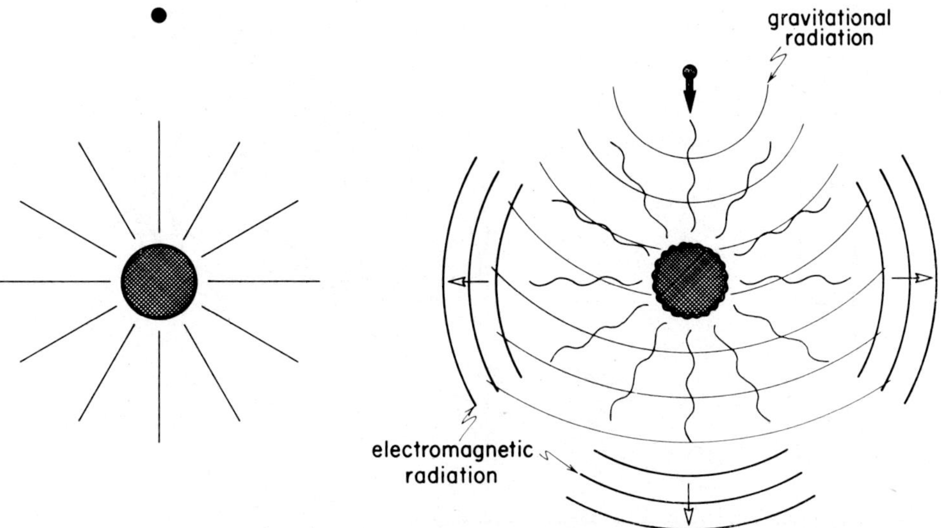

Fig. 13.   Schematic representation of the gravitationally induced electromagnetic radiation. For the sake of example, we have considered the problem of electromagnetic and gravitational radiation emitted in the infall of an *uncharged* test particle of mass m in the field of a charged black hole (Reissner-Nordstrøm) of mass M and charge Q. Similar effects should be expected to occur in a more general situation in which the collapsed object is endowed with a more complex electromagnetic field structure (generalize Kerr solution with magnetosphere [17]). The infalling particle perturbs the background electromagnetic field both through the effect of the gravitational field of the test particle and the scattering of gravitational waves. The key and most important result can be simply summarized:

$$(\Delta E)_{\text{elec}} \sim (Q/M)\,(\Delta E)_{\text{grav}}.$$

The conversion of gravitational into electromagnetic radiation can reach a 100% efficiency in the limit $Q \rightarrow M$. Details in Reference 57. The inverse phenomena, the electromagnetic induced gravitational radiation, has also been considered and details given in Reference 58.

the shape of the burst (see Figure 10), and the energetics (see Figure 11) of a burst of gravitational radiation emitted by a particle falling or projected into a Schwarzschild black hole have been analyzed. The analysis has as well been carried out for the much discussed problem of the emission of electromagnetic radiation from a charge infalling on a Schwarzschild or a Reissner-Nordstrøm black hole [53, 54].

An entire program has also been directed toward the exploration of possible beaming effects similar to the well known synchrotron radiation of classical electrodynamics, into the field of a collapsed object (see Figure 12).

It has been pointed out by Partridge and Ruffini [55] how the enormous difference of cross section between electromagnetic and gravitational detectors make most promising an experimental search for coupled bursts of gravitational and electromagnetic radiation. The theoretical analysis of this coupling has generated a third major field of research. The first fully relativistic analysis of this coupling process is based on the generalization [56] of the Regge-Wheeler [42]-Zerilli [43] formalism to the case in which the background field is endowed with an electromagnetic structure. Two major new effects have been pointed out as occurring in this regime: The

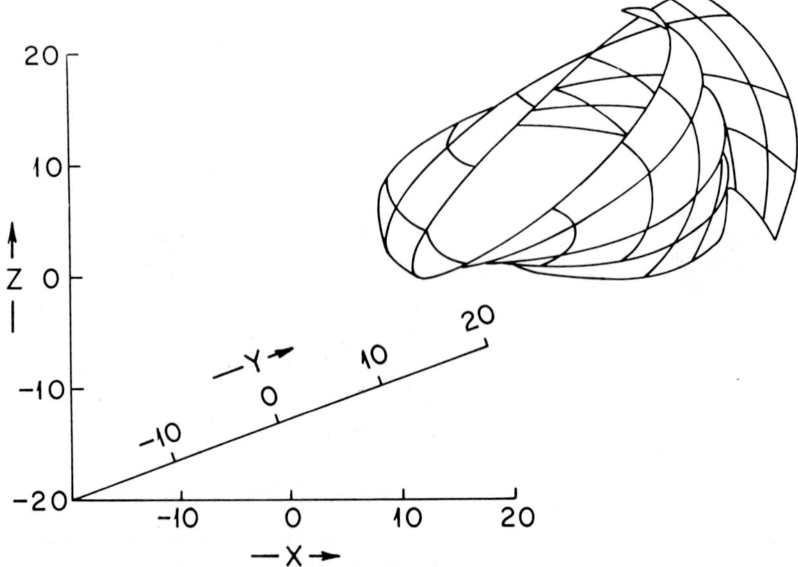

Fig. 14.   Motion of an uncharged cloud of test particles of mass μ, each endowed with an angular momentum $p_\phi/\mu m = 2$, corotating around an extreme spinning black hole of mass m. The vertical lines
are isochronous points as seen from infinity. The typical 'Klein bottle' behavior of this cloud of particles is a direct consequence of the 'Wilkins effect' [62, 63]. In the field of a Kerr solution a self
gravitating orbiting cloud of particles acquires an additional periodicity in the θ direction and is confined to an equatorial region. It is very difficult for a particle to spiral toward the poles of the rotation
          axis (see References 62 and 63). (From M. Johnston and R. Ruffini, *Phys. Rev.* **D10**, 1974.)
             Published by American Physical Society; © 1974, American Institute of Physics.

gravitationally induced electromagnetic radiation [57] and the electromagnetically
induced gravitational radiation [58, 54]. Both these analyses focus on the possibility
of converting electromagnetic into gravitational radiation and vice-versa in the field
of a collapsed object (see Figure 14). In a specific example [57] it has been pointed out
how this conversion can reach 100% efficiency. The very large amount of gravitational
radiation expected to be emitted during the process of gravitational collapse together
with the clear possibility that the collapsed object be endowed with an electromagnetic
field [59] make this analysis most interesting and attractive.

It has, of course, to be clear that this analysis describes an extremely idealized
physical situation and much more complex electromagnetic processes certainly occur
simultaneously to the ones here considered in a realistic process of gravitational
collapse.

An experimental search for bursts of electromagnetic radiation at radio and infrared wavelength is currently being made [60].

All these researches on the ultrarelativistic radiation processes in the field of collapsed objects have also naturally led to an in depth analysis of the properties of
gravitational waves detectors (see Section 4) in order to check the feasibility of observing some of these theoretical predictions.

If we turn now to perhaps the most interesting problem connected with the direct detectability of a black hole, namely the radiation emitted by an accreting plasma, much work still has to be done in order to prepare the ground for a detailed analysis. The detailed structure of the electromagnetic fields of a black hole *in vacuum* has been given by Christodoulou and Ruffini [61]. The trajectories of clouds of charged and uncharged test particles in the most general background geometry have been analyzed [62, 63] (see Figures 14 and 15). Finally, the effect of a black hole on a uniform magnetic field has also been studied [64] (see Figure 16). It is, however, now clear that these are only preliminary works. To approach the complete problem of plasma accreting about a rotating black hole, it is necessary to solve the entire self-consistent

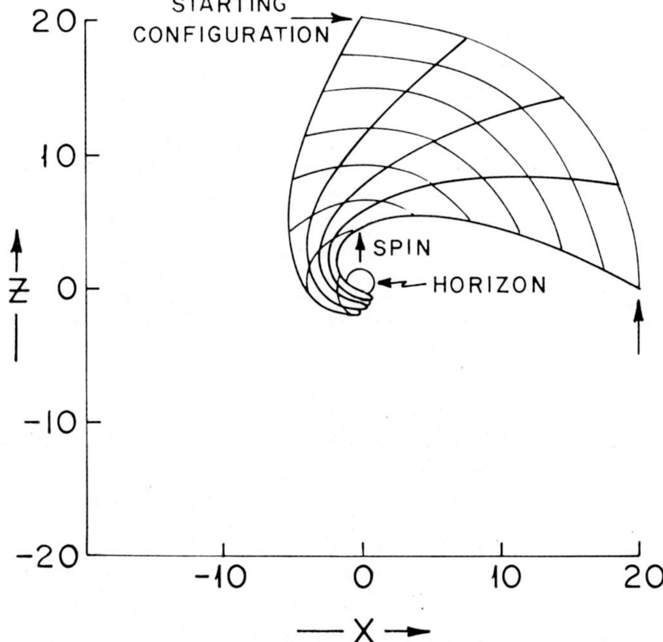

Fig. 15.   This figure beautifully exemplifies the concept of angular velocity of a black hole introduced, in Reference 61. A cloud of particles with initial zero angular momentum along the ϕ direction ($P_\phi/\mu m = 0$) acquires an angular velocity along the ϕ direction which is simply given by:

$$\omega = a/(r_+^2 + a^2) \tag{1}$$

as the cloud approaches the horizon $r_+ = m + (m^2 - a^2)^{1/2}$. Here, as usual, a is the angular momentum per unit mass of the black hole $a = L/m$. After this angular velocity of the black hole was introduced, J. A. Wheeler suggested that it should have been possible to obtain this same result from the formula governing the energetic of black holes [33]:

$$m^2 = \left(m_{ir} + \frac{e^2}{4m_{ir}}\right)^2 + \frac{L^2}{4m^2_{ir}}. \tag{2}$$

Indeed we have

$$\omega = \frac{\partial m}{\partial L}. \tag{3}$$

This result greatly contributes to a deeper understanding of the meaning and significance of the formula (2). (From M. Johnston and R. Ruffini, *Phys. Rev.* **D10**, 1974.) Published by American Physical Society; ©1974, American Institute of Physics.

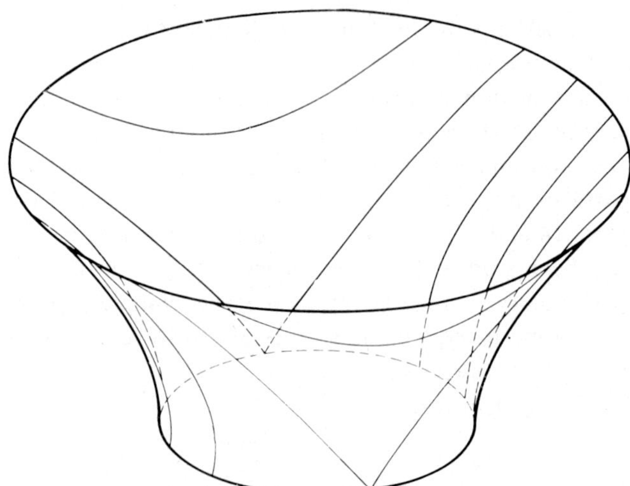

Fig. 16. The effects of a Schwarzschild black hole on a uniform magnetic field are here displayed in an embedding diagram. To solve this problem, the fully covariant Maxwell equations have been solved in a Schwarzschild background with the boundary conditions of a uniform field at infinity. We have adopted for the lines of force the definition given in References 61 and 17. This analysis is of the greatest relevance for the study of the accretion of plasma into a black hole since we expect from the usual condition of infinite conductivity, that the plasma will flow toward the black hole along the magnetic field lines and the final configuration of the magnetic field near the horizon will largely resemble the one here reproduced if, of course, the black hole is not rotating. (Details in Reference 64.)

solution of the background electromagnetic field and of the accreting plasma in the magnetosphere. Preliminary results clearly show that in this process very large currents and electromagnetic fields should be expected to exist in the magnetosphere of such a black hole [65, 66]. Much work has therefore been done in recent years ranging from the analysis of the structure of black holes, to trajectories of particles in a given background geometry of a black hole, to the analysis of energy extraction from black holes, to the processes of accretion around collapsed objects to finally the electrodynamics processes near the surface of black holes. Thanks to this work it has become clear that each kind of black hole appears to have its own characteristic signature which could in principle be detected and observed. This entire program can be carried out only with a very strong set of new results in the experimental field in order to have feedback in the theoretical field not just for verification of the theoretical predictions but for new information in this drastically new field of physics.

4. Gravitational Radiation Detectors

The detailed analysis of the processes of emission of gravitational radiation in extreme relativistic regimes has naturally led to the study of the cross section, directionality, and polarization response of detectors of gravitational radiation [39, 67, 68]. The cross section of a gravitational wave antenna characterized by a resonant frequency ω_0 is given for frequencies near resonance by a generalization of the Breit-

Wigner formula [70, 69]

$$\sigma(\omega) = \frac{\pi \lambda^2}{2} \frac{A_{\text{grav}} A_{\text{diss}}}{(\omega - \omega_0)^2 + (A_{\text{diss}}/2)^2} \ \text{cm}^2 \ \text{s}^{-1} \ \text{Hz}^{-1}$$

with the value for the integrated cross section

$$\int_{\text{res}} \sigma(v) \, dv = (\pi/2) \, \lambda^2 A_{\text{grav}},$$

where $A_{\text{grav}} = -(dE/dt)_{\text{grav}}/E$ is the rate of damping of the detectors caused by radiation of gravitational waves and $A_{\text{diss}} = -(dE/dt)_{\text{diss}}/E$ is the rate of damping caused by all form of dissipation other than gravitational radiation and $\lambda = \lambda/2\pi$ is the reduced wavelength of gravitational radiation at resonance. Under any realistic circumstance we have of course $A_{\text{diss}} \gg A_{\text{grav}}$. If we then indicate by $I(v)$ (erg cm^{-2} Hz^{-1}) the spectrum of the gravitational radiation pulse we obtain for the energy E absorbed by the detector

$$E = \int I(v) \, \sigma(v) \, dv \ \text{erg} \ \text{s}^{-1}.$$

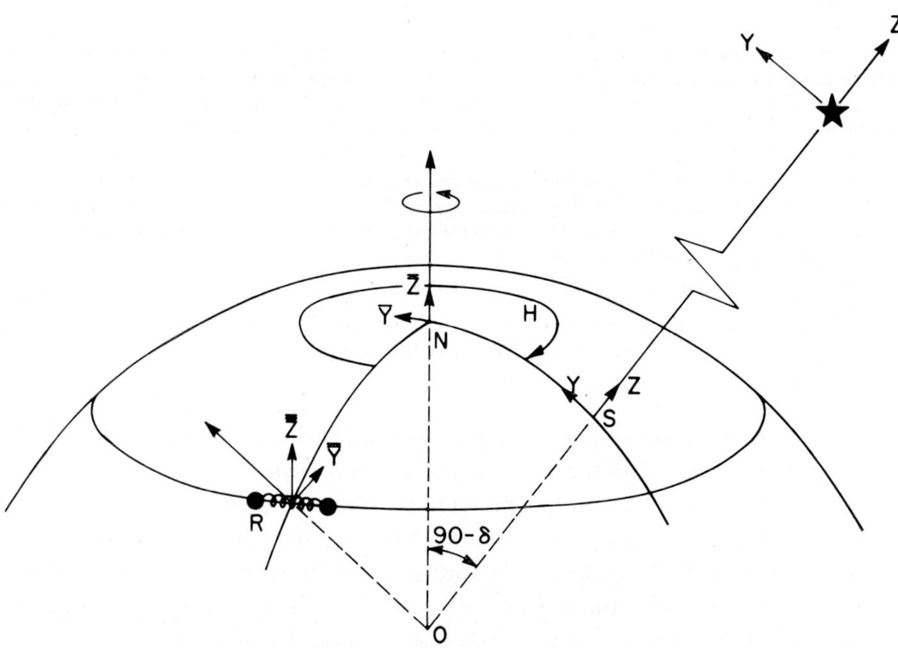

Fig. 17. Idealized detector of gravitational waves, R, on the surface of the earth is driven by a source on a far away star. The coupling between the gravitational radiation emitted at the source and the receptor on the Earth surface have been analyzed in Reference 39 by transforming tensorial components from the laboratory frame (double barred coordinates) to a frame at the north pole (barred coordinates) and then to a frame at S. If the source of gravitational radiation has random polarization and is located at declination δ and hour angle H the response factor of the detector is given by

$$W(H, \delta) = (\cos^2 H - \sin^2 \delta \sin^2 H)^2 + (\sin \delta \sin 2H)^2,$$

see also Figure 18. (Figure reproduced from Reference 39 with permission of the authors.)

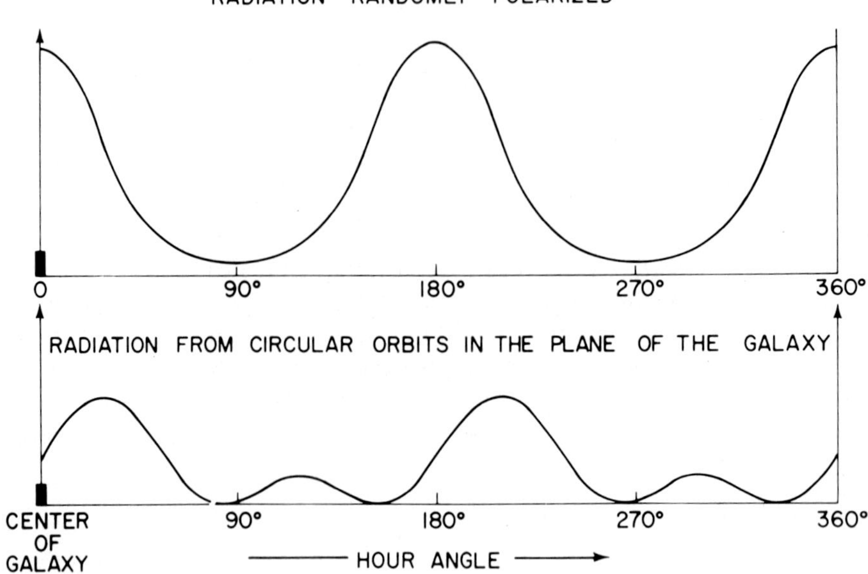

Fig. 18.   The response factor of a gravitational wave detector depends drastically upon the polarization of the gravitational radiation signal. In this figure we compare and contrast the response factor
$W(H, \delta)$ of a detector of gravitational radiation aligned east–west (see Figure 17) to a source of
gravitational radiation located at hour angle H and declination $\delta = -28.9$ (center of the galaxy).
Compared and contrasted are the two examples in which the radiation is randomly polarized (upper
part of the figure) and the radiation is 100% polarized and originates from circular orbits in the plane
of the galaxy (details in Reference 51). The intensity as well as the peak of the response function are
markedly different in the two cases. The response function of a detector directed east–west to a
100% polarized source of radiation is given by

$$W(H, \delta) = [(\cos^2 H - \sin^2 H \sin^2 \delta) \cos 2\alpha + \sin 2\alpha \sin \delta \sin 2H]^2$$

the angle α is the angle between the plane of the polarization of the source and the plane $y-z$ in
Figure 17. (Details in Reference 51.)

Any estimate of the power required for a source of gravitational radiation to deposit
a fixed amount of energy E in the detector will strongly depend on the spectrum of the
radiation. In the case of a broad spectrum we have $I(v_0) = E/\int_{\text{res}} \sigma(v)\, dv$. If the
spectrum is flat from $v = 0$ up to $v = v_0$ we have for the total energy in the pulse
$\int I(v)\, dv = I(v_0)\, v_0$. If the pulse has a spectrum of the form $I(v) = I(v_0)\, (\Delta\omega/2)^2/$
$[(\omega - \omega_0)^2 + (\Delta\omega/2)^2]$ we obtain $\int I(v)\, dv = \Delta\omega\, I(v_0)/4$. If $\Delta\omega \gg A_{\text{diss}}$; $\int I(v)\, dv =$
$= A_{\text{diss}}\, I(v_0)/2$ if $\Delta\omega = A_{\text{diss}}$ and $\int I(v)\, dv = A_{\text{diss}}\, I(v_0)/4$ if $\Delta\omega \ll A_{\text{diss}}$.

To evaluate explicitly the value of the integrated cross section for a given detector
we have first to evaluate the A_{grav} for the mode at which the gravitational wave detector operates. Considering e.g. the detectors used by Weber [71], an alluminum cylinder
of 153 cm in length and 66 cm in diameter, and limiting our considerations to the
modes of longitudinal vibration of the form [39, 69]

$$\xi = \xi_0 \sin(n\pi\, x/L) \sin(\omega t)$$

we have

$$A_{\text{grav}} = -\langle(dE/dt)\rangle_{\text{av}}/E = \frac{64}{15}\frac{G}{c^5}\frac{Mv^4}{L^2}$$

and for the integrated cross section over randomly polarized radiation [67]

$$\int_{\text{res, random}} \sigma(v)\,dv = \left(\frac{\pi}{2}\right)\lambdabar^2 A_{\text{grav}} = \frac{32}{15\pi}\frac{G}{c}\frac{v^2}{c^2}\frac{M}{n^2}.$$

Here $\eta = 1, 3, 5\ldots$ correspond to the even vibrational modes, v is the speed of sound, and M the mass of the cylinder. The cross section for different modes of vibration has been also considered [72]. The directional properties of a detector of the kind used by Weber [71] as well as its response factor to gravitational radiation for selected polarization have been studied by idealizing the detector to a system of two masses m coupled by a spring of length L and resonance frequency ω_0 (see Figures 17 and 18).

5. Observations and Criteria to Differentiate Between Neutron Stars and Black Holes

A direct comparison of the physical size and some of the parameters characterizing a neutron star and a black hole clearly summarize the similarities and the difficulties in distinguishing between these two possible outcomes of gravitational collapse (see Figure 19). The value of the angular velocity as well as the magnitude of the magnetic fields and the radius of these two different kinds of collapsed objects can be extremely similar (see Figure 19). One fundamental difference, however is the value of their

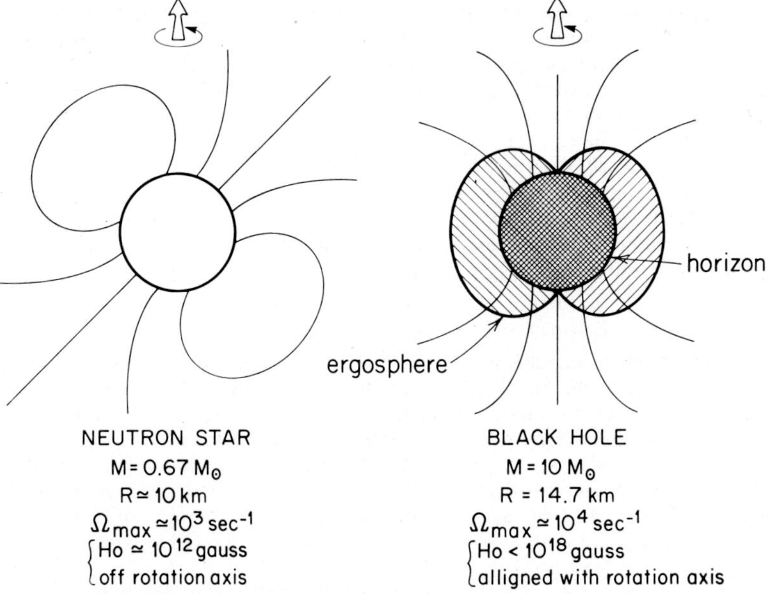

Fig. 19. Neutron stars and black hole compared and contrasted. The magnetic field of a black hole must always be aligned along the rotation axis. (For details see Reference 17.)

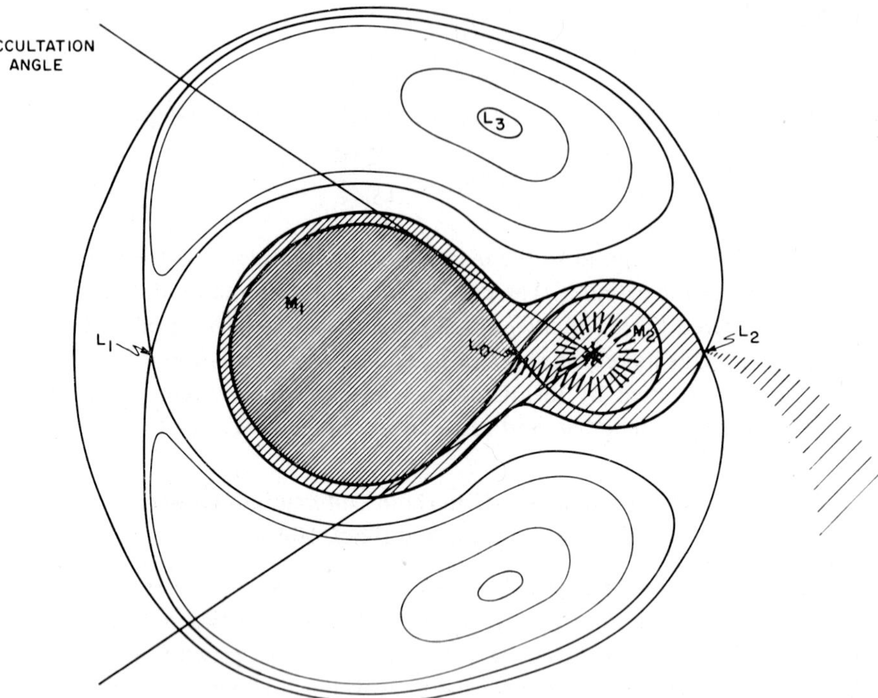

Fig. 20. Roche lobes and equipotential surfaces of a binary system formed by a normal star M_1, and a compact collapsed companion M_2. L_0, L_1, and L_2 are the Lagrange points of the system. X-rays are emitted by the matter from the main star being accreted into the very strong gravitational field of the collapsed companion star and, if the inclination of the orbit is high enough (see Table I), occultation will occur with the characteristic period of the binary. The equipotential surfaces here traced have been computed under the following simplifying assumptions of the Roche model: (a) The gravitational field generated by the two stars are computed as if the masses of the stars were concentrated in two points (b) the orbit of the two stars are assumed to be circular orbits around the center of mass (c) the axis of rotation of the two stars is perpendicular to the orbital plane and (d) the period of intrinsic rotation of the star M_1 is assumed to be the same as the one of orbital revolution (corotation). Information on the masses of the system can be immediately acquired by an application of the Kepler law. We have $G(M_1 + M_2) = (2\pi/T)^2 a^3$, where a is the separation between the center of mass of the two stars and T is the period of the binary system. If we know the projected velocity of both stars $(v \sin i)_1$ and $(v \sin i)_2$ and the orbital period T, or $\omega = 2\pi/T$, we can derive

$$M_2 \sin^3 i = [(v \sin i)_2 + (v \sin i)_1]^2 \, (v \sin i)_1/\omega G$$
$$M_1 \sin^3 i = [(v \sin i)_2 + (v \sin i)_1]^2 \, (v \sin i)_2/\omega G \, .$$

By assuming that the main star fills its Roche lobe and material outflow through the Lagrangian point L_0, from the value of the binary period and the occultation angle $\Phi = (T_{\mathrm{occ}}/T_{\mathrm{orb}}) \times 180°$, it is possible to obtain for selected values of the inclination angle i the value of the ratio $q = M_2/M_1$ (see Table I). Profoundly different from pulsars in which the energy source is due to the rotational energy of the neutron star, the energetics of these binary X-ray sources is totally determined by the gravitational binding energy of the infalling material. The luminosity of the X-rays is controlled by the Eddington limit [81] $L_{\mathrm{E}} = 4\pi \, GMc/\sigma_c$ where as usual σ_c is the Compton cross section. Ruffini and Wilson [80] have pointed out that accretion processes can be conceived in which

$$\left(\frac{\mathrm{d}E}{\mathrm{d}t}\right)_{\mathrm{accr}} = \frac{GM_2}{r_0} \frac{\mathrm{d}M}{\mathrm{d}t} \gg \frac{4\pi GM_2 c}{\sigma_c} = L_{\mathrm{E}} \sim 10^{38} \; \mathrm{erg \; s^{-1}} \; \mathrm{for} \; M_2 \sim M_\odot$$

due to the fact that a large fraction of the energy of the accreting material $(\mathrm{d}E/\mathrm{d}t)_{\mathrm{accr}}$ could be emitted masses, neutron stars can exist only for masses smaller than the critical mass against gravitational collapse ($< 3.2\ M_\odot$) and black holes can exist only for values of the mass larger than this critical value. To form a black hole from a star with mass smaller than the critical mass, enough kinetic energy should be given to the collapsing material in order to tunnel through the barrier of the neutron star equilibrium configurations (see Figure 3). Another fundamental difference between these two families of collapsed objects follows from the structure of their electromagnetic fields. In a neutron star the magnetic field can have any inclination with respect to the rotation axis, and explicit solutions have been given by Deutsch for such configurations [10]. In particular, the existence of an off-axis magnetic field in a neutron star can explain most directly the very regular pulsation and the lengthening in pulsational period observed in pulsars. In a black hole the magnetic field always has to be aligned with the rotation axis in order to have a stationary metric at infinity [17].

Therefore, although we can expect very short time structure in the signal emitted by material falling into black holes with the characteristic time constant given by the revolution period of material orbiting down to the last stable circular orbit, this signal will last at most, a few revolution periods [17]. In no way, therefore, can we expect that a regular signal of the kind observed in a pulsar can be emitted from a black hole.

To allow further progress in this entire field of research, a large amount of data was needed to infer not only the mass and the angular velocity of the collapsed objects, but also the structure of their magnetospheres as well as, in the case of neutron stars, the details of their internal constitution. This was impossible to do on the basis of the data acquired from pulsars for two reasons: if we exclude a few glitches and microglitches which affect very slightly their pulsational period, pulsars are extremely steady in their emission processes and no variation occurs to give a hint on their internal structure. Moreover, no possibility exists of a direct measure of the mass of the pulsar since, of the 120 pulsars observed, none has been found to be a member of a multiple system [30]. The detection of an isolated black hole in space would have been even more hopeless: "No light comes directly from it. It cannot be seen by its lens action or other effect on a more distant star. It is difficult enough to see Venus, 12000 km in diameter, swimming across the disc of the Sun; looking for a 15 km object moving across a far-off stellar light source would be unimaginably difficult!" [73] Following the work of Zel'dovich and Guseynov [74], Shklovsky [75], Zel'dovich and Novikov [76], and Schwartzman [77], the emphasis was directed in 1971 not to isolated systems, but binary systems: "The possibility of capitalizing on double star system is most favorable when the black hole is so near to a normal star that it draws

---

as neutrinos through to the reaction $\gamma \rightleftarrows e^- + e^+ \rightarrow v_e + v_e$. The accretion rate into the neutron star, far from being then constrained by the Eddington luminosity to the value $(\mathrm{d}M/\mathrm{d}t)_{\mathrm{accr}} \lesssim 10^{-9}\ M$ yr^{-1}, could reach much higher values e.g. $\mathrm{d}M/\mathrm{d}t \sim 10^{-6}\ M_\odot$yr^{-1}. This accretion rate would give rise to neutrino fluxes at $L_\nu \sim 8.5 \times 10^{39}$ erg s^{-1} or to neutrino and antineutrino fluxes at the surface of the Earth $\Phi_\nu = 22.2$ cm^{-2} s^{-1} assuming the source 1000 parsec away. During these processes (as pointed out in Reference 80) the X-ray luminosity of the X-ray source is still equal to the Eddington luminosity. Details on the structure of the accretion disk can be found in References 82–88.

in matter from its companion. Such a flow from one star to another is well known in close binary systems, but no unusual radiation emerges. When one of the components is a neutron star or a black hole, a strong emission in the X-ray region is expected" [73] (see Figure 20).

The discovery made by the team lead by Riccardo Giacconi [78] through the observations from the Uhuru satellite and the joint observations made from the ground in the optical and radio wavelength have given irrefutable evidence for the discovery and direct observation inside our own Galaxy of a very large number of short period binary systems ($P_0 \lesssim 5$ days) with a normal star and a collapsed object as components. It is hard to overemphasize the relevance of this experimental discovery for the entire field of the astrophysics and for the physics of collapsed objects. For the first time, we are now in the position not only of measuring the masses of collapsed objects with great accuracy, but of also obtaining, from the observation of the detailed features and short time structure of the radiation emitted by material accreting into the collapsed object under a variety of conditions and regimes, an accurate description of the magnetosphere both of neutron stars and black holes. More important even the analysis of very short time variability allows having information from regions more and more near the surface of a collapsed object.

As in the physics of elementary particles where we can never 'see' an elementary particle but we can 'infer' its structure and form factor through an analysis of scattering experiments, similarly here we can never see the surface or the internal structure of a black hole, but we certainly can infer its 'form factor', through the large-scale scattering experiment originated by the accretion of matter in the field of the collapsed object. Finally, the reason that the collapsed object is continuously accreting material from the normal star implies that at least in principle we should be able to observe dynamical changes and increase in the mass of the collapsed object and consequently a direct 'neutronization' of matter in a neutron star or an expansion of the horizon of the black hole.

One of the most impressive features of the binary X-ray sources consists in a sharp differentiation in the kind of X-ray spectrum they emit: In one family of sources the X-rays are emitted in pulses of great regularity recalling many features of pulsars, in the other, although variability down to time scale of a few milliseconds are observed, no regular pulsations are present in the spectrum with the possible exception of train of pulses of radiation. The difference between these two families of sources is exemplified better by a direct look at Figure 21 than by any further word.

In 1972 [17, 79] a classification was proposed to identify all the pulsating binary X-ray sources with neutron stars, and the bursting sources, with short time variability but no regular pulsation, with black holes.

Today this classification appears to be supported by experimental evidence (see Sections 6 and 7). The crucial point of this classification consists in the clear possibility of determining from direct observations the value of the critical mass of a neutron star against gravitational collapse: The pulsating sources are expected to have masses *up* to this critical value, and *all* the bursting sources masses larger than this critical value.

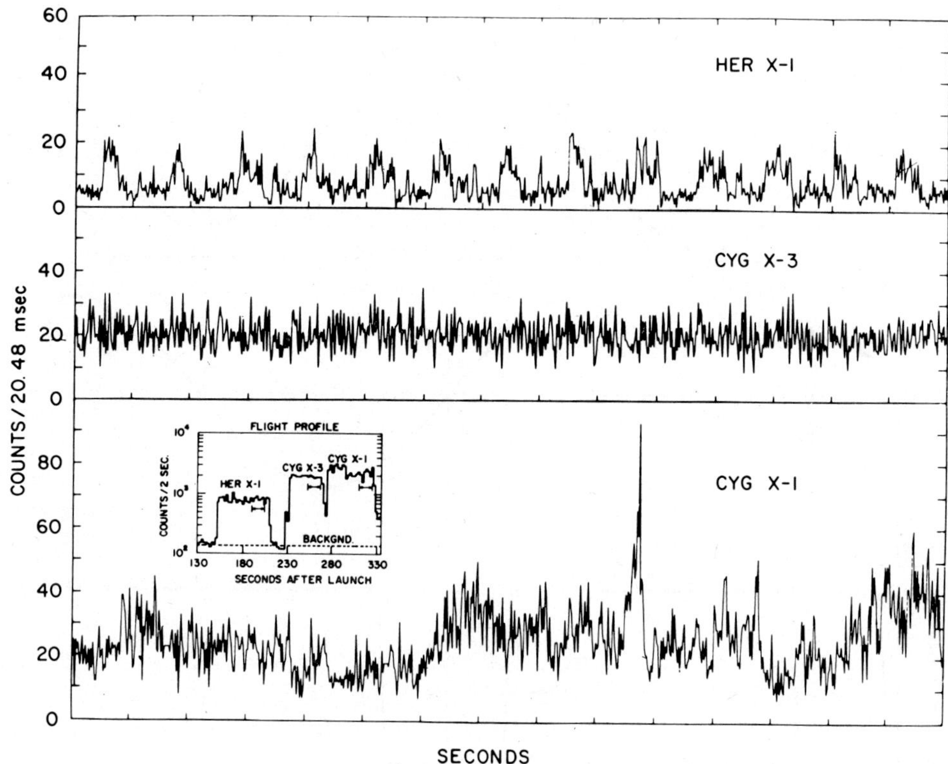

Fig. 21. X-rays data of the rocket flight from Goddard Space Center flown by Rothschild *et al.* [109]. The flight lasted 330 s with ~ 70 s observations of the pulsating X-ray sources Hercules X1, ~ 40 s on Cygnus X3 and 60 s on Cygnus X1. A direct comparison between the data of Hercules X1, Cygnus X3 and Cygnus X1 clearly shows the marked differences in the X-ray signals in the three sources; regularly pulsating with a period of ~ 1.2 s the ones in Hercules X1 bursting with time scales down to a few milliseconds the ones in Cygnus X1 (for details see Sections 5 and 6, and reference mentioned there, also see Reference 17). (Courtesy of R. Rothschild, Goddard Space Flight Center.)

6. Regularly Pulsating Binary X-Ray Sources

There are two binary X-ray sources which have a sharply defined pulsational period in their X-ray emission in the range 1–20 kev: Hercules X1 [91] has a pulsational period $P_0 \sim 1.23$ s, Centaurus X3 [92] $P_0 \sim 4.84$ s. Their binary nature is most clearly shown by their occultation in the X-ray emission and by the Doppler shift in their intrinsic pulsation period P_0 due to the orbital motion of the X-ray source (see Figure 22). In the case of Hercules X1 the companion star has been identified with the star Hz Hercules [93] while the optical identification of the normal companion star of Centaurus X3 is still tentative [94].

The main arguments leading to the identification of these sources with rotating neutron stars, members of binary systems, leads to the simple understanding of three main experimental facts:

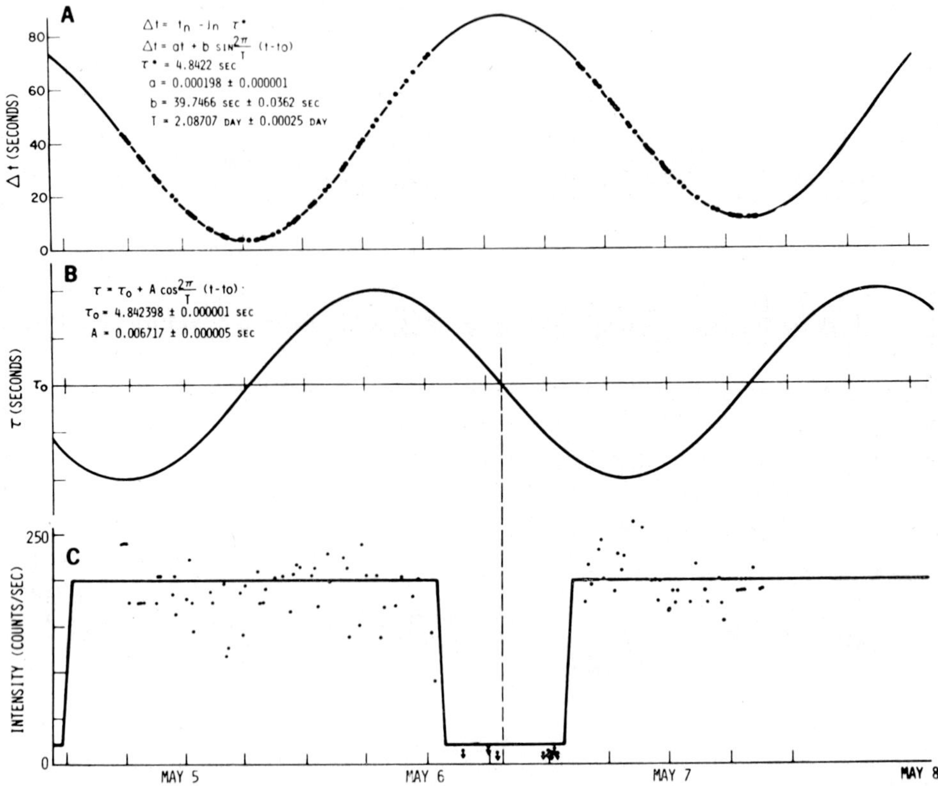

Fig. 22.   X-rays data from Centaurus X3 [92]. The clear occultation of the X-ray source by the normal star in shown in Part c. The binary period is $T = 2.087$d and the eclipse or low state 0.55d, the transition from the high to the low state takes place in ~ 0.04d (In the case of Hercules X1 [91] the orbital period $T = 1.7$d, the eclipse last 0.24d, the transition between the high intensity state and the eclipse occurs in less than 12 min). The intrinsic period of the X-ray source is modulated by two different effects: A Doppler effect, Part b, and an arrival time delay, Part a. Due to its velocity in the orbital motion and to the Doppler effect the intrinsic period appears to have a sine wave modulation, the amplitude being proportional to the projected velocity of the X-ray source along the line of sight. As a direct consequence the velocity of the X-ray source in the circular orbit can be estimated to be $v_x \sin i = 415.1$ km s^{-1}, i being the inclination of the orbit. (The corresponding velocity for Hercules X1 is $v_x \sin i = 169$ km s^{-1}.) As a consequence of the fact that the X-ray source is moving in a circular orbit the arrival time of the pulse is delayed 39.7 s at the center of the occultation and is 39.7 s early at the center of the high state. (The corresponding time delay of Hercules X1 gives 13.2 s.) The delay time of 39.7 s gives a direct measurement in light seconds of the radius of the orbit of the X-ray source about the center of mass of the system as projected into the observing plane. The nearly sinusoidal feature of the curves in (a) and (b) allow to place a limit to the eccentricity of the orbit to $\varepsilon < 0.05$ ($\varepsilon < 0.05$ also in the case of Hercules X1). If we adopt the Roche model for the accretion of matter in the binary system [17, 79] we can evaluate parameters of the binary system from the value of the period T, of the occultation time and the projected orbital velocity of the masses. For Centaurus X3 for selected value of the inclination we have:

i	$m_x/M_\odot$	$M/M_\odot$	$a/R_\odot$	$R/R_\odot$	v_x	v_M
90°	0.275	16.0	17.4	12.7	415.1	7.14
80°	0.250	16.6	17.7	13.3	421.5	6.32
60°	0.194	24.2	19.9	15.5	479.3	3.83

(1) An amount of energy of the order of $\sim 10^{37}$ erg s^{-1} is emitted by these sources in the X-rays band.

(2) This radiation is emitted in sharply defined pulses with a period $P \sim 1$ s.

(3) The intrinsic pulsational period of the X-ray source decreases with time (see Figure 24).

As in the case of pulsars, the identification of these pulsating X-ray sources with rotating neutron stars endowed with an off axis magnetic field allows a simple explanation for the pulsating nature of the X-ray signal. However these X-rays sources depart from pulsars in a very important respect. There the pulsational period – increases with time $(\mathrm{d}P/\mathrm{d}t > 0)$, and the energetics of the system are most easily explained by the loss of rotational energy of the neutron star. In our case, instead, the period of the X-ray sources decreases with time. This clearly implies that the X-ray source is, in fact, gaining rotational energy! (see Reference 78 and particularly Figure 7 in H. Gursky and E. Schreier report).

Both the energetics of the system and the gain of rotational energy are most easily explained if we account for the accretion of matter from the main star into the collapsed object. The infalling material then imparts angular momentum to the neutron star while the energetic can be explained by the conversion into X-rays of the gravitational binding energy of the infalling material at the surface of the neutron star (up to 10% of its rest mass).

Many of the features of the X-ray emission both of Hercules X1 and Centaurus X3 still present outstanding difficulties for their detailed explanation. However, we can emphasize that for the first time we shall be able in the near future to obtain from these two systems an accurate direct measure of the mass of a neutron star. Assuming that the main star fills its Roche lobe, then from the velocity of the neutron star in its binary orbit, the length of the occultation, and the binary period, we can directly estimate the neutron star mass for selected values of the inclination as shown in Figure 22. We can then conclude that in the case of Hercules X1, it is most likely that the neutron star has a mass larger than the value of the critical mass as computed by Oppenheimer and Volkoff [13].

and in the case of Hercules X1

i	$m_X/M_\odot$	$M/M_\odot$	$a/R_\odot$	$R/R_\odot$	v_X	v_M
90°	1.20	2.1	8.9	3.8	169.0	96.3
80°	0.78	1.8	8.2	3.7	171.6	73.8
60°	0.17	1.6	7.2	4.1	195.2	20.5

Here we have indicated by $m_X/M_\odot$ and $M/M_\odot$ the mass of the X-ray source and of the main star, by $a/R_\odot$ and $R/R_\odot$ the separation between the center of masses of the two stars the radius of the main star and by v_X and v_M the velocity of the X-ray source and of the main star. It is interesting to remark that both these estimates are very interesting: the neutron star in Cen X3 appears to have a very small mass while the one in Hercules X1 has a mass which for a suitable inclination is larger than the critical value of neutron star as computed from an equation of state neglecting the nuclear interactions. Both results should be confirmed by a more model independent derivation possibly by the observation of the velocity of the companion star. (Figure reproduced with the kind permission of Giacconi [95].)

If this result is confirmed by an analysis of the Doppler shift of the main star (Hz Hercules) associated with the X-ray source then we will have the first clear experimental evidence that the contribution of strong interactions in the description of the neutron star material has to be taken into serious account and cannot be neglected.

A detailed monitoring of the intrinsic pulsational period and of the binary period of the X-ray source can give important informations both on the amount of material being transfered in the binary system and the one accreting onto the neutron star. We can write the following general formula [89].

$$I\frac{d\omega_0}{dt} + \omega\frac{dI}{dm}\frac{dm}{dt} = \left(\frac{dJ}{dt}\right)_{\text{diss.}} + \left(\frac{dJ}{dt}\right)_{\text{accr.}}.$$

Here $(dJ/dt)_{\text{diss}}$ is always negative and takes into account all the loss of angular momentum due to dissipative processes from electromagnetic or gravitational radiation emitted from magnetic fields or changing gravitational quadrupole moments of the neutron star [90]. $(dJ/dt)_{\text{accr}}$ is the angular momentum transferred to the neutron star by the accreting matter. dI/dm determines the change in the moment of inertia of the neutron star as a function of accreting mass which is drastically dependent upon the details of the equation of state of the neutron star material (see Figure 5). We can then conclude that to determine with great accuracy the change of the moment of inertia of a neutron star as a function of its change in mass can lead to basic informations about the equation of state of matter at nuclear and supranuclear densities in neutron star material. If we assume [89] for dI/dm the value computed from selected equations of state as given in Figure 5 and $dm/dt \sim 10^{-8} M_\odot \text{ yr}^{-1}$, then in both the case of Hercules X1 and Centaurus X3 we have $I(d\omega/dt) \gg \omega(dI/dm)(dm/dt)$. It is however conceivable that in some other binary X-ray sources not yet detected, or during some phases of the accretion process, the quantity $(dJ/dt)_{\text{accr}}$ is so small as to make observable both the change of moment inertia and the dissipative terms in the loss of angular momentum. It is interesting here to remark that the change of the intrinsic period of pulsation also allows obtaining information on the structure of the accreting disc of material around the neutron star. If we assume that the accretion on the neutron star occurs from a disc in which angular velocity is removed by viscous stresses [88] then, the accreting material transfer to the neutron star the Keplerian angular momentum of the inner edge of the disc R

$$\frac{dm}{dt}(GMR)^{1/2} = I\,d\omega/dt. \tag{2}$$

Here we indicate by M and I the mass and the moment of inertia of the neutron star and by dm/dt and $d\omega/dt$ the rate of mass accretion on the neutron star and the change of angular velocity, respectively. If we substitute for $d\omega/dt$ the observed values we can then obtain an absolute upper limit from (2) to the rate of matter accretion: $dm/dt \lesssim 1.1 \times 10^{-10} M_\odot \text{ yr}^{-1}$ for Hercules X1. This value is much smaller than the absolute lower limit on the rate of accretion obtainable on purely energetic grounds: assuming that up to ten percent of the rest mass of the accreting material could be transformed into X-rays we would obtain for a source intensity of $\sim 10^{37}$ erg s^{-1} an absolute lower limit of $dm/dt \sim 1.5 \times 10^{-9}$ $M_\odot$ yr^{-1}. We can then conclude that the disc structure, if existing at all, is drastically modified by the presence of the magnetic field of the rotating neutron star. Additional information on the dynamics of the binary system can be acquired if we notice that the binary period also changes with time. In the case of Centaurus X3 the binary period T has been observed to decrease of $\Delta T / T \sim 3.5 \times 10^{-5}$ over one year in 1971. It has been shown how this variation can be explained [89] with a very large outflow of matter ($dM/dt \gtrsim 10^{-3}$ $M_\odot/$yr^{-1}) from the binary system. It is most likely, therefore, that some of the low states of Centaurus X3 (see Reference 78 and particularly Figure 7 in H. Gursky and E. Schreier report) are indeed due to the absorption of the X-rays by this very large outflow of matter it is also very important to correlate changes either of the intrinsic pulsational period or of the binary period with changes in the intensity of the X-ray emission. This correlation appears to be most promising for the understanding of the accretion processes.

7. Bursting Binary X-Ray Sources

The characteristics of these sources are very similar to those presented in the previous paragraph: they are members of binary systems and the energy they radiate in X-rays is $dE/dt \lesssim 10^{38}$ erg s^{-1}. They drastically differ, however, from the ones presented there in one important respect: the X-rays are not emitted in regular pulses but they present only short intensity variations and flare like phenomena in the X-rays with intensity changes by a factor two or more on a time scale down to a few milliseconds. Since we are dealing again with close binary systems the most direct explanation for the strong X-ray emission is, as in the previous case, accretion of matter from a normal star into a compact collapsed object. The main reasons for requiring that the object on which the accretion occurs be a collapsed object are made both on energetic grounds (we need a deep potential well in order to transform enough gravitational energy into electromagnetic energy) and on the grounds of the irregular variations in the X-ray intensity observed in some sources to extend down to a few milliseconds. This last experimental result clearly implies that the region of X-ray emission has to be very compact.

The absence of a regular pulsation in the X-rays can be ascribed to the fact that the collapsed object is either a black hole (see Section 5) or a neutron star deprived of an off axis magnetic field of such an intensity as to modulate the X-ray emission of the accreting material [79]. The identification with a neutron star is clearly impossible if the collapsed object proves to have a mass larger than the absolute upper limit of the neutron star critical mass [29].

There are several binary X-ray sources which have these common features in their X-ray emission: Cygnus X3, 2U 0115−73 or SMX1, 2M 0900−40 or Vel XR1, 2U 1700−37, 3U 1516−56 or Circinus X1 and finally Cygnus S1. A detailed description of these systems can be found in Reference 78. In the following we shall mainly focus on Cygnus X1 and we will give reasons why we consider this sytem the one of the greatest physical interest. The first detailed observations of Cygnus X1 were obtained by Giacconi *et al.* [96, 97] and Rappaport *et al.* [98]. From the distance of the source [99, 100] and the observed flux and spectrum of the X-rays it was possible to infer that Cygnus X1 had to emit $dE/dt \sim 10^{37}$ erg s^{-1}. The X-ray intensities were observed to have very large changes on a time scale of less than 50 ms [98]. Very high energy flux and the short time variability in the X-ray intensity most naturally lead to the assumption that Cygnus X1 had to be an accreting collapsed object and a member of a binary system. However, the identification of this source with a binary system appeared very problematic from the beginning. The major 'signature' characterizing an X-ray source member of a binary system (see Figure 20) was missing in this case, namely, the X-ray source was not regularly occulted by the main star with the regular binary period of a few days. Moreover the absence of an intrinsic pulsational period, or for that matter of any regular long lasting structure in the intensity variations did not allow the use of the Doppler effect to infer the orbital motion of the X-ray source as in Centaurus X3 or Hercules X1.

The absence of regular occultation of the X-rays could still be made consistent with a binary system model if it was assumed that the angle between our line of sight and the orbital plane of the binary was larger than a critical amount ($i < 40°$, see Table I). From the pure geometrical features of the Roche accretion model we should in fact expect that out of N_{obs} observed binary X-ray sources a number $N \sim 4\,N_{obs}/9$ should *not* present any occultation if indeed the orientation of their orbital plane is, as it should be, completely random. Clearly this figure should be taken with the due caution since is based on a direct application of the idealized case of a Roche model

TABLE I

Occulation angle $\Phi = (T_{occ}/T_{orb}) \times 180°$ for selected values of the inclination of the orbit and selected value of the ratio $q = M_1/M_2$. The case $i = 90°$ corresponds to the line of sight in the orbital plane of the binary system. It has been assumed in these computations that the main star fills its Roche lobe and that the X-ray source can be considered point-like. For every value of q there exists a critical value of the inclination angle i_{crit} for which no occultation is possible if $i < i_{crit}$

$q = \dfrac{m}{M}$	Ω_0	$\phi_{i=90°}$	$\phi_{i=80°}$	$\phi_{i=70°}$	$\phi_{i=60°}$	$\phi_{i=50°}$	$\phi_{i=40°}$	$\phi_{i=30°}$	$\phi_{i=20°}$
1.0	3.7500	22.00	19.81	10.46	–	–			
0.8	3.41697	23.30	21.29	13.25	–	–			
0.6	3.06344	25.03	23.22	14.40	–	–	no occulation		
0.4	2.67810	27.56	26.00	20.41	–	–	possible		
0.3	2.46622	29.42	28.00	23.09	9.58	–			
0.2	2.23273	32.09	30.85	26.69	17.15	–	–	–	–
0.15	2.10309	34.00	32.88	29.16	21.16	–	–	–	–
0.1	1.95910	36.72	35.73	32.52	26.04	10.77	–	–	–
0.05	1.78886	41.32	40.52	37.97	33.15	24.50	–	–	–
0.02	1.65702	47.16	46.54	44.60	41.13	35.58	26.70	–	–
0.01	1.59911	51.29	50.77	49.17	46.37	42.10	35.92	27.01	12.59
0.005	1.56256	55.13	54.69	53.35	51.04	47.63	42.98	36.97	29.83
0.001	1.52148	62.87	62.55	61.63	60.06	57.87	55.07	51.83	48.54

which at least in some sources (see e.g. 2U 1700−37) is proven [79] not to fit the experimental results. In any case, since $N_{obs} \sim 8$, the fact that two sources, Cygnus X1 and Cygnus X3, do not have occultations should be considered in perfect agreement with the general expectation and explainable as an effect of the orientation of their orbital plane.

In sharp difference from the other X-ray sources, Cygnus X1 presents very little low energy cutoff due to absorption (see Figure 23)· This experimental fact gave the first confirmation that, most likely, we were looking at Cygnus X1 at a small inclination angle. The reason is simply explained: we expect in the accretion processes that the majority of the X-rays are emitted in a region very near the surface of the collapsed objects. From the geometry of the system we should then expect large absorption in

$$\frac{d\epsilon}{dE} = CE^{-\alpha} \exp\left\{-\left(\frac{E_A}{E}\right)^{\frac{8}{3}}\right\}$$

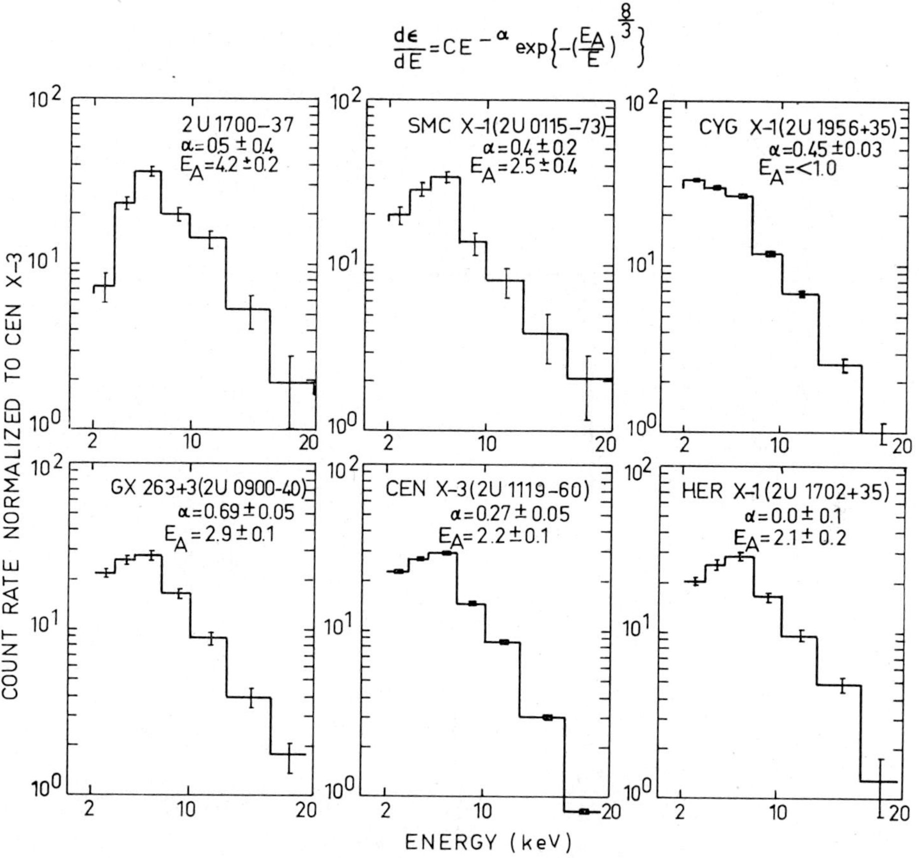

Fig. 23. Spectra of six binary X-Ray sources in the range of energy 2 keV $\leqslant E \leqslant$ 20 keV. In the first row from left to right are 2U 1700 − 37, SMC X-1, Cyg X-1, in the second row again from left to right 2U 0900 − 40 Cen X3 and Her X1. All these sources present a drastic cutoff in the range of energies 2 keV $\leqslant E \leqslant$ 6 keV due to absorption with the only clear exception of Cygnus X1. This result was extremely important in reaching the conclusion that we are observing the binary source in Cygnus X1 at a small inclination angle and therefore avoiding the absorption of the accreting matter. (Figure reproduced from Reference 78 with kind permission of the author.)
© 1974, International Astronomical Union.

the X-ray spectrum the more the line of sight will approach the orbital plane of the binary system.

In this sense Cygnus X1 will most likely be the most interesting system to examine: only in the case of small inclination angle will we be able to see the processes occurring near the surface of the collapsed object. Moreover the more the emission processes will occur near the surface of the collapsed object the shorter should their time varia-

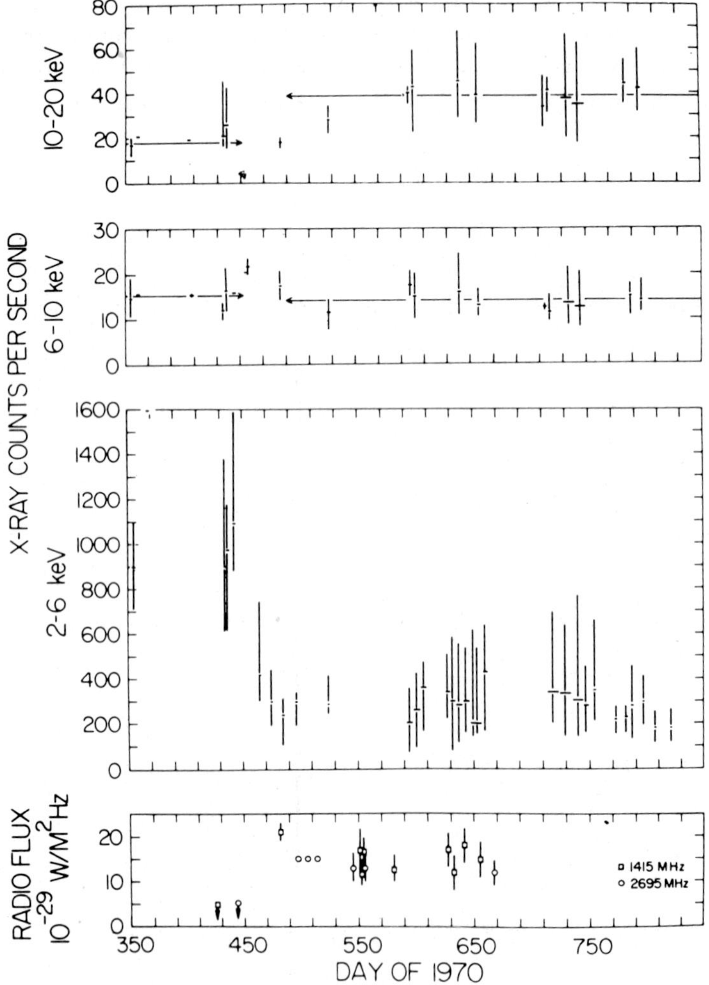

Fig. 24.   Transition occurred in Cygnus X1 between 12 March and 28 April 1971. The spectrum drastically changed with an increase of intensity in the high energy (10–20 keV upper figure) of a factor 2 and a decrease of intensity in the low energy (2–6 keV) of a factor 4. Hjellming [101] has reported that sometime between March 22 and March 31 a radio source first appeared and remained at a level of 0.02 fu in the X-ray error box (lower figure). No radio source was present in the X-ray error box before this transition down to a level of 0.01 fu. This result allowed to give a much improved position for the X-ray source and led to the identification of the optical companion star HDE226868 (see Figure 25). (Figure reproduced from Reference 78 with kind permission of the author.)

bility be expected to be. We can then conclude that we should notice a correlation in X-ray sources between the absorption and their intensity changes: The smaller the absorption, the more structure we should find at shorter time scale.

To further analyze the nature of Cygnus X1 the region of the sky around this source was carefully analyzed for the possible existence of radio emission. A detailed examination was made at 11 cm by Hjellming [101] who could not find any source down to the limit of 0.01 fu. The motivation of this detailed analysis was dictated from a straightforward consideration: if Cygnus X1 was indeed a collapsed object and if its formation had occurred through the usual supernova process then we should have expected a radio remnant around the collapsed object. The absence of a radio remnant together with the absence of the disruption of the binary system [102, 103] allowed the general conclusion that the formation of Cygnus X1 had occurred through a different process than the usual supernova explosion (see Section 9).

The key result which led at once to the clear identification of Cygnus X1 as a mem-

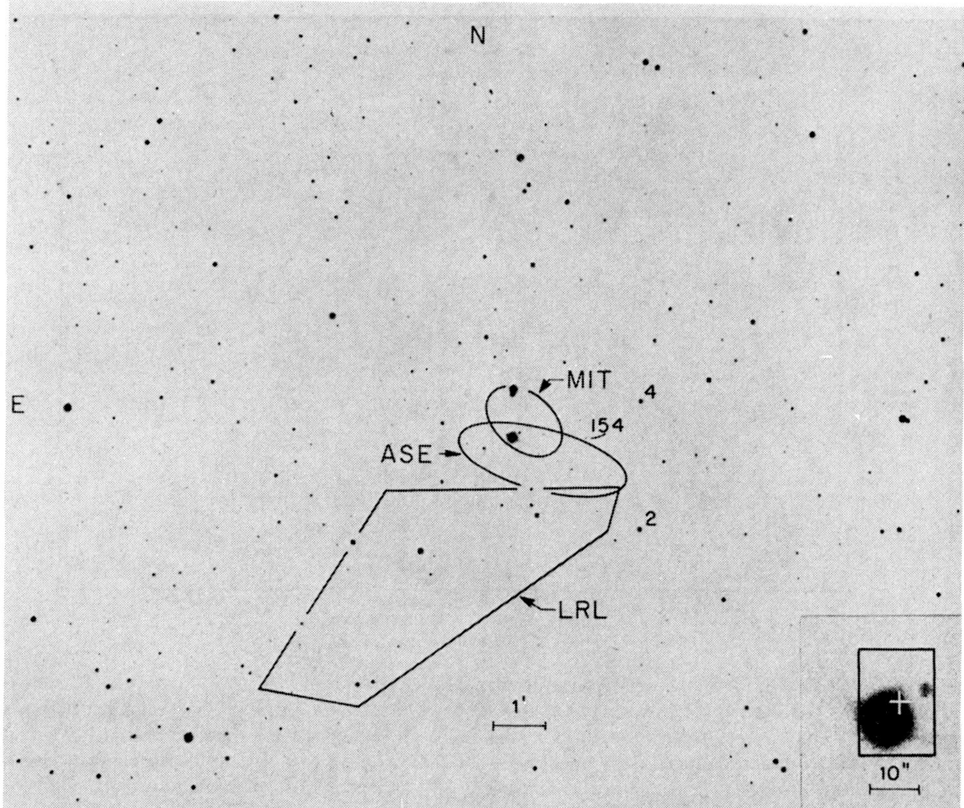

Fig. 25.   Optical plate showing the star HDE226868. The error boxes indicate the X-ray position of the source as from rocket flights and Uhuru satellite data (ASE). In the corner is a magnification of the central region (10″) with the star HDE226868 and the radio error box (white cross). The analysis of the spectral lines of the star have clearly shown the sine wave shift typical of a star member of a binary system. (Figure reproduced from Reference (95) with kind permission of the author.)
© 1974, International Astronomical Union.

ber of a binary system and to the identification of the companion normal star of the system with the B0 supergiant HDE226868 [101, 105] has been due to the direct observation of an abrupt change in the X-ray spectrum occurred between 22 March and 28 April 1972. For still unexplained reasons correlated with this change of spectrum, a radio source appeared (0.02 fu) in the X-ray error box of Cygnus X1 (see Figure 24).

The location of the radio source allowed the determination with great accuracy of the position of Cygnus X1 and the immediate optical identification by Webster and Murdin [104] and Bolton [105] of the other member of the binary system with the B0 supergiant HDE226868 (see Figure 25). The further evidence for the binary nature of the system was proved by Bolton through a detailed analysis of the Doppler shift of the line of the star due to its orbital motion (see Figure 26).

Evidence for the accretion and infall of material from the main star into the X-ray source, and a direct estimate of the inclination of the orbital plane of the binary system with respect to the line of sight have been given by Hutching $et\ al.$ [106] $(i \sim 27°)$

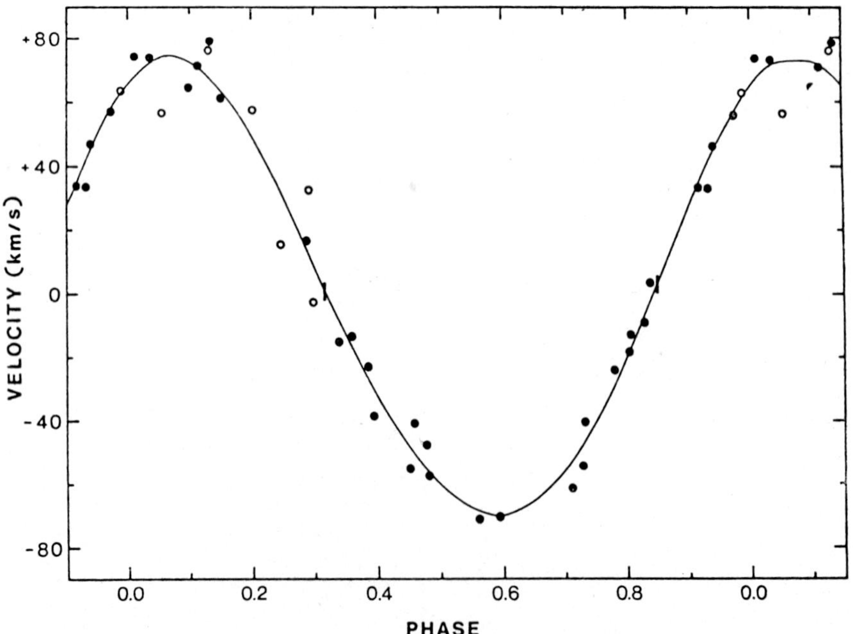

Fig. 26.   Velocity curve of the absorption lines of HDE226868 obtained from 12 Å mm^{-1} spectrograms taken by Bolton [111] with the 74 in. telescope of the David Dunlap Observatory. The open circles are $\frac{1}{2}$ weight points and the vertical stick marks on the velocity curve indicate times of conjunction. The orbital period deduced is $P = 5.599823 \pm 0.000037$ days, the velocity of the center of mass of the system is $v_0 = -1.7 \pm 0.5$ km s^{-1}. The projected velocity of the star $k = v \sin i = 0.8$ km s^{-1} the eccentricity of the orbit $\varepsilon = 0.061 \pm 0.11$ the time of phase zero $T = $ JD 244 1562.520 ± 0.305 the projected value of the semi axis $a \sin i = 5.549 \pm 0.061 \times 10^6$ km and finally the mass function $f(m) = 0.217 \pm 0.007\ M_0$. T. Bolton quotes for the mass of the components of the system values which fall inside an 'error box' whose corners in the M_1 (mass of HDE226868) M_2 (mass of X-ray source) plane are $(M_1, M_2)/M_\odot$ (12.7, 9), (21, 11.5) (35, 18.5), (19.5, 14.9). We thank T. Bolton for allowing to reproduce these data and for many enlightening discussions. Published by the University of Chicago Press; © 1974, American Astronomical Society.

By the analysis of the emission line He II $\lambda4686$, first discovered by Bolton [105] and Brucato and Kristian [107], Hutching *et al.* were able to show that these emissions line were originated in a stream of material from the main star to the X-ray source their velocity curve being 120° out of phase from the lines of HDE226868 (see Figure 27).

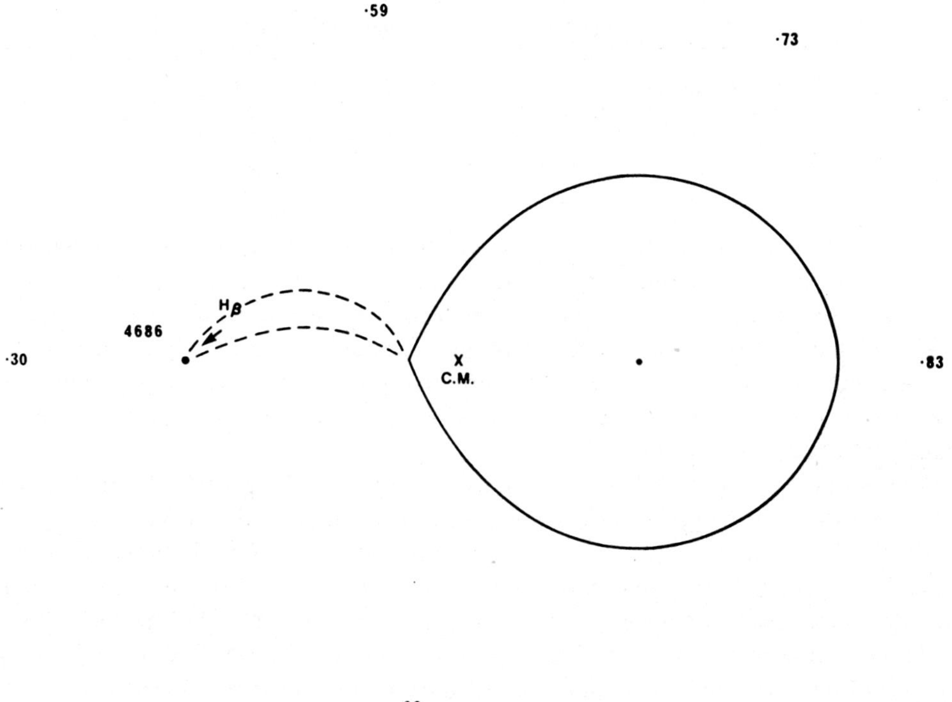

Fig. 27. Profile of matter leaving the main star and accreting on the collapsed object as derived from the observation of He II $\lambda4686$ by Hutching *et al.* [106]. The He II $\lambda4686$ ($v\sin i \sim 100$ km s^{-1}) are 120° out of phase from the lines of HDE 226868 ($v\sin i \sim 80$ km s^{-1}). These observations were essential in strengthening both the argument of the binary nature of the system containing Cygnus X1 and the role of accretion in the generation of the X-ray emission. (Details in Reference 106. Figure reproduced from Reference 106 with kind permission of the authors.) Published by the University of Chicago Press; © 1973, American Astronomical Society.

This entire set of observations by Bolton and Hutchings lead to the following clear conclusions:

(a) Cygnus X1 is indeed a member of a binary system,

(b) accretion is occurring between the main star and the X-ray source,

(c) the absence of an occultation of the X-ray source is a direct consequence of the high inclination of the orbit ($i \sim 26°$).

The experimental data of Cygnus X1 have allowed a direct estimate of the masses of the system see Table II. It is important to remark here that in all cases the estimate of the mass of the X-ray source is much larger than the absolute upper limit to the neutron star mass of 3.2 $M_\odot$ [29].

TABLE II

Estimates of the masses of Cygnus X1 and HDE 226868 as given
by Brucato-Kristian [103], Hutchings *et al.* [106], Sunyaev *et al.*
[110], and Bolton [105]. In all these estimates the mass of Cygnus X1 is well above the absolute upper limit of a neutron
star [29]

	Cygnus X1	HDE 226868
Brucato-Kristian	$M_2/M_\odot > 5.5$	$M_1/M_\odot \sim 22$
Hutchings *et al.*	$10 \leqslant M_2/M_\odot \leqslant 18$	$16 \leqslant M_1/M_\odot \leqslant 23$
Sunyaev *et al.*	$7.8 \leqslant M_2/M_\odot \leqslant 17$	$10 \leqslant M_1/M_\odot \leqslant 22$
Bolton	$10 \leqslant M_2/M_\odot \leqslant 20$	$25 \leqslant M_1/M_\odot \leqslant 35$

The three main arguments therefore [108]: (1) very large emission of energy
$dE/dt \gtrsim 10^{37}$ erg s^{-1}, (2) intensity variations on a time scale as short as 50 ms, and
(3) mass larger than 3.2 $M_\odot$, all point to the clear evidence of a completely collapsed
object or black hole in Cygnus X1 accreting material from the companion star. As
pointed out in reference 17, the only alternative possibility is that a violation of a
fundamental law of physics (either violation of general relativity or violation of
causality) occurs in this system.

The discovery of a black hole in nature would prove to be completely sterile if we
could not observe the processes occurring near its horizon. The reason is simply explained: the most clear and important relativistic effects (gravitational redshift,
dragging of inertial frames and fully relativistic electrodynamical processes) give large
and detectable effects only within a few kilometers from the surface of the black hole.

To probe the most novel predictions of this kind of physics a line of attack very
similar to the one usually adopted in elementary particle physics is needed: the only
way to infer the near field structure of a collapsed object is to proceed with scattering
experiments. Similarly we define the entire set of the electromagnetic and gravitational
multipole moments of a black hole as the 'form factor' of the black hole. To determine
this 'form factor' we have therefore to consider emission processes which occur more
and more near the surface of the black hole or equivalently with shorter and shorter
time scales [17, 79]. It is already clear that as a direct consequence of the finite radius
of the collapsed object we should also find a clear cutoff for signal with time constant
shorter than the characteristic time of the last stable orbit of revolution of a particle
around a black hole $(P \sim GM/c^3)$.

Only in the case of Cygnus X1 in which the observations can be made at a small
inclination angle we can expect to infer the properties of the processes occurring near
the horizon of the collapsed object.

In this respect the most important results have been the observations made by
Rothschild *et al.* [109] from a rocket flight flown on October 4th, 1973 (see Figures 28
and 29). It is not our goal to go into a detailed explanation of the origin of the train of
pulses observed from Cygnus X1 during this flight. Here, it is important only to stress
the relevance of these results for black holes physics. If we look at the width of the last

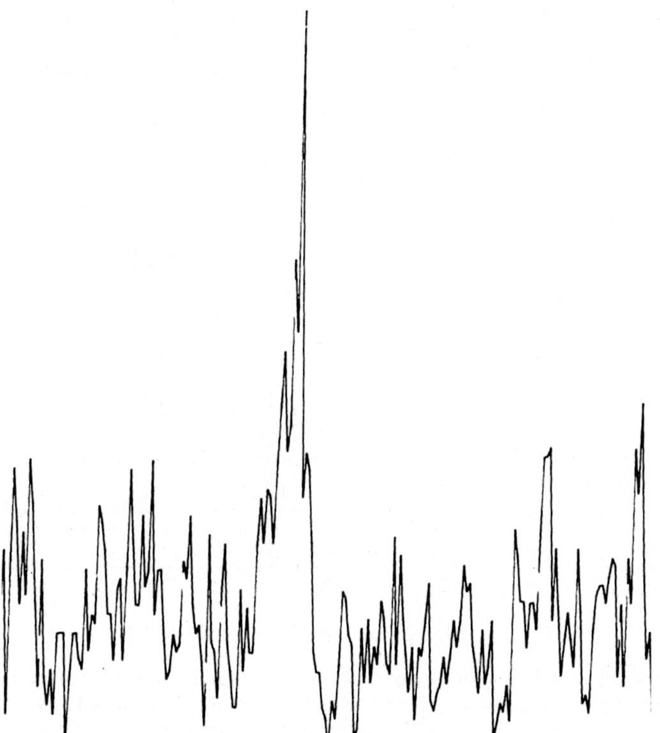

Fig. 28.   Enhancement in the data recorded between 318 and 319 s after launch of the rocket flight by Rothschild *et al.* [109]. During this time the counting rate increased from the overall mean value of 1274 counts s^{-1} (see Figure 21) to 2188 counts s^{-1} averaged over 409.6 ms. The statistical significance of this enhancement is discussed in Reference 109. The interval of 409.6 ms was further divided into 320 bins each of 1.28 ms. Eight bursts were found in this set of data. (See also Figure 29). (Figure reproduced from Reference 109 with kind permission of the authors.) Published by the University of Chicago Press; © 1974, American Astronomical Society.

three pulses in the train (~ 1.28 ms, see Figure 22) and examine their relative separation we find that the first two pulses are 4.48 ms and the last two 8.32 ms apart. Assuming for the black hole a mass $M \sim 10\, M_{\odot}$ we obtain for the last stable circular orbit of a particle around a Schwarzschild black hole $P = 12\pi(6)^{1/2}\, GM/c^3 \sim 4.5$ ms and in the case of an extreme Kerr black hole using the definition of angular velocity given by Christodoulou and Ruffini [61]

$$\Omega = a/(r_+^2 + a^2) \quad \text{with} \quad r_+ = m + (m^2 - a^2)^{1/2} \quad \text{and} \quad a = L/m = m$$

we then have

$$P = 4\pi\, GM/c^3 \sim 0.64 \text{ ms}.$$

It is therefore clear that these observations are of the greatest relevance and of the right time variability for a deeper understanding of the 'form factor' of a black hole. We need more and continuous observations of this kind with accurate statistical analy-

                                    REMO RUFFINI

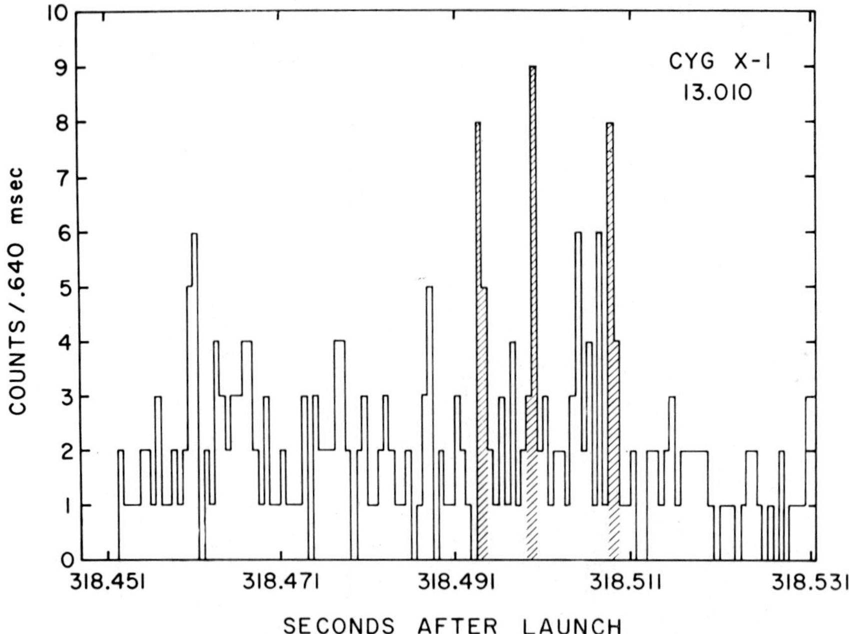

Fig. 29.   Eighty milliseconds of exposure of Cygnus X1 containing the peak of the enhancement near 318 s after launch. The count rates are binned every 0.64 ms. Bursts with $\geqslant 12$ counts per 1.28 ms are shaded (see text and Reference 109). (Figure reproduced from Reference 109 with kind permission of the authors.) Published by the University of Chicago Press; © 1974, American Astronomical Society.

sis and with as much intensity of X-rays as possible (larger collecting areas in telescopes!). Only through a direct analysis of similar data and a continuous feed back with the theoretical predictions we will be able to know if and how general relativity applies in these very strong ultrarelativistic regimes.

8. White Dwarfs in Contact Binary Systems

In the previous paragraphs we have emphasized how a direct observation of both families of binary X-ray sources and of their short time variability will most likely determine in the near future some of the major features characterizing the physics of collapsed objects. This task will be greatly simplified by the discovery of more sources inside our own Galaxy, by a continuous monitoring of the existing sources and possibly by the detection of sources in nearby galaxies. As a by product of these observations we should also be able to determine the experimental value of the critical mass against gravitational collapse.

Thanks to the work of Paczinsky [112] and Arnett [113] it has become more and more clear in recent years the fundamental role that white dwarf stars play in the processes leading to the formation of collapsed objects. Paczińsky has given a detailed treatment of the evolution of Population I stars with masses 0.8, 1.5, 3, 4, 10 and 15 $M_\odot$.

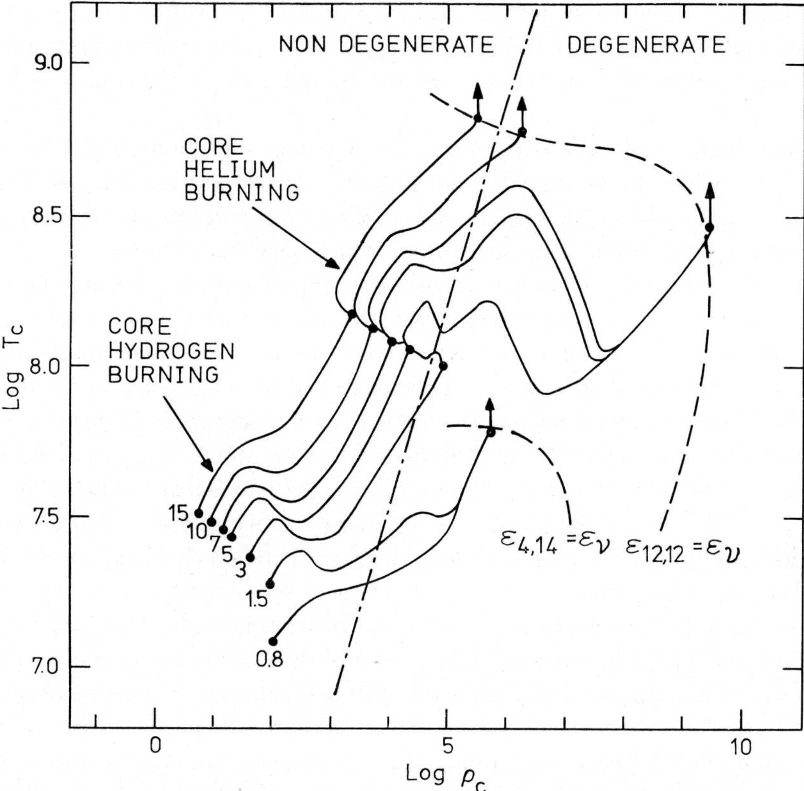

Fig. 30. Evolutionary track for stellar models proceeding toward the endpoint of thermonuclear evolution. The temperature at the center of the star is given as a function of the central density. At the beginning of each track the value of the mass of the star is given in units of the solar mass. The large dots indicate the position of the centers of the models on the main sequence and at the time helium and carbon ignition occur. The cores of the 3, 5 and 7 $M_\odot$ models are smaller and do not contract too rapidly after the helium exhaustion. Neutrino emission cools them down and carbon ignition takes place when the density af the center reaches 3×10^9 g cm^{-3}. Along the broken lines neutrino energy losses balance either nitrogen + helium burning or carbon burning. The ignition of carbon as suggested by Arnett [113] could be explosive. However, in these computations the effects of crystallization, electron capture, and general relativity were not taken into account. Some of the nuclear reactions governing carbon burning rate are also uncertain. The effects of these uncertainties could be so large as to make the system unable to cause a thermonuclear explosion and the core would collapse directly to neutron star density. Certainly the most striking feature in this model is the convergence of the evolutionary tracks for $3\,M_\odot \lesssim M \lesssim 7\,M_\odot$ into a common track after the exhaustion of helium in the core has taken place. (Figure and caption based on the paper in Reference 124.)

© 1970, Pergamon Press, Inc.

In all these computations the evolutionary tracks of the center of stellar models were computed and some of the results are here reproduced in Figure (30). The most striking aspect of these computations is the formation of a standard size degenerate core of material (white dwarf material) after the exhaustion of helium in the center of the star has occurred. The size of this core for initial configurations with $3 \lesssim M/M_\odot \lesssim 7$ is always the same: $M_{\mathrm{core}} \sim 1.39\,M_\odot$. The remaining mass of the star, distributed in a large envelope, is expected to be expelled when the core becomes unstable and undergoes gravitational collapse if the star is single [113], or drive the entire dynamical aspects and transfer of material between the components, if the star is in a binary system [114].

If indeed these results are confirmed, then a deeper understanding of the physics governing the white dwarf stars will not only be relevant in itself but will also be of basic relevance for the understanding of the physical conditions existing at the onset of gravitational collapse and of the evolution of binary star systems.

In this light, it is of the greatest importance to obtain a direct experimental verification not only of the main parameters of white dwarfs but also to try to infer as much information as possible about their internal constitution, and to particularly obtain an experimental value for their critical mass against gravitational collapse [115]. Masses of white dwarfs have indeed been measured with great accuracy in binary systems: the best known example being certainly the classical observations of Sirius a and Sirius b [116]. Measurements of this kind, though relevant for a mass measurement, are practically of no value to infer the physics or the internal constitution of white dwarfs.

The situation has been drastically changed by the detailed analysis done by Warner *et al.* [117] of cataclysmic variable stars. These sources are in many respects very similar to the ones considered in the previous two paragraphs. They are very short period binaries ($P_0 \lesssim 5$ h) and they have as one of the component of the system not a neutron star or a black hole but a white dwarf. These binaries, however, are *not* strong X-ray sources. Also, in these systems matter is transferred from the main normal star into the white dwarf. The gravitational field is, however, not strong enough to produce X-ray emission from the accreting plasma. Some of the best known systems are given with their typical parameters in Table III.

By far, the system most studied in this class is DQ Hercules [118]. This system has a binary period $P_0 = 4^{\mathrm{h}}39$ min which is observed to change at a rate $(\mathrm{d}P_0/\mathrm{d}t)/P_0 = 3.484 \times 10^{-10}$ days^{-1}. The white dwarf luminosity is observed to pulsate with a period $P_1 = 71$ s with an amplitude in 1971 of 0.016 mag. [117]. Both the period and

TABLE III

Parameters of cataclysmic binaries reproduced from References 117 M_1 is here the mass of the white dwarf companion star, a_1 the semi major axis and i the orbital inclination (details in References 117)

Star	P $(10^4$ s$)$	$\dfrac{2\pi a_1}{P}\sin i$ (km s^{-1})	q	$M_2/M_\odot$	$M_1/M_\odot$
RV Peg	3.21	137	1.15	1.3	1.1
Z Cam	2.51	144	0.73	1.0	1.4
SS Cyg	2.38	122	0.86	1.0	1.2
DQ Her	1.67	150	0.72	0.70	1.0
ss Aur	1.56	85	0.64	0.65	1.0
V603 Agl	1.20	37	0.46	0.50	1.1

the amplitude of the pulsation of the white dwarf are observed to decrease with time $(\mathrm{d}P_1/\mathrm{d}t \sim -20\ \mu\mathrm{s}\ \mathrm{yr}^{-1}$, amplitude of the pulsation 0.026 mag, in 1961 and 0.016 mag. in 1971). An attractive and self consistent model has been advanced by Warner [117] in order to explain the major features of this system. The onset of the pulsation of the white dwarf in this model is triggered by the nova explosion which occurred in 1934 in DQ Hercules. The change of the binary period is simply explained as due to the outflow of mass from the system which gives $(\mathrm{d}M/\mathrm{d}t)_{\mathrm{out}} \sim 10^{-7}\ M_\odot\ \mathrm{yr}^{-1}$. The observed decrease in the pulsational period is explained by Warner as due to an accretion rate of $(\mathrm{d}M/\mathrm{d}t)_{\mathrm{acc}} \sim 10^{-7}\ M_\odot\ \mathrm{yr}^{-1}$ on the white dwarf: to an increase in mass corresponds an increase in density and therefore a decrease in period of pulsation since $\omega_{\mathrm{Puls}} \sim$ $\sim (\pi G\varrho)^{1/2}$. Finally a direct extrapolation of the observed binary period today to the 1934 value, compared with the binary period before the nova outburst shows a discontinuity which would imply an output of mass during the nova outburst of $\sim 10^{-3}$ $M_\odot$. From this the conclusion that at the current accretion rate the system DQ Hercules would undergo a nova explosion once every $\sim 10^3$ yr. From this model [117] follows a considerable difference between systems like DQ Hercules and the binary pulsating X-ray sources. The intrinsic variation of luminosity (71 s in DQ Her, 4.8 s in Cen X3, 1.2 s in Her X1) is due to rotation in the regularly pulsating binary X-ray sources and to non radial pulsation in the present case [119].

It should be expected that some features of the accretion disk far away from the collapsed objects and some of the dynamical processes governing the transfer of material from the main star in binary X-ray sources are very similar to the ones encountered in these cataclysmic binaries. Moreover, a detailed analysis and understanding of the ignition processes occurring in nova processes in cataclysmic binaries could be of the greatest interest in order to explore the possibility of the occurrence of similar phenomena on the surface of a collapsed object. This possibility appears to be very promising in order to explain burst like phenomena as observed in Cygnus X3 and Cygnus X1 (see Section 7) or as an explanation of the newly discovered X-ray bursts [120].

A direct comparison of these three families of binary systems, with a direct measurement of the masses of the collapsed stars and of the white dwarfs, as well as a detailed interpretation of the short time structures of their intensity variations, will not only give a definite understanding of all the possible equilibrium configurations to be found at the endpoint of thermonuclear evolution of a star but also a detailed knowledge of the physical processes governing these equilibrium configurations (see Figure 31).

It is now clear that if the general results of the evolution of a star given by Paczińsky [112] and Arnett [113] will be confirmed by further analysis, then the abundance of neutron stars to black holes in nature is mainly determined by the ratio

$$(M_{\mathrm{crit}})^{\mathrm{W.D.}}/(M_{\mathrm{crit}})^{\mathrm{N.S.}}.$$

If this ratio is larger than one, $(M_{\mathrm{crit}})^{\mathrm{N.S.}} < 1.39\ M_\odot$, than the majority of gravitational collapse processes will lead to formation of black holes. In this case neutron stars could be formed uniquely if enough matter will be expelled during the process of

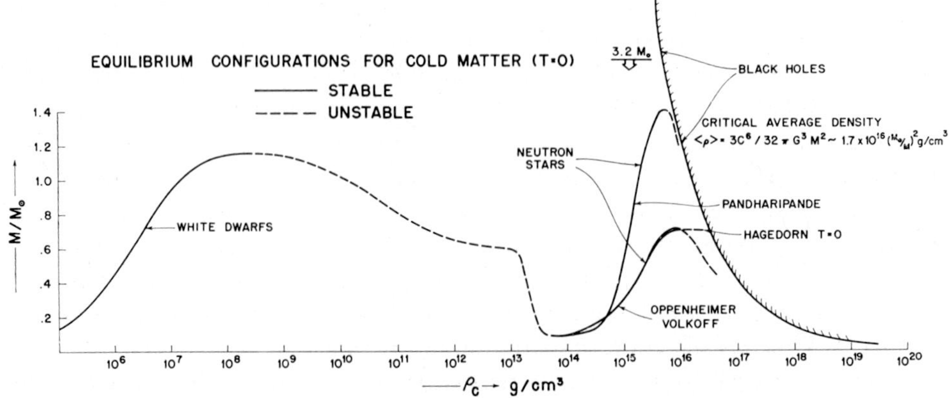

Fig. 31.   Configurations of equilibrium of degenerate matter. The masses measured in units of solar mass are plotted here as a function of the central density of the configuration of equilibrium. In the case of black holes we have plotted an 'effective' average density as given by $\langle \varrho \rangle = M/\frac{4}{3}\pi(2GM/c^2)^3$, strictly speaking the black hole never reaches an equilibrium but approaches asymptotically (as seen from us far-away) its own horizon with a time constant $\tau \sim GM/c^3$ ($\tau = 4.9 \times 10^{-3}$ s for a 10 $M_\odot$ black hole!) [39]. It is clear that the entire dynamics and outcome of gravitational collapse are totally governed by only two critical values: The critical mass of white dwarfs (Chandrasekhar limit), the critical mass of neutrons stars (largely unknown but certainly smaller than 3.2 $M_\odot$ [29]). The experimental determination of the value of the white dwarf critical mass is extremely important as a consequence of the central role this mass appears to have as the 'standard' starting point for gravitational collapse to occur. The value of the critical mass for neutron stars will drastically influence the ratio between neutron stars and black holes to be found in nature.
(See text and caption to Figure 30.)

gravitational collapse. If instead this ratio is smaller than one then the formation of neutron stars will be largely enhanced. The formation of a black hole, since the critical mass of a neutron star must always be smaller than 3.2 $M_\odot$, will still be possible either by an implosion with enough kinetic energy to overcome the potential barrier of the equilibrium configuration of neutron stars or by a multiple step process through further accretion of material on the neutron star.

The fact that in all the known cataclysmic variable stars, the masses of the white dwarf is near the theoretical value of the critical mass, see Table III and the very high accretion rate expected in these systems ($M \gtrsim 10^{-7}\ M_\odot\ \mathrm{yr}^{-1}$) has led Ruffini and Treves to advance the hypothesis [121] that indeed we should be able to observe binary systems which have evolved from cataclysmic variable stars into binary X-ray sources. This model is particularly appealing for an explanation of Cygnus X3. Initially it is assumed that Cygnus X3 was a contact binary of short period one of the components having evolved to a white dwarf [122]. Due to accretion of material the white dwarf would have then collapsed to a black hole which is now orbiting inside the envelope of the main star. This could explain both the sine behavior in the intensity of the X-ray emission as well as the very large infrared luminosity measured by Becklin *et al.* [123]. (For a detailed description of the properties of Cygnus X3, see Reference 78.)

A direct comparison and intercorrelation of the systems presented in Sections 6, 7, and 8 is strongly and urgently needed.

9. The Moment of Gravitational Collapse

Since the first days of general relativity it has been clear that one of the major new predictions of Einstein's theory of gravitation was the fact that gravitational interactions had to propagate with a finite velocity, equal to the velocity of light, and that waves carrying gravitational energy should exist in nature. It was, however, also very clear that the amount of energy radiated by any conceivable source inside our solar system was totally negligible and its effect well below any realistic limit of detectability. The main reason is simply given by a direct estimate of the gravitational waves energy radiated by a source as computed in the slow motion approximation [125]

$$\left(\frac{\mathrm{d}E}{\mathrm{d}t}\right)_{\mathrm{rad}} = \frac{G}{45c^5}\,\dddot{Q}^{ij}\dddot{Q}_{ij}$$

with $i, j = 1, 3$, G the gravitational constant, c the speed of light, and $\dddot{Q}^{ij}$ the third time derivative of the quadrupole moment of the system. The only way to counterbalance the smallness of the factor G/c^5 is to consider very large systems ($M \sim M_\odot$) and rapidly varying quadrupole moments (velocity of sound $\sim c$) [126].

It was suggested as far back as 1952 by Dyson [127] that similar regimes could be reached inside our own galaxy during supernovae explosions.

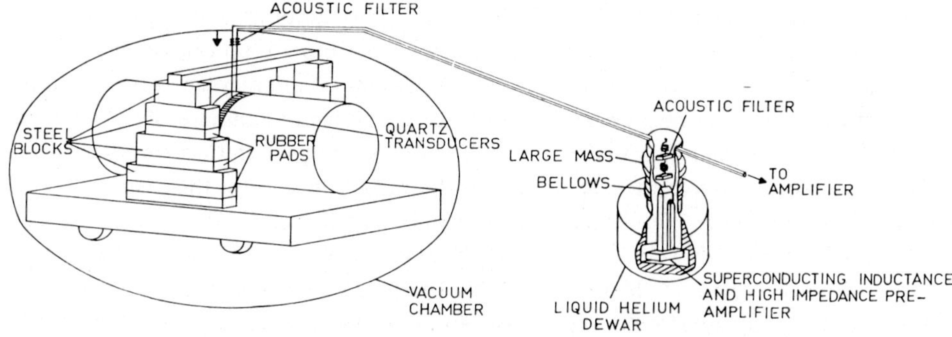

Fig. 32.   Schematic diagram of the detector of gravitational radiation conceived, constructed and operated by Weber [71]. The detector isolated in a vacuum chamber consisted of an aluminum bar 153 cm in length and 61 cm in diameter acoustically and thermally isolated. The stress in the cylinder were measured by piezoelectric crystals distributed around the surface of the cylinder and allow to measure displacements as small as 2×10^{-14} cm which allow to detect the thermal vibrations of the cylinder ($\xi^2 = kT/m\omega^2$, k as usual is the Boltzmann's constant, T the temperature, m the mass of the cylinder and ω is angular frequency, $T \sim 300$k). The electronic and the amplifier are tuned at the resonance frequency of the bar for its lowest mode of stretching vibration which is $v = 1660$ cyc s^{-1}. Similar observations have been carried out by other groups: Tyson [128] (Bell Telephone Lab. Murray Hill, New Jersey) and Douglas [129] (University of Rochester, New York); and Billing *et al.* [130] (Max Planck Institute – Munich and E.S.R.I.N., Frascati, Italy); Braginsky [131] (University of Moscow); Garwin [132] (I.B.M. Watson Research Labs., New York). (From J. Weber, *Phys. Rev. Letters* **17**, 1966.) Published by American Physical Society; © 1966, American Institute of Physics.

Weber [71] was the first to conceive the possibility of detecting gravitational waves from extraterrestrial sources and to build and operate the first detector to observe bursts of this radiation. The detector constructed by Weber (see Figure 32) consists of an aluminum cylinder 153 cm in length, 66 cm in diameter with a mass of 1.4×10^6 g and a resonance frequency in its lowest mode of stretching vibration of $v = 1660$ cyc s^{-1}. The Q for this mode is estimated to be $Q \sim 10^5$ and the bandwidth of the detector $\Delta\omega = \omega/Q \sim 10^{-1}$ rad s^{-1}. Following the considerations presented in Figure 13 and Reference 39 we can then evaluate the total cross section of this detector

$$\int_{\text{resonance}} \sigma(v)\,dv = 1.0 \times 10^{-21} \text{ cm}^2 \text{ Hz}.$$

To detect the stresses of the bar, Weber uses piezoelectric crystals which enable him to detect displacements as small as 10^{-14} cm, corresponding to vibration energy in the bar $\sim kT$. To discriminate between locally produced noise and gravitational waves signal, Weber uses two identical detectors, one at Argonne National Laboratory (Chicago) and one at the University of Maryland and he looks for coincidences in the signals in the two widely separated detectors. Despite early reports of positive finding we can today conclude on the ground of further work of Weber himself, Tyson [128], Douglas [129], Levine and Garwin [132], Braginsky [131], Billing *et al.* [130] that no signals of gravitational radiation have been seen of such an intensity as to deposit in the detectors an amount of energy $\sim kT$.

This conclusion allows to estabish a direct limit to the strength of a gravitational waves signals during the period of observation. A signal of $\sim kT$ represents an uptake of energy by the detector of $\Delta E \sim 4 \times 10^{-14}$ erg. This amount of energy divided by the cross section of the detector (see Section 4) gives an upper limit on the energy $I(v_0)$ of an incoming pulse of gravitational radiation per cm^2 of intercepted surface and per unit frequency near the resonance frequency of the detector, namely, $\Delta E = 4.0 \times 10^7$ erg cm^{-2} Hz^{-1}. To establish limits on the flux of incoming radiation we have still to make some assumptions on the spectrum of the radiation and on the distance of the source. If we assume for the spectum of the radiation a width equal to the detector bandwidth (of course this possibility is highly unlikely) we would obtain an incoming flux $\Delta E \sim 2 \times 10^6$ erg cm^{-2}. If we assume for the pulse a flat spectrum from $v = 0$ up to the frequency of the detector $v = v_0$ we obtain a flux of $\Delta E \sim 5.5 \times 10^{10}$ erg cm^{-2}. If we then assume the source to be located at the center of the galaxy ($D = 8.2$ kpc) we have to multiply the above fluxes by $4\pi D^2$ (No possibility of beaming gravitational radiation see Section 8). We then obtain the *lower* limit for the output of gravitational waves from a source located at the center of the galaxy in order to produce a signal $\sim kT$ in our detector to be respectively 1.6×10^{51} erg and 4.4×10^{55} erg for the two spectra previously considered. These bursts would correspond respectively to an amount of mass m which would have to be in gravitational radiation (assuming 100% efficiency of conversion!) of $m = 8 \times 10^{-4} \, M_\odot$ or $m = 22 \, M_\odot$ respectively.

In the light of these results we can reach the following two general conclusions on the Weber type detectors:

(1) There exist no evidence for signals of gravitational radiation which have deposited in the detectors an amount of energy comparable to the thermal energy of the detector itself.

(2) To generate signals as small as the thermal energy of the detectors it would require an enormous amount of gravitational radiation, of the order of a few solar masses radiated into gravitational radiation!

This was the status of this field of research only a few years ago. These two conclusions gave rise to two major reactions:

(1) In the theoretical research it became clear that major emphasis should be given to a fully relativistic analysis of processes in which a sizable fraction of a solar mass can be radiated into gravitational radiation in time scales of a few milliseconds.

(2) In the experimental research it became self evident that the only hope to launch a realistic program for the detection of gravitational radiation was heavily based

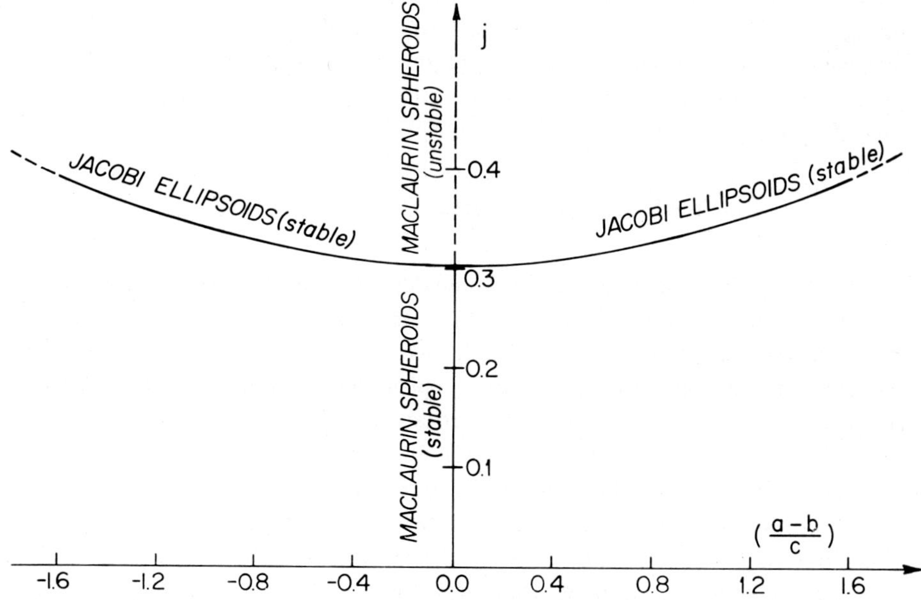

Fig. 33. Equilibrium configurations of an incompressible fluid turning at a uniform angular velocity. The solid lines indicate stable equilibrium configurations the dotted lines unstable. If the total angular momentum is larger than the critical value $L = 0.303751 L_0$ with $L_0 = (GM^3)^{1/2} (abc)^{1/6}$, abc being the semiaxis of the ellipsoid of equilibrium, then the stable equilibrium configuration is given by Jacobi ellipsoids [140]. The horizontal scale gives the difference of semiaxis of the ellipsoid in the equatorial plane divided by the semiaxis along the axis of rotation. A neutron star with an equation of state with an effective polytropic index $\gamma > 2.2$ [69] will also, if endowed with large enough angular momentum, have equilibrium configurations given by a Jacobi ellipsoid. It will then radiate a very large amount of gravitational radiation with a characteristic signature. A neutron star of $M \sim 0.65 \, M_\odot$ and $\bar{R} \sim 15$ km rotating with a period of $P \sim 1.4$ ms will have an eccentricity in the equatorial plane $\varepsilon \sim 0.87$ and will radiate $dE/dt \sim 0.6 \times 10^{51}$ erg s^{-1} with the period starting from $P \sim 1.4$ ms (extreme Jacobi configuration) decreasing in a time $\tau \sim 1.4$ s to a value of 1.3 ms. Details of this process as well estimates of the amount of gravitational radiation emitted by vibrating neutron stars are given in References 69 and 67.

upon the possibility of drastically improving the sensitivity of existing detectors.

Natural candidates of powerful sources of gravitational waves are then neutron stars and black holes.

A neutron star either pulsating in non radial modes or rotating with an asymmetrical distribution around the rotation axis is a very strong emitter of gravitational radiation. Let us consider a neutron star of mass $M \sim 0.6\ M_\odot$, $R \sim 15$ km rotating with a period $P \sim 1.5$ ms. As a consequence of the very large angular momentum the configuration of equilibrium of the neutron star is expected to be not axially symmetric (Jacobi ellipsoid [133], see Figure 33) will be endowed with an equatorial eccentricity $\varepsilon \sim 0.87$. The power of gravitational radiation will then be given by [90]

$$\frac{dE}{dt} \sim -\ 288 G I^2 \varepsilon^2 \omega^6 / 45 c^2 = 0.6 \times 10^{51}\ \text{erg s}^{-1},$$

emitted at twice the characteristic rotational frequency of the neutron star with a damping time $\tau \sim 0.4$ s. If we turn to the pulsation of neutron stars their eigen frequency is given by $\omega \sim (\pi G \varrho)^{1/2}$. Assuming a neutron star of mass $M \sim 0.7\ M_\odot$ radius $R = 9$ km, central density $\varrho_c \sim 6 \times 10^{15}$ g cm^{-3} the power, dE/dt, emitted in gravitational radiation is given by [69]:

$$-\frac{dE}{dt} / (\delta R / R)^2 \sim 3.0 \times 10^{53}\ \text{erg s}^{-1},$$

where $(\delta R / R)^2$ is the mean squared fractional departure of the surface from sphericity. The characteristic damping time is $\tau \sim 0.2$ s. In both cases the energy radiated is proportional to the square of the amplitude either of the asymmetrical pulsation or of the effective eccentricity. We should then expect that a freshly formed neutron star still endowed with large asymmetries originating during the process of gravitational collapse would emit into gravitational radiation a few percent of its rest mass. Vice versa an 'old' neutron star would be a very unlikely source of a large amount of radiation. The reason is simply explained: either the rotation is very slow (the power emitted in gravitational radiation is proportional to the sixth power of the frequency!) or the required eccentricity or asymmetry are energetically extremely difficult to be generated as a consequence of the very strong gravitational fields.

The second natural candidates of powerful sources of gravitational waves are given by black holes. There are two different reasons for this: (a) large masses of material are indeed observed to accrete in these collapsed objects (see Sections 6 and 7) and (b) the infalling matter reaches relativistic velocities during the accretion processes as a consequence of the very strong gravitational fields. Thanks to the detailed fully relativistic treatments presented in Section 3. we can then estimate the energy emitted during these accretion processes.

The following general conclusions can be reached on the energy emitted by particle radially imploding into a Schwarzschild black hole:

(1) the peak of the radiation is emitted at a frequency $\omega_{\text{peak}} \sim c^3 / GM$ uniquely

characterized by the mass of the black hole ($\omega_{\text{peak}} \sim 2 \times 10^4$ rad s^{-1} for a 10 $M_\odot$ black hole),

(2) the amount of gravitational radiation emitted at infinity during the infall of a mass m is given by (see Figures 8 and 10),

$$\Delta E \sim 0.01 mc^2 (m/M)$$

(3) the gravitational radiation is emitted in lobes of $\sim 70°$ (see Figure 9).

Although these considerations apply only to the idealized case of a particle infalling radially into a Schwarzschild metric we do expect their validity to apply also in the case of more general orbits of implosion as well as in the case of a Kerr metric. The major difference will occur only in the value of the numerical constants in front of the dimensional factors mc^2 (m/M) (for the energy radiated) and c^3/GM (for the peak radiation). Moreover we do not expect these numerical factors to vary drastically. It is therefore clear that the maximum efficiency in a radiation process is reached in the case $m \sim M$. On energetic grounds we can conclude also that these processes can turn out to be of physical interest only if $m \sim M_\odot$.

If we turn now to particles in circular orbits at a radius r around a black hole of mass M we can conclude that gravitational radiation will be emitted:

(1) at a frequency $\omega \sim (GM/r^3)^{1/2}$,

(2) with an integrated intensity essentially proportional to the gravitational binding energy of the orbit. Stable circular orbits can exist only down to a minimum radius $r_{\text{min}} = 6\, GM/c^2$ in the case of a Schwarzschild black hole and $r_{\text{min}} = r_{\text{horizon}} = = m + (m^2 - a^2)^{1/2}$ in the case of the Kerr metric. We can conclude that the total amount of radiation emitted by a particle in circular orbit is simply proportional to the gravitational binding energy of the last stable circular orbit in the case of an extreme Kerr black hole $E_{\text{bind}} \sim 0.42\, mc^2$ and in the case of a Schwarzschild metric $E_{\text{bind}} \sim 0.051\, mc^2$.

We can then conclude that both in the case of radial infall and of circular orbits the black hole uniquely characterizes the frequency of the outgoing radiation. If we keep in mind the technological possibilities of gravitational wave detectors we must then conclude that the only realistic situation in which a burst of gravitational waves can have enough intensity at the right frequency for detectability is reached in the special case where $m \approx M \approx M_\odot$!

It is also clear, however, that all the formulae and the treatment we have been speaking about applies only if the size of the infalling particle can be considerd negligible by comparison to the size of the black hole. The constraints imposed by realistic astrophysical situations are then even stronger! If we look at the radius of a black hole of 10 $M_\odot$ we have $R \sim 29.4$ km (in the case of extreme Kerr solution $\frac{1}{2}$ as much), and the radius of a main sequence star of a solar mass is $R \sim 10^5$ km! If the pointlike assumption on the imploding mass cannot be adopted then tidal deformations have to be considered. A white dwarf imploding into a black hole will be disrupted by tidal forces at a distance [69]

$$R_{\text{limit}} \sim 2.45\, (\varrho_{\text{BH}}/\varrho_{\text{WD}})\, R_{\text{B.H.}}$$

from the black hole surface. The disruption of the infalling star by tidal forces will totally spoil the effect of the fully relativistic asymmetric implosion. In the case of a normal star imploding into a black hole, since its radius is much larger than the radius of the black hole, the amount of gravitational radiation emitted will be totally negligible: the star will flow inside the black hole with a highly symmetric flow on a very long time scale essentially governed by the Eddington limit and, as a consequence the third derivative of the quadrupole moment of the system will be negligible. The general consideration and limitations here presented for black holes apply as well to the case of neutron stars with the only limitation that the emission processes will be in this case less efficient as a consequence of the less relativistic gravitational fields.

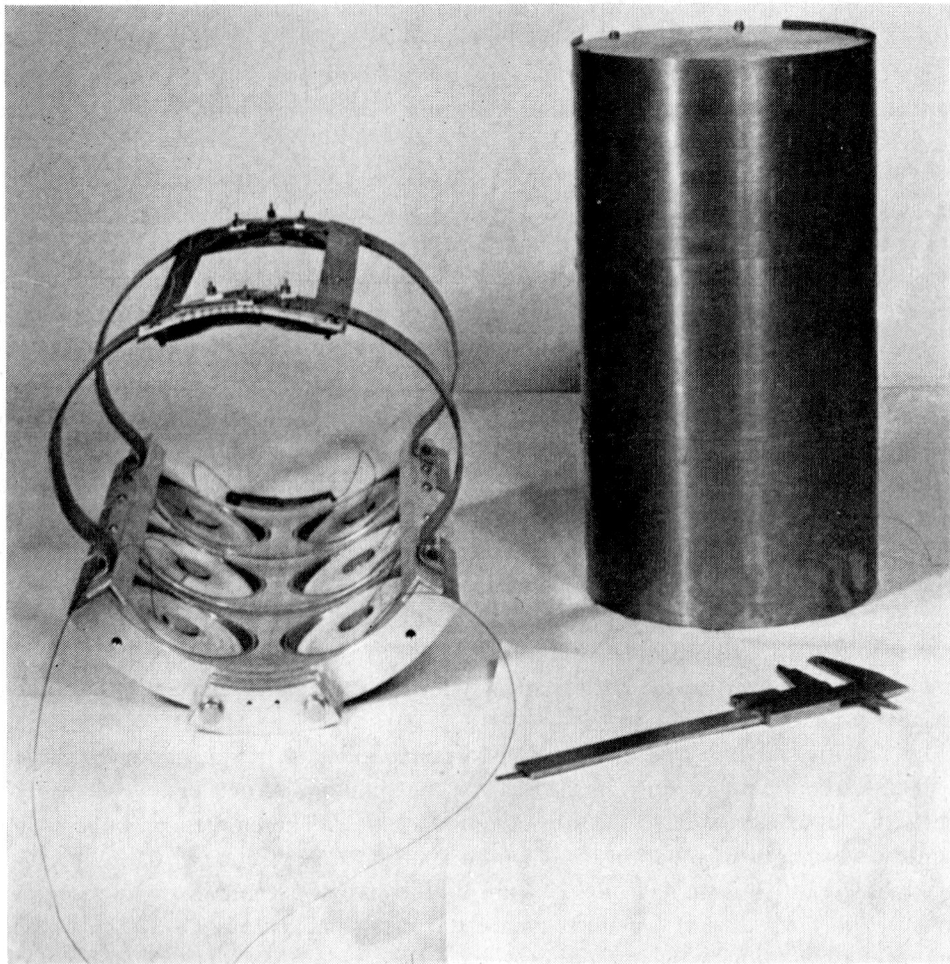

Fig. 34. Superconducting coils being developed at the University of Rome to magnetically levitate the gravitational wave antenna. The final antenna, an aluminum cylinder 3 meters in length, 0.9 meters in diameter and 5 tons in weight, will be coated with a 0.4 mm foil of NbTi. Similar magnetic suspension systems have been developed at Stanford University and at Louisiana State University.

We can therefore conclude that under no realistic circumstance can a process of accretion either on a neutron star or a black hole be of any interest in the production of a detectable amount of gravitational radiation.

In conclusion as far as we can see today there exists only *one* regime in which, for certain, all the optimal conditions for the emission of a detectable amount of gravitational radiation can be reached at once: at the moment of gravitational collapse.

One of the possible ways of estimating the number of gravitational collapses to be expected in the galaxy is to count the number of visible supernovae [5]. On this ground we would expect a gravitational collapse every thirty years. This number of supernovae would give a population of $\sim 10^9$ collapsed objects in our own galaxy, in substantial

Fig. 35.   Gravitational wave antenna being assembled at Louisiana State University (Baton Rouge). Identical detectors are being assembled at the University of Rome and at Stanford University. The final goal is to reach a sensitivity 10^6 times larger than existing detectors capitalizing, among other approaches, on reducing the temperature of the detectros (an aluminum bar of $\sim 2 \times 10^6$ g) to 3×10^{-3} K. The displacements at the end of the bar are being measured by a resonant diaphragm [138]. Displacements as smail 4.9×10^{-18} cm are expected to be measureable at the final temperature of 3×10^{-3} K. The magnetic suspension has already been tested at a temperature of 2 K both at Stanford (aluminium bar of 1.5 m and 30 cm in diameter covered with Nb Ti foil) and in Rome (aluminum bar 16 cm in diameter and 30 cm in length covered with Nb Ti foil). The first detectors are expected to be operational at 2 K at the end of 1975. The eigenmodes of the detectors are currently being studied theoretically and experimentally [141]. (Photograph courtesy of W. O. Hamilton, Stanford University.)

agreement with the number of pulsars observed if their lifetime and the geometrical effects of possible beaming are taken into proper account.

However, there are today reasons to believe that processes of gravitational collapse can occur under different regimes than the ones encountered in supernovae explosions.

We have emphasized in the previous paragraph how a detailed knowledge is being gained on the two different families of collapsed objects inside our own Galaxy thanks to the data acquired from binary X-ray sources. In no way can we escape the necessity that these objects have to be formed in a process of gravitational collapse (see Figure 31). The very general arguments lead inescapably to the result that during the process of gravitational collapse up to 4–$5 \times 10^{-2}\, Mc^2$ of the mass M of the collapsing star is emitted in a sharp burst of radiation. If the collapse ends in a neutron star additional energy can still be extracted in the form of gravitational radiation either from the rotational energy or the vibrational energy of the freshly formed neutron star during the few seconds following the occurrence of the gravitational collapse itself. [134]

Colgate [102] has pointed out that these collapsed objects have to be formed through a different process than usual suprenovae, since the binary system would be disrupted by the relativistically expanding shell. It is therefore likely that the formation of a collapsed object in a binary system from a white dwarf core [114] will have a totally different behavior than in a supernova. Most likely the difference could be due to the gradual transfer of mass in the binary system from the initially more massive star to its companion [114]. Although no final understanding of these processes exists, we stress the conclusion that today there is some evidence that we should expect a larger number of gravitational collapses inside our own Galaxy than the one obtained from a direct count of supernovae.

To start a realistic program for the detection of gravitational radiation it is then necessary to drastically augment the sensitivity of existing gravitational wave detectors in order to observe all the gravitational wave bursts originating in gravitational collapses inside our own Galaxy and possibly in nearby galaxies. With this aim in mind two different kinds of experiments have been conceived and are in advanced stage of realization: one is a joint effort [135, 136] by the Universities of Louisiana (Baton Rouge), Rome (Italy), Stanford (California), the other is led by Dr Braginsky at the Moscow State University [137] (U.S.S.R.).

The key idea beyond these two experiments is simply explained. The mininum energy detectable by a gravitational wave detector is limited by the information which can be extracted from its thermal noise. If Nyquist's theorem can be applied to analyze the noise of the detector then we find that the minimum energy flux detectable by the gravitational wave antenna is given by [138]

$$(E)_{\mathrm{grav}} \times \int_{\mathrm{res}} \sigma(v)\, \mathrm{d}v = kT\, \tau/\tau_{\mathrm{D}},$$

where $\int_{\mathrm{res}} \sigma(v)\, \mathrm{d}v$ is the cross section of the detector integrated over the resonance,

T is the temperature of the detector, and $\tau_D = 2Q/\omega_0$, where ω_0 is the resonance frequency of the detector and τ the time over which the signal is recorded. This formula applies uniquely to the case $\tau \ll \tau_D$ or in the limit that the time over which the observations are made is much shorter than the relaxation time of the antenna. There are only two ways of increasing the sensitivity of a gravitational wave antenna of a given cross section: (a) to lower its temperature and (b) to increase its Q.

The Louisiana-Rome-Stanford experiment is mainly directed at exploiting the gain coming from the reduction of the temperature of the gravitational wave detector. The antenna, as in the case of J. Weber, is an aluminum cylinder of a mass $\sim 2 \times 10^6$ g. Instead of being suspended at the center by a cable the antenna is in this experiment magnetically suspended (see Figure 34) coated of foils of NbTi. The antenna floats in magnetic fields generated by superconducting current loop. The system is thermally isolated (see Figure 35) and a final temperature of the antenna of 3×10^{-3} K is expected to be reached. Instead of using the piezoelectric crystals in order to detect the strains of the bar a vibrating superconducting diaphragm is used (see Figure 36). The three experiments are in advanced stages of preparation and should be operational at a temperature of 2 K by the end of 1975. The final expected gain in sensitivity with respect to existing detectors is a factor $\sim 10^6$.

The major effort of the Moscow State University experiment [137] is, instead, mainly directed at exploiting all possible gains obtainable from increasing the Q of

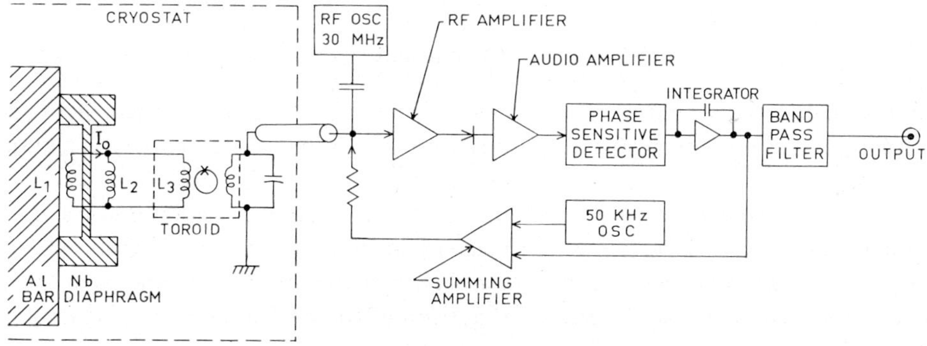

Fig. 36. To detect the very small displacements down to the Brownian motion of the antenna itself ($\sim 3 \times 10^{-10}$ cm for an antenna of mass $m = 2.5 \times 10^6$ g and $\omega_0 = 2\pi \times 900$ Hz) a new kind of detector has been conceived by H. J. Paik at Stanford University. It consists of a circular diaphragm of mass m much smaller than the mass of the bar M but whose fundamental mode has the same resonant frequency as the fundamental mode of the longitudinal oscillation of the bar. The antenna and the detector will then act as a coupled oscillator and the vibrational energy of the bar will be transferred to the diaphragm which will oscillate with an amplitude larger by a factor ($M^{1/2}/m$) than the bar itself. Two inductors L_1 and L_2 wound with 300 turns of 5×10^{-3} cm in diameter niobium wire in a pancake shape are placed 10^{-2} cm away from the two sides of the diaphragm. A supercurrent I_0 is stored in the closed loop made up of L_1 and L_2 and a third superconducting inductor L_3 is connected in parallel with L_1 and L_2. The current i_3 in L_3 is zero when the diaphragm is in the equilibrium position. When the diaphragm is displaced, the value of the inductances L_1 and L_3 is changed and a current i_3 appears in L_3 proportional to this displacement. The magnetic flux generated in L_3 is then detected by a Josephson Junction magnetomater. (Diagram reproduced from Reference 138 with kind permission of the author.) © 1974, Academic Press.

the system neglecting, at least in a first stage, the possible gain from low temperature techniques. A sapphire monocrystal cylinder 152 cm long, 45 mm in diameter, with a mass of $\approx 2 \times 10^3$ g and an eigenfrequency of 34016 Hz was first used for detection. A $Q = 4 \times 10^7$ was reached at room temperature; however, it has been suggested that with a better polishing of the surface of the crystal and an improvement in the suspension, the Q could be increased of 2–3 orders of magnitude [137].

Both experiments are in an advance stage of realization and hopefully they will be operational in the next few years. If, indeed, since it is theoretically achievable, we do gain the six orders of magnitude on existing detectors then we should be able to see for the first time with clear evidence a signal of gravitational radiation originating from gravitational collapse. It is not clear if this signal will be detected first as a clear signal over the thermal background of the detector by the Louisiana-Rome-Stanford experiment or by the information extracted from the noise in the Moscow State University experiment. What is clear, however, is that either way we will have for the first time a record of one of the most important physical processes occurring during the moment of gravitational collapse.

Most likely this gravitational wave pulse will be emitted together with an electromagnetic signal (see Section 3). The enormous difference of cross section between electromagnetic and gravitational wave detectors [139] then make the burst of electromagnetic radiation associated with gravitational collapse the ideal event to look for coincidences in signals of the gravitational wave detectors.

10. Conclusions

For a long time the attention of astrophysicists has been directed towards the understanding of the nuclear processes occurring in the evolution and energy generation of stars.

Only recently it has become apparent that the late stages of evolution of a star is uniquely characterized from the energetics point of view by gravitational and rotational energy, and that the most violent and energetically relevant moments in the life of a star indeed take place after the exhaustion of the nuclear sources of energy has occurred.

Gravitational interaction either slowly generating the luminosity of white dwarfs through their continuous contraction, or producing up to $10^5 L_\odot$ in the accretion processes in X-ray sources or, again, completely determining the physical processes during gravitational collapse or in the description of the totally collapsed objects ensuing from gravitational collapse, appear to be more and more the back-bone, the only fundamental field theory, of this drastically new domain of physics. In this sense the in depth analysis of a relativistic theory of gravity has become in the recent years not only much more easy due to the existence in nature of collapsed objects, but also much more relevant to our understanding of a very large number of physical processes.

During the next five or six years, if the resonance between so many different branches of astronomy (radio, X-ray, optical, infrared) will continue in close contact with the theoretical work we should be able to clearly determine all the details of the curves in Figure 33 with direct experimental data.

If we turn to the long range program of research we have seen in the recent years the preparation of an entire new chapter of astrophysics, what we could call 'Burst Astronomy'. Detectors of gravitational radiation, of neutrino bursts, and of possible associated electromagnetic radiation are all rapidly improving in sensitivity and sophistication while in the theoretical field basic advances are made in the analysis of fully relativistic short time phenomena. In the next few years we should be able to reach in all these different experimental fields the limits of detectability theoretically predicted.

The direct observation through different techniques of the moment of gravitational collapse appears to be of the greatest interest. More we learn of the physics characterizing the configurations of equilibrium of cold catalyzed matter (see Figure 33), the more we see the need of processes of gravitational collapse to occur under a variety of regimes. It is also clear that it is unavoidable that all collapsed objects we are today observing either in pulsars or in binary X-ray sources had to be formed through this process.

The observation of the moment of gravitational collapse will disclose, among others, one of the most fundamental predictions of general relativity namely that gravity, as any other long range interaction has to propagate with a finite velocity equal to the speed of light and that gravitational energy can be carried by waves. The direct technological advance brought by these observations will most likely be very limited, the influence however for our understanding of nature will certainly be enormous.

The moment of gravitational collapse appears by now to be the most energetic process occurring inside a galaxy second in the entire universe to the formation of the Universe itself.

This brings us to another domain of physics also dominated by a fully relativistic theory of gravity: Cosmology. The greatest success of our research on gravitational collapse will be reached if not only we can describe using different techniques and detailed theoretical work the physical processes occurring in this very short phenomena ($\tau < 1$ s) but if we can apply the enormous amount of knowledge we are acquiring in this research to Cosmology.

Acknowledgement

This work was partially supported by N.S.F. Grant #GP 30799X to Princeton University.

References

1. Landau, L.: *Phys. Z. Soviet* **1**, 285 (1932).
2. Chandrasekhar, S.: *Phil. Mag.* **7**, 63 (1929). Also *Monthly Notices Roy. Astron. Soc.* **91**, 456 (1931).
3. Baade, W. and Zwicky, F.: *Phys. Rev.* **45**, 138 (1934).
4. Staelin, D. H. and Reifenstein, E. C.: *Science* **162**, 1481 (1968).
5. For a detailed account see e.g. Shklovski, J. S.: *Supernovae*, New York, 1968.

6. Locke, W. J., Disney, M. J., and Taylor, D. J.: *Nature* **221**, 525 (1969).
7. Gold, T.: *Nature* **218**, 731 (1968).
8. Fritz, G., Henry, R. C., Meekins, J. F., Chubb, T. A., and Friedman, H.: *Science* **164**, 709 (1969).
9. Richards, D. W. and Comella, J. M.: *Nature* **222**, 551 (1969).
10. For a detailed treatment of the magnetic field of a rotating star, see Deutsh, A.: *Ann. Astrophys.* **18**, 1 (1955). These general equations have been applied to a rotating neutron star by Pacini, F.: *Nature* **219**, 145 (1968).
11. For an early suggestion of the relevance of the rotational energy of a neutron star in the energetics of the Crab Nebula, see Wheeler, J. A.: *Ann. Rev. Astron. Astrophys.* **4**, 393 (1966). See also Finzi, A. and Wolf, R. A.: *Astrophys. J. Letters* **155**, L107 (1969).
12. Oppenheimer, J. R. and Serber, R.: *Phys. Rev.* **54**, 540 (1938).
13. Oppenheimer, J. R. and Volkoff, G. M.: *Phys. Rev.* **55**, 374 (1939).
14. Oppenheimer, J. R. and Snyder, H.: *Phys. Rev.* **56**, 455 (1939).
15. This point has been extensively considered and a complete proof of the validity of using an equation of state has been given by Ruffini, R. and Bonazzola, S.: *Phys. Rev.* **187**, 1767 (1969). Instead of using the fluid approximation, the neutron star is there described as a system of many self gravitating fermions at zero temperature by solving a second quantized Dirac field in the curved background generated by the average matter distribution. This self consistent Hartree-Fock approach is proved to be completely equivalent to the approximation of a perfect fluid with locally determined equation of state in the range of nuclear densities considered in neutron star models. A summary on the possible effects of alternative theories of gravity in the computation of the equilibrium configuration of a neutron star has been given in Ruffini, R. and Wheeler, J. A.: in H. V. Hardy (ed.), *Cosmology from Space platform*, ESRO book SP52, Paris, 1971. Recently Wagoner, R. V. and Malone, R. C.: *Astrophys. J. Letters* **189**, L75 (1974), have purported large effects of the gravitational theory used in the computation of the equilibrium configuration of a neutron star. The validity of their treatment, however, is cast in doubt since it leads to a direct violation of the conservation law of the momentum and energy of an isolated system and also they are mistakes in the field equations they have used.
16. See e.g. Ambartsumyan, V. A. and Saakyan, G. S.: *Soviet Astron. AJ* **5**, 601 (1962) and **6**, 601 (1961). See also Cameron, A. G. W., Cohen, J. H., Langer, W. D., and Rosen, L. R.: *Astrophys. Space Sci.* **6**, No. 2, 228 (1970) and references mentioned there.
17. See e.g. Ruffini, R. in B. and C. DeWitt (ed.), *Black Holes*, Gordon and Breach Publ., New York-London, 1973.
18. See e.g. Pandharipande, V. R. in *Proceedings of the 1973 Solvay Meeting in Astrophysics*.
19. Hagedorn, R.: *Nuovo Cimento Suppl.* **3**, 147 (1965).
    Hagedorn, R.: *Nuovo Cimento* **56 A**, 1027 (1968).
    Hagedorn, R.: *Astron. and Astrophys.* **5**, 184 (1970).
    Hagedorn, R.: CERN Lecture Notes – 71–72.
20. Fermi, E.: *Progr. Theor. Phys.* **5**, 570 (1950).
21. Ruderman, M.: *Phys. Rev. Letters* **27**, 1306 (1971) and references mentioned there.
22. Dyson, F.: *Ann. Phys.* **63**, 1, 1971. See also Witten, T.: *Astron, Astrophys. J.* **188**, 615 (1974).
23. Baym, G. and Pines, D.: *Ann. Phys.* **66**, 816 (1971) and references mentioned there.
24. Ruderman, M.: *Nature* **225**, 838 (1970).
25. Smoluchowsky, R.: *Phys. Rev. Letters* **24**, 923 (1970).
26. Rhoades, C.: Ph. D. Thesis, Princeton University, 1971, unpublished.
27. Observed change in the period of the Vela Pulsar, PSR0833-45 of $\Delta P = 2.08 \times 10^{-7}$ s, Reichley P. E. and Douris G. S.: *Nature* **222**, 229 (1969).
28. Observed change in the period of the Crab Nebula Pulsar, NP0532 of $\Delta P \sim 0.3 \times 10^{-4}$ s, Boynton, P. E., Groth, E. J., Hutchinson, D. P., Nanos, G. R., Partridge, R. B., and Wilkinson, D. T.: *Astrophys. J. Letters* **175**, 217 (1972).
29. Rhoades, C. and Ruffini, R.: *Phys. Rev. Letters* **32**, 324 (1974).
30. See Groth. E.: this volume, p. 121.
31. For an up to date presentation of the status of this topic see Bardeen, J., Carter, B., Gursky, H., Hawking, S., Novikov, I., Ruffini, R., and Thorne, K.: in B. and C. DeWitt (eds.), *Black Holes*, Gordon and Breach, 1973.
32. Christodoulou, D.: *Phys. Rev. Letters* **25**, 1596 (1970).
33. Christodoulou, D. and Ruffini, R.: *Phys. Rev.* **4**, 3442 (1971).

34. That a black hole can never increase its parameters L and e such as to have $L^2/m^2 + e^2 > m^2$ by accretion of test particles has explicitly been shown by Christodoulou, D.: Ph. D. Thesis, Princeton University, 1971, unpublished.
35. Denardo, G. and Ruffini, R.: *Phys. Letters* **45B**, 259 (1973).
36. Denardo, G., Hively, C., and Ruffini, R.: *Phys. Letters* **49B**, 185 (1974).
37. Christodoulou, D. and Ruffini, R.: *On the Energetics of Magnetic Black Hole*, Preprint Institute for Advances Study, February 1971.
38. Penrose, R. and Floyd, R.: *Nature* **229**, 193 (1971).
39. Ruffini, R. and Wheeler, J. A.: in H. V. Hardy (ed.), *Cosmology from Space Platform*, ESRO book SP52, Paris, 1971.
40. Hawking, S.: *Phys. Rev. Letters* **26**, 1344 (1971).
41 Hanni, S. and Ruffini, R.: *Phys. Rev.* **8**, 3259 (1973).
42. Regge, T. and Wheeler, J. A.: *Phys. Rev.* **108**, 1063 (1957).
43. Zerilli, F.: *Phys. Rev.* **D2**, 2141 (1970).
    Zerilli, F.: *Phys. Rev. Letters* **24**, 737 (1970).
    Zerilli, F.: *J. Math. Phys.* **11**, 2203 (1970).
44. Davis, M. and Ruffini, R.: *Nuovo Cimento Letters* **2**, 1165 (1971).
45. Davis, M., Ruffini, R., Press, W., and Price, R.: *Phys. Rev. Letters* **27**, 1466 (1971).
46. Davis, M., Ruffini, R., and Tiomno, J.: *Phys. Rev.* **5**, 2932 (1972).
47. Ruffini, R.: *Phys. Rev.* **D7**, 972, (1973).
48. Misner, C. W.: *Phys. Rev. Letters* **28**, 994 (1972).
49. Misner, C. W., Breuer, R. A., Brill, D. R., Chrzanovski, P. L., Hughes III, H. G., and Pereira, C. M.: *Phys. Rev. Letters* **28**, 998 (1972).
50. Davis, M., Ruffini, R., Tiomno, J., and Zerilli, F.: *Phys. Rev. Letters* **28**, 1352 (1972).
51. Ruffini, R. and Zerilli, F.: submitted to *Phys. Rev.* See also Appendix 2.5 in Reference 17.
52. Haxton, W. and Ruffini, R.: submitted to *Annals of Phys.*
53. Ruffini, R.: *Phys. Letters* **41B3**, 334 (1972).
54. Johnston, M., Ruffini, R., and Zerilli, F.: *Phys. Letters* **B49**, 185 (1974). See also Reference 58.
55. Partridge, R. B. and Ruffini, R.: *Gravitational Waves and a Search for the Associated Microwave Radiation*, Gravity Research Foundation, Third Award Essay on Gravitation, 1970.
56. Zerilli, F.: *Phys. Rev.* **9**, 860 (1974).
57. Johnston, M., Ruffini, R., and Zerilli, F.: *Phys. Rev. Letters* **31**, 1317 (1973).
58. Johnston, M., Ruffini, R., and Peterson, M.: *Nuovo Cimento Letters* **9**, 217 (1974).
59. Ruffini, R. and Treves, A.: *Astrophys. Letters* **13**, 109 (1973).
60. See Partridge, R. B.: this volume, p. 29.
61. Christodoulou, D. and Ruffini, R.: 'On the Electrodynamics of Collapsed Objects', see Appendix 4.2 in Reference 17.
62. For the effective potentials of uncharged particles in Kerr and Schwarzschild geometry, see Reference 39. See also Wilkins, D.: *Phys. Rev.* **D5**, 814 (1972). For the trajectory of charged particle, see Reference 17 and 39.
63. Johnston, M. and Ruffini, R.: 'On the Generalized Wilkins Effect', *Phys. Rev.* **D 10**, 2324 (1974).
64. Hanni, R. S. and Ruffini, R.: submitted to *Phys. Rev.* (1974).
65. Wilson, J.: in vited talk delivered at the seventh Texas symposium on relativistic astrophysics, to appear in the proceedings, 1974.
66. Ruffini, R.: in vited talk delivered at the seventh Texas symposium on relativistic astrophysics, to appear in the proceedings, 1974.
67. Ruffini, R. and Wheeler, J. A.: in L. Radicati (ed.), *Proceedings of the Cortona Symposium on Weak interactions*, Academia Nationale dei Lincei, Rome, 1971.
68. Ruffini, R. and Zerilli, F.: 'Polarization of Gravitational and Electromagnetic Radiation etc.', Appendix A 3.7 in Reference 17, 1973.
69. Rees, M., Ruffini, R., and Wheeler, J. A.: *Black Holes, Gravitational Waves and Cosmology*, Gordon and Breach, New York, 1974.
70. Breit, G. and Wigner, E. P.: *Phys. Rev.* **49**, 519 (1936).
71. Weber, J.: *Phys. Rev.* **117**, 306 (1960).
    Sinsky, J. and Weber, J.: *Phys. Rev. Letters* **18**, 795 (1967).
    Weber, J.: *Phys. Rev. Letters* **17**, 1228 (1966).

Weber, J.: *Phys. Rev. Letters* **21**, 395 (1968).
Weber, J.: *Phys. Rev. Letters* **24**, 276 (1970).
Weber, J.: *Phys. Rev. Letters* **22**, 1320 (1969).
Weber, J.: *Phys. Rev. Letters* **18**, 498 (1967).
Weber, J.: *Phys. Rev. Letters* **25**, 180 (1970).
Weber, J.: *Nuovo Cimento* **4B**, 197 (1971).
Weber, J.: *Nature* **240**, 28 (1972).
72. d'Anna, E., Pizzella, G., and Trevese, D.: *Nuovo Cimento Letters* **9**, 231 (1974).
73. Ruffini, R. and Wheeler, J. A.: *Physics Today* **24**, 30 (1971).
74. Zel'dovich, Ya. B. and Guseynov, O. Kh.: *Dokl. Akad. Nauk USSR* **162**, 791 (1965).
75. Shklovsky, I. S.: *Astrophys. J.* **148**, L1 (1967).
76. Zel'dovich, Ya. B. and Novikov, I. D.: *Soviet Phys. Dokl.* **9**, 246 (1964).
77. Schwartzman, V. F.: *Astron. Zh.* **47**, 824 (1970).
78. Giacconi, R., see e.g., Talk delivered at the *64th Symposium of the International Astronomical Union*, Warsaw 1973. To appear in the proceedings.
    See also Gursky, H. and Schreier, E.: this volume, p. 175.
79. Leach, R. and Ruffini, R.: *Astrophys. J. Letters* **180**, L15 (1973).
80. Ruffini, R. and Wilson, J.: *Phys. Rev. Letters* **32**, 324 (1974).
81. See e.g. Cameron, A. G. W. and Mock, M.: *Nature* **215**, 264 (1967).
82. Zel'dovich, Ya. B. and Novikov, I. D.: *Relativistic Astrophysics*, Vol. 1, Izdatel'stvo Nauka, Moscow, 1967.
83. Prendergast, K. and Burbidge, G.: *Astrophys. J. Letters* **151**, L83 (1968).
84. Schwartzman, V. F.: *Astron. Zh.* **48**, 438 (1971).
    Schwartzman, V. F.: *Soviet Astron.* **15**, 343 (1971).
85. Pringle, J. E. and Rees, M.: *Astron. Astrophys.* **21**, 1 (1972).
86. Ostriker, J. and Davidson, K.: in H. Bradt and R. Giacconi (eds.), *X-Ray and Gamma Ray Astronomy*, D. Reidel, Dordrecht, 1973.
87. Lamb, F. K., Pethick, C. J., and Pines, D.: *Astrophys. J.* **184**, 271 (1973).
88. Shakura, N. and Sunyaev, R.: *Astron, Astrophys.* **24**, 337 (1973).
89. Baker, K. and Ruffini, R.: submitted to *Astron. Astrophys.*
90. Ferrari, A. and Ruffini, R.: *Astrophys. J. Letters* **158**, L71 (1969).
91. Giacconi, R., Gursky, H., Kellog, E., Schreier, E., and Tananbaum, H.: *Astrophys. J. Letters* **167**, L7 (1971).
92. Giacconi, R., Gursky, H., Kellog, E., Levinson, R., Schreier, E., and Tananbaum, H.: *Astrophys. J.* **184**, 227 (1973).
    Tananbaum, H., Gursky, H., Kellog, E. M. Levinson, R., Schreier, E., and Giacconi R.: *Astrophys. J. Letters* **174**, L143 (1972).
93. Bahcall, J. N. and Bahcall, N. A.: *IAU Circ.*, Nos. 2427 and 2428 (1972).
    Bahcall, J. N. and Bahcall, N. A.: *Astrophys. J. Letters* **178**, L1 (1972).
    Boynton, P. E., Canterna, R., Crosa, L., Deeter, J., and Gerend, D.: *Astrophys. J.* **186**, 617 (1973)
    Crampton, D. and Hutchings, J. B.: *Astrophys. J. Letters* **178**, L65 (1972).
    Davidson, A., Henry, J. P., Middleditch, J., and Smith, H. E.: *Astrophys. J. Letters* **177**, L97 (1972).
    Forman, W., Jones, C. A., and Liller, W.: *Astrophys. J. Letters* **177**, L103 (1972).
    Jones, C. A., Forman, W., and Liller, W.: *Astrophys. J. Letters* **182**, L109 (1973).
    Petro, L. and Hiltner, W. A.: *Astrophys. J. Letters* **181**, L39 (1973).
94. Liller, W.: *IAU Circ.*, No. 2517 (1973).
95. Giacconi, R.: talk delivered at the *1973-Solvay Conference on Relativistic Astrophysics*. To appear in the proceedings, in press.
96. Oda, H. Gorenstein, P., Gursky, H., Kellog, E., Schreier, E., Tananbaum, H., and Giacconi, R.: *Astrophys. J. Letters* **166**, L1 (1971).
97. Schreier, E., Gursky, H., Kellog, E., Tananbaum, H., Giacconi, R.: *Astrophys. J. Letters* **170**, L21 (1971).
98. Rappaport, S., Zaumen, W., Doxsey, R.: *Astrophys. J. Letters* **168**, L17 (1971).
    Shulman, S., Fritz, G., Meekins, J. F., Friedmann, H., and Meidav, M.: *Astrophys. J. Letters* **168**, L49 (1971).
    Rappaport, S., Doxsey, R., and Zaumen, W.: *Astrophys. J. Letters* **168**, L43 (1971).

99. Margon, B., Bowyer, S., Stone, R. P. S.: *Astrophys. J. Letters* **185**, L113 (1973).
100. Bergman, J., Butler, D., Kemper, E., Koski, A., Kraft, R. P., and Stone, R. P.: *Astrophys. J. Letters* **185**, L117 (1973).
101. Hjellming, R. M., Wade, C. M., Hughes, V. A., and Woodsworth, A.: *Nature* **234**, 138 (1971). Hjellming, R. M.: *Astrophys. J. Letters* **182**, L29 (1973).
102. Colgate, S.: *Nature* **225**, 247 (1970).
103. Ruffini, R.: see Reference 17.
104. Webster, B. L. and Murdin, P.: *Nature* **235**, 37 (1972).
105. Bolton, C. T.: *Nature Phys. Sci.* **240**, 124 (1972).
106. Hutchings, J. B., Crampton, D., Glaspey, J., and Walker, G. A. H.: *Astrophys. J.* **182**, 549 (1973).
107. Brucato, R. and Kristian, J.: *Astrophys. J. Letters* **179**, L129 (1973).
108. Ruffini, R.: 'X-Ray Sources: Neutron Stars or Black Holes', invited talk at the *Texas Meeting of Relativistic Astrophysics*, New York, 1972, unpublished. See Reference 17.
109. Rothschild, R. E., Boldt, E. A., Holt, S. A., and Serlemitsos, P. F.: *Astrophys. J. Letters* **189**, L13 (1974).
110. Lyutiy, V. M., Sunyaey, R. A., and Cherepashchuk, A. M.: *Astron. Zh.* **50**, 3 (1973).
111. Bolton, T.: to be published.
112. Paczińsky, B.: *Ann. Rev. Astron. Astrophys.* **9**, 183 (1971) and references mentioned there.
113. Arnett, W. D.: *Nature* **219**, 1344 (1968). Arnett, W. D.: *Astrophys. Space Sci.* **5**, 180 (1969). See also Colgate, S.: this volume, p. 13.
114. The special role of this evolutionary computations in the case of a binary system has been emphasized by Paczińsky (see e.g. Reference 112) by Kippenhahn and Weigert see e.g. Kippenhahn, R.: *Astron. Astorphys.* **3**, 83 (1969). Kippenhahn, R., Kohl, K., and Weigert, A.: *Z. Astrophys.* **66**, 58 (1967). Kippenhahn, R. and Weigert, A.: *Z. Astrophys.* **65**, 251 (1967). By Giannone *et al.*, see e.g. Giannone, P. and Giannuzzi, M. A.: *Astron. Astrophys.* **6**, 309 (1971). Giannone, P. and Giannuzzi, M. A.: *Astron. Astrophys.* **18**, 111 (1972); see also **19**, 298 (1972). Giannone, P., Refsdal, S., and Weigert, A.: *Astron. Astrophys.* **4**, 428 (1970). And by Van den Heuvel in the special case of binary X-ray sources see e.g. Van den Heuvel, E.: *Bul. Astron. Inst. Neth.* **19**, 432 (1968). Van den Heuvel, E.: *Nature Phys. Sci.* **242**, 71 (1973). Van den Heuvel, E. and Heise, J.: *Nature Phys. Sci.* **239**, 67 (1972). Van den Heuvel, E. and Ostriker, J.: *Nature Phys. Sci.* **245**, 99 (1973). See also Kraft, R. P.: this volume, p. 235.
115. For the mass-radius relation and much of the details of white dwarfs stars, see the classical work of Chandrasekhar, S.: *Stellar Evolution*, Dover Publ. Inc., 1957.
116. See e.g. Van de Kamp, P.: *Handbuch der Physik* **50**, 187 (1958).
117. Warner, B.: *Monthly Notices Roy. Astron. Soc.* **160**, 35 (1972). Warner, B.: *Monthly Notices Roy. Astron. Soc.* **162**, 189 (1973) and references mentioned there. See also Warner, B. and Robinson, E. L.: *Nature Phys. Sci.* **239**, 2 (1972) and references mentioned there.
118. Walker, M. F.: *Publ. Astron. Soc. Pacific* **66**, 230 (1954).
119. This pulsation appears to be very abnormal for a white dwarf star of reasonable mass. Although an explanation of this abnormal period has been advanced by Channugan, G.: *Nature* **236**, 3 (1972) in term of non radial pulsations more work is needed to explore the full significance of this very important experimental result and especially to explore the possibility that the pulsation in the luminosity of these systems, be explainable as associated to the rotational period of a white dwarf.
120. See Strong, I. B.: this volume, p. 47. See also Strong, I. B., Klebesadel, R. W., and Olson, R. A.: *Astrophys. J. Letters* **188**, L1 (1974).
121. Ruffini, R. and Treves, A.: submitted to *Astrophys. Letters*.
122. See e.g. *The Evolution Computations of Binary Systems* by Paczińsky Weigert and Kippenhan, Giannone *et al.*, Van den Heuvel (see Reference 130).
123. Becklin, E. E., Kristian, J., Neugehauer, G., Wynn Williams, C, G.: *Nature* Phys. Sci **239**, 130 (1972).

124. Paczińksy, B.: *Acta Astronomica* **20**, 2 (1970).
125. For derivation of this formula see Landau, L. and Lifschitz, E.: *Theorie des champs*, editions M.I.R., Moscow, 1970.
126. For an estimate of a large set of sources of gravitational radiation with their characteristic signature and energetics, see Reference 69.
127. Dyson, F. J.: Chapter 12 in A. G. W. Cameron (ed.), *Interstellar Communication*, W. A. Benjamin, New York, 1963.
128. Tyson, J. A.: *Phys. Rev. Letters* **31**, 326 (1973).
129. Douglas, D. H.: see e.g. Talk delivered at the *Sixth Cambridge Conference On Exprimental Relativity, MIT*, 1974, unpublished.
130. Billing, H. *et al.*: see e.g. Talk delivered at the *Sixth Cambridge Conference on Experimental Relativity, MIT*, 1974, unpublished.
131. Braginsky, V.: *J.E.T.P. Letters* **16**, 108 (1972).
132. Levine, J. L. and Garwin, R. L.: *Phys. Rev. Letters* **31**, 173 (1973).
     Garwin, R. L. and Levine, J. L.: *Phys. Rev. Letters* **31**, 176 (1973).
133. Ruffini, R.: in R. D. Davies and F. G. Smith (eds.), 'The Crab Nebula', *IAU Symp.* **46**, 382, (1971).
134. For a detailed estimate of the energy and frequency of gravitational radiation emitted by a vibrating neutron star, see Table XVIII in Reference 39.
135. Fairbank, W. M., Boughn, S. P., Paik, H. J., McAshan, M. S., and Opfer, J. E., and Taber, R. C., Hamilton, W. O., Pipes, B., and Bernat, T.: in B. Bertotti (ed.), *Varenna 1972 'Enrico Fermi' Summer School*, Academic Press, 1974.
136. Amaldi, E.: *Rapporto al Comitato Fisica del C.N.R.* (1973). See also Pizzella, G.: 'Low Temperature Detectors of Gra. rad.', Nota Interna. Univ. of Rome, 1974; and Carelli, P., Cerdonio, M., Giovannardi, U., Lucano, G., and Modena, I.: 'Low Temperature Gravitational Radiation Antenna – A Progress Report', *Colloque Int. Ondes Gravitationelle*, Paris, 1973, to be published.
137. Braginsky, V. B. and Rudenko, B. N.: *Uspekhi* **100**, 395 (1970).
     Bagdasarov, H. S., Braginsky, B. B., and Mitrafanov, V. P.: preprint I.T.P. 73-93E, Kiev, 1973.
138. Paik, H. J.: 'Displacement Detection for the Temperature Gravitational Wave Detector', in B. Bertotti (ed.), *Varenna 1972 'Enrico Fermi' Summer School*, Academic Press, 1974.
139. See Reference 50. See also Field, G. B., Rees, M. J., and Sciama, D. W.: *Comments Astron. Space Sci.* **1**, 187 (1969).
140. See e.g. Chandrasekhar, S.: *Ellipsoidal Figures of Equilibrium*, Yale University, 1969.
141. De Gasperis, A. and Pietronero, L.: to be published.
142. Ruffini, R.: 'Theoretical and Experimental Progress in the Detection of Gravitational Radiation', *La Recherche*, in press (1974).

OBSERVATIONAL PROPERTIES OF PULSARS

EDWARD J. GROTH

Joseph Henry Laboratories, Physics Dept., Princeton University,
Princeton, N.J. 08540, U.S.A.

1. Introduction

In the six years since the announcement [129] of the discovery of the first pulsar, a total of 105 pulsars have been discovered and a great deal has been learned of their properties. Pulsars are slowly rotating (~ 1 Hz) neutron stars with huge magnetic fields ($\sim 10^{12}$ G) which convert rotational energy into energy of (radio) radiation and relativistic particles. The pulsed radiation presumably arises from a beaming effect associated with the magnetic field.

In this chapter, we will discuss the properties of pulsars from an observational viewpoint. The basic parameters of pulsars are presented in Section 2. Propagation of the pulses through the interstellar medium is dealt with in Section 3 and in Section 4, pulsar supernova remnant associations are discussed. Distance estimates are presented in Section 5 and flux density spectra are discussed in Section 6. The structure of the pulses is summarized in Section 7 while Section 8 discusses timing of the pulses. Concluding remarks are given in Section 9.

2. Basic Parameters

Table I lists the basic parameters of the 105 pulsars insofar as they are known as of June, 1974. Generally, only the published measurement with the smallest quoted error is listed. Some authors have not given errors, others give 1σ errors, others give 2σ errors, etc., so the errors listed in Table I are not a uniform assessment of the errors in the parameters.

The first column lists the name of the pulsar with the convention PSRhhmm $\pm$ dd, where hhmm is the hours and minutes of right ascension and $\pm$ dd is the degrees of declination (equinox 1950). As better coordinates become known, some of the names in Table I will change. In the early days of pulsar astronomy, pulsars were often designated by a letter standing for the discovering observatory, P for pulsar, and the hours and minutes of right ascension, e.g. CP1919 $=$ pulsar at $19^{\mathrm{h}}19^{\mathrm{m}}$ discovered at Cambridge. References (indicated by R.) in column (1) cite the discovery publication. Columns (2) and (3) give the 1950 right ascension and declination of the pulsar. In most cases these positions are determined by finding the coordinates of maximum intensity. More accurate positions may be obtained with timing observations spanning several years. Timing positions are indicated by T. Galactic coordinates are given in columns (4) and (5). Periods, rates of change of period, and epoch of period measurement are listed in columns (6), (7), and (8). Periods are given in A1 seconds. Where

H. Gursky and R. Ruffini (eds.), Neutron Stars, Black Holes and Binary X-Ray Sources, 119–173. All Rights Reserved.
Copyright © 1975 by D. Reidel Publishing Company, Dordrecht-Holland.

epochs were not given, the epochs are estimated from the dates of observation and enclosed in parentheses. If dates of observation were not given, no epoch is listed. Column (9) lists the timing age, $\frac{1}{2}P/(dP/dt)$ for those pulsars with measured period derivatives. The discussion in Section 8 indicates that these ages are probably within a factor of two of the true age of the pulsar. Dispersion measures are tabulated in column (10). Full widths at half maximum of the pulse, measured near 400 MHz, are given in column (11). In four cases only an equivalent width (energy per unit area per unit frequency contained in the pulse divided by peak flux density) was available and is denoted by e. Widths enclosed in parentheses were measured from published pulse shapes. An estimate of the pulsating flux density (energy per unit area per unit frequency contained in the pulse divided by period) at an observing frequency near 400 MHz is given in column (11). Data in columns (11) and (12) are in some cases very rough estimates strongly affected by pulse shape and intensity fluctuations. Column (13) contains distance estimates which are probably uncertain by a factor of three. Those marked P are taken from References 232 or 234. Those marked S or H are obtained from association with a supernova remnant at known distance or from 21 cm data. Unmarked estimates are calculated from the dispersion measure assuming $\langle n_e \rangle = 0.04\ \mathrm{cm}^{-3}$. Details are given in Section 5. Finally column (14) gives a partial list of references where a plot of the pulse shape may be found.

3. Propagation of the Pulses

Observed pulses are considerably distorted by their propagation through the interstellar medium. Dispersion and Faraday rotation are due to the average properties of the interstellar medium, while scintillation and pulse broadening are due to inhomogeneities.

Pulses are dispersed by the conducting interstellar medium. A broadband pulse arrives later at lower observing frequencies than at higher frequencies. The form of the dispersion law (Reference 141, p. 227) yields,

$$t_d = \frac{e^2}{2\pi m_e c v^2} \int n_e\, dl,\tag{1}$$

where t_d is the time delay at observing frequency v, n_e is the electron number density, l is the path length along the line of sight to the pulsar, and the other symbols have their usual meanings. Equation (1) ignores the ionic contribution to the plasma frequency, which leads to a small uncertainty in the relation between t_d and $\int n_e\, dl$. Dispersion is usually given by the dispersion measure, DM, which is $\int n_e\, dl$ in units of $\mathrm{pc\ cm}^{-3}$. Another unit is the dispersion constant, DC, defined by

$$t_d = DC/v^2.\tag{2}$$

DC is the directly measured quantity and is usually given in units of s Hz^2. The relation between DM and DC is,

$$DM \ (\text{pc cm}^{-3}) = 2.41 \times 10^{-16} \ DC \ (\text{s Hz}^2). \tag{3}$$

Equation (1) results from an expansion of the group velocity to lowest order in the plasma frequency. The v^{-2} dependence is accurately satisfied [59, 92, 93, 299]. The absence of v^{-4} terms sets limits on interstellar electron density fluctuations and magnetic field strength. These limits are generally much higher than can be obtained by other means but they were important in establishing that the dispersion arises in the interstellar medium and not in the source where high densities and field strengths would be expected. With Equation (1) it is found that the pulses are emitted simultaneously (~ 1.5 ms Reference 93) over a wide range of wavelengths. In fact, the radio and optical pulses of PSR0531+21 are emitted simultaneously to within 400 μs, the measurement uncertainty [234]. Except for PSR0531+21, discussed separately below, dispersion measures have been found to be independent of time [93, 189]. Because the v^{-2} interstellar dispersion law satisfactorily describes the data and because the dispersion measure is constant, it is usually assumed that there is no dispersion intrinsic to the pulsars.

It is interesting that a massive photon yields a dispersion relation identical to Equation (1). By assuming that all the dispersion in PSR0531+21 (which has a known distance) is due to effects of a massive photon, one finds $m_\gamma < 10^{-44}$ g (Reference 79). This limit applies to real photons while somewhat better limits obtained from measurements of the Earth's magnetic field apply to virtual photons.

Equation (1) shows that in all observations the pulse will be smeared out by differential time delay across the reciever bandwidth. If Δt is the amount of smearing in ms, Δv the bandwidth in MHz, and v the observing frequency in MHz,

$$\Delta t = 8.3 \times 10^6 \ DM \ \Delta v / v^3. \tag{4}$$

Dispersion broadening can make a substantial contribution to the observed pulse width for high dispersion pulsars. It can be eliminated by the use of narrow bandwidths or dedispersing techniques.

Dispersion permits the pulsar distance to be estimated if the average electron density along the line of sight to the pulsar is known. A detailed discussion of pulsar distances is deferred until Section 5. Here we note that estimates of $\langle n_e \rangle$ hover around $0.04 \, \text{cm}^{-3}$ and that there are large fluctuations about this value.

The interstellar magnetic field produces Faraday rotation of the plane of polarization of the pulses by the amount (Reference 141, p. 229),

$$\Delta \phi = \frac{e^3 \lambda^2}{2\pi m_e^2 c^4} \int B_l n_e \, dl, \tag{5}$$

where $\Delta \phi$ is the amount of rotation, λ is the wavelength, B_l is the line of sight component of the magnetic field and the other symbols have their usual meanings. The rotation measure has units of rad m^{-2} and is defined by

$$\Delta \phi = RM \lambda^2. \tag{6}$$

Pulsar Faraday rotation is found to be accurately described by Equation (6), is time independent*, and shows no variation across the pulse [189, 235]. It is therefore assumed that the observed Faraday rotation is entirely due to the interstellar medium and that pulsars have no intrinsic Faraday rotation.

Together, rotation and dispersion measure provide an estimate of the galactic magnetic field in the directions of the pulsars. From the definition of RM and DM,

$$\langle B_l \rangle = 1.24 \; RM/DM, \tag{7}$$

where $\langle B_l \rangle$, measured in μG, is the average line of sight magnetic field weighted by electron density. Pulsar Faraday rotation measurements of $\langle B_l \rangle$ have several advantages over other methods of estimating $\langle B_l \rangle$. The principal ones are that there is no intrinsic Faraday rotation and the electron density is easily determined via the dispersion measure [190].

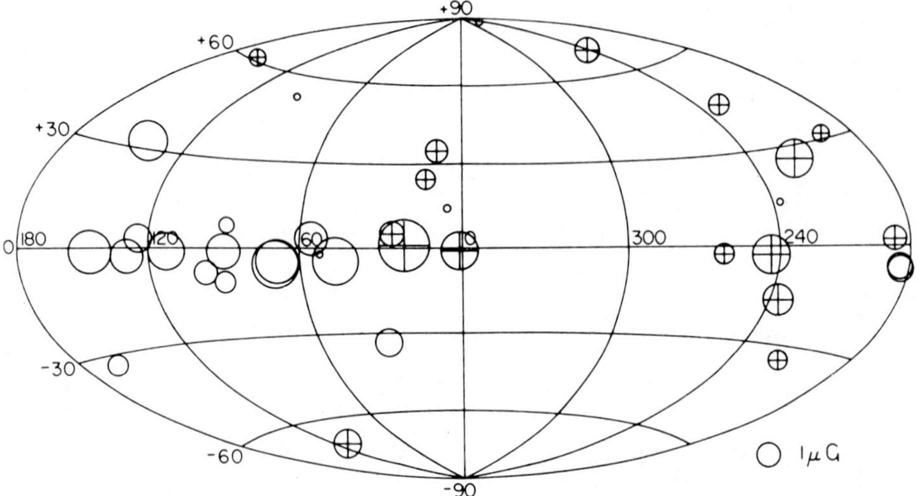

Fig. 1. Line of sight magnetic field components in galactic coordinates, reproduced from Reference 190 by permission. The areas of the circles are proportional to the field strength. Circles with crosses indicate a field directed towards the observer. Published by the University of Chicago Press; © 1974, American Astronomical Society.

Rotation measures exist for 38 pulsars and are summarized in Reference 190. The derived magnetic fields are shown in Figure 1. From these data the galactic magnetic field is found to be approximately homogeneous out to about 1 kpc from the Sun. Its strength is $2.2 \pm 0.4 \; \mu$G and it is directed towards galactic longitude $94° \pm 11°$. Isolated irregularities of about the same strength also occur.

Scattering of pulsar radiation by electron density inhomogeneities produces a diffraction pattern at the Earth. The relative motions of the pulsar, Earth, and interstellar medium cause the diffraction pattern to sweep across the Earth, yielding a time

---

* A decrease of about 1.5% over two years has been reported for PSR0329 + 54. These observations have not yet been confirmed. If real, the decrease may be due to motion of H I clouds containing strong frozen in fields [190].

varying flux density – scintillation. Pulsars are very small angular diameter sources and the modulation is very high. The effect is analogous to the ordinary twinkling of stars.* The scattered radiation arrives at the Earth slightly later than the unscattered radiation. Thus, pulses are asymmetrically broadened by the addition of an exponential tail. Unlike dispersive broadening, scattering broadening (often called multipath broadening) cannot be removed by using narrow bandwidths.

The theory of scintillations and pulse broadening is in satisfactory shape although there are some areas requiring more study. The basic theory is presented in References 251, 261, 262, 269, 293, 328, 329, and 330, among others. A summary of the theory and fit to observations may be found in Reference 162 and provides the basis for the following discussion.

Scintillation theories usually employ the thin screen approximation in which all the inhomogeneities are confined to a thin screen midway between the source and observer. In the limit of geometric optics, the rms scattering angle is (Reference 162, Equation (2)),

$$\theta_{\text{Scat}} = 4 \times 10^7 \left(\frac{D}{a}\right)^{1/2} \frac{\langle \Delta n_e^2 \rangle^{1/2}}{v^2}, \tag{8}$$

where D is the distance to the source, $\langle \Delta n_e^2 \rangle^{1/2}$ is the rms fluctuation in electron density on the scale a, and v is the observing frequency. Scintillations of nearby frequencies are correlated over a 'decorrelation bandwidth' given by (Reference 162, Equation (3))

$$f_v = \frac{7 \times 10^{-5} a v^4}{\langle \Delta n_e^2 \rangle D^2} \propto v^4 DM^{-2}, \tag{9}$$

where the proportionality is obtained by assuming that $\langle \Delta n_e^2 \rangle^{1/2}$ is proportional to $\langle n_e \rangle$. The scattered radiation travels a longer path and arrives later than the unscattered radiation. The observed pulse shape is then the convolution of the emitted shape with an exponential broadening function of the form $\exp(-t/\Delta t)$, $t > 0$, $\Delta t = f_v^{-1}$. The movement of the diffraction pattern causes intensities at a given frequency to be correlated over a time (Reference 162, Equation (6)),

$$\tau_v = 10^3 \left(\frac{a}{D}\right)^{1/2} \frac{v}{v \langle \Delta n_e^2 \rangle^{1/2}} \propto vDM^{-1}, \tag{10}$$

where v is the velocity of the scattering irregularities perpendicular to the line of sight.

The relations given above are approximately satisfied and the following conclusions have been reached [162]: (i) $\langle n_e \rangle \approx 0.03 \text{ cm}^{-3}$; (ii) $\langle \Delta n_e \rangle^{1/2} \approx 3 \times 10^{-4} \text{ cm}^{-3}$; (iii) $a \approx 10^{11} \text{ cm}$; (iv) $30 \text{ km s}^{-1} \lesssim v \lesssim 200 \text{ km s}^{-1}$; and (v) the constant of proportionality in Equation (9) is 2–$5 \times 10^{-9} \text{ pc}^2 \text{ MHz}^{-3} \text{ cm}^{-6}$. It should be emphasized that these relations are only approximately satisfied.

---

* The first pulsar was discovered by an experiment to detect quasistellar radio sources by their scintillation. Although scintillation did not play a role in the discovery, it was the first time that a large area of the sky had been surveyed with high enough sensitivity and short enough time constant to detect a pulsar.

The relation $f_v \propto DM^{-2}$ is not well established. For some of the more distant pulsars (i.e. high dispersion measure) the pulse broadening is a factor of ten to a hundred times higher than predicted by Equation (9) [163]. It has been suggested [293] that $f_v \propto$ $\propto DM^{-4}$. These results probably indicate that $\langle \Delta n_e^2 \rangle^{1/2}$ is not well correlated with $\langle n_e \rangle$. Another problem with the simple theory is that with scattering irregularities distributed along the line of sight, the broadening function will have a slightly rounded leading edge [328]. For most pulsars this effect is quite small.

The estimate of $\langle n_e^2 \rangle^{1/2}$ given above refers to lengths of 10^{11} cm. This length is not necessarily the correlation length of the density fluctuations but may be only the length of the first Fresnel zone [142]. In fact, as will be seen in Section 5, the rms fluctuations in $\langle n_e \rangle$ are the same order as $\langle n_e \rangle$ and therefore the true correlation length is probably much larger than 10^{11} cm.

Decorrelation bandwidths as a function of frequency have been measured by several observers [51, 77, 162, 250]. (Beware – definitions differ, but see Reference 293 for a summary.) All find $f_v \propto v^4$. In one set of observations [138], a considerably weaker v dependence was found. However, these results are in disagreement with others and have been questioned on the grounds of improper calibration [293].

Figure 2 shows an example of the frequency and time structure in radiation from PSR0329+54. Traces are successive, average spectra centered on 408 MHz. Randomly placed features with a characteristic size of 150 kHz by four minutes are clearly seen and provide a most vivid example of scintillation.

Pulse broadening of a number of pulsars has been investigated and found to agree with the v^{-4} law [162, 163]. Pulse broadening of PSR0833−45 is shown in Figure 3 together with the expected λ^4 dependence at long wavelengths [1]. In another paper [150] it is shown that convolution of the high frequency (1410 MHz) Stokes parameters with an exponential broadening function reproduces the low frequency (down to 300 MHz) Stokes parameters quite well. These results, in addition to verifying the pulse broadening theory, also show that there is no intrinsic change in pulse shape over the range 300 to 1400 MHz. If the pulse broadening width becomes larger than

---

Notes to Table I

PSR0355 + 54	Announced independently in Reference 304.
PSR0531 + 21	Crab nebula pulsar; period irregular, see Section 8; dispersion variable, see Section 3.
PSR0611 + 22	Probably associated with IC443, see Section 4.
PSR0740 − 28	FWHM and flux density at 327 MHz.
PSR0833 − 45	Vela supernova remnant pulsar; period irregular, see Section 8; width at 1420 MHz.
PSR0904 + 77	Width and flux density at 112 MHz.
PSR0943 + 10	Width at 111 MHz.
PSR1133 + 16	Has detectable proper motion, epoch of coordinates same as period, see Section 8.
PSR1451 − 68	Width at 11 cm.
PSR1541 − 09	Flux density at 327 MHz.
PSR1749 − 28	Position from lunar occultation.
PSR1858 + 03	Width at 327 MHz.
PSR1953 + 29	Width at 327 MHz.
PSR2020 + 28	Width and flux density at 327 MHz.

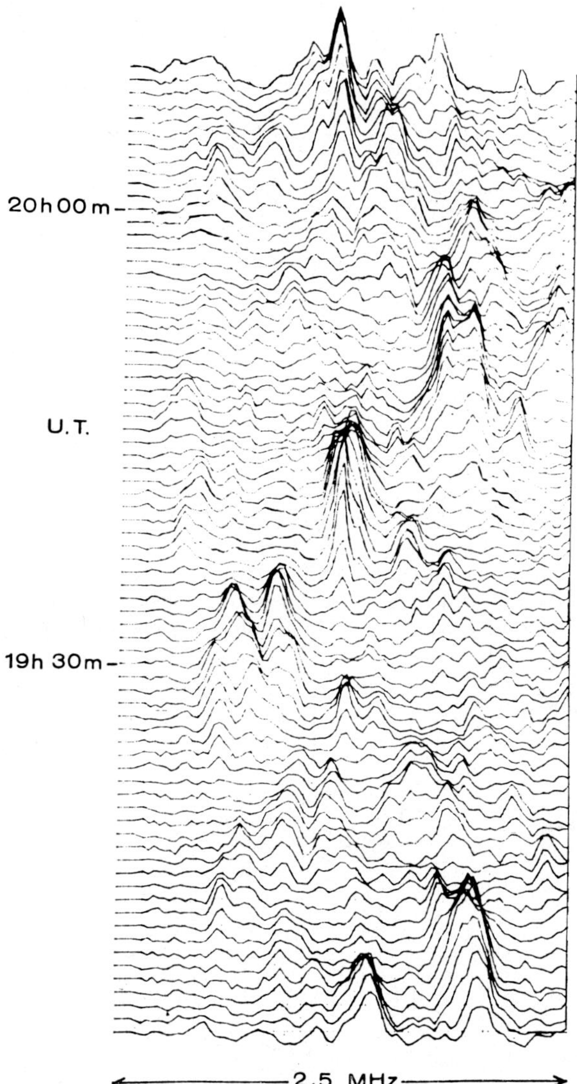

Fig. 2. Frequency and time structure of scintillation in PSR0329 + 54, reproduced from Reference 250 by permission. Each trace is a 2.5 MHz segment of the spectrum averaged over 50 seconds. © 1969, MacMillan Journals, Ltd.

the pulse period, the pulses will be smeared out to steady emission. A point source at 80 MHz has been found coincident with PSR0833 − 45. The flux density of this source agrees with the extrapolated high frequency pulsar spectrum. Thus it is natural to suppose that this source is simply smeared out pulsar radiation [130]. A similar effect is well established in the case of PSR0531 + 21 (see below and Section 5).

PSR0531 + 21 is the only known pulsar for which the exponential broadening function does not provide adequate results. A broadening function resulting from the

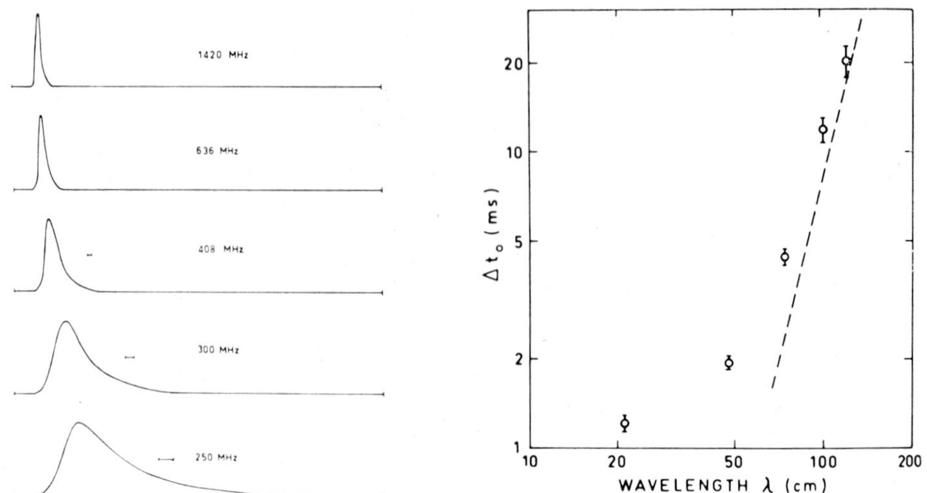

Fig. 3.   Pulse broadening in PSR0833 − 45, reproduced from Reference 1 by permission. On the left are average pulse shapes at several frequencies. Each trace covers an entire period. The short bars indicate the amount of dispersion broadening where significant. On the right is pulse width vs wavelength together with the expected λ^4 long wavelength dependence. Published by the University of Chicago Press; © 1974, American Astronomical Society.

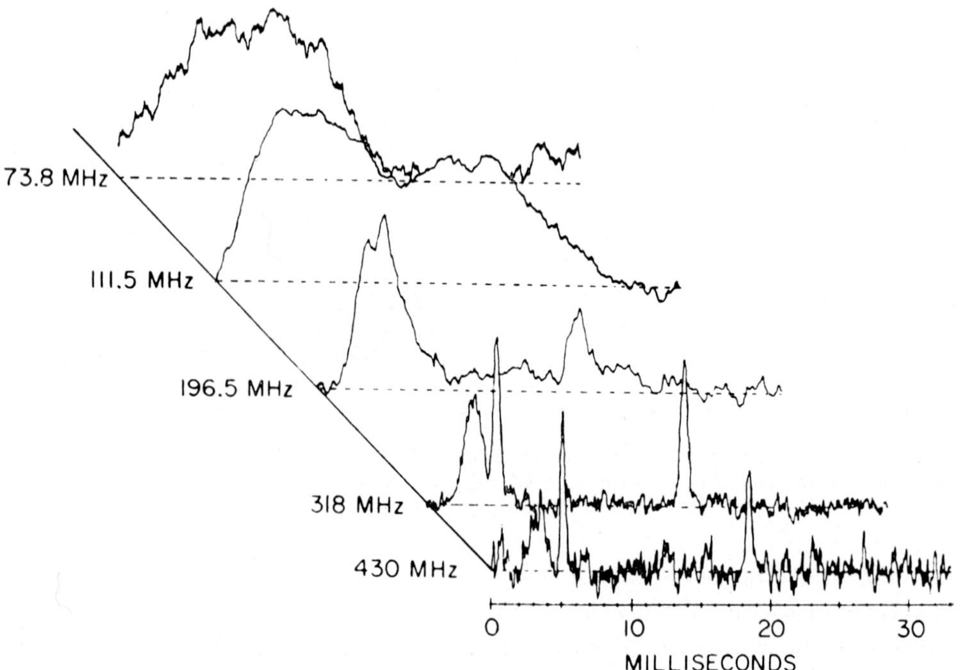

Fig. 4.   Average pulse shape of PSR0531 + 21 as a function of frequency, reproduced from Reference 239 by permission. Horizontal alignment and intensity normalization are arbitrary. These shapes vary with time due to variations in the amount of scattering. Published by the University of Chicago Press; © 1970, University of Chicago Press.

convolution of two exponential functions is required. Furthermore, the dispersion measure and amount of scattering vary on a time scale of months [239, 240]. Figure 4 gives the pulse shape of PSR0531+21 as a function of frequency. At 430 MHz, a narrow main pulse, a narrow interpulse, and a broad precursor preceding the main pulse are clearly seen. As the frequency decreases to 73.8 MHz the pulses are smeared out to essentially a sine wave. The precursor is not seen at frequencies higher than 600 MHz [186, 242], indicating a very steep spectral index for the precursor above 400 MHz. However, the good agreement of predicted and observed pulse shapes below 400 MHz indicates that all components have about the same spectral index in this region [240]. Good agreement is obtained only with the double exponential broadening function. Reports to the contrary [292, 296] have been based on instrumental resolution inadequate to distinguish between a single or double exponential scattering function [240]. The broadening function also accurately predicts the apparent size of the compact low frequency source in the center of the Crab nebula, providing strong evidence that this source is nothing more than smeared out pulsar radiation.

Figure 5 shows the variation of dispersion *constant* and S_1 and S_2 with time. (Here, the scattering width is $\Delta t = S v^{-4}$.) These time variations are supported by very long

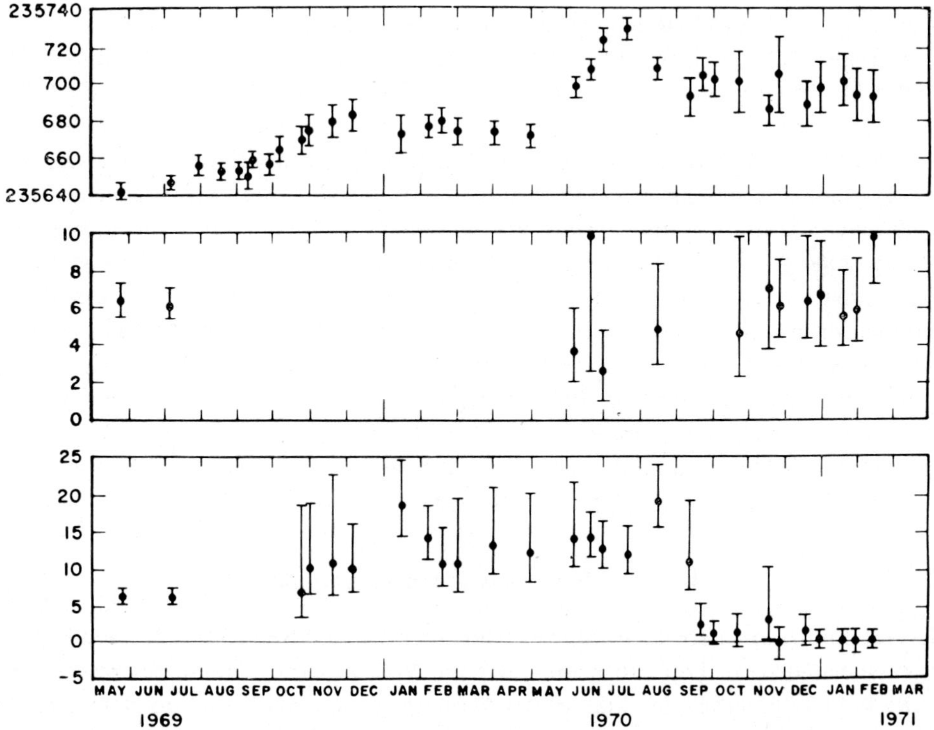

Fig. 5.   Time variations of dispersion constant and scattering measures of PSR0531+21, reproduced from Reference 240 by permission. Published by the University of Chicago Press; © 1973, American Astronomical Society.

baseline interferometry observations which show that the sum of the pulsed and steady (smeared out pulses) flux densities remains constant while the individual components vary [312]. S_1 is identified with scattering due to the interstellar medium and all measurements are consistent with a constant value of 6.2×10^5 s MHz.[4] S_2 is thought to be due to scattering within the central regions of the Crab nebula and varies from 0 to 3 times the interstellar scattering measure. The minimum value of the dispersion measure (56.80 pc cm^{-3}) is identified with the interstellar medium while the time varying component is thought to be dispersion within the Crab nebula [240]. These identifications are not unreasonable as considerable optical activity on the time scale of months is observed in the central regions of the nebula [266]. However, attempts to correlate dispersion and scattering variations with optical activity in the nebula or timing irregularities in the pulsar period were unsuccessful [240].

Scintillation pattern velocities may be measured directly by simultaneous observations at two widely separated observatories. Cross correlation of the intensities yields a maximum at a lag which is the baseline divided by the projection of the pattern velocity along the baseline. In addition, a large, narrow maximum is usually observed at zero lag, indicating that the pulse to pulse fluctuations are intrinsic to the source [49, 284].

The first results were obtained for PSR1133+16 and indicated a velocity in the range 22 to 92 km s^{-1} (Reference 166). Later results are $v = 370 \pm 90$ km s^{-1} in position angle $293° \pm 15°$ for PSR0329+54 (Reference 86), $16 < v < 400$ km s^{-1} for PSR0834+06 (Reference 252), and 35 and 121 km s^{-1} for PSRs 1929+10 and 2045$-$ -16, respectively (Reference 283). However, recent observations of PSR1133+16 not only disagreed with the earlier results but the velocity changed sign on a time scale of days [252]. A similar effect was found for PSR1642$-$03 which gave velocities of $+40$ and -52 km s^{-1} during observations separated by only three days [283]. The velocity changes have been interpreted as a breakdown of the thin screen scattering approximation [252, 283] and cast considerable doubt on the assocation of the scintillation pattern velocity with the actual velocity of the pulsar through the interstellar medium. In addition, it has been pointed out that intrinsic fluctuations on the same time scale as the scintillation fluctuations will systematically shift the maximum of the cross correlation function towards zero lag, thereby overestimating the pattern velocity [283]. Thus, scintillation measurements appear to neither support nor contradict the hypothesis [102] that pulsars are high velocity runaway stars.

4. Pulsar Supernova Remnant Associations

Several pulsars are associated with supernova remnants. PSRs 0531+21 and 0833$-$45 are associated with the Crab nebula and the Vela supernova remnant, respectively. PSRs 0611+22, 1154$-$62, and 2021+51 may be associated with remnants IC443, G296.8$-$0.3 and HB21, respectively. Finally, it has been suggested that PSR1919+21 is associated with G55.7+3.4 but this identification is doubtful.

On the general subject of supernova remnants there are recent reviews by Shklovsky

[277] and Woltjer [331], among others, so in what follows we will list only the main points relevant to the identification of pulsar and remnant.

Phased television photography has shown that the pulsar in the Crab nebula (= NGC1952, M1, Tau X-1, Tau A, 3C144) is the south preceding component of the central double star of the nebula [207]. The nebula contains two components – filaments emitting line radiation and an amorphous region emitting synchrotron radiation (see e.g. References 277 or 311). The nebula is the remnant of a supernova observed and recorded in summer, 1054 A.D. by Chinese astrologers [71, 199]. The timing age of PSR0531+21 is 1200 yr, in good agreement with its actual age of 920 yr.

Various arguments [307] indicate that the nebula is at a distance of 2 kpc although it might be somewhat closer (~ 1.5 kpc). The expansion velocity of the optical nebula is about 1500 km s^{-1} (Reference 307) and on this basis it has been argued [213] that the nebula is not a remnant of a type I supernova which have velocities ~ 20000 km s^{-1} during the supernova outburst. On the other hand, the light curve reconstructed from ancient observations does not disagree with that expected from a type I supernova [212]. Also, it has been shown that the present day remnant of Kepler's supernova, a well established type I supernova, has an expansion velocity ~ 1400 km s^{-1}, similar to that of the Crab [311]. Thus, the supernova producing the nebula was probably of type I although this assignment is still uncertain.

Photographic measurements [307] of PSR0531+21 yield a proper motion of $\mu_\alpha = -0\rlap{.}''009$ yr^{-1} and $\mu_\delta = -0\rlap{.}''002$ yr^{-1} with uncertainty $\pm 0\rlap{.}''003$ yr^{-1}. At 2 kpc the transverse velocity of PSR0531+21 is 87 km s^{-1}. The velocity expected from differential galactic rotation and the Sun's peculiar velocity is only about 25 km s^{-1} (Reference 270). The relative transverse velocity of pulsar and nebula may be as large as 130 km s^{-1} (Reference 307). Thus, the pulsar appears to have acquired a large velocity during the supernova event.

When the expansion of the nebula is extrapolated backwards, it is found that the nebula has been accelerated since the supernova event. Assuming uniform acceleration, the current energy input is 2–4×10^{38} erg s^{-1} (Reference 308). The total synchrotron radiation produced by the nebula amounts to 7–12×10^{37} erg s^{-1} at the assumed distance of 2 kpc [116, 277]. The rotational energy loss rate of a 'standard' neutron star rotating with the period of PSR0531+21 and slowing down at the same rate as the pulsar is $\sim 10^{38}$ erg s^{-1} (References 81 and 89). It was the discovery of the hitherto unexplained energy source of the Crab nebula that layed to rest the white dwarf hypothesis of pulsars and established the rotating neutron star hypothesis. In fact, it may be possible to use the energy requirements of the Crab nebula to obtain information on the equation of state of matter at nuclear densities. In particular, very soft equations of state seem to be ruled out [28]

Further details on the Crab nebula and PSR0531+21 may be found in Reference 66.

The evidence associating PSR0833−45 with the Vela supernova remnant is reviewed in Reference 155. The identification rests mainly on positional and distance agreement. Also, the timing age of the pulsar is 11000 yr, the same order of magnitude as the estimated age of the remnant, 30000–50000 yr (Reference 278). The distance to the remnant is about 500 pc and its estimated mass is characteristic of a type II super-nova event [209]. The radio remnant has a shell-like structure similar to most other remnants, but different from the Crab nebula which is brightest at the center. Optical nebulosity associated with the remnant consists entirely of filaments [311], again dissimilar from the Crab nebula but similar to most other remnants. X-rays have been detected from the remnant [145, 274] but it is not yet clear whether pulsed X-rays from PSR0833−45 have been detected (cf. Section 6).

The pulsar is $0°\!.6$ from the center of the nebula which implies a relative transverse velocity of ~ 450 km s^{-1} if the distance and age given above are correct.

The remnant appears to be immersed in a decaying H II region – the Gum nebula – created by the supernova event [6, 32]. The distance to the center of the nebula is about 460 pc, its radius is about 360 pc, and it subtends about $90°$ in the sky. The Gum nebula is responsible for an anomalously high electron density (~ 0.18 cm^{-3} within the nebula) in the direction of PSR0833−45.

PSR0611+22 is $0°\!.6$ from the center of IC443 and its timing age is 90000 yr. IC443 is a remnant of type II [277] with an estimated age of 60000 yr [213]. The estimated distance of IC443 is 1.4 kpc. If PSR0611+22 is at this distance, the required value of $\langle n_e \rangle$ is 0.07 cm^{-3}, somewhat larger than usually assumed. However, there is an H II region in this direction [64, 276], so the discrepancy is probably not serious. IC443 is an asymmetric radio shell with optical filaments coinciding with maximum radio brightness [211, 311]. PSR0611+22 is located just beyond the edge of the shell on the side opposite the optical filaments, suggesting an asymmetric supernova event. The radio brightness contours of the shell show an indentation near the pulsar suggesting that the pulsar has pushed out a cavity around itself [271]. There is only one piece of evidence which argues against the association of PSR0611+22 and IC443. It has been suggested [311] that IC443 is associated with HD43836. If this were true its distance would be only 500 pc, its age would be much less than 60000 yr, and it would probably not be associated with PSR0611+22. However, in view of all the other evidence, it seems more likely that IC443 is associated with the pulsar rather than HD43836. With the parameters given above, the relative transverse velocity of the pulsar and remnant is about 150 km s^{-1}.

On somewhat more shaky ground is the association of G296.8−0.3 with PSR1154−62. As far as I am aware, the only discussion of this identification is contained in Reference 169. G296.8−0.3 is too faint to appear in the catalogs of supernova remnants but is shown to have the shelll structure and non-thermal spectrum characteristic of supernova remnants. The distance and age of G296.8−0.3 are 4 kpc and 25000 yr with large uncertainties. PSR1154−62 is slightly outside the shell and $0°\!.2$ from the center. An asymmetry in the shell points approximately towards the pulsar. The distance to the pulsar is estimated as 3 to 6 kpc. The timing age has not been measured. The estimated probability of a chance positional agreement is 10%. If the association is real then the transverse velocity is ~ 550 km s^{-1}. The uncertainties here are so large, however, that more data, particularly the age of PSR1154−62, are needed before the association can be regarded as established.

The association of PSR2021+51 with HB21, a type II remnant [277], is discussed in Reference 319. The pulsar has 21 cm absorption at the same velocity as an emission feature originating in HB21 and must be behind the remnant. It is argued that the pulsar cannot be much farther than HB21 on the basis of its dispersion measure. The angular separation of the remnant and pulsar is $3°5$ and their ages are $\sim 10^6$ and $\sim 3 \times 10^6$ yr. The transverse velocity implied by these data is ~ 25 km s^{-1}. While no discrepancies are apparent in these data, the large angular separation makes the probability of a chance superposition very high (~ 1 if it is remembered that supernova remnants and pulsars are concentrated towards the galactic plane). More data are needed to confirm this association.

The association [35] of PSR1919+21 with the extended source G55.7+3.4 appears unlikely. The age and distance of G55.7+3.4 are 5×10^4 yr and 7 kpc (References 101 and 159). The timing age of PSR1919+21 is 1.6×10^7 yr. At 7 kpc, the mean electron density to PSR1919+21 would be more than an order of magnitude smaller than the standard value (cf. Section 5). The galactic latitude of PSR1919+21 is $3°5$ so it is hard to see how the electron density could be so small.

In addition to the specific pulsar supernova remnant pairs dicussed above, 28 pairs were found in a search of known remnants and pulsars [309]. However, it was shown that this is just the number expected on the basis of chance superpositions [48, 169].

To summarize, there are two definite pulsar supernova remnant associations, one probable association and two possible associations. Three of the associations are with type II remnants, one with a type I remnant, and one remnant type is unknown. The three possible and probable associations suggest high velocities were acquired by the pulsar during the supernova event. These associations are with young ($\lesssim 10^5$ yr) pulsars. Of the pulsars not associated with a supernova remnant and having a measured timing age, only PSR1747–27 is younger than PSR0611+22. Thus it is not surprising that more associations haven't been found. What is surprising is that of the half dozen or so galactic supernovae (mostly type I) recorded in the last 2000 yr (see e.g. Reference 277) the only remnant to contain a pulsar is the Crab nebula. Perhaps not all supernova make pulsars – certainly not observable pulsars.

That neutron stars should be formed by supernovae was suggested [13] as long ago as 1934 and with the discovery of PSRs 0531+21 and 0833−45, this idea has become entrenched in the conventional wisdom. However, the exact details of the process of neutron star formation are still uncertain (see chapter by S. Colgate in this volume) and it is not clear that all types of supernovae can make pulsars. Hence it is important to ask which type of supernovae produce pulsars.

Pulsars are concentrated towards the galactic plane (Figure 6) with an angular distribution similar to that of supernova remnants and OB stars, but with a somewhat larger scale height [96, 109, 231]. Longer period (presumably older) pulsars have a slightly broader distribution in galactic latitude (Figure 6) than do short period pulsars [109, 222]. A possible explanation of these distributions is that pulsars arise from both type I and type II supernovae [222]. However, the estimated birth rate of pulsars in the Galaxy is about one per thirty years [109] in agreement with the

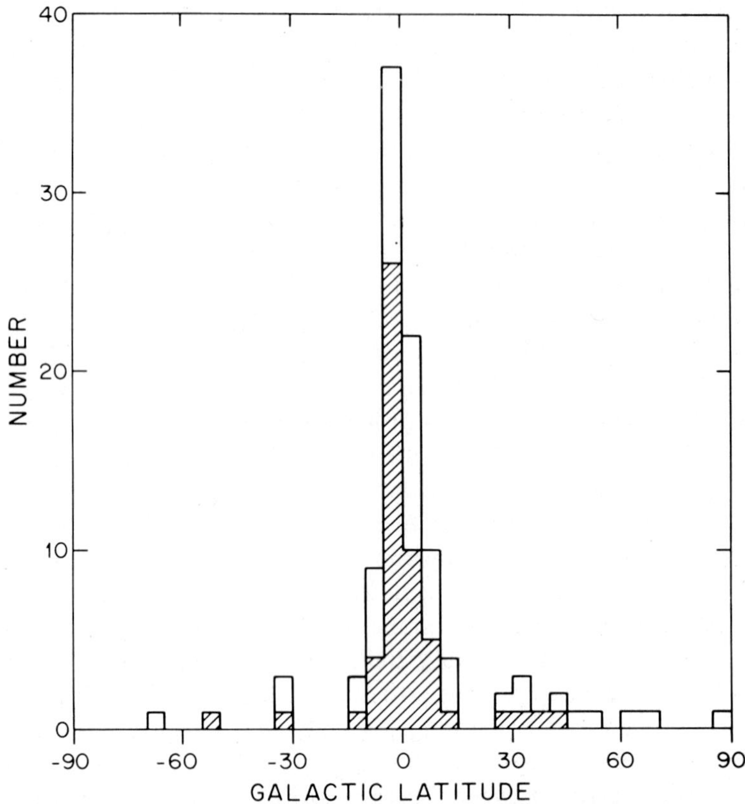

Fig. 6.   Distribution of pulsars in galactic latitude. Shaded area represents 53 pulsars with period
less than 0.655 s. Clear area represents the 52 pulsars with period greater than 0.655 s.

estimated rate of type II supernovae and several times the rate of type I supernovae
[144, 277, 298]. Thus most pulsars (PSR0531 + 21 is an exception) may be produced
in type II supernovae. Under this hypothesis the height distribution and period-
latitude correlation can be understood on the basis of the high velocities ($\gtrsim 100$ km-
s^{-1}) acquired by the pulsar during the supernova event [109, 231]. The high velocities
also fit well with observations of runaway stars [231] and the fact that no pulsars are
in binary systems (Section 8).

5. Pulsar Distances and Distributions

Figure 7 shows the angular distribution of pulsars in galactic coordinates. Pulsars
are concentrated towards the plane and there are no strong concentrations in latitude.
Hence, most pulsars are more than several hundred parsecs and less than several
kiloparsecs distant.

The usual method of obtaining pulsar distances is to assume that dispersion mea-

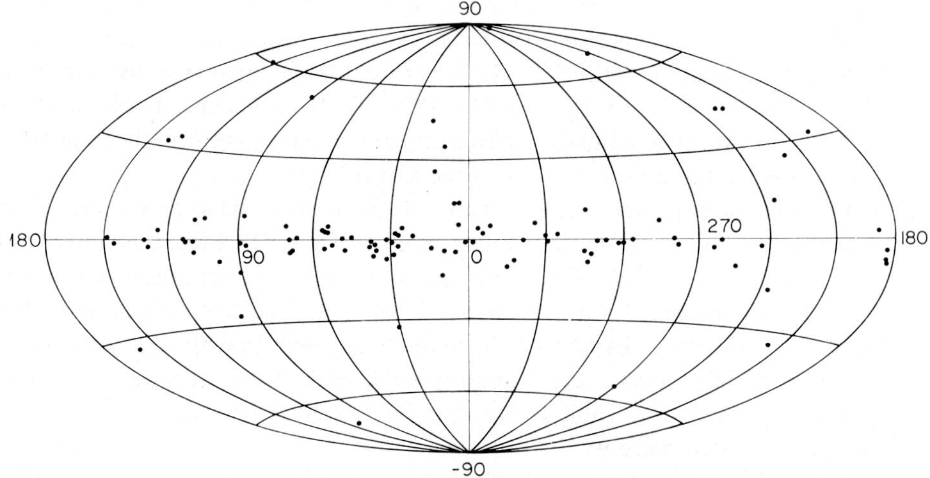

Fig. 7.   Galactic distribution of pulsars.

sure and distance are proportional, with $\langle n_e \rangle$ the proportionality constant. Unfortunately, there are large variations in $\langle n_e \rangle$ so this method is semi-statistical at best.

In a few cases pulsars are associated with supernova remnants (Section 4) at known distance and in other cases 21 cm absorption, together with a model for galactic rotation can provide distance limits. These pulsars are listed in Table II where the refer-

TABLE II

Neutral hydrogen absorption distances

PSR	Dist. (kpc)	DM (pc cm^{-3})	$\langle n_e \rangle$ (cm^{-3})	Ref.
$0531 + 21$	2	56.8	0.028	
$0611 + 22$	1.4	96.7	0.069	
$0833 - 45$	0.5	63	0.126	
$0329 + 54$	>1	26.8	<0.027	98
$0355 + 54$	>1.5	57.0	<0.038	96
$0740 - 28$	>1.5	73.8	<0.018	97
$1718 - 32$	>1	120	<0.120	96
$1749 - 28$	>1	50.9	<0.051	98
$2016 + 28$	>1	14.2	<0.014	75
$2020 + 28$	>2	24.6	<0.012	96
$2021 + 51$	>1.1	22.6	<0.021	97
$0329 + 54$	<1	26.8	>0.027	108
$0355 + 54$	<2.5	57.0	>0.023	96
$0740 - 28$	<2.5	73.8	>0.030	97
$1749 - 28$	<1	50.9	>0.051	108
$1818 - 04$	<1.5	84.5	>0.056	97
$1822 - 09$	<1.5	19.3	>0.013	96
$1929 + 10$	<1	3.2	>0.003	95
$2021 + 51$	<1	22.6	>0.023	95

ence column indicates the source of the 21 cm distance limit. The first group are those pulsars probably or certainly associated with supernova remnants. The second group of pulsars show 21 cm absorption and the lower limit is the kinematic distance of the absorbing feature. The limit for PSR2021 + 51 comes from placing it behind HB21. Pulsars in the last group failed to show absorption by a known feature; the upper limit comes from placing the pulsar in front of the feature.

In addition to those pulsars listed in Table II, many more have been searched for 21 cm absorption (see any of the references listed in Table II). Either the pulsar was not detected at 21 cm because of its steep spectrum, or no absorption or absorption only at the local standard of rest was observed. This fact has led to the proposal that the absorbing hydrogen is distributed in small clouds such that the probability of a particular line of sight intersecting a cloud is small [99]. If this hypothesis is correct the distance upper limits in Table II are meaningless. Arguments against this hypothesis are given in Reference 96.

Even if the third group of pulsars in Table II is ignored, the message is clear: there is a wide variation in $\langle n_e \rangle$. Probably the best that can be expected from dispersion measure distances is an accuracy of a factor of three.

In the early days of pulsar astronomy, the electron density was taken to be 0.1 to 0.2 cm^{-3}. These high values were first questioned in Reference 110 where 0.01 cm^{-3} was suggested. While this value is probably too low, much early work indicated that the density was considerably less than 0.1 cm^{-3} (References 56, 106, 208, 233, and 280). Considerations [57] of ionization and heating within H I regions, absorption of low frequency radiation, and the distribution of pulsars yielded $\langle n_e \rangle = 0.03$ cm^{-3}. The same result was obtained [96] from an investigation of data similar to that given in Table II.

Distance estimates for 54 pulsars have been obtained from a detailed study [232, 234] of the contributions to $\langle n_e \rangle$ from H II regions around galactic clusters and OB stars and from electrons in H I regions. The average and median values of density obtained in this study were 0.05 and 0.04 cm^{-3}, respectively. These distances appear to be the most accurate in the literature and are listed in Table I where available. For pulsars in Table II, the distance to the remnant or the 21 cm distance has been listed in Table I. (Where both an upper and lower limit exist, the average was used.) All other distances in Table I were calculated with $\langle n_e \rangle = 0.04$ cm^{-3}. Considerable extrapolation is involved since most recently discovered pulsars have high dispersion measure.

With 105 known pulsars it is possible to preform some crude statistical investigations. Earlier results may be found in References 109 and 167, among others. The present discussion is based on the data in Table I and does not closely follow the earlier investigations.

Figure 8 is a logarithmic plot of the cumulative number of pulsars versus dispersion measure. If dispersion measure is strictly proportional to distance and if all pulsars at a given distance are detected, then the $\log N - \log DM$ plot should have a slope of two or three if pulsars are uniformly distributed in a disk or sphere, respectively [128].

Lines with these slopes are indicated in Figure 8. Below $DM = 16$ pc cm^{-3} the statistics are very poor and the data are consistent with either slope. If the break in the curve near $DM = 16$ pc cm^{-3} is due to the fact that an increasing number of higher dispersion pulsars are too faint to be detected and if pulsars are distributed in a disk, then the present sample of pulsars is complete out to $DM = 16$ pc cm^{-3} or 400 pc. Within this radius there are 16 pulsars, and taking the radius of the Galaxy as 15 kpc, a total of 22 000 pulsars would be contained in the Galaxy. Assuming that

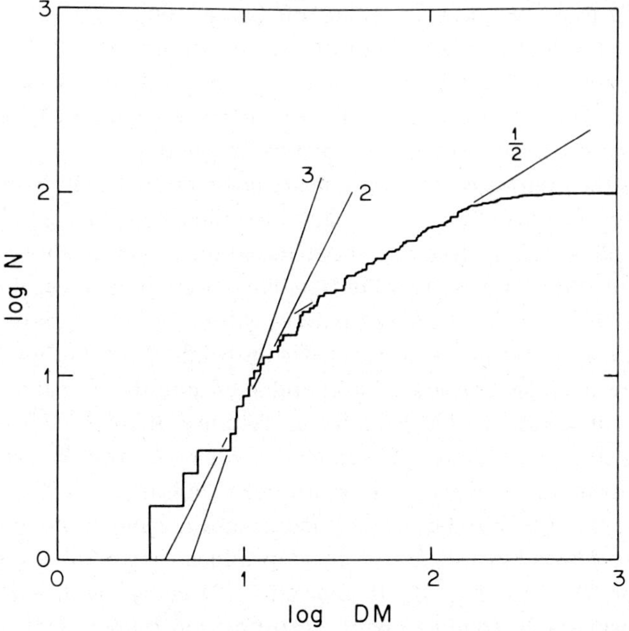

Fig. 8.   Log N − log DM relation. The cumulative number of pulsars with dispersion measure less than DM is plotted against DM.

pulsar radiation is beamed in a cone, this number should be increased by about a factor of ten to account for pulsars we are unfavorably placed to observe [120]. If a pulsar is observable for 10^7 years (roughly the largest age estimates in Table I), a galactic birth rate of one per 45 years is obtained. Considering all the uncertainties, this number agrees remarkably well with the rate of one per 30 years obtained in a previous investigation [109].

Returning to Figure 8, the data with dispersion measure between 17.4 and 158 pc cm^{-3} have a slope of one-half. If this slope arises because a greater proportion of intrinsically faint pulsars are missed as the distance increases, then the luminosity function is $dN/dL \propto L^{-7/4}$, where L is the intrinsic radio luminosity. Because the efficiency of conversion of rotational energy loss into radio emission varies widely among different pulsars (see Section 6) it is not clear how the radio luminosity function should be related to the evolution of pulsars.

If the breaks in the $\log N - \log DM$ curve near 17.4 and 158 pc cm^{-3} arise because most pulsars nearer than the first break are bright enough to be detected and most farther than the second break are too faint to be detected then the range of most pulsar luminosities is confined to two orders of magnitude. (Pulsars beyond 158 pc cm^{-3} may be exceptionally bright, may have anomalously high dispersion, or may have been detected during a particularly active period.) From Table I, it is seen that pulsars are detected if their flux density is above 10 to 100 mfu. Thus the maximum radio luminosity is 3 to 30×10^{28} erg s^{-1}. Here it is assumed that the spectral index is -2 above 100 MHz, that the spectrum is cut off below 100 MHz and the radiation is beamed into one-tenth of a sphere. With the same assumptions, PSR0531+21 is the most luminous pulsar in Table I and has a radio luminosity of 6×10^{29} erg s^{-1}. Next in order are PSRs 1831−03 (4×10^{29} erg s^{-1}) and 0329+54 (3×10^{29} erg s^{-1}). Thus, Figure 8 appears to present a consistent picture of the data.

Figure 6 gives the impression that pulsars are concentrated slightly below the plane. Earlier, this effect was more pronounced and attributed to a strongly dispersing layer centered on the plane [171]. However, the required pulsars lying above the plane have never been found and as more distant pulsars have been discovered, the effect has decreased. In fact, the average latitude of pulsars within 20° of the plane is $-0.6\pm0°.6$. The effect (if it exists at all) may be attributed to the height, Z, of the Sun above the plane. In this case, the average latitude of a cylinder of pulsars at distance R is $-Z/R$. Weighting the contribution of a cylinder at distance R by $R^{-1/2}$ as indicated by Figure 8, the average latitude is $-Z/(R_1 R_2)^{1/2}$ where R_1 and R_2 are the minimum and maximum distances at which pulsars are seen. Pulsars with dispersion measure less than 17.8 pc cm^{-3} tend to be at high galactic latitude and do not contribute to the average latitude. Those beyond 158 pc cm^{-3} produce only a small correction. With these limits for R_1 and R_2, $Z = 0.56$ pc cm^{-3}. Finally, with $\langle n_e \rangle = 0.04$ cm^{-3}, $Z = 14$ pc, in surprisingly good (probably fortuitous) agreement with 10 to 20 pc obtained from studies of the local mass distribution [74]. With a sample of 41 pulsars and a different technique, a similar result was obtained in Reference 109.

While necessarily very crude because of the limited data available, the above discussions indicate that the present data form a consistent picture and the adopted electron density, 0.04 cm^{-3}, is not grossly in error even though large pulsar to pulsar variations exist.

6. Pulsar Flux Density Spectra

The measurement of pulsar flux density spectra is complicated by scintillation as well as intrinsic variability. Also at low frequencies, pulse boadening may completely wash out the pulse. Ideal measurements would consist of observations over a wide frequency range for times much longer than the scintillation time scale and with bandwidths much wider than the decorrelation bandwidth.

To date, only thirteen pulsars have been simultaneously observed at different frequencies for the purpose of determining their spectra [19, 22, 180, 239, 256]. The largest set of observations [22] contains measurements of ten pulsars at six frequencies between 250 and 8085 MHz.

In addition to these observations, all available flux density measurements for 27 pulsars have been compiled, generous errors estimated, and phenomenological spectra derived [281]. The results of the simultaneous observations as well as this latter compilation may be summarized as follows: Near 400 MHz, the spectrum may be described as a simple power law with spectral indices ranging from -0.7 to -3.3 and average spectral index -1.62. Some pulsars show a low frequency cut-off which typically occurs between 100 and 200 MHz. At frequencies between 1 and 2 GHz, the spectrum may steepen. The average spectral indices below and above the high frequency break are about -1 and -2, respectively [281].

As an example, the spectrum of PSR1929+10 and the phenomenological best fit spectrum are shown in Figure 9. The spectrum is cut-off below 100 MHz, has spectral index -1 between 100 MHz and 2.8 GHz, and has spectral index -1.8 above 2.8 GHz. Since the dispersion measure of PSR1929+10 is only 3 pc cm^{-3}, the low frequency cut-off is not due to pulse broadening but must be intrinsic to the source. The low frequency cut-off is not observed in all pulsars. For example, PSR0950+08 has been observed down to 53 MHz with no evidence of a cut-off [281].

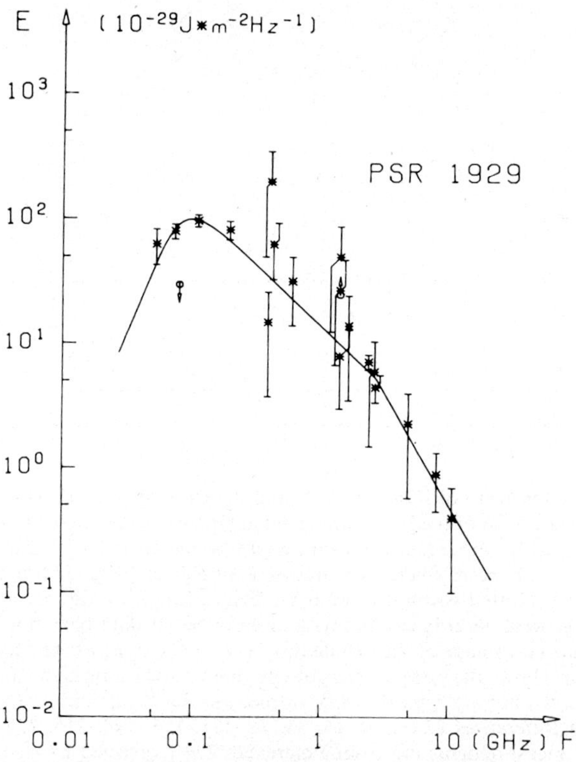

Fig. 9.   Flux density spectrum of PSR1929 + 10, reproduced from Reference 281 by permission. The units of the vertical scale are 10^{-29}Jm^{-2}Hz^{-1} and may be converted into milliflux units by dividing by the pulsar period, 0.2265 s. © 1973, Springer and Verlag Publ.

PSR0531 + 21 is remarkable in that it has been observed from 10 MHz up to about 2 GeV. The flux density spectrum of this pulsar is shown in Figure 10 along with the spectrum of the Crab nebula. Original sources are cited in the figure caption. Below about 100 MHz, the pulsed component decreases due to pulse smearing by scattering in the interstellar medium (Section 3). The lower of the two spectra below 100 MHz refers only to the pulsed component, while the upper spectrum refers to the compact source which is coincident with the pulsar and believed to be smeared out pulsar radiation. Data on the compact source were obtained with very long baseline interfero-

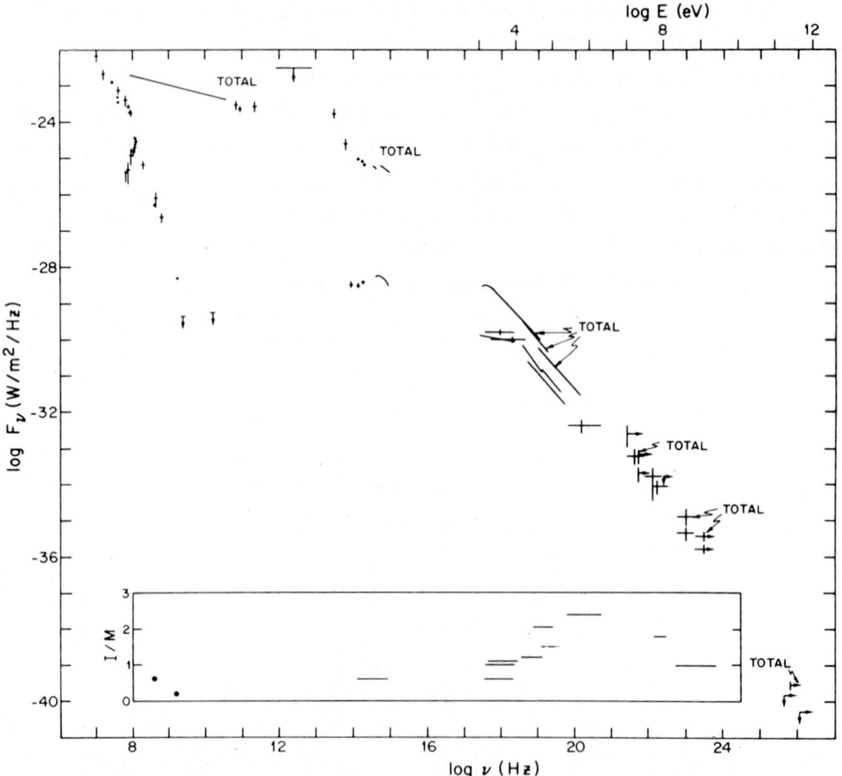

Fig. 10. Flux density spectrum of PSR0531 + 21 and the Crab nebula. Measurements marked total refer to the combined emission of the pulsar and nebula. Other measurements refer to the pulsar only. In the sub-millimeter, X-ray, and γ-ray regions the length of the horizontal bar indicates the region of sensitivity of the detector. In cases where measurements were given by specifying the best fit spectrum, that spectrum has been plotted without error bars. Typical uncertainties are 10% to 20% in both height and slope. In general, data in conflict with more or better data have not been used, nor have pulsed fluxes given as a percentage of the nebula flux been used except where necessary to fill a gap in the spectrum. The inset shows the ratio of interpulse to main pulse energies as a function of frequency. Error bars are not shown but are typically 20% or more in the X-ray and γ-ray regions. Radio data has been taken from References 7, 11, 14, 26, 33, 76, 131, 186, 198, 215, 239, 312, 332, and 333. Optical and infrared measurements are from References 224 (corrected to $V = 16.5$), 2, 24, 25, 85, 221, and 223 and have been corrected for 1.6 mag. of visual absorption with a standard reddening curve (References 206 and 324). X-ray and γ-ray data are from References 3, 31, 34, 36, 38, 69, 78, 82, 84, 88, 100, 116, 119, 147, 174, 200, 225, 229, 244, and 287.

metry or interplanetary scintillation measurements. Data from 90 MHz to 1660 MHz are consistent with a spectral index of -3.6 while below 90 MHz the spectral index is in the range -1.6 to -1.8. The radio spectrum is discussed in greater detail in Reference 76 where a spectral index of -2.5 fits all the data if some curvature is allowed.

The pulsar emission peaks in the optical or near infrared. The absorption correction is uncertain by ±0.2 magnitudes [206] so the location of the maximum is somewhat uncertain. The optical emission is smooth, showing no absorption or emission lines [178, 224]. The deep minimum between the radio and optical regions in the pulsar spectrum probably indicates that at least two emission mechanisms are operating. On the other hand, the smooth nebula spectrum is produced by synchrotron radiation over its entire range [277].

In the soft X-ray region, 2–10 keV, the pulsar spectral index is -0.2 ± 0.1 (Reference 84). The turnover below 2 keV is due to absorption by interstellar matter with a density of 0.5 to 0.8 hydrogen atoms per cm^3, assuming standard abundances. This density is somewhat higher than that found by 21 cm measurements [38, 84].

Higher energy measurements are summarized in Reference 200. Spectral indicies of -1.08 ± 0.03 and -1.18 ± 0.03 are found to describe the pulsar and nebula plus pulsar emission from 10 keV to 2 GeV. The upper limits above 200 GeV indicate that a break must occur between 2 and 200 GeV.

The inset in Figure 10 shows the ratio of energies contained in the interpulse and main pulse. The ratio is about 0.6 in the optical and soft X-ray regions, increases to about 2.4 at 1 MeV and decreases at higher energies. These variations have been questioned [200] on the grounds of poor signal-to-noise, although inspection of published pulse shapes does seem to indicate a more prominent interpulse at hard X-ray energies.

PSR0531$+$21 is the only pulsar to have been detected at other than radio frequencies. The second fastest pulsar, PSR0833$-$45 might be thought to offer the best chance of optical or X-ray detections. Upper limits in the optical have been obtained by several groups [83, 127, 155, 322], and the best limit is $m_v > 24.8$ (Reference 37). A report [114] of detection of soft X-ray pulsations has not been confirmed [55, 216]. Reported [115] detection of hard X-ray pulsations is doubtful because the observed period did not agree with the radio period. For several other radio pulsars, upper limits on optical and X-ray emission are given in References 156, 244, and 249, among others.

While there are no widely accepted theories of the pulse emission mechanism, a few general remarks may be in order.

Figure 9 indicates that the flux received from PSR1929$+$10 is 3.2×10^{-15} erg cm^{-2} s^{-1}. Using the distance in Table I and assuming the emission is beamed into 10% of the sphere, the luminosity of PSR1929$+$10 is 9×10^{26} erg s^{-1}, near the lower limit of the range discussed in Section 5. If the pulse width is larger than the light travel time across the emission region, then the emissivity is greater than 3×10^{10} erg cm^{-2} s^{-1}, the electric field is greater than 2×10^5 V m^{-1} and the radio brightness temperature at 90 MHz is greater than 10^{22} K. Actually, the emitting region could be as small as a few kilometers or less in which case these numbers would increase by 10^6. Thus, not only is the emission nonthermal, it is highly coherent [182]. That is, it must be produced by groups, often called bunches, of charges, all moving coherently. The wavelength of the high frequency break near 1 GHz in pulsar spectra has been interpreted [90, 182] as the coherence length or bunch size, about 30 cm.

If the moment of inertia [258] of PSR 1929 + 10 is 2×10^{44} g cm^2, then the period and slowing down rate in Table I imply a rotational energy loss rate of 8×10^{32} erg s^{-1}, about 10^6 times more power than is going into the radio pulses. This power may go into very low frequency electromagnetic radiation [226] or into acceleration of charged particles to relativistic velocities [91]. In fact, it was suggested quite early that pulsars should provide a copious supply of cosmic rays [89], and we have already seen (Section 4) how this suggestion is confirmed in the case of PSR 0531 + 21.

7. Pulses

When attention is directed to the pulses themselves, a tremendous variety of phenomena is encountered. This section is primarlily a listing of these phenomena.

A distinction must be made between individual and average pulses. A sequence of pulses shows great variability from one pulse to the next, but when several hundred pulses are averaged, a stable pulse shape energes. Average pulse shapes may range from simple, symmetric pulses (e.g. PSRs $1642 - 03$, $1749 - 28$, Figure 11) to two component pulses (e.g. PSRs $0525 + 21$, $1133 + 16$) to very complex shapes with at least five components (e.g. PSR $1237 + 25$).

While the average pulse shape is generally stable there are a few exceptions to this rule. In what is known as mode switching, the average pulse changes shape for a few hundred periods and then reverts back to its former shape. This phenomena has been observed in only two pulsars, PSRs $1237 + 25$ (References 18 and 179) and $0329 + 54$ (References 124, 125, and 179). In PSR $1237 + 25$, intervals between the abnormal mode are several thousand periods while the abnormal mode lasts from tens to hundreds of periods. The switch between modes is very abrupt, apparently occurring within a few periods. The normal and abnormal modes are shown in Figure 12. The abnormal mode is characterized by a change in the relative intensity of the five components. Changes in polarization are not nearly as drastic as intensity changes. A change in the intensity ratio of the 'outriders' (Figure 11) from $I_I/I_{IV} \approx 0.5$ to $I_I/I_{IV} \approx 2$, characterizes the abnormal mode of PSR $0329 + 54$ at 408 MHz. (The leading outrider is component I, the trailing outrider is component IV.) These changes are illustrated in Figure 13 which indicates that the abnormal mode occurs about 15% of the time. At higher frequencies, the normal mode is similar to that at 408 MHz, but the abnormal mode shows considerable variation with frequency. At 2.7 GHz, a new component V appears between the primary component (III) and component IV [124], while at 10.7 GHz the abnormal mode contains only components III and IV with about equal

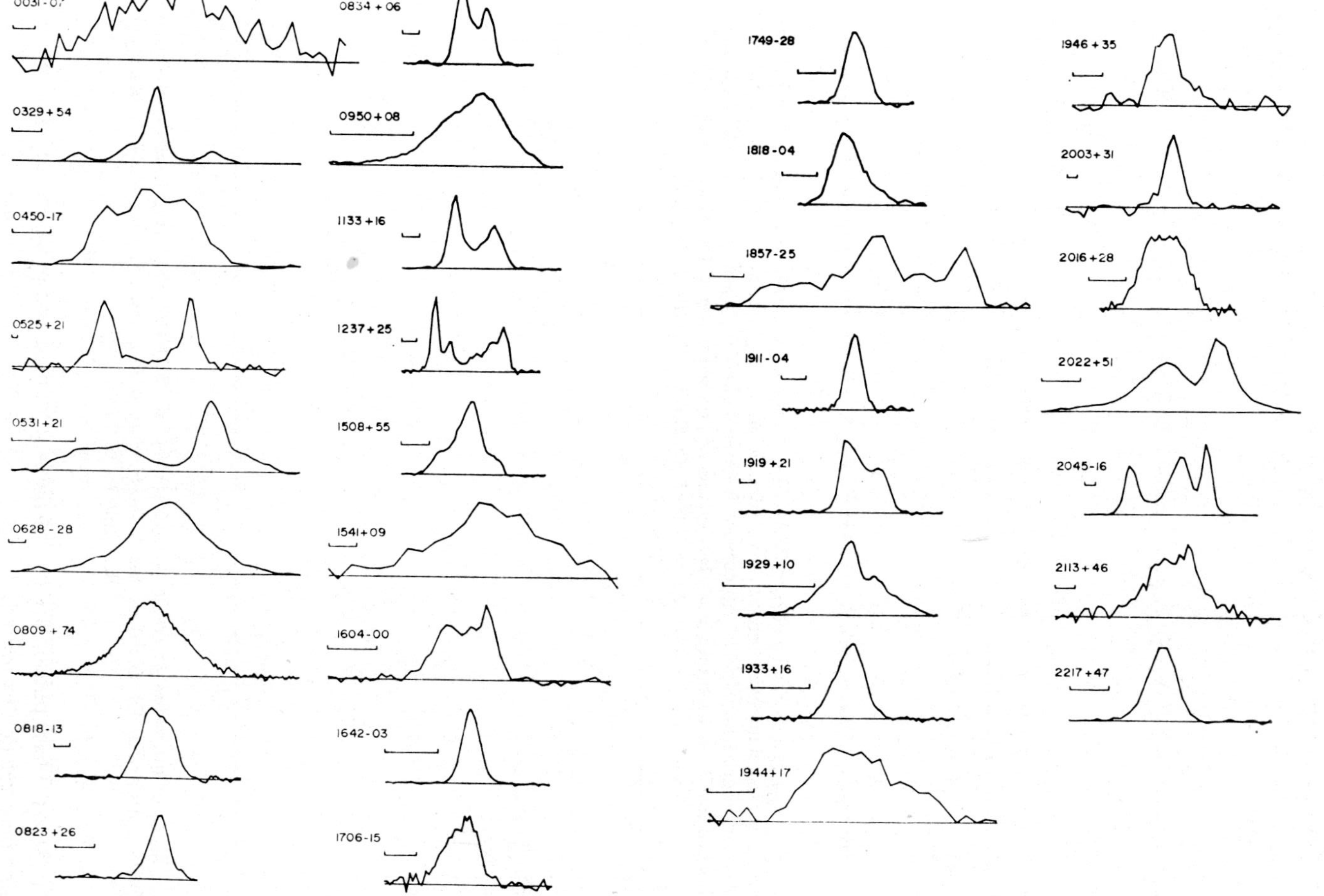

Fig. 11. Assorted average pulse shapes, reproduced from Reference 184 by permission. In each case, the same length along the horizontal scale is the same fraction of the period. The horizontal bar indicates 1 ms for PSR0531 $+21$ and 10 ms for all other pulsars. The signal to noise ratio varies from pulsar to pulsar and may be estimated from the scatter in the wings. Note that some pulsar names have changed slightly since this figure (and others) was constructed. Published by the *Monthly Notices Roy. Astron. Soc.*; © 1971, Blackwell Scientific Publ., Ltd.

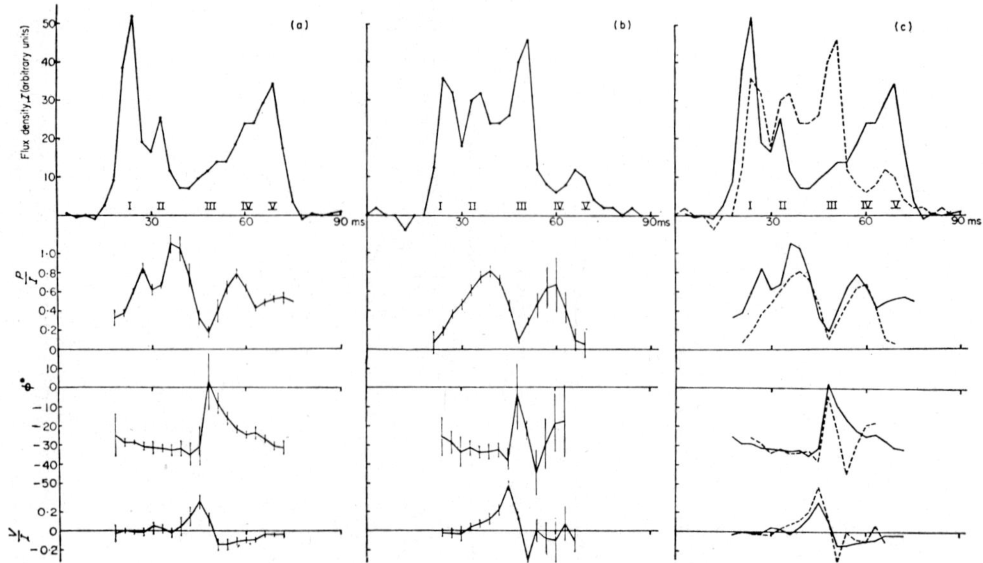

Fig. 12. Normal and abnormal modes of PSR1237 + 25, reproduced from Reference 179 by permission. Top curve is total intensity, second curve is fractional linear polarization, third curve is position angle of linear polarization, and fourth curve is fractional circular polarization. Normal mode is at left, abnormal mode in the center, and the two are superposed at right. Published by the *Monthly Notices Roy. Astron. Soc.*; © 1971, Blackwell Scientific Publ., Ltd.

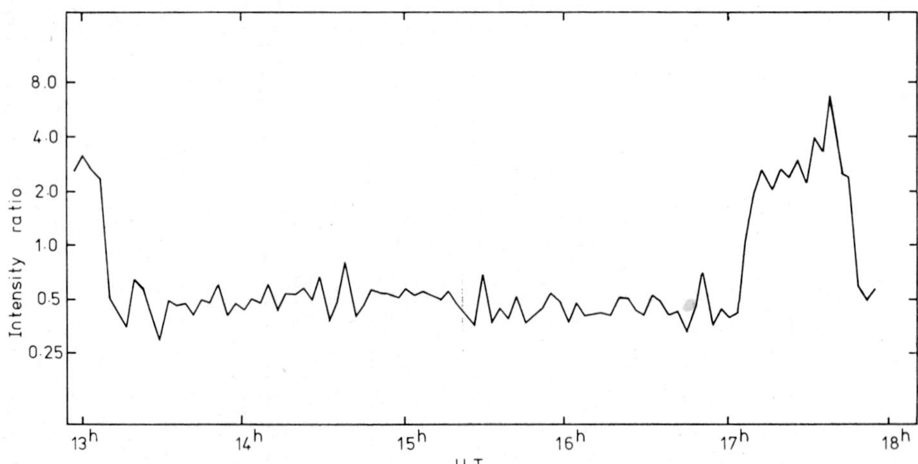

Fig. 13. Ratio of intensities of components I and IV of PSR0329 + 54, reproduced from Reference 179 by permission. Published by the *Monthly Notices Roy. Astron. Soc.*; © 1971, Blackwell Scientific Publ., Ltd.

intensity [125]. These pulsars are two of the best studied and more examples of mode switching may yet be discovered.

Pulse width vs pulse period is plotted in Figure 14 for the 85 pulsars with measured pulse width (Table I). The lines indicate widths of 1% and 10% of the period. Most

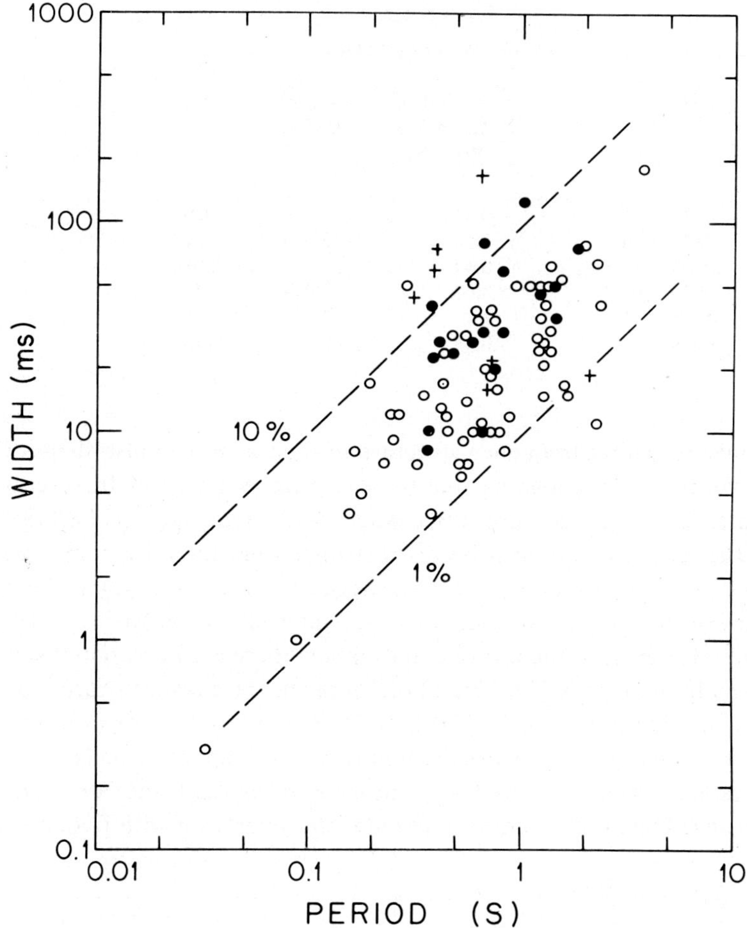

Fig. 14.   Pulse width vs pulse period. Open circles, $DM < 100$ pc cm^{-3}; filled circles, $100 < DM$ < 200; crosses. $200 < DM$. At high dispersion measure, substantial pulse broadening may occur.

pulsars fall between these lines and one may take 3% as a representative value. The lack of dependence of duty cycle on pulse period has been interpreted in favor of the rotating neutron star model of pulsars [89].

While the emission of most pulsars is confined to a small fraction of the period, five and possibly six pulsars are known to have a weak interpulse approximately midway between successive main pulses. Data on these pulsars are summarized in Table III.

Another example of emission outside the main pulse is weak emission extending between the main pulse and interpulse as in the optical pulse of PSR0531+21 (Figure 15) or between the interpulse and main pulse as in PSR0950+08 where the emission between the pulses is never less than 0.1% of peak emission [21]. Still another example of extended emission is the broad wings of the main pulse of PSR1541+09 which extend over half the pulse period [21].

                        EDWARD J. GROTH

TABLE III

Pulsars with interpulses

PSR	Separation M to I (Periods)	Strength I/M	Ref.
$0531 + 21$	0.41	~ 0.7	239
$0823 + 26$	0.50	~ 0.005	21
$0904 + 77$	0.44	~ 0.2	301
$0950 + 08$	0.60	~ 0.018	253
$1055 - 51$	?	?	316
$1929 + 10$	0.52	~ 0.02	188

Some pulsars exhibit frequency structure in their average pulse shapes (Not to be confused with pulse broadening due to scattering, Section 3.) In several cases the relative intensity of the components changes with frequency. In PSR1919+21, for example, the second component becomes stronger and the third component weaker, relative to the first component, as the frequency increases [202]. The frequency dependence of the main pulse, precursor, and interpulse of PSR0531+21 has already been noted. However, in the infrared and optical there are no variations greater than 3% of the main pulse peak [24, 321, 323]. There may be a color variation of about one percent in the interpulse region [219]. In PSR1133+16, the separation of the two components decreases with increasing frequency [51], approximately as $v^{-1/4}$. These phenomena have been observed in a number of pulsars and are summarized in Reference 184. Data on variation of component separation with frequency for seven

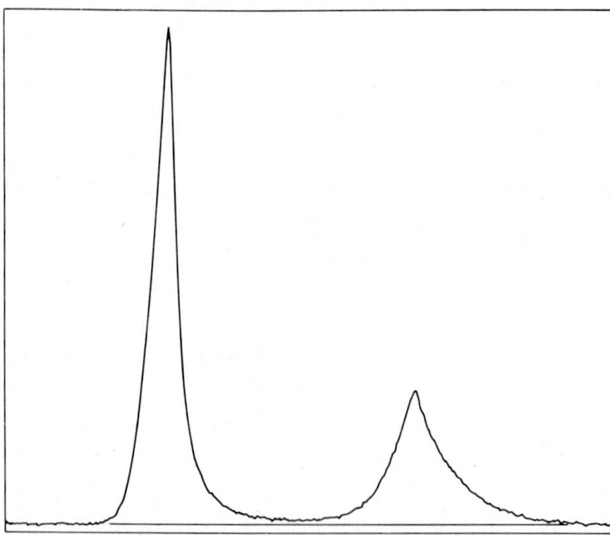

Fig. 15.   Optical pulse shape of PSR0531+21. The horizontal axis is one period long.

pulsars are shown in Figure 16. The separation depends weakly on v and may either increase or decrease with increasing v.

The polarization properties of the average pulses may be described as follows. Many pulsars show a high degree of linear polarization – up to 100% in some cases. Often the amount and position angle of linear polarization vary with position across the pulse. Sometimes these variations are smooth, in other cases, chaotic. Average circular polarization is not so common, and when observed is usually less strong than the linear polarization. In some cases the circular polarization changes hand during the pulse.

The average polarization properties of PSR0833−45 at 400 MHz are shown in Figure 17. The structure in Figure 17 is constant over the frequency range 300 to

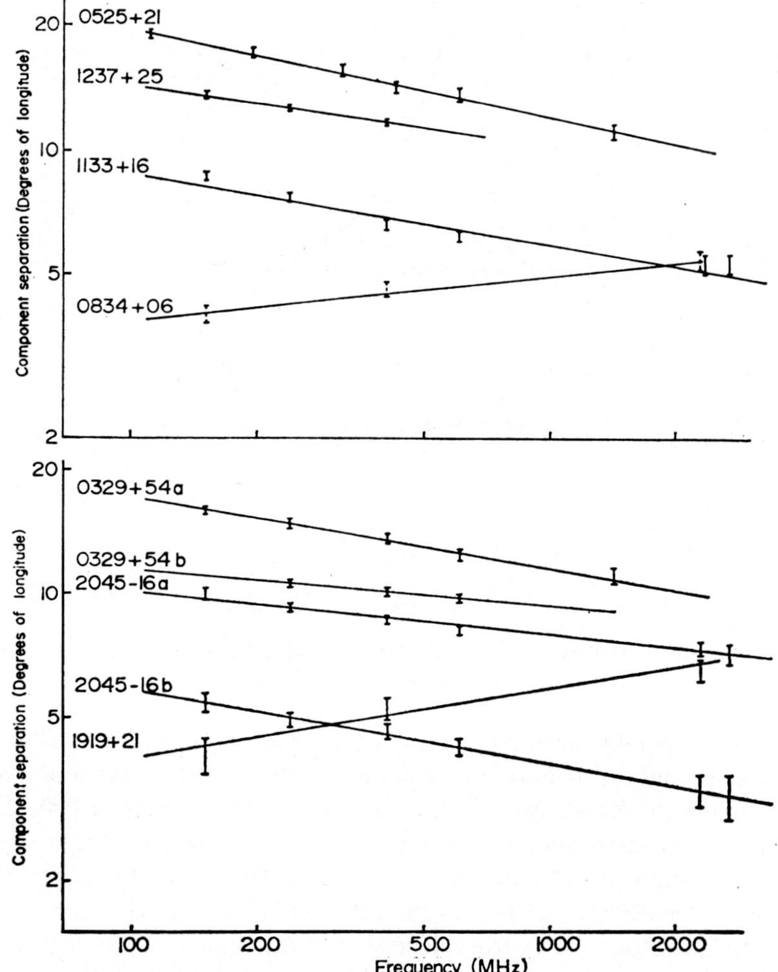

Fig. 16.   Component separation versus frequency for several pulsars, reproduced from Reference 184 by permission. One period = 360° of longitude. Published by the *Monthly Notices Roy. Astron. Soc.*; © 1971, Blackwell Scientific Publ., Ltd.

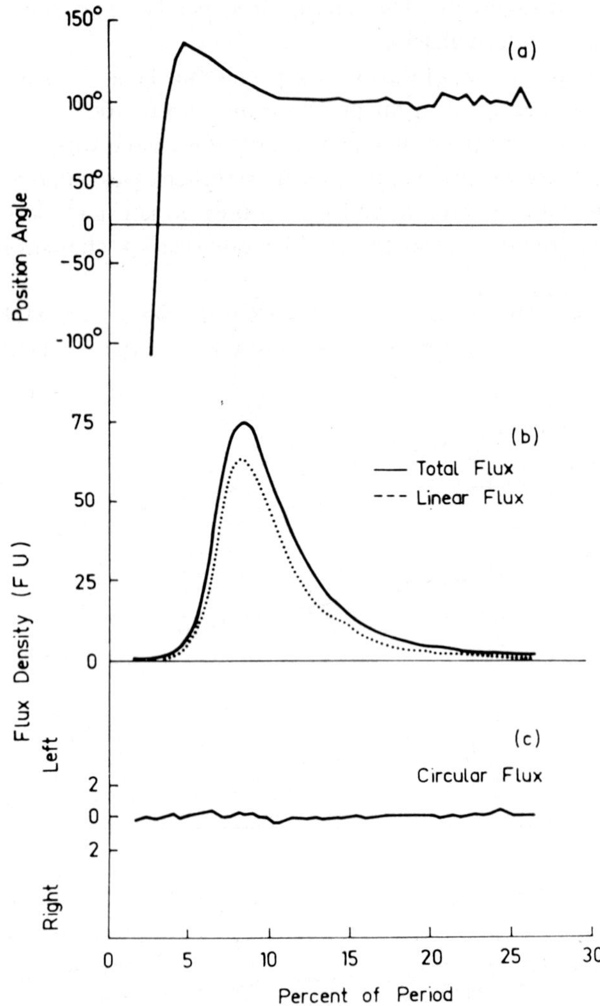

Fig. 17.   Polarization of PSR0833 − 45 at 400 MHz, reproduced from Reference 201 by permission.
© 1972, Gordon and Breach.

1400 MHz. The smooth variation of position angle after the maximum is interpreted
by a number of authors as indicating rotation of the magnetic field component perpendicular to the line of sight as the pulsar rotates beneath the observer [148, 235, 236].
On the basis of the smooth rotation, it was suggested that the line of sight must intersect the pulsar within $10°$ of a magnetic pole [235]. The angle of polarization of the
optical pulses of PSR0531 + 21 rotates smoothly and about the same amount in both
pulses. The polarization varies from near zero at the centers of the pulses to about
20% in the wings. Circular polarization is less than 0.1% [43]. With the simple model
mentioned above, it is found that the rotation axis must be very nearly in the plane of
the sky [42, 158, 321]. In the radio, the precursor of PSR0531 + 21 is almost 100%

polarized while the degree of polarization of the main pulse and interpulse is the same order as the optical polarization [188]. Apparently, no rotation of position angle has yet been observed at radio frequencies.

The simple model of position angle rotation can be fit to several other pulsars [153, 217]. However, not all pulsars show a smooth rotation of position angle: two counter-examples are shown in Figure 18.

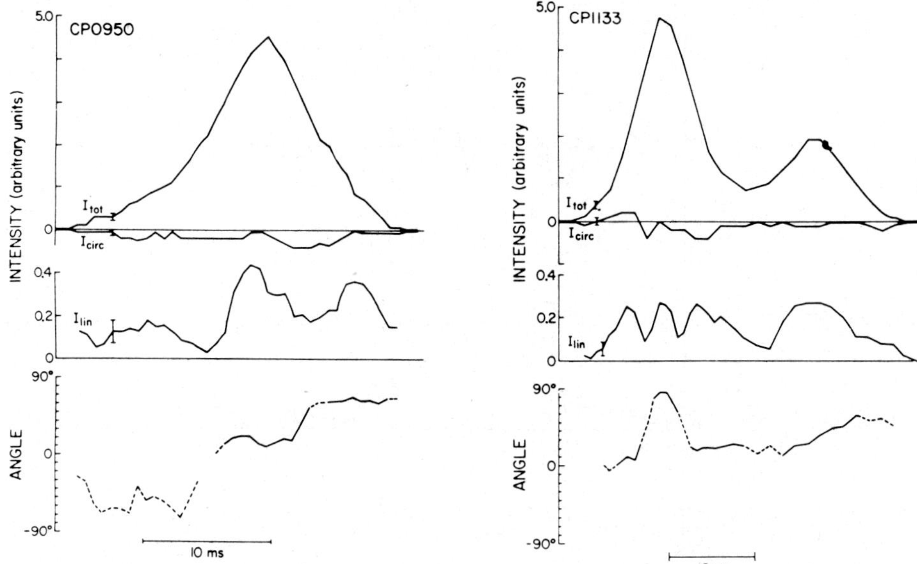

Fig. 18.   Polarization of PSRs 0950 + 08 and 1133 + 16, reproduced from Reference 73 by permission. Position angles indicated by a solid line have error less than 10° while those indicated by a dashed line have errors between 10° and 30°. Published by the University of Chicago Press; © 1969, University of Chicago Press.

Average circular polarization is less strong than linear polarization, but in a survey of 20 pulsars, 12 had circular polarization greater than 10% at some time during the pulse [104]. Quite often the circular polarization changes hand during the pulse as shown in Figure 19. PSR1749 − 28 is an example of a pulsar in which the polarized component is predominantly circular and reaches 30% [73].

In some cases, the degree of polarization decreases with increasing frequency [273]. For four pulsars [196] the polarization is independent of frequency up to a critical frequency, beyond which it decreases as v^{-1}. Typical critical frequencies are of the order of 1 GHz. More observations are needed before it is known how widespread this phenomenon is.

Before turning our attention to individual pulses where we will encounter a host of time variable phenomena, it should be pointed out that no definite time variations have ever been observed in the average pulse shape or polarization properties of any

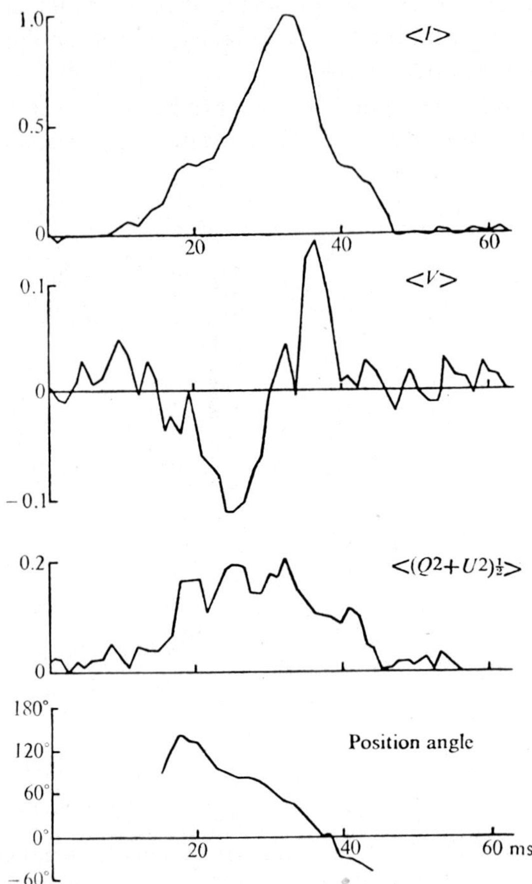

Fig. 19. Polarization of PSR1508 + 55, reproduced from Reference 104 by permission. V is the circularly polarized component and $(Q^2 + U^2)^{1/2}$ is the linearly polarized component.
© 1971, MacMillan Journals, Ltd.

pulsar.* (We ignore mode switching – when the pulsar returns to the normal mode it is the same normal mode as before.) Even the spin-ups (Section 8) in PSRs 0531 + 21 and 0833 − 45 did not cause a detectable change in the average pulse shape or polarization of these pulsars [29, 148, 158, 235]. Many authors (References 188, 192, and 235, among others) argue that the great stability of the average shape and polarization indicates that the emission region is near the surface of the neutron star where the magnetic field configuration is fixed by the solid stellar surface.

For strong pulsars it is possible to record and study individual pulses. Typically

---

* Early reports [40, 41] of time variations in the optical pulse shape of PSR0531 + 21 were due to incorrect experimental technique. A report [42] of small changes in the average polarization of the optical pulse of PSR0531 + 21 with no change in average shape, between 1969 and 1970, is unconfirmed. Data in the two years were obtained with different techniques and later measurements [80] agree with the 1969 measurements. Until more data is available, this question is still open.

these subpulses* are narrower than the average pulse shape, symmetrical, and highly elliptically polarized. They may occur at various positions in the 'window' defined by the average pulse shape and they are highly variable from one pulse to the next. In some cases the variability is ordered (marching subpulses – see below). In other cases the variability is more or less random – but not completely random because the Stokes

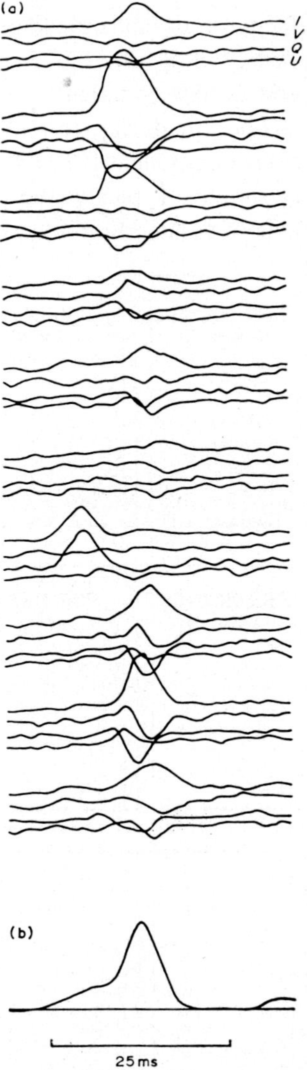

Fig. 20.   Successive subpulses from PSR0329 + 54, reproduced from Reference 184 by permission. The Stokes parameters *I*, *V*, *Q*, and *U* are the intensity, circularly polarized intensity and two orthogonal linearly polarized intensities, respectively. The curve at the bottom is the average pulse shape. Published by the *Monthly Notices Roy. Astron. Soc.*; © 1971, Blackwell Scientific Publ., Ltd.

* Here subpulse refers to the individual pulse(s). Some authors have used the term subpulse to refer to the components of the average pulse shape.

parameters of the subpulses must average to those of the average pulse. Typically, subpulses from a given pulsar have a characteristic width which is not strongly dependent on frequency. Subpulse phenomena are discussed by a number of authors [4, 39, 67, 182, 238]. A summary is given in Reference 184. The latter authors regard subpulses as the basic unit of pulsar emission and believe the average pulse shape is imposed after the radiation is formed. They are led to a model in which the beaming is provided by the relativistic velocity of an isotropic source corotating near the light cylinder. This model is not specified in detail.

Subpulses from PSR0329+54 (which apparently do not march) are shown in Figure 20. Most of the subpulses occur during component III of the average pulse shape but one occurs in the weaker component II just preceding the main component. All subpulses are qualitatively similar. Typically the degree of linear and circular polarization is high and the plane of polarization may rotate during the subpulse. An

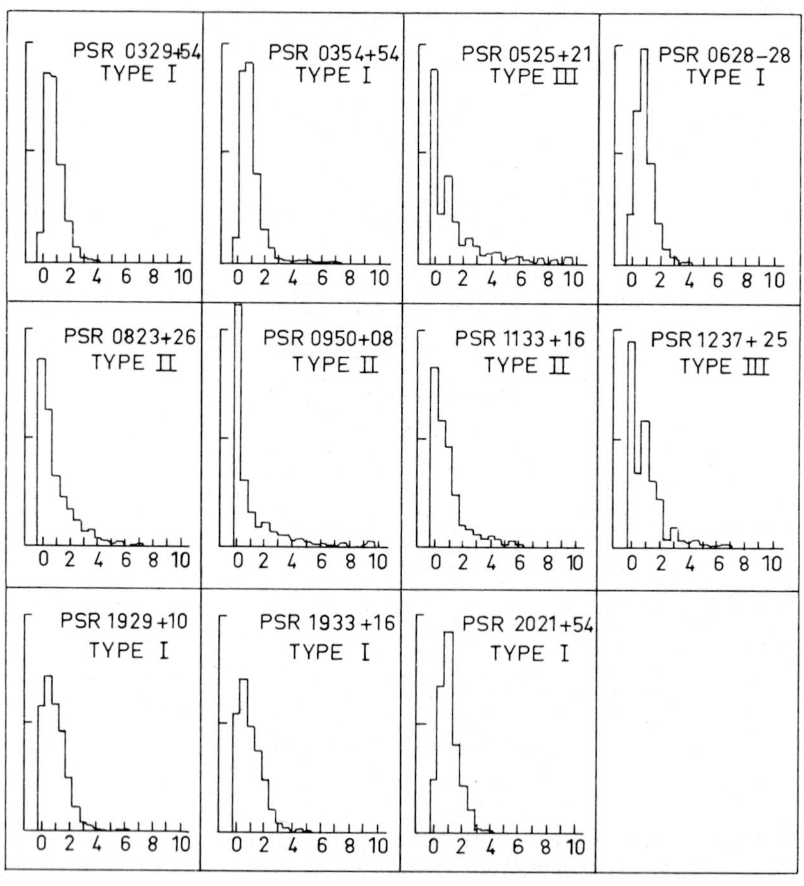

Fig. 21.   Pulse energy histograms, reproduced from Reference 126 by permission.
© 1974, Springer and Verlag, Publ.

example (PSR0809 + 74, Figure 23) exists in which the planes of polarization of the subpulses and average pulse rotate in opposite directions [290].

The strong pulse to pulse intensity variations can be seen in Figure 20. In many pulsars the radiation occurs in bursts lasting from a few to a few hundred pulses. These variations are made more apparent by histogramming the number of occurrences of a given pulse energy versus energy in a pulse energy histogram [10, 126, 143, 289, 325, 326] as shown in Figure 21. By separately normalizing histograms of a few hundred pulses and summing the individual histograms, the effects of scintillation are largely removed. In this way, three basic types of pulse energy histograms are observed. Type I has a maximum at positive intensity, type II has a maximum at zero intensity and type III has maxima at both zero and positive intensities [126]. In most cases a long tail extending to high intensities is observed. There is some indication that histogram type is correlated with period and with pulse morphology. For example, there is some tendency for type I histograms to be associated with short period pulsars, type II with intermediate periods, and type III with long periods [126].

PSR0531 + 21 provides an extreme example of the long tail in the pulse energy histogram. Usually, integration times of several minutes are required before this pulsar can be seen above the Crab nebula background. However, about once every five minutes (10000 periods) the pulsar emits a 'giant' pulse about 1000 times stronger

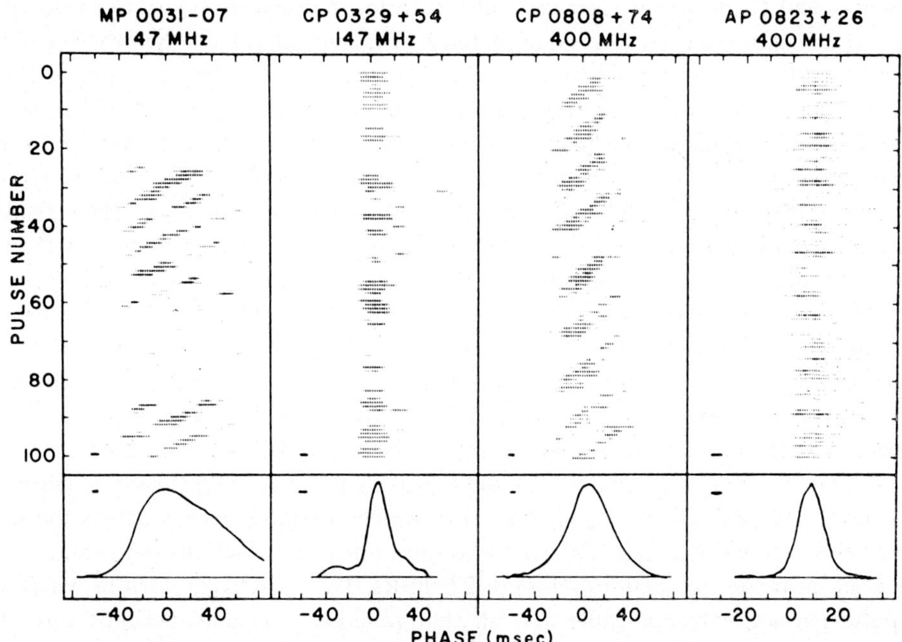

Fig. 22.   Marching subpulses reproduced from Reference 302 by permission. The density in each horizontal line shows the intensity during one pulse of the pulsar. Successive horizontal lines show successive pulses. PSRs 0031 − 07 and 0809 + 74 have well defined subpulse paths. PSRs 0329 − 54 and 0823 + 26 do not have marching subpulses but have a quasi-periodic intensity modulation. The curves at the bottom show the average pulse shapes and the bars indicate the time resolution.
Published by the University of Chicago Press; © 1971, University of Chicago Press.

than the average pulse. Giant pulses are weakly linearly polarized ($\sim 20\%$); about 92% occur during the main pulse and 8% during the interpulse [9, 103, 241]. Giant pulses with hand changing circular polarization have been observed [117]. Giant pulses are strongly correlated over 1 MHz bandwidths and weakly correlated (if at all) over 100 MHz bandwidths [118]. Since 1 MHz is much larger than the scintillation decorrelation bandwidth, giant pulses are probably not a propagation effect. Giant pulses are not present in the optical radiation [157].

We now turn to a very interesting phenomenon– that of marching (or drifting, depending on one's temperament) subpulses, first discovered [68] in PSR1919+21. In the simplest version, a subpulse appears at the trailing edge of the average pulse window. One period later, the subpulse has moved earlier in time, and the next period, still earlier in time, and the next period, still earlier, and so on, until it marches through the entire pulse. Some examples are shown in Figure 22. Note that subpulse structure and intensity are highly correlated within a subpulse sequence but vary markedly from one sequence to the next. It appears that subpulses within a sequence are produced by emission from the same group of particles, while successive sequences involve different groups.

The subpulse sequences are approximately periodic and are characterized by two periods. Referring to Figure 22, P_2 measures the horizontal separation of the subpulse sequences and P_3 measures their vertical separation.* When necessary to avoid confusion, the usual pulsar period is designated by P_1. Typically, P_2 is the order of milliseconds and P_3 is the order of several P_1. P_1, P_2, and P_3 are generally incommensurate.

A common method of analyzing marching subpulses is to record a time series of pulse intensities, Fourier transform the time series, and produce its power spectrum, which has come to be known as a fluctuation spectrum (References 15, 20, 121, 143, 160, 161, 176, 204, 272, 282, 285, 286, 297, 302, and 305). In more sophisticated analyses, separate spectra are formed for several phases of the average pulse window. Since the sample rate is one sample per P_1, fluctuation spectra provide information on periodic components with periods longer than $2P_1$, i.e. P_3, and no information (ignoring aliasing) on periods shorter than $2P_1$, i.e. P_2. However, analysis [20] of the phases of the Fourier transforms as a function of position in the pulse may provide a drift rate which together with P_3 can give information on P_2.

About twenty pulsars are known to exhibit marching subpulses or some form of periodic modulation of the pulses. The modulation can range from 'linear' (Figure 22, PSRs 0031−07 and 0809+74; Figure 23) in which subpulses march across the window in either direction in a straight path, to 'non-linear' in which the paths may curve and even bifurcate, to 'pattern' (Figure 22, PSRs 0329+54, 0823+26) in which no subpulse paths are recognizable and all that remains is a quasi-periodic intensity fluctuation. The terminology used here is that suggested in Reference 20.

PSR1237+25 provides an example of a nonlinear subpulse path [21]. The average pulse shape (Figures 11 and 12) has five components designated I through V starting

---

* Some pulsars are modulated at several P_3. Some authors have used P_4 to designate the second period. Most authors seem to prefer multiple P_3 rather than P_3 and P_4.

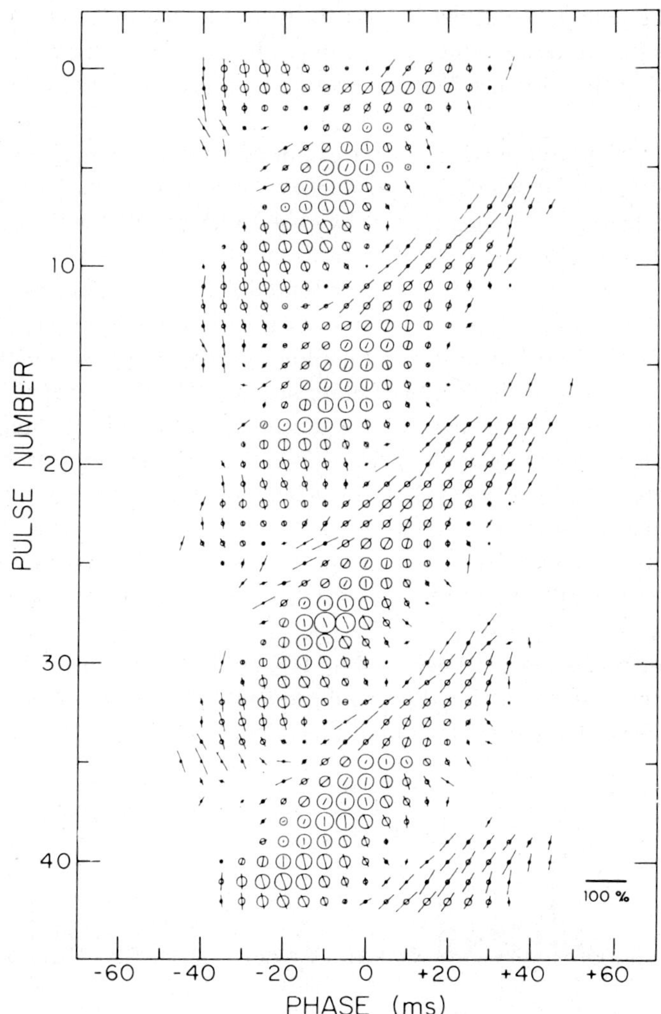

Fig. 23.   Marching subpulse polarization in PSR0809 + 74, reproduced from Reference 303 by permission. Diameter of circles is proportional to intensity, length of line is proportional to percentage linear polarization, and orientation of line is the position angle of polarization.
© 1971, Gordon and Breach.

at the leading edge. Subpulses first appear in the middle of component I. They then split in two with one marching toward the leading edge of component I while the other marches backwards through components I and II, disappears in component III, reappears in component IV, and marches to the trailing edge of component V. P_3 associated with this sequence is $2.82\,P_1$. A different sequence of subpulses marches forward in component III, with $P_3 = 48P_1$.

Polarization can be correlated with the subpulse sequence. Data on PSR0809 + 74 are shown in Figure 23. The polarization of the subpulses is quite large but the average pulse is only 6% polarized [303]. In this case, the polarization is correlated with the subpulses rather than the average pulse. Not only does the polarization rotate with time during the subpulse but also from one appearance of the subpulse to the next.

The stability of the subpulse modulation can be very low with $Q \approx 1$, or very high with $Q \approx 100$–1000. For example, the fluctuation spectrum [20] of PSR2016+28 shows a broad feature extending from 0.1 to 0.4 cycles per P_1. On the other hand, the marching subpulses of PSR0809+74 are quite stable [44, 227] with $Q \approx 100$. In fact, in this pulsar an extremely stable P_3 is observed for several hundred periods, after which a pulse drops out (a null, see below), the phase jumps by almost a full cycle, an P_3 is stable again until the next null. With allowance for the phase jumps, Q is greater than 700 and is limited by the length of observation [227]. Broad peaks in the fluctuation spectra may show fine structure. For example, the peak in the fluctuation spectrum of PSR0834+06 at 0.46 cycles per P_1 has $Q \approx 40$. Higher resolution [286] reveals two closely spaced peaks with $Q \approx 250$. It has been suggested that there are two fundamental frequencies with the modulation switching from one to the other [20]. An effect that may be related is observed in PSR0031$-$07. Typically $P_3 \approx 7P_1$ and $P_2 \approx 56$ ms. Occasionally P_3 increases to $13P_1$ or decreases to $4P_1$ while P_2 remains constant [139].

In a phenomenon known as pulse nulling [16, 161] one or more pulses in an otherwise strong burst may be missing. Pulse nulls in PSR1133+16 are illustrated in Figure 24, where the burst character of the radiation is clearly seen. About every forty periods,

Fig. 24. Pulse nulls in PSR1133+16, reproduced from Reference 161 by permission. Height of the vertical bars is proportional to pulse intensity. Published by the University of Chicago Press; © 1970, University of Chicago Press.

six pulses drop out. In several pulsars, the nulls are known to be quasi-periodic with $P_3 \sim 100P_1$. Nulls may be responsible for the zero intensity peaks in pulse energy histograms.

Parameters describing the subpulse phenomena show no change over long time scales or over widely separated observing frequencies [20].

Several authors [20, 302] have attempted to classify pulsars on the basis of pulse and subpulse morphology and to relate these classifications to other properties. For example, pulsars with single component average pulse shapes and no subpulse modulation tend to have short periods ($\lesssim 0.75$ s). Single component pulsars with linear marching subpulses tend to have longer periods. It is still too early to draw any firm conclusions from these trends as the current classifications are rather loose and may not even be internally consistent.

More details on marching subpulses and periodic modulation of pulse intensities may be found in the summary given in Reference 20.

As we have seen, pulsars exhibit large variations on time scales ranging from pulse to pulse through minutes and pulsar radiation often occurs in bursts. On longer time scales, minutes to hours, interstellar scintillation plays a major role in the intensity fluctuations. We now turn to very short and very long time scales.

The first report [52] of sub-millisecond structure in PSRs 0950+08 and 1133+16 was later shown to be noise produced by an improper choice of receiver bandwidth and time constant [72]. However, this second paper did find evidence for sub-milli-second (i.e. sub-subpulse) structure in these two pulsars. High time resolution [113] reveals 1.4 ms wide notches in the average pulse shape of PSR 1919+21.

Recently, it has been shown that features of characteristic width ~ 175 μs and ~ 600 μs appear in the subpulses of PSRs 0950+08 and 1133+16, respectively [112]. In addition, it has been claimed [111, 112] that these pulsars exhibit subpulse structure which is unresolved at 7 μs. However, I believe the reality of the 7 μs features is still in doubt. The problem is similar to that mentioned above, namely, the bandwidth was 125 KHz and the RC time constant was 7 μs yielding a bandwidth time constant product of about 1. Hence the intensity ($\propto E^2$) in each 7 μs interval has a chi-squared distribution with two degrees of freedom. It is well known that such a distribution has a very long tail and one expects to see unresolved spikes. The question then becomes, are there more spikes than expected or are they larger than expected. To answer, one must know the expected value of the intensity at the location of the spike. The spikes are superposed on the sub-millisecond structure discussed above and it is difficult to evaluate just what mean level should be used in a statistical test. Arguments that the spikes disappear if the signal is not dedispersed do not appear relevant because the sub-millisecond structure also disappears. Thus, the existence of 7 μs structure appears to be an open question.

At the other extreme, pulsars exhibit intensity variations on time scales of days to years [45, 122, 123, 137]. As an example, the intensity of PSR 0950+08 was enhanced by a factor of 200 for a few days [230]. Observations of five pulsars over several years revealed fluctuations on all time scales up to the observation period. Characteristic times, defined as the width of the zero lag peak of the autocorrelation function, ranged from one to two months and typical maximum to minimum intensity ratios were about ten to one [137]. No periodic structure is observed in the long term fluctuations [122, 137]. These fluctuations are usually regarded as an intrinsic effect rather than a scin-tillation effect because the requirements on the electron density irregularities are too severe. A large cloud might produce the fluctuations via a large scale refraction effect. However, a refraction effect is probably ruled out because fluctuations at 81 and 151 MHz are well correlated [123].

8. Pulse Timing

Pulse timing observations consist of measuring as accurately as possible the pulse arrival times. Topocentric arrival times are converted to barycentric arrival times with a solar system ephemeris. The barycentric arrival times are then directly related to the rotation frequency of the pulsar along with any Doppler shift due to the relative motion of the pulsar and the solar system.

The phase, or pulse arrival time measured in units of the pulse period, is given by,

$$\phi_1 - \phi_2 = \int_{t_1}^{t_2} v(t)\, dt, \tag{11}$$

where ϕ is the phase and $v = P^{-1}$ is the pulse repetition frequency.* If t_1 and t_2 are measured arrival times, then $\phi_2 - \phi_1$ must be exactly an integer. Usually v is specified as a function of t depending on several adjustable parameters which are determined by a fit of the function to the data via Equation (11).

In order to apply Equation (11) successfully, observations must be sufficiently closely spaced that there is no ambiguity in determining which integer goes with which arrival time. Occasionally, cycle miscounting has led to erroneous period estimates.

If the pulsar coordinates are in error, the barycentric reduction introduces a spurious one year sinusoidal term in the arrival times with a maximum amplitude of 240 μs per 0.1 arcs position error. Hence, pulsar positions may be determined to a few tenths of an arc second by timing observations. Measurements [197] of PSR1133+16 over four years have detected a proper motion of $\mu_\alpha = 0''.00 \pm 0''.10$ yr^{-1}, $\mu_\delta = 0''.58 \pm 0''.22$ yr^{-1} yielding a transverse velocity of 360 km s^{-1} at 130 pc and again supporting the idea that pulsars are high velocity objects.

After the discovery of pulsars, it was hoped that timing observations might be used to measure the gravitational redshift of Earth based clocks as they move in and out of the Sun's field due to the Earth's elliptical orbit [132, 146]. Another suggestion was that pulsars could be used to measure the gravitational path delay effect near the Sun [50]. Neither of these hopes has been realized. In the first case, the timing data cannot be used to separate a redshift effect from a position error effect. Independent methods of estimating the position are not sufficiently accurate to provide a significant test of the equivalence principle. In the second case, radar bounce measurements are more accurate than pulsar measurements. Hence, both effects, as predicted by standard general relativity, are usually included in the barycentric reductions.

After correction for position errors, the phases of most pulsars have a quadratic time dependence, indicating that v is changing at a constant rate. In all cases v decreases – evidence in favor of rotation, rather than vibration or revolution, as the basic clock mechanism.

The timing positions, accurate periods, and period rates of change listed in Table I

---

* In this section, v is the pulse repetition frequency rather than the observing frequency.

have been determined by the methods outlined above. The first report of slowing down is contained in Reference 60 and more recent results may be found in References 141, 191, 203, and 247.

Doppler shifts due to motion in a binary orbit with period less than four years would be easily detected in the timing data. None have been found and pulsars are not members of binary systems.

The expected relation between frequency and rate of change of frequency is

$$\frac{\mathrm{d}v}{\mathrm{d}t} = - Kv^{n}, \tag{12}$$

where, in simple models K and n, the 'braking index', are constants related to the details of the mechanism of angular momentum loss. For example, in the magnetic dipole radiation model [226], the braking index is three and K depends on the magnetic moment and moment of inertia of the neutron star.

Integration of Equation (12) yields t_0, the time since the pulsar was spinning very rapidly,

$$t_0 = \frac{-1}{n-1} \; v\left/\frac{\mathrm{d}v}{\mathrm{d}t}\right. = \frac{1}{n-1} \; P\left/\frac{\mathrm{d}P}{\mathrm{d}t}\right. \tag{13}$$

If pulsars were born spinning only twice as fast as their present rate, if $n=3$, and if effects not described by Equation (12) are ignored, then t_0 differs from the true age of the pulsar by only 25%, a difference comparable to the uncertainty produced by the uncertainty in n. As most pulsars have periods of the order of one second and there is a pulsar spinning thirty times faster, the ages listed in Table I are probably accurate to better than a factor of two.

One way to estimate n is to plot $\mathrm{d}v/\mathrm{d}t$ vs v as shown in Figure 25. While there is considerable scatter in the data, the solid line with slope 3.6 approximately fits the data. The braking index of PSR0531+21 is 2.5 (see below) and the dashed line shows the track this pulsar will follow according to Equation (12). While there is no accepted explanation of the discrepancy between the measured index of PSR0531+21 and that derived from Figure 25, the data appear to indicate that n is in the range 2 to 4 and that Equation (12) may be adopted as a working hypothesis.

Timing data on PSR0531+21 indicate that ϕ in Equation (11) is a cubic function of time and PSR0531+21 has a measurable second derivative related to the braking index by,

$$n = v \frac{\mathrm{d}^2v}{\mathrm{d}t^2}\left/\left(\frac{\mathrm{d}v}{\mathrm{d}t}\right)^2\right. . \tag{14}$$

Published values of the braking index range from 0 to 5 (References 29, 30, 70, 133, 220, 228, 248, and 255). These results are not due to measurement errors as all observers are in substantial agreement [134]. Rather, different segments of data are described by different cubic polynomials and hence different braking indices.

Actually, on a fine scale, the data are not described at all well by a cubic polynomial

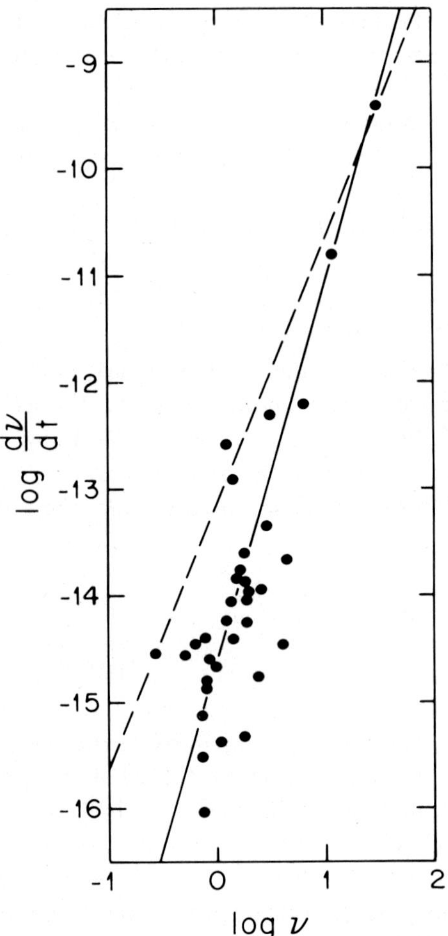

Fig. 25. dv/dt vs v for 34 pulsars with measured slowing down rates. Solid line corresponds to a braking index of 3.6 and approximately fits the data. Dashed line shows the path of PSR0531 + 21 if its braking index continues at the presently observed 2.5.

as seen in Figure 26. The one to ten millisecond structure in Figure 26, while large compared with measurement error ($\sim 10\ \mu s$), is small compared to the cubic term (~ 2 s after one year). The fine structure (called timing irregularities, restless behavior, or wobbling by some authors) is responsible for the scatter in published values of n, particularly those obtained from short segments of data.

In Reference 29, the fine structure is qualitatively interpreted as a random process superposed on an otherwise smooth slowing down with intrinsically constant braking index. My own unpublished work quantitatively confirms these conclusions. The random process acts on the pulse frequency with an ω^{-2} power spectrum. A model for such a process is a random walk in the pulse frequency. The data are not consistent with a random walk in either ϕ or dv/dt. The strength of the random walk and the

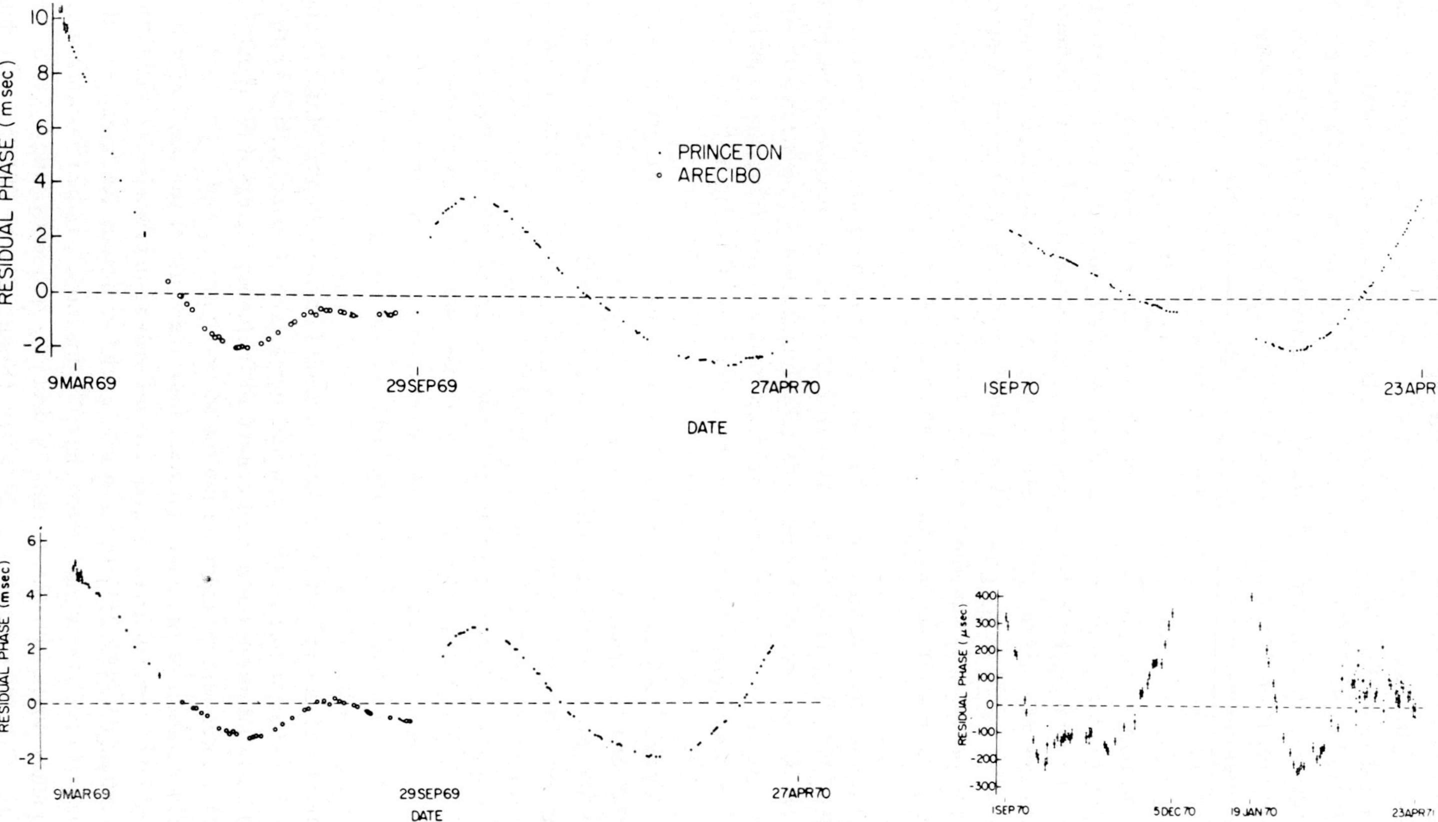

Fig. 26. Timing data on PSR0531 + 21, reproduced from Reference 29 by permission. The plotted points are the residuals from a cubic fit to the phase. For example, a residual of 2 ms indicates that the pulse arrived 2 ms earlier than predicted by the cubic function. The top curve is a fit to the entire two years of data while the bottom curves are separate fits to the first and second years. The large discontinuity near 29SEP69 is the glitch. The remaining structure is due to the random walk discussed in the text. Published by the University of Chicago Press; © 1972, American Astronomical Society.

braking index have not varied by more than the measurement accuracy ($\sim 300\%$ and $\sim 10\%$ for six month intervals) over five years of observation. The strength of the random walk is described by the mean rate of events times the second moment of their amplitude, $R\langle \Delta v^2 \rangle = 10^{-22}$ Hz^2 s^{-1}. The best estimate of the braking index is 2.5. Errors in these numbers are about 50% and 1%, respectively. While the analysis gives no information on R or Δv separately, observation of structure on scales at least as small as one week indicates $R > 1.6 \times 10^{-6}$ s^{-1} and $\Delta v < 8 \times 10^{-9}$ Hz.

In Reference 220 six months of data could not be adequately described by a cubic function, so the data were divided into four sections with a frequency jump between each section. Braking indices from 0 to 5 were obtained. Such an interpretation is consistent with the random model described above although the rate and size of events are much smaller and much larger, respectively, than estimated above. Furthermore, this interpretation provides no natural explanation of the braking index variations, which if intrinsic, would imply gross changes in the radiation mechanism. A more reasonable interpretation is that the frequency structure has not been time resolved and the model discussed in the preceding paragraph provides a better description of the data.

Fine structure has very recently been observed [194] in timing data from two slow pulsars, PSRs $1642-03$ and $0329+54$. Available data are not sufficient to establish the random walk character of this fine structure, but are not inconsistent with this interpretation. $R\langle \Delta v^2 \rangle\, v^{-3}$, the mean square fractional change in frequency or period, per period is a dimensionless measure of the random walk strength. It has values 3.6×10^{-27}, 3×10^{-27} and 7×10^{-28} for PSRs $0531+21$, $1642-03$ and $0329+54$, respectively. Unpublished data on PSR$0833-45$ obtained by the JPL group apparently show similar fine structure.

While the fine structure may be unresolved frequency steps, there are several instances of resolved frequency jumps. Two glitches, as they are now commonly called, have been observed [237, 245, 246] in PSR$0833-45$, one [29, 228, 255] in PSR$0531+21$, and possibly one [194] in PSR$1508+55$. In all cases, the frequency increased during the glitch. In the PSR$0833-45$ glitches, the pulsar was observed to be spinning faster than it had been about two weeks earlier. In the other glitches, the pulsars were never observed to be spinning faster but the increase was apparent after allowance for the normal slowing down.

The PSR$0833-45$ glitches occurred between 1969 February 24 and March 3 and 1971 August 21 and September 4. The fractional increases in frequency were 2.3×10^{-6} and 2.0×10^{-6}. In addition, the rate of change of frequency increased by about 1% during the first glitch and probably during the second glitch as well.

In one interpretation of the glitch phenomenon, the crust of the neutron star is suddenly sped up by some mechanism and is then slowly spun down by frictional drag from the superfluid core [23, 257]. In this way, the relative sizes of the changes in v and dv/dt may be explained. Three parameters describe this model: τ, the relaxation time; Δv, the frequency jump; and Q, the ratio of decaying frequency jump to total frequency jump. The 1969 September 29 glitch (Figure 27) in PSR$0531+21$ had

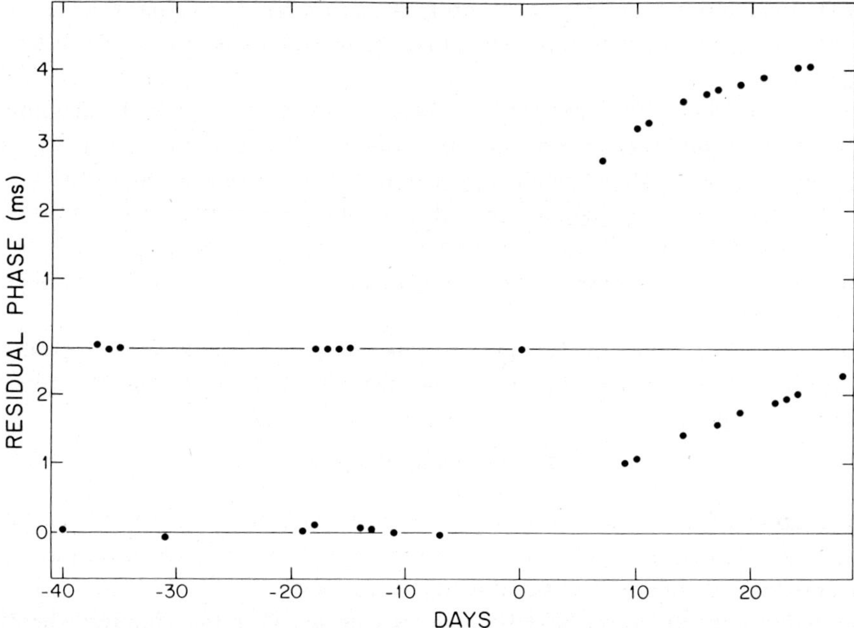

Fig. 27. Comparison of the PSR0531 + 21 glitch of 29SEP69 (top) and the alleged glitch of 26OCT71 (bottom). Units on the horizontal axis are measured from the glitch. The plotted points are residuals from the cubic function which describes the preglitch data. Note that the slope is proportional to the change in frequency. Structure the size of the alleged second glitch occurs throughout data (see lower right curve of Figure 26). Better coverage in the vicinity of 26OCT71 may be found in Reference 175.

$\tau = 4 \pm 1$ days, $\Delta v / v = (9 \pm 2) \times 10^{-9}$, and $Q = 0.93 \pm 0.03$. These results are from a fit to a length of data sufficiently short that the contribution of the random walk is negligible. It should be noted that the data, while consistent with this model, certainly do not establish the model. Apparently it is not possible to carry out a similar analysis for the PSR0833 − 45 glitches because the relaxation time is so long (~ 1 yr) that the fine structure introduces appreciable error.

The glitch in PSR0531 + 21 has been associated with apparent changes in the braking index [268], with changes in dispersion measure (Section 3), with activity in the Crab nebula [267], and recently, with γ-ray bursts [107]. While only one glitch has occurred, each associated event has been observed on several occasions. Thus, the suggested associations are not very convincing.

A second glitch in PSR0531 + 21 has been reported [175]. This event occurred on 1971 October 26, was much smaller than the first glitch ($\Delta v / v \sim 1.7 \times 10^{-9}$), and did not show the relaxation found in the first glitch. The two events are compared in Figure 27. Events the size of the alleged second glitch occur throughout the data and are just a manifestation of the random walk described earlier. The second glitch is probably not a distinct time resolved frequency step. The structure associated with the first glitch is huge compared to the structure elsewhere in the data and occurs on a much shorter time scale. This is the primary reason for regarding it as a distinct event. It is unlikely, but not impossible, that the first glitch is also a manifestation of the random walk.

The possible glitch [194] in PSR1508 + 55 occurred in late 1972, had a magnitude of $\Delta v/v \sim 2 \times 10^{-10}$, and was accompanied by a fractional change in dv/dt of -6×10^{-3}. The change in slowing down rate is opposite to that observed in PSRs 0531 + 21 and 0833 − 45. However, more data are probably required before it is known whether this glitch is a manifestation of a random walk or a distinct event. Should it turn out to be a distinct event, the presence of a random walk could still affect the measured change in dv/dt.

Suggested mechanisms for the random walk and glitches range from superfluid instabilities to starquakes to magnetospheric instabilities to accretion. Reference 258 contains a review.

9. Concluding Remarks

In this chapter, I have tried to present an overview of observational pulsar astronomy. While some may disagree with my choice of topics for emphasis, as far as I am aware, every known phenomenon has been at least mentioned.

Theoretical considerations have been entirely neglected in this chapter. Theories of neutron star structure and formation are particularly relevant to the material presented in Sections 4, 6, and 8. Neutron star structure is understood in a general sense but there is disagreement about the specifics. For reviews, see Chapter 5 of this volume and References 258 and 275. Theories of the pulse emission mechanism are relevant to Sections 6, 7, and 8. No theory yet exists which can quantitatively account for more than a small fraction of the observed phenomena. Qualitative theories abound, but no single one has attracted a large following. In general, pulsar theorists may be divided into two camps: those who believe the emission occurs near the stellar surface and those who believe it occurs near the light cylinder. Probably the entire region between the stellar surface and the light cylinder and even beyond must be understood before a quantitative theory emerges. For reviews, see References 87, 254, 259, 288, and 306.

Acknowledgements

I acknowledge helpful conversations with a great number of my colleagues and thank the National Science Foundation and Alfred P. Sloan Foundation for support.

Notes added in proof. Since this article was prepared the following papers have appeared.
1. Hulse, R. A. and Taylor, J. H.: *Astrophys. J. Letters 191*, L59 (1974).
   This paper announces the discovery of eleven new pulsars with a search technique about 10 times more sensitive than previous searches. The new pulsars are all close to the plane and have $39° \langle l^{II} \langle 52°$. All but three have DM greater than 100 pc cm^{-3}. One pulsar may be associated with the supernova remnant W51. The total number of pulsars is now at least 116.

2. Winkler, P. F. Jr. and Clark, G. W.: *Astrophys. J. Letters 191*, L67 (1974).
This paper reports the detection of soft X-rays from IC443. Interpretted in terms of a shock wave model, the X-ray data lead to an age of 3400 yrs for IC443. If true, this result makes the identification of PSR0611=22 with IC443 much less convincing.

J. H. Taylor and R. A. Hulse have discovered (*IAU Circ.* No. 2704) a 59 ms period pulsar in a binary system with an orbital period of 0.323 days and eccentricity 0.61. Potential results of studying this system include the measurement of the mass of a neutron star and the testing of general relativity.

References

1. Ables, J. G., Komesaroff, M. M., and Hamilton, P. A.: *Astrophys. Letters* **6**, 147 (1970).
2. Aitken, D. K. and Polden, P. G.: *Nature Phys. Sci.* **233**, 45 (1971).
3. Albats, P., Frye, G. M., Jr., Zych, A. D., Mace, O. B., Hopper, V. D., and Thomas, J. A.: *Nature* **240**, 221 (1972).
4. Alekseev, Y. I., Vitkevitch, V. V., and Malofeev, V. M.: *Nature Phys. Sci.* **235**, 167 (1972).
5. Alekseev, Y. I., Vitkevitch, V. V., Zhuravlev, V., and Shitov, Yu. P.: *IAU Circ.*, No. 2123 (1968).
6. Alexander, J. K., Brandt, J. C., Maran, S. P., and Stecher, T. P.: *Astrophys. J.* **167**, 487 (1971).
7. Andrew, B. H., Branson, N. J. B. A., and Wills, D.: *Nature* **203**, 171 (1964).
8. Antonova, T. D., Panadzhyan, V. G., and Pynzar', A. V.: *Soviet Astron. AJ* **15**, 13 (1971).
9. Argyle, E.: *Astrophys. J.* **183**, 973 (1973).
10. Argyle, E. and Gower, J. F. R.: *Astrophys. J. Letters* **175**, L89 (1972).
11. Armstrong, J. W., Coles, W. A., Kaufman, J. J., and Rickett, B. J.: *Astrophys. J. Letters* **186**, L141 (1973).
12. Assousa, G. E. and Erkes, J. W.: *Astron. J.* **78**, 885 (1973).
13. Baade, W. and Zwicky, F.: *Phys. Rev.* **45**, 138 (1934).
14. Baars, J. W. M. and Hartsuijker, A. P.: *Astron. Astrophys.* **17**, 172 (1972).
15. Backer, D. C.: *Nature* **227**, 692 (1970).
16. Backer, D. C.: *Nature* **228**, 42 (1970).
17. Backer, D. C.: *Nature* **228**, 752 (1970).
18. Backer, D. C.: *Nature* **228**, 1297 (1970).
19. Backer, D. C.: *Astrophys. J. Letters* **174**, L157 (1972).
20. Backer, D. C.: *Astrophys. J.* **182**, 245 (1973).
21. Backer, D. C., Boriakoff, V., and Manchester, R. N.: *Nature Phys. Sci.* **243**, 77 (1973).
22. Backer, D. C. and Fisher, J. R.: *Astrophys. J.* **189**, 137 (1974).
23. Baym, G., Pethick, C., Pines, D., and Ruderman, M.: *Nature* **224**, 872 (1969).
24. Becklin, E. E. and Kleinmann, D. E.: *Astrophys. J. Letters* **152**, L25 (1968).
25. Becklin, E. E., Kristian, J., Matthews, K., and Neugebauer, G.: *Astrophys. J. Letters* **186**, L137 (1973).
26. Bell, S. J. and Hewish, A.: *Nature* **213**, 1214 (1967).
27. Boclet, D., Brucy, G., Claisse, J., Durouchoux, P., and Rocchia, R.: *Nature Phys. Sci.* **235**, 69 (1972).
28. Börner, G. and Cohen, J. M.: *Astrophys. J.* **185**, 959 (1973).
29. Boynton, P. E., Groth, E. J., Hutchinson, D. P., Nanos, G. P., Jr., Partridge, R. B., and Wilkinson, D. T.: *Astrophys. J.* **175**, 217 (1972).
30. Boynton, P. E., Groth, E. J., Partridge, R. B., and Wilkinson, D. T.: *Astrophys. J. Letters* **157**, L197 (1969).
31. Bradt, H., Rappaport, S., Mayer, W., Nather, R. E., Warner, B., MacFarlane, M., and Kristian, J.: *Nature* **222**, 728 (1969).
32. Brandt, J. C., Stecher, T. P., Crawford, D. L., and Maran, S. P.: *Astrophys. J. Letters* **163**, L99 (1971).
33. Bridle, A. H.: *Nature* **225**, 1035 (1970).

34. Browning, R., Ramsden, D., and Wright, P. J.: *Nature Phys. Sci.* **232**, 99 (1971).
35. Caswell, J. L. and Goss, W. M.: *Astrophys. Letters* **7**, 141 (1970).
36. Cavani, C., Frontera, F., Fuligni, F., and Brini, D.: *Nature Phys. Sci.* **233**, 153 (1971).
37. Chiu, H.-Y., Lynds, R., and Maran, S. P.: *Astrophys. J. Letters* **162**, L99 (1970).
38. Clark, G. W., Bradt, H. V., Lewin, W. H. G., Markert, T. H., Schnopper, H. W., and Sprott, G. F.: *Astrophys. J.* **179** 263 (1973).
39. Clark, R. R. and Smith, F. G.: *Nature* **221**, 724 (1969).
40. Cocke, W. J., Disney, M. J., and Gehrels, T.: *Nature* **223**, 576 (1969).
41. Cocke, W. J., Disney, M. J., and Taylor, D. J.: *Nature* **221**, 525 (1969).
42. Cocke, W. J., Ferguson, D. C., and Muncaster, G. W.: *Astrophys. J.* **183**, 987 (1973).
43. Cocke, W. J., Muncaster, G. W., and Gehrels, T.: *Astrophys. J. Letters* **169**, L119 (1971).
44. Cole, T. W.: *Nature* **227**, 788 (1970).
45. Cole, T. W., Hesse, H. K., and Page, C. G.: *Nature* **225**, 712 (1970).
46. Cole, T. W. and Pilkington, J. D. H.: *Nature* **219**, 574 (1968).
47. Colla, G., Salter, C., and Sutton, J.: *IAU Circ.*, No. 2374 (1971).
48. Combe, V. and Large, M. I.: *Astrophys. Letters* **14**, 59 (1973).
49. Conklin, E. K., Howard, H. T., Craft, H. D., Jr., and Comella, J. M.: *Nature* **219**, 1238 (1968).
50. Counselman, C. C., III and Shapiro, I. I.: *Science* **162**, 352 (1968).
51. Craft, H. D., Jr. and Comella, J. M.: *Nature* **220**, 676 (1968).
52. Craft, H. D., Jr., Comella, J. M., and Drake, F. D.: *Nature* **218**, 1122 (1968).
53. Craft, H. D., Jr., Lovelace, R. V. E., and Sutton, J. M.: *IAU Circ.*, No. 2100 (1968).
54. Craft, H. D., Jr., Sutton, J. M., Comella, J. M., and Lovelace, R. V. E.: *Nature* **219**, 1237 (1968).
55. Culhane, J. L., Cruise, A. M., Rapley, C. G., and Hawkins, F. J.: *Astrophys. J. Letters* **190**, L9 (1974).
56. Davidson, K. and Terzian, Y.: *Nature* **221**, 729 (1969).
57. Davidson, K. and Terzian, Y.: *Astron. J.* **74**, 849 (1969).
58. Davies, J. G.: *IAU Circ.*, No. 2107 (1968).
59. Davies, J. G., Horton, P. W., Lyne, A. G., Rickett, B. J., and Smith, F. G.: *Nature* **217**, 910 (1968).
60. Davies, J. G., Hunt, G. C., and Smith, F. G.: *Nature* **221**, 27 (1969).
61. Davies, J. G. and Large, M. I.: *Monthly Notices Roy. Astron. Soc.* **149**, 301 (1970).
62. Davies, J. G., Large, M. I., and Pickwick, A. C.: *Nature* **227**, 1123 (1970).
63. Davies, J. G., Lyne, A. G., and Seiradakis, J. H.: *IAU Circ.*, No. 2436 (1972).
64. Davies, J. G., Lyne, A. G., and Seiradakis, J. H.: *Nature* **240**, 229 (1972).
65. Davies, J. G., Lyne, A. G., and Seiradakis, J. H.: *Nature Phys. Sci.* **244**, 84 (1973).
66. Davies, R. D. and Smith, F. G. (eds.): 'The Crab Nebula', *IAU Symp.* **6** (1971).
67. Drake, F. D.: *Science* **160**, 416 (1968).
68. Drake, F. D. and Craft, H. D., Jr.: *Nature* **220**, 231 (1968).
69. Ducros, G., Ducros, R., Rocchia, R., and Tarrius, A.: *Astron. Astrophys.* **7**, 162 (1970).
70. Duthie, J. G. and Murdin, P.: *Astrophys. J.* **163**, 1 (1971).
71. Duyvendak, J. J. L.: *Publ. Astron. Soc. Pacific* **54**, 91 (1942).
72. Ekers, R. D. and Moffet, A. T.: *Nature* **220**, 756 (1968).
73. Ekers, R. D. and Moffet, A. T.: *Astrophys. J. Letters* **158**, L1 (1969).
74. Elvius, T.: *Stars and Stellar Systems* **5**, 41 (1965).
75. Encrenaz, P. and Guelin, M.: *Nature* **227**, 476 (1970).
76. Erickson, W. C., Kuiper, T. B. H., Clark, T. A., Knowles, S. H., and Broderick, J. J.: *Astrophys. J.* **177**, 101 (1972).
77. Ewing, M. S., Batchelor, R. A., Friefeld, R. D., Price, R. M., and Staelin, D. H.: *Astrophys. J. Letters* **162**, L169 (1970).
78. Fazio, G. G., Helmken, H. F., O'Mongain, E., and Weekes, T. C.: *Astrophys. J. Letters* **175**, L117 (1972).
79. Feinberg, G.: *Science* **166**, 879 (1969).
80. Ferguson, D. C., Cocke, W. J., and Gehrels, T.: *Astrophys. J.* **190**, 375 (1974).
81. Finzi, A. and Wolf, R. A.: *Astrophys. J. Letters* **155**, L107 (1969).
82. Fishman, G. J., Harnden, F. R., Jr., Johnson, W. N., III, and Haymes, R. C.: *Astrophys. J Letters* **158**, L61 (1969).
83. Freeman, K. C., Rodgers, A. W., Rudge, P. T., and Lyngå, G.: *Nature* **222**, 459 (1969).

84. Fritz, G., Meekins, J. F., Chubb, T. A., Friedman, H., and Henry, R. C.: *Astrophys. J. Letters* **164**, L55 (1971).
85. Furniss, I., Jennings, R. E., and Moorwood, A. F. M.: *Nature Phys. Sci.* **236**, 6 (1971).
86. Galt, J. A. and Lyne, A. G.: *Monthly Notices Roy. Astron. Soc.* **158**, 281 (1972).
87. Ginzburg, V. L.: in C. de Jager (ed.), *Highlights of Astronomy 1970*, D. Reidel, Dordrecht, 1971, p. 20.
88. Glass, I. S.: *Astrophys. J.* **157**, 215 (1969).
89. Gold, T.: *Nature* **221**, 25 (1969).
90. Goldreich, P.: *Publ. Astron. Soc. Pacific* **83**, 599 (1971).
91. Goldreich, P. and Julian, W. H.: *Astrophys. J.* **160**, 971 (1970).
92. Goldstein, R. M.: *Science* **161**, 44 (1968).
93. Goldstein, S. J., Jr. and James, J. T.: *Astrophys. J. Letters* **158**, L179 (1969).
94. Goldstein, S. J., Jr. and Meisel, D. D.: *Nature* **224**, 349 (1969).
95. Gómez-González, J., Falgarone, E., Encrenaz, P., and Guélin, M.: *Astrophys. Letters* **12**, 207 (1972).
96. Gómez-González, J. and Guélin, M.: preprint (1974).
97. Gómez-González, J., Guélin, M., Falgarone, E., and Encrenaz, P.: *Astrophys. Letters* **13**, 229 (1973)
98. Gordon, C. P., Gordon, K. J., and Shalloway, A. M.: *Nature* **222**, 129 (1969).
99. Gordon, K. J. and Gordon C. P.: *Astrophys. Letters* **5**, 153 (1970).
100. Gorestein, P., Kellogg, E. M., and Gursky, H.: *Astrophys. J.* **160**, 199 (1970).
101. Goss, W. M. and Schwarz, U. J.: *Nature Phys. Sci.* **234**, 52 (1971).
102. Gott, J. R., III, Gunn, J. E., and Ostriker, J. P.: *Astrophys. J. Letters* **160**, L91 (1970).
103. Gower, J. F. R. and Argyle, E.: *Astrophys. J. Letters* **171**, L23 (1972).
104. Graham, D. A.: *Nature* **229**, 326 (1971).
105. Graham, D. A. and Hunt, G. C.: *Nature Phys. Sci.* **242**, 86 (1973).
106. Grewing, M., Mebold, U., and Rohlfs, K.: *Nature* **221**, 751 (1969).
107. Grindlay, J. E. and Fazio, G. G.: *Astrophys. J. Letters* **187**, L93 (1974).
108. Guélin, M., Guibert, J., Huchtmeier, W., and Weliachew, L.: *Nature* **221**, 249 (1969).
109. Gunn, J. E. and Ostriker, J. P.: *Astrophys. J.* **160**, 979 (1970).
110. Habing, H. J. and Pottasch, S. R.: *Nature* **219**, 1137 (1968).
111. Hankins, T. H.: *Astrophys. J.* **169**, 487 (1971).
112. Hankins, T. H.: *Astrophys. J. Letters* **177**, L11 (1972).
113. Hankins, T. H.: *Astrophys. J. Letters* **181**, L49 (1973).
114. Harnden, F. R., Jr. and Gorenstein, P.: *Nature* **241**, 107 (1973).
115. Harnden, F. R., Jr., Johnson, W. N., III and Haymes, R. C.: *Astrophys. J. Letters* **172**, L91 (1972).
116. Haymes, R. C., Ellis, D. V., Fishman, G. J., Kurfess, J. D., and Tucker, W. H.: *Astrophys. J. Letters* **151**, L9 (1968).
117. Heiles, C., Campbell, D. B., and Rankin, J. M.: *Nature* **226**, 529 (1970).
118. Heiles, C. and Rankin, J. M.: *Nature Phys. Sci.* **231**, 97 (1971).
119. Helmken, H. F., Fazio, G. G., O'Mongain, E., and Weekes, T. C.: *Astrophys. J.* **184**, 245 (1973).
120. Henry, G. R. and Paik, H.-J.: *Nature* **224**, 1188 (1969).
121. Hesse, K. H.: *Nature* **225**, 837 (1970).
122. Hesse, K. H.: *Nature Phys. Sci.* **231**, 54 (1971).
123. Hesse, K. H.: *Nature Phys. Sci.* **235**, 27 (1972).
124. Hesse, K. H.: *Astron. Astrophys.* **27**, 373 (1973).
125. Hesse, K. H., Sieber, W., and Wielebinski, R.: *Nature Phys. Sci.* **245**, 57 (1973).
126. Hesse, K. H. and Wielebinski, R.: *Astron. Astrophys.* **31**, 409 (1974).
127. Hesser, J. E., Lasker, B. M., Bochonko, D. R., and Mook, D. E.: *Nature* **223**, 485 (1969).
128. Hewish, A.: *Ann. Rev. Astron. Astrophys.* **8**, 265 (1970).
129. Hewish, A., Bell, S. J., Pilkington, J. D. H., Scott, P. F., and Collins, R. A.: *Nature* **217**, 709 (1968).
130. Higgins, C. S., Komesaroff, M. M., and Slee, O. B.: *Astrophys. Letters* **9**, 75 (1971).
131. Hobbs, R. W., Corbett, H. H., and Santini, N. J.: *Astrophys. J. Letters* **155**, L87 (1969).
132. Hoffmann, B.: *Nature* **218**, 667 (1968).
133. Høg, E. and Lohsen, E.: *Nature* **227**, 1229 (1970).
134. Horowitz, P., Papaliolios, C., Carleton, N. P., Nelson, J., Middleditch, J., Hills, R., Cudaback, D., Wampler, J., Boynton, P. E., Groth, E. J. Partridge R. B., Wilkinson, D. T., Duthie, J. G., and Murdin, P.: *Astrophys. J. Letters* **166**, L91 (1971).

135. Huguenin, G. R. and Taylor, J. H.: *IAU Circ.*, No. 2084 (1968).
136. Huguenin, G. R. and Taylor, J. H.: *IAU Circ.*, No. 2128 (1969).
137. Huguenin, G. R., Taylor, J. H., and Helfand, D. J.: *Astrophys. J. Letters* **181**, L139 (1973).
138. Huguenin, G. R., Taylor, J. H., and Jura, M.: *Astrophys. Letters* **4**, 71 (1969).
139. Huguenin, G. R., Taylor, J. H., and Troland, T. H.: *Astrophys. J.* **162**, 727 (1970).
140. Hunt, G. C.: *Monthly Notices Roy. Astron. Soc.* **153**, 119 (1971).
141. Jackson, J. D.: *Classical Electrodynamics*, Wiley, New York, 1962.
142. Jokipii, J. R. and Hollweg, J. V.: *Astrophys. J.* **160**, 745 (1970).
143. Jones, B. and Wielebinski, R.: *Astrophys. Letters* **5**, 17 (1970).
144. Katgert, P. and Oort, J. H.: *Bull. Astron. Inst. Neth.* **19**, 239 (1967).
145. Kellogg, E., Tananbaum, H., Harnden, F. R., Jr., Gursky, H., Giacconi, R., and Grindlay, J.: *Astrophys. J.* **183**, 935 (1973).
146. Keswani, G. H.: *Nature* **220**, 148 (1968).
147. Kinzer, R. L., Noggle, R. C., Seeman, N., and Share, G. H.: *Nature* **229**, 187 (1971).
148. Komesaroff, M. M., Ables, J. G., and Hamilton, P. A.: *Astrophys. Letters* **9**, 101 (1971).
149. Komesaroff, M. M., Ables, J. G., Morris, D., Cooke, D. J., Schwarz, U. J., and Hamilton, P. A.: *IAU Circ.*, No. 2201 (1970).
150. Komesaroff, M. M., Hamilton, P. A., and Ables, J. G.: *Australian J. Phys.* **25**, 759 (1972).
151. Komesaroff, M. M., Hamilton, P. A., McCulloch, P. M., Ables, J. G., and Cooke, D. J.: *IAU Circ.*, No. 2505 (1973).
152. Komesaroff, M. M., Hamilton, P. A., McCulloch, P. M., Ables, J. G., and Cooke, D. J.: *IAU Circ.*, No. 2563 (1973).
153. Komesaroff, M. M., Morris, D., and Cooke, D. J.: *Astrophys. Letters* **5**, 37 (1970).
154. Krishna Mohan, S., Balasubramanian, V., and Swarup, G.: *Nature Phys. Sci.* **234**, 151 (1971).
155. Kristian, J.: *Astrophys. J. Letters* **162**, L103 (1970).
156. Kristian, J.: *Astrophys. J. Letters* **162**, L173 (1970).
157. Kristian, J.: *Publ. Astron. Soc. Pacific* **82**, 456 (1970).
158. Kristian, J., Visvanathan, N., Westphal, J. A., and Snellen, G. H.: *Astrophys. J.* **162**, 475 (1970).
159. Kundu, M. R. and Velusamy, T.: *Nature Phys. Sci.* **234**, 54 (1971).
160. Lang, K. R.: *Astrophys. J. Letters* **158**, L175 (1969).
161. Lang, K. R.: *Astrophys. J. Letters* **161**, L133 (1970).
162. Lang, K. R.: *Astrophys. J.* **164**, 249 (1971).
163. Lang, K. R.: *Astrophys. Letters* **7**, 175 (1971).
164. Lang, K. R.: *IAU Circ.*, No. 2381 (1972).
165. Lang, K. R. and Bosque, B.: *IAU Circ.*, No. 2137 (1969).
166. Lang, K. R. and Rickett, B. J.: *Nature* **225**, 528 (1970).
167. Large, M. I.: in R. D. Davies and F. G. Smith (eds.), 'The Crab Nebula', *IAU Symp.* **46**, 165 (1971).
168. Large, M. I. and Vaughan, A. E.: *Nature* **220**, 43 (1968).
169. Large, M. I. and Vaughan, A. E.: *Nature Phys. Sci.* **236**, 117 (1972).
170. Large, M. I., Vaughan, A. E., and Mills, B. Y.: *Nature* **220**, 340 (1968).
171. Large, M. I., Vaughan, A. E., and Wielebinski, R.: *Nature* **220**, 753 (1968).
172. Large, M. I., Vaughan, A. E., and Wielebinski, R.: *Nature* **223**, 1249, (1969).
173. Large, M. I., Vaughan, A. E., and Wielebinski, R.: *Astrophys. Letters* **3**, 123 (1969).
174. Leray, J. P., Vasseur, J., Paul, J., Parlier, B., Forichon, M., Agrinier, B., Boella, G., Maraschi, L., Treves, A., Buccheri, L., Cuccia, A., and Scarsi, L.: *Astron. Astrophys.* **16**, 443 (1972).
175. Lohsen, E.: *Nature Phys. Sci.* **236**, 70 (1972).
176. Lovelace, R. V. E. and Craft, H. D., Jr.: *Nature* **220**, 875 (1968).
177. Lovelace, R. V. E. and Sutton, J. M.: *IAU Circ.*, No. 2135 (1969).
178. Lynds, R.: *Astrophys. J. Letters* **157**, L11 (1969).
179. Lyne, A. G.: *Monthly Notices Roy. Astron. Soc.* **153**, 27P (1971).
180. Lyne, A. G. and Rickett, B. J.: *Nature* **218**, 326 (1968).
181. Lyne, A. G. and Rickett, B. J.: *Nature* **219**, 1339 (1968).
182. Lyne, A. G. and Smith, F. G.: *Nature* **218**, 124 (1968).
183. Lyne, A. G. and Smith, F. G.: *Monthly Notices Roy. Astron. Soc.* **157**, 15P (1972).
184. Lyne, A. G., Smith, F. G., and Graham, D. A.: *Monthly Notices Roy. Astron. Soc.* **153**, 337 (1971).
185. Manchester, R. N.: *Nature* **225**, 1124 (1970).

186. Manchester, R. N.: *Astrophys. J. Letters* **163**, L61 (1971).
187. Manchester, R. N.: *Astrophys. J. Letters* **167**, L101 (1971).
188. Manchester, R. N.: *Astrophys. J. Suppl.* **23**, 283 (1971).
189. Manchester, R. N.: *Astrophys. J.* **172**, 43 (1972).
190. Manchester, R. N.: *Astrophys. J.* **188**, 637 (1974).
191. Manchester, R. N. and Peters, W. L.: *Astrophys. J.* **173**, 221 (1972).
192. Manchester, R. N., Tademaru. E., Taylor, J. H., and Huguenin, G. R.: *Astrophys. J.* **185**, 951 (1973).
193. Manchester, R. N. and Taylor, J. H.: *Astrophys. Letters* **10**, 67 (1972).
194. Manchester, R. N. and Taylor, J. H.: *Astrophys. J. Letters* **191**, L63 (1974).
195. Manchester, R. N., Taylor, J. H., and Huguenin, G. R.: *Nature Phys. Sci.* **240**, 74 (1972).
196. Manchester, R. N., Taylor, J. H., and Huguenin, G. R.: *Astrophys. J. Letters* **179**, L7 (1973).
197. Manchester, R. N., Taylor, J. H., and Van, Y. Y.: *Astrophys. J. Letters* **189**, L119 (1974).
198, Matveenko, L. I. and Pynzar' A. V.: *Soviet Astron. AJ* **13**, 433 (1969).
199. Mayall, N. U. and Oort, J. H.: *Publ. Astron. Soc. Pacific* **54**, 95 (1942).
200. McBreen, B., Ball, S. E., Jr., Campbell, M., Greisen, K., and Koch, D.: *Astrophys. J.* **184**, 571 (1973).
201. McCulloch, P. M., Hamilton, P. A., Ables, J. G., and Komesaroff, M. M.: *Astrophys. Letters* **10**, 163 (1972).
202. McCulloch, P. M., Hamilton, P. A., Komesaroff, M. M., and Cooke, D. J.: *Proc. Astron. Soc. Austalia* **1**, 225 (1969).
203. McCulloch, P. M., Komesaroff, M. M., Ables, J. G., Hamilton, P. A., and Rankin, J. M.: *Astrophys. Letters* **14**, 169 (1973).
204. McCutcheon, W. H. and Shuter, W. L. H.: *Astrophys. Letters* **9**, 201 (1971).
205. McNamara, B. J.: *Publ. Astron. Soc. Pacific* **83**, 491 (1971).
206, Miller, J. S.: *Astrophys. J. Letters* **180**, L83 (1973).
207. Miller, J. S. and Wampler, E. J.: *Nature* **221**, 1037 (1969).
208. Mills, B. Y.: *Nature* **224**, 504 (1969).
209. Milne, D. K.: *Australian J. Phys.* **21**, 201 (1968).
210. Milne, D. K.: in R. D. Davies and F. G. Smith (eds.), 'The Crab Nebula', *IAU Symp.* **46**, 248 (1971).
211. Milne, D. K.: *Australian J. Phys.* **24**, 429 (1971).
212. Minkowski, R.: *Stars and Stellar Systems* **7**, 923 (1968).
213. Minkowski, R.: in R. D. Davies and F. G. Smith (eds.), 'The Crab Nebula', *IAU Symp.* **46**, 241 (1971).
214. Mohanty, D. K., Balasubramanian, V., and Swarup, G.: *IAU Circ.*, No. 2356 (1971).
215. Montgomery, J. W., Epstein, E. E., Oliver, J. P., Dworetsky, M. M., and Fogarty, W. G.: *Astrophys. J.* **167**, 77 (1972).
216. Moore, W. E., Agrawal, P. C., and Garmire, G.: *Astrophys. J. Letters* **189**, L117 (1974).
217. Morris, D., Schwarz, U. J., and Cooke, D. J.: *Astrophys. Letters* **5**, 181 (1970).
218. Morris, D., Schwarz, U. J., and Slee, O. B.: *Astrophys. Letters* **5**, 187, (1970).
219. Muncaster, G. W. and Cocke, W. J.: *Astrophys. J. Letters* **178**, L13 (1972).
220. Nelson, J., Hills, R., Cudaback, D., and Wampler, J.: *Astrophys. J. Letters* **161**, L235 (1970).
221. Neugebauer, G., Becklin, E. E., Kristian, J., Leighton, R. B., Snellen, G., and Westphal, J. A.: *Astrophys. J. Letters* **156**, L115 (1969).
222. Notni, P., Oleak, H., and Schmidt, K.-H.: *Astrophys. Letters* **6**, 61 (1970).
223. O'Dell, C. R.: *Astrophys. J.* **136**, 809 (1962).
224. Oke, J. B.: *Astrophys. J. Letters* **156**, L49 (1969).
225. Orwig, L. E., Chupp, E. L., and Forrest, D. J.: *Nature Phys. Sci.* **231**, 171 (1971).
226. Ostriker, J. P. and Gunn, J. E.: *Astrophys. J.* **157**, 1395 (1969).
227. Page, C. G.: *Monthly Notices Roy. Astron. Soc.* **163**, 29 (1973).
228. Papaliolios, C., Carleton, N. P., and Horowitz, P.: *Nature* **228**, 445 (1970).
229. Parlier, B., Agrinier, B., Forichon, M., Leray, J. P., Boella, G., Maraschi, L., Buccheri, R., Robba, N. R., and Scarsi, L.: *Nature Phys. Sci.* **242**, 117 (1973).
230. Pilkington, J. D. H., Hewish, A., Bell, S. J., and Cole, T. W.: *Nature,* **218**, 126 (1968).
231. Prentice, A. J. R.: *Nature* **225**, 438 (1970).
232. Prentice, A. J. R., Buckee, J. W., and ter Haar D.: *Nature* **228**, 452 (1970).

233. Prentice, A. J. R. and ter Haar, D.: *Nature* **222**, 964 (1969).
234. Prentice, A. J. R. and ter Haar, D.: *Monthly Notices Roy. Astron. Soc.* **146**, 423 (1969).
235. Radhakrishnan, V. and Cooke, D. J. *Astrophys. Letters* **3**, 225 (1969).
236. Radhakrishnan, V., Cooke, D. J., Komesaroff, M. M., and Morris, D.: *Nature* **221**, 443 (1969).
237. Radhakrishnan, V. and Manchester, R. N.: *Nature* **222**, 228 (1969).
238. Rankin, J. M., Campbell, D. B., and Backer, D. C.: *Astrophys. J.* **188**, 609 (1974).
239. Rankin, J. M., Comella, J. M., Craft, H. D., Jr., Richards, D. W., Campbell, D. B., and Counselman, C. C., III: *Astrophys. J.* **162**, 707 (1970).
240. Rankin, J. M. and Counselman, C. C., III: *Astrophys. J.* **181**, 875 (1973).
241. Rankin, J. M. and Heiles, C.: *Nature* **227**, 1330 (1970).
242. Rankin, J. M., Heiles, C., and Comella, J. M.: *Astrophys. J. Letters* **163**, L95 (1971).
243. Rao, A. P. and Krishna Mohan, S.: *Nature Phys. Sci.* **238**, 69 (1972).
244. Rappaport, S., Bradt, H., and Mayer, W.: *Nature Phys. Sci.* **229**, 40 (1971).
245. Reichley, P. E. and Downs, G. S.: *Nature* **222**, 229 (1969).
246. Reichley, P. E. and Downs, G. S.: *Nature Phys. Sci.* **234**, 48 (1971).
247. Reichley, P. E., Downs, G. S., and Morris, G. A.: *Astrophys. J. Letters* **159**, L35 (1970).
248. Richards, D. W., Pettengill, G. H., Counselman, C. C., III, and Rankin, J. M.: *Astrophys. J. Letters* **160**, L1 (1970).
249. Ricker, G. R., Gerassimenko, M., McClintock, J. E., Ryckman, S. G., and Lewin, W. H. G.: *Astrophys. J. Letters* **186**, L111 (1973).
250. Rickett, B. J.: *Nature* **221**, 158 (1969).
251. Rickett, B. J.: *Monthly Notices Roy. Astron. Soc.* **150**, 67 (1970).
252. Rickett, B. J. and Lang, K. R.: *Astrophys. J.* **185**, 945 (1973).
253. Rickett, B. J. and Lyne, A. G.: *Nature* **218**, 934 (1968).
254. Roberts, D. H., Sturrock, P. A., and Turk, S. J.: *Ann. New York Acad. Sci.* **224**, 206 (1973).
255. Roberts, J. A. and Richards, D. W.: *Nature Phys. Sci.* **231**, 25 (1971).
256. Robinson, B. J., Cooper, B. F. C., Gardiner, F. F., Wielebinski, R., and Landecker, T. L.: *Nature* **218**, 1143 (1968).
257. Ruderman, M.: *Nature* **223**, 597 (1969).
258. Ruderman, M.: *Ann. Rev. Astron. Astrophys.* **10**, 427 (1972).
259. Ruderman, M. A. and Sutherland, P. G.: *Astrophys. J.* **190**, 137 (1974).
260. Rudnick, L.: thesis, Princeton University (1974).
261. Salpeter, E. E.: *Astrophys. J.* **147**, 433 (1967).
262. Salpeter, E. E.: *Nature* **221**, 31 (1969).
263. Salter, C. J.: *IAU Circ.*, No. 2287 (1970).
264. Salter, C. J.: *IAU Circ.*, No. 2295 (1970).
265. Salter, C. J. and Facondi, S. R.: *Monthly Notices Roy. Astron. Soc.* **152**, 5P (1971).
266. Scargle, J. D.: *Astrophys. J.* **156**, 401 (1969).
267. Scargle, J. D. and Harlan, E. A.: *Astrophys. J. Letters* **159**, L143 (1970).
268. Scargle, J. D. and Pacini, F.: *Nature Phys. Sci* **232**, 144 (1971).
269. Scheuer, P. A. G.: *Nature* **218**, 920 (1968).
270. Schmidt, M.: *Stars and Stellar Systems* **5**, 513 (1965).
271. Schönhardt, R. E.: *Nature Phys. Sci.* **243**, 62 (1973).
272. Schönhardt, R. E. and Sieber, W.: *Astrophys. Letters* **14**, 61 (1973).
273. Schwarz, U. J. and Morris, D.: *Astrophys. Letters* **7**, 185 (1971).
274. Seward, F. D., Burginyon, G. A., Grader, R. J., Hill, R. W., Palmieri, T. M., and Stoering, J. P.: *Astrophys. J.* **169**, 515 (1971).
275. Shaham, J., Pines, D., and Ruderman, M.: *Ann. New York Acad. Sci.* **224**, 190 (1973).
276. Sharpless, S.: *Astrophys. J. Suppl.* **4**, 257 (1959).
277. Shklovsky, I. S.: *Supernovae*, Wiley, New York, 1968.
278. Shklovsky, I. S.: *Nature* **225**, 252 (1970).
279. Shklovsky, I. S.: *Astrophys. Letters* **8**, 101 (1971).
280. Shuter, W. L. H., Venugopal, V. R., and Mahoney, M. J.: *Nature* **220**, 356 (1968).
281. Sieber, W.: *Astron. Astrophys.* **28**, 237 (1973).
282. Slee, O. B.: *Proc. Astron. Soc. Australia* **1**, 223 (1969).
283. Slee, O. B., Ables, J. G., Batchelor, R. A., Krishna Mohan, S., Venugopal, R., and Swarup, G.: *Monthly Notices Roy. Astron. Soc.* **167**, 31 (1974).

284. Slee, O. B., Komesaroff, M. M., and McCulloch, P. M.: *Nature* **219**, 342 (1968).
285. Slee, O. B. and Mulhall, P. S.: *Proc. Astron. Soc. Australia* **1**, 322 (1970).
286. Slee, O. B. and Mulhall, P. S.: *Astrophys. Letters* **8**, 5 (1971).
287. Smathers, H. W., Chubb, T. A., and Sadeh, D.: *Nature Phys. Sci.* **232**, 120 (1971).
288. Smith, F. G.: *Monthly Notices Roy. Astron. Soc.* **149**, 1 (1970).
289. Smith, F. G.: *Monthly Notices Roy. Astron. Soc.* **161**, 9P (1973).
290. Smith, F. G.: *Monthly Notices Roy. Astron. Soc.* **167**, 43P (1974).
291. Staelin, D. H. and Reifenstein, E. C., III: *IAU Circ.*, No. 2110 (1968).
292. Staelin, D. H. and Sutton, J. M.: *Nature* **226**, 69 (1970).
293. Sutton, J. M.: *Monthly Notices Roy. Astron. Soc.* **155**, 51 (1971).
294. Sutton, J. M., Craft, H. D., Jr., Lovelace, R. V. E., and Comella, J. M.: *IAU Circ.*, No. 2093 (1968).
295. Sutton, J. M., Salter, C., and Colla, G.: *IAU Circ.*, No. 2386 (1972).
296. Sutton, J. M., Staelin, D. H., and Price, R. M.: in R. D. Davies and F. G. Smith (eds.), 'The Crab Nebula', *IAU Symp.* **46**, 97 (1971).
297. Sutton, J. M., Staelin, D. H., Price, R. M., and Weimer, R.: *Astrophys. J. Letters* **159**, L89 (1970).
298. Tamman, G. A.: *Astron. Astrophys.* **8**, 458 (1970).
299. Tanenbaum, B. S., Zeissig, G. A., and Drake, F. D.: *Science* **160**, 760 (1968).
300. Taylor, J. H. and Huguenin, G. R.: *IAU Circ.*, No. 2120 (1968).
301. Taylor, J. H. and Huguenin, G. R.: *Nature* **221**, 816 (1969).
302. Taylor, J. H., and Huguenin, G. R.: *Astrophys. J.* **167**, 273 (1971).
303. Taylor, J. H., Huguenin, G. R., Hirsch, R. M., and Manchester, R. N.: *Astrophys. Letters* **9**, 205 (1971).
304. Taylor, J. H., Huguenin, G. R., and Manchester, R. N.: *IAU Circ.*, No. 2435 (1972).
305. Taylor, J. H., Jura, M., and Huguenin, G. R.: *Nature* **223**, 797 (1969).
306. ter Haar, D.: *Phys. Rep.* **3C**, No. 2 (1972).
307. Trimble, V.: *Astron. J.* **73**, 535 (1968).
308. Trimble, V. L. and Rees, M. J.: *Astrophys. Letters* **5**, 93 (1970).
309. Tsarevsky, G. S.: *Astrophys. Letters* **10**, 71 (1972).
310. Turtle, A. J. and Vaughan, A. E.: *Nature* **219**, 689 (1968).
311. van den Bergh, S., Marscher, A. P., and Terzian, Y.: *Astrophys. J. Suppl.* **26**, 19 (1973).
312. Vandenberg, N. R., Clark, T. A., Erickson, W. C., Resch, G. M., Broderick, J. J., Payne, R. R., Knowles, S. H., and Youmans, A. B.: *Astrophys. J. Letters* **180**, L27 (1973).
313. Vaughan, A. E. and Large, M. I.: *Proc. Astron. Soc. Australia* **1**, 220 (1969).
314. Vaughan, A. E. and Large, M. I.: *Nature* **225**, 167 (1970).
315. Vaughan, A. E. and Large, M. I.: *Monthly Notices Roy. Astron. Soc.* **156**, 25P (1972).
316. Vaughan, A. E. and Large, M. I.: *Monthly Notices Roy. Astron. Soc.* **156**, 27P (1972).
317. Vaughan, A. E., Large, M. I., Wielebinski, R.: *Nature* **222**, 963 (1969).
318. Vaughan, A. E. and McAdam, W. B.: *Nature Phys. Sci.* **241**, 138 (1973).
319. Verschuur, G. L.: *Astrophys. J. Letters* **183**, L9 (1973).
320. Vitkevich, V. V. and Shitov, Yu. P.: *Nature*, **226**, 1235 (1970).
321. Wampler, E. J., Scargle, J. D., and Miller, J. S.: *Astrophys. J. Letters* **157**, L1 (1969).
322. Warner, B. and Nather, R. E.: *Nature* **222**, 254 (1969).
323. Warner, B., Nather, R. E., and MacFarlane, M.: *Nature* **222**, 233 (1969).
324. Whitford, A. E.: *Astron. J.* **63**, 201 (1958).
325. Wielebinski, R.: *Nature* **219**, 1135 (1968).
326. Wielebinski, R.: *Proc. Astron. Soc. Australia* **1**, 226 (1969).
327. Wielebinski, R., Vaughan, A. E., and Large, M. I.: *Nature* **221**, 47 (1969).
328. Williamson, I. P.: *Monthly Notices Roy. Astron. Soc.* **157**, 55 (1972).
329. Williamson, I. P.: *Monthly Notices Roy. Astron. Soc.* **163**, 345 (1973).
330. Williamson, I. P.: *Monthly Notices Roy. Astron. Soc.* **166**, 499 (1974).
331. Woltjer, L.: *Ann. Rev. Astron. Astrophys.* **10**, 129 (1972).
332. Wrixon, G. T., Gott, J. R., III, and Penzias, A. A.: *Astrophys. J.* **174**, 399 (1972).
333. Zabolotny, V. F., Sholomitsky, G. B., and Slysh, V. I.: *Astrophys. Letters* **8**, 1 (1971).
334. Zeissig, G. A.: *Nature* **226**, 536 (1970).

THE GALACTIC X-RAY SOURCES

HERBERT GURSKY and ETHAN SCHREIER

Center for Astrophysics,
Harvard College Observatory, Smithsonian Astrophysical Observatory,
Cambridge, Mass., U.S.A.

1. Introduction

When X-ray sources in the galaxy were discovered in 1962 (Giacconi *et al.*, 1962) it was only possible to speculate on their nature, which centered on supernova and cosmic-ray phenomena since these were the only very energetic events known. In particular, the discovery of an X-ray source associated with the Crab Nebula led to the idea that the emission might be the thermal radiation from the surface of a hot, neutron star, However, it was soon demonstrated (Bowyer *et al.*, 1964) that a neutron star could not be responsible for the bulk of the X-radiation from the Crab, and it was not possible to exclude highly pathological conditions in otherwise ordinary stellar systems as being responsible for the X-ray sources (c.f., Hayakawa and Matsuoka, 1964).

Since that time, there have been fundamental advances in astronomy and in the physics of compact stars. The quasars and the pulsars have been discovered and the generation of enormous quantities of non-thermal power on a galactic scale is clearly demonstrated. Partially stimulated by the discovery of X-ray sources and of pulsars, substantial progress has been made in the physics of neutron stars and black holes the physical conditions within these objects and in their vicinity, the manner in which they are formed and their likely appearance in an astronomical setting. In an apparently unrelated activity, astronomers became concerned with how the evolution of stars is affected by their being in close binary systems where mass transfer might be important as seemed to be the case in certain stellar systems such as the old novae. These studies are now believed to be an important consideration in the evolution of neutron stars and black holes.

Until the last few years, little of this work could be tied directly to observational X-ray astronomy. However, beginning in late 1970, with the launch of NASA's X-ray satellite 'Uhuru', a wealth of new observational material has been accumulated. The quantity of data has produced a qualitative change in our understanding of the field. The change has been so great that it is almost as if a new field of astronomy has been created. In particular, it is possible to account for many of the galactic X-ray sources as originating from neutron stars and black holes which are in close proximity to a more normal star. The reason for this view is that,

(1) The short time variability of the X-ray emission requires a compact emission region.

(2) Many of the sources are positively confirmed as being in binary systems.

H. Gursky and R. Ruffini (eds.), Neutron Stars, Black Holes and Binary X-Ray Sources, 175–220. All Rights Reserved.
Copyright © 1975 by D. Reidel Publishing Company, Dordrecht-Holland.

(3) Mass accretion onto a neutron star or black hole from its nearby companion should be an extremely efficient means of generating X-rays.

The field has come full circle in the sense that the supernova remnants, which provided the early incentive for associating X-ray emission with compact stars, may be the least interesting of the sources. Except for sorting them out, they will not be discussed further. The prominent galactic X-ray sources are listed in Table I. As in the other branches of non-optical astronomy, the sources with identified optical counterparts have an unusual status since for these it is possible to determine fundamental

TABLE I

Outstanding galactic X-ray sources

3U Designation	Name	Binary period	X-ray characteristics	Optical characteristics
$0115 - 37$	SMC X-1	3.9 days	X-ray Eclipse extended lows	13 mag. BOIb
$0900 - 40$	Vela XR-1	8.9 days	X-ray Eclipse slow (h) flares	6 mag. B0.5Ib
$1118 - 60$	Cen X-3	2.1 days	X-ray Eclipse 4.8 s pulse period extended lows	13 mag. BOIb
$1617 - 15$	Sco X-1	–	Slow (min-h) flares	12–13 mag. non-stellar
$1653 + 35$	Her X-1	1.7 days	X-ray Eclipse 1.2 s pulse period 35 day high/low cycle	13–15 mag. – late A
$1700 - 37$		3.4 days	X-ray Eclipse irregular variability to 0.1 s	6 mag. 07f
$1956 + 35$	Cyg X-1	5.6 days	Irregular variability to 1 min	9 mag. 09.7Ib
$2030 + 40$	Cyg X-3	4.8 h	$\sim$ sinusoidal 4.8h variation radio outbursts	IR source
$2142 + 38$	Cyg X-2	–	Irregular variability to min	14 mag. G

characteristics such as distance, mass, age, etc. However, because of restrictions of both X-ray and optical observations, only nine optical identifications have been made.

Seven of the 9 identified sources are binary systems; however, as will be discussed further, the binary character is particularly easy to distinguish observationally. Nevertheless, it is possible the preponderance of binaries in Table I reflects the preponderance of binaries among all the galactic X-ray sources. In fact, it can be argued that all the X-ray sources are binaries.

One group of the identified sources is associated with a very particular kind of star – a late O or early B supergiant, a star which is very massive and very luminous. The relative rarity of this kind of star probably means that the star itself is a necessary ingredient in the presence of the X-ray emission in these systems. Other X-ray sources are clearly not associated with O–B stars, but rather with much later spectral type stars – stars which are much more like the Sun in temperature, luminosity and mass.

Thus, two very different kinds of stellar systems contain the X-ray sources. The X-ray data themselves do not reveal any particular homogeneity in their characteristics. The temporal structure of the radiation is found to be highly variable on every time scale from millisecond to years, and the variability is both random and periodic. One of these X-ray sources, Cyg X-1, is most credibly interpreted as a black hole. Its X-ray emission is very highly variable on a time scale down to 1 ms, and its mass is significantly greater than can be accommodated by other compact stars; namely, white dwarfs or neutron stars. Two other X-ray sources, Cen X-3 and Her X-1, are likely to be rapidly rotating neutron stars. Their X-ray emission is almost precisely periodic with periods of 4.8 and 1.2 s, respectively.

This paper presents the current observational evidence on these objects both from an astrophysical and astronomical point of view. We discuss at first the distributional properties of the sources, where they appear in the Galaxy, and certain average characteristics. In this way, certain properties of the X-ray sources can be deduced which are not apparent in the study of single objects. We go on to describe the properties of individual X-ray sources.

The hope is that we can learn about neutron stars and black holes, their physical properties, their origin and evolution, and their influence on other galactic phenomena. Thus, we focus on those elements of data which appear to have the most bearing on these questions.

2. General Characteristics of the Galactic X-Ray Sources

Astronomy encompasses both the study of the properties of individual objects and the study of the average characteristics of these objects. The latter kind of study has frequently proven to be extremely fruitful. For example, the Hertzsprung-Russell diagram establishes an average relation between luminosity and surface temperature of stars which has been the fundamental tool in studying stellar evolution. In the case of X-ray Astronomy, the 'average' property most studied has been the distribution of the sources around the Milky Way and their average intensity and spectral characteristics.

Even though observations have been conducted from about 0.2 to several hundred keV, the galactic sources described here are seen most clearly in the 1–10 keV energy range. At higher energies, rapidly decreasing photon fluxes make source detection more difficult. Below 1 keV, interstellar absorption and self-absorption severely limits observability.

The general feature of the distribution of the galactic sources; namely, a strong concentration of sources at low galactic longitude, was established during the earliest rocket experiments. The first observations that approached the angular resolution of current instruments and showed that there were a large number of discrete sources strung along the Milky Way were those of Fisher *et al.* (1966), and of Friedman *et al.*

(1967). The latter study also included an attempt to enumerate the number of sources based on guessing the location of sources within specific spiral arms. Gursky *et al.* (1967) attempted to demonstrate a specific correlation with spiral arms based on the similarity of the longitude distribution of the X-ray sources and 21 cm radio emission, H II regions and O-B associations. The latter are the accepted spiral arm indicators. Ryter (1970) and Setti and Woltjer (1970) made an important contribution to this kind of study by pointing out the importance of a galactic 'ridge', (an excess flux of radiation along the Milky Way which could be attributed to unresolved sources) in estimating the average distance to the resolved sources and the space-density of sources.

With the availability of Uhuru data, these distributional studies could be made more quantitative than they had in the past as was shown by Salpeter (1973) and Gursky (1973). Salpeter analyzed the latitude and longitude distribution of the sources in the central region of the Galaxy and the consequences for a ridge. His principal conclusion was that the bright sources at low galactic longitude were not Population I. Gursky analyzed the same data and argued that the sources outside the central region still showed a correlation with the spiral arms.* The general conclusions from these studies was as follows:

(1) The luminosity of the sources is within the range 10^{35-38} erg s^{-1}.

(2) The number of sources in the Galaxy is not large, perhaps no more than 100.

(3) The X-ray luminosity is concentrated in the range 2–10 keV. Possibly most of the X-ray sources are surrounded by an absorbing envelope of order 10^{22} atoms cm^{-2} of hydrogen. In particular, the various low energy (<1 keV) rocket surveys (c.f. Seward *et al.*, 1972) fail to reveal a significant number of sources not seen at higher energy.

(4) The sources in the outer regions of the Galaxy are found in the spiral arms and are thus of extreme Population I.

(5) The bright sources in the central region are not of extreme Population I, but rather seem to be distributed as are the ordinary stars.

The newly prepared 3U Catalog (Giacconi *et al.*, 1974) can be used to extend these conclusions. However, the really new development has been the abundance of information available on single X-ray sources. For a significant number of objects we have good indication of distance, mass and stellar type. Thus, provided we can assure ourselves that the identified sources represent a proper sample of the X-ray sources, many of the properties sought after by studying the sources as a whole are at hand.

2.1. Results from the 3U catalog

The 3U Catalog lists 161 discrete sources ranging in intensity from the most intense Sco X-1 at 3×10^{-7} erg cm^{-2}-s to the faintest at about 10^4 times weaker. The distribution of all these sources is shown in Figure 1 which is a plot in galactic coordinates of

---

* The significance of identifying the X-ray sources with a specific population is that it gives a clue as to the mass, age, and chemical composition of the stellar progenitors. In general, it is found that the spatial characteristics of the stars in the Milky Way correlate with these properties (c.f. Blaauw, 1965).

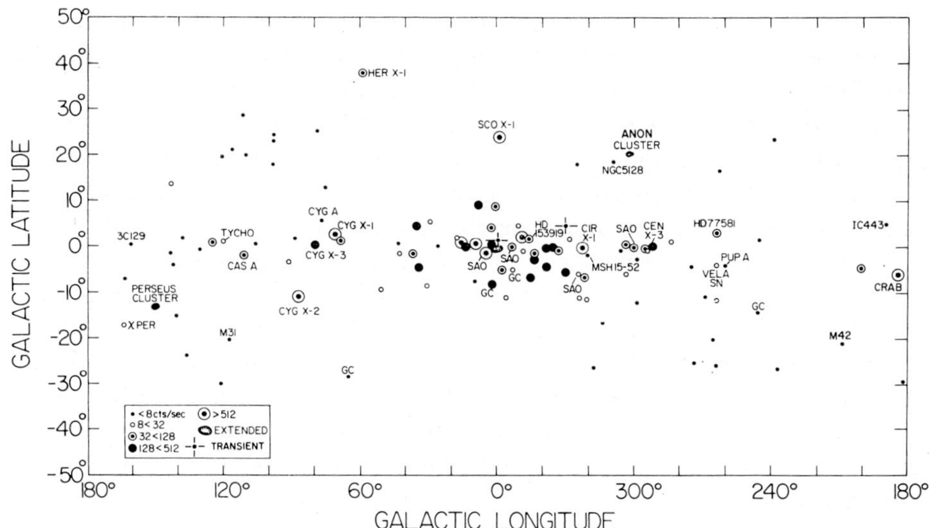

Fig. 1. The distribution of X-ray sources from the 3U Catalog galactic coordinates. Only low latitude sources are plotted.

the source positions. The sources are divided into intensity classes based on the photon counting rate observed by the Uhuru instruments in the 2–6 keV energy range (a count rate of 1 count s^{-1} is approximately 1.7×10^{-11} erg cm^{-2}-s).

The striking feature about the distribution is the concentration of strong sources along the Milky Way, especially at low galactic longitude (l). The sources at high galactic latitude (b) are much weaker. In fact, many of them can be identified with specific extragalactic objects, and it can be argued based on their distribution that most of them must be extragalactic. The only exceptions to this are identified objects; i.e., Her X-1 and Sco X-1, which are both galactic sources. It is the strong sources which concern us here. They make up the great majority of the identified galactic sources and contain those individual objects which have come to exemplify the field as a whole. There is also a class of weak (< 8 counts s^{-1}) sources associated with the Galaxy. These can be seen only at large angular separation from the galactic center and along the Milky Way where the density of strong sources is not great; in particular, in the interval 80–160° l. Within the band $\pm 5°$ of latitude, there are 16 3U sources < 8 counts s^{-1} whereas only 5 are expected based on the all sky density of such weak sources.

The galactic X-ray sources are listed in the Appendix. The first portion of the list includes those sources with $I > 8$ counts s^{-1}, which are not identified with well known galactic objects such as supernova. These sources are listed in order of galactic longitude beginning at 320°. It is likely that all these are galactic. As will be discussed below, the region 320–40° delineates the central region of the Galaxy. We list separately the single sources seen in the Magellanic Clouds and, for completeness, the sources < 8 counts s^{-1} which are within 5° of the galactic plane. It is likely that many of these are extragalactic. Also listed separately are the transient sources, those objects which were seen for one period of time but not otherwise. In addition to galactic coordinates, the appendix lists the 3U name, which is the right ascension (hours and minutes) and declination (degrees) of the source, the intensity in Uhuru counts and the variability, being the ratio between the maximum and minimum daily average intensity with which the source was observed. A comments column lists other significant features regarding the X-ray emission, including optical candidates.

The distinguishing characteristic of the strong galactic sources is that they can not be identified with well defined classes of stars. They are not old novae, planetary nebula, Wolf-Rayet stars, flare stars, or any of the classes of peculiar stars which seem to be the seat of extraordinary activity. However, the identified X-ray sources give the essential information regarding the nature of these sources. As noted earlier, one important group are binaries in which the companion star is a B0 supergiant; several others appear also to be binaries, but the companion appears to be a more common, late spectral type star.

2.2. Evidence for Extreme Population I Sources

The X-ray sources associated with massive B0 supergiants, such as Cyg X-1 and 2U0900−40, are clearly of extreme Population I. The significance of this is their youth – it can be no more than about 10^7 years since they have condensed out of the interstellar medium. These objects are presumably examples of the class of sources associated with spiral arms and exhibiting a very small dispersion in galactic latitude.

The distributional evidence for an association with spiral arms is actually very weak. The Cygnus region is frequently cited as an example of such an association. As can be seen from the Appendix, only 3 sources >8 counts s^{-1} lie in the range 70–90° longitude which delineate that arm; however, of those sources, Cyg X-3 is now known to be about 10 kpc distant, far outside the limits of this arm, and Cyg X-2 will be cited below as belonging to a different population. Thus, the 'association' reduces to the single object Cyg X-1, which is hardly significant.

The Centaurus Arm, the region 280–300° longitude, seems to be a better delineated feature since it contains 7 X-ray sources and is followed (304–320°) by a significant region devoid of sources.

The X-ray sources that may be of extreme Population I are listed in Table II. The average latitude of the identified sources is $\sim 2°$, and the average distance from the galactic plane is about 90 pc based on the distance estimate to the companion stars. This is as expected for an extreme Population I. The spread in X-ray luminosity is from $3 \times 10^{35-38}$ erg s^{-1}.

Further evidence for extreme Population I objects came from the statistical association of X-ray sources with bright B stars found by Gursky (1972) using the selected sources from the 2U Catalog. Such an association was first suggested by Liller (1972). When this is done with the 3U Catalog, out of 41 unidentified sources with positional uncertainty less than 0.02 deg^2, ten are found to be associated with stars listed in the

TABLE II

Extreme Population I X-ray sources

3U Designation	l, b	Z (pc)	I (counts s^{-1})	Distance (kpc)	L (erg s^{-1})
Positively Identified					
0900 − 40	(263, 4)	90	100	1.3	3.2×10^{37}
1118 − 60 (Cen X-3)	(292, 0)	60	350	10	6×10^{37}
1700 − 37	(348, 2)	65	102	1.7	6×10^{37}
1956 + 35 (Cyg X-1)	(71, 3)	130	1175	2.5	1.4×10^{37}
0115 − 37 (SMC X-1)	–		20	60	1.4×10^{38}
Possible Identifications					
0352 + 30	(163, −17)		20		X-Perseus
1145 − 61	(296, 0)		72		B0, 9.2m; B2, 8.6m
1556 − 60	(324, −5)		17		B9, 8.9m
1624 − 49	(335, 0)		50		B8, 9m
1727 − 33	(354, 0)		65		B5, 6.7m; B8, 9.4m

Old population X-ray sources

Positively Identified					
1617 − 15 (Sco X-1)	(359, 24)	80–800	17000	0.2–2	10^{36}–10^{38}
1653 + 35 (Her X-1)	(58, 38)	3700	100	6	7×10^{36}
2030 + 40 (Cyg X-3)	(80, 1)	120	194	10	4×10^{37}
2142 + 38 (Cyg X-2)	(87, −11)	120	540	0.600	4×10^{35}
Possible Identifications					
1746 − 37	(354, −5)		31		Glob. Cluster, NGC6441
1820 − 30	(3, −8)		250		Glob. Cluster, NGC6624

SAO Catalog whereas the number of accidental associations is expected to be fewer than three. Of the ten, 5 are associated with B spectral type stars and, surprisingly for two of these, 3U1727−33 and 3U1145−61, two B stars are within the X-ray area positional uncertainty. These associations are listed as possible candidates in Table II. These plus other possible stellar associations are listed in the Appendix in the comments on individual sources.

The probability of having two B stars within a single X-ray error box in two instances is extremely small. This may be a reflection of the tendency of B stars to be found in clusters; thus, the B stars are not randomly distributed but rather tend to group together. In fact, one X-ray source 3U1624−49 may be within the star cluster NGC6134.

2.3. Evidence for Old Population Sources

Several X-ray sources are known not to be associated with massive B0 stars. These sources are listed in Table II. In particular, the optical companion of Her X-1 is late

A, that of Cyg X-2 is either of F or G spectral type, and no stellar companion can be detected for Sco X-1.* Thus, these objects could be common, low mass stars, much like the Sun, either of Population I or Population II. It is noteworthy that these three sources lie at the largest latitudes – 38°, 11°, 24°, respectively – of any of the galactic X-ray sources. Cyg X-3 is the fourth object with a low mass companion, based on the assumption that it is a binary system with a 0.2 day period. A massive star cannot be a member of such a short period binary system. The range in luminosity of these sources is 4×10^{35}–4×10^{37} erg s^{-1}.

One piece of evidence favoring a Population II halo origin is that two X-ray sources, 3U1746−37 and 3U1820−30, are coincident with the globular clusters NGC6441 and NGC6624, which is statistically a very unlikely event.

2.4. Very low luminosity sources

The lowest luminosity of an identified source is $\sim 4 \times 10^{35}$ erg s^{-1}, however, there is evidence for sources of lower luminosity. One source, 3U0352+30 is coincident with the star X-Perseus. This star is cited as possibly being a runaway from the Perseus OB2 association of O–B stars, Blaauw (1961). On this basis, its distance is about 300 pc and the luminosity of the X-ray source, about 4×10^{33} erg s^{-1}.

Another piece of evidence for low luminosity is the excess of weak (<8 counts s^{-1}) sources found along the Milky Way in the interval 80–160° longitude. If their distance is the order of 5 kpc, their luminosity is about 4×10^{35}, which is at the lowest end of luminosity of the identified sources. Thus, they could be merely more distant examples of known sources.

2.5. Are all the X-ray sources in close binary systems?

We can estimate the probability of seeing a binary system in eclipse by making assumptions on the mass of the system and the radius of the larger star. The probability of eclipse is simply,

$$P \text{ (eclipse)} = r/(R_1 + R_2),$$

where

$r =$ radius of the star
$R_1 + R_2 =$ separation of star and X-ray source,

and from Kepler's Laws,

$$R_1 + R_2 = (T/2\pi)^{2/3} \left[G(m_1 + m_2) \right]^{1/3},$$

where

$T =$ binary period
$m_1, m_2 =$ mass of the binary components.

---

* Sco X-1 is seen as +13 m at minimum intensity. In order for there to be present an equal intensity of light from a B0 supergiant (which could hardly go undetected) the distance to Sco X-1 would need to be ~ 50 kpc, which is well outside our Galaxy.

Then
$$P \text{ (eclipse)} = 0.19\, r/T^{2/3}\, m^{1/3}$$

where now r is in units of solar radii, T is in units of days, and m is in units of solar masses.

Three X-ray sources of the extreme Population I category ($0900-40$, Cen X-3, $1700-37$) are seen to eclipse with periods between 2.1 and 8.9 days. Assuming the total mass to be 20 $m_\odot$, the probability of eclipse of these objects varies between 16 and 43%, if the radius of the B star is taken to be $10 r_\odot$ and between 32 and 85% if the radius is taken to be $20 r_\odot$. Having seen these three in eclipse, we would expect there to be between 5 and 10 systems of this kind in the Galaxy with the same strength.

In the case of the two eclipsing sources of the older population (Her X-1 and Cyg X-3) we assumed the total mass to be 2 $m_\odot$ and take the radius to be either 1 or $2 r_\odot$. Again, we expect between 5 and 10 systems of this kind making a total of 10–20. In fact, 25 galactic X-ray sources are seen with $I > 100$ counts s^{-1}. Considering the fact that longer period binaries are less likely to eclipse and more difficult to discover, the data are consistent with all the X-ray sources being in binary systems of the kind seen in eclipse. However, these number also are consistent with $\sim \frac{1}{3}$ to $\frac{1}{2}$ of the X-ray sources not being in binary systems at all.

2.5. SPACE DENSITY AND LUMINOSITY OF THE X-RAY SOURCES

As discussed above, the identified X-ray sources lie at distances ranging from 500 pc to 10 kpc and have luminosity from 10^{35}–10^{38} erg s^{-1}. This information by itself

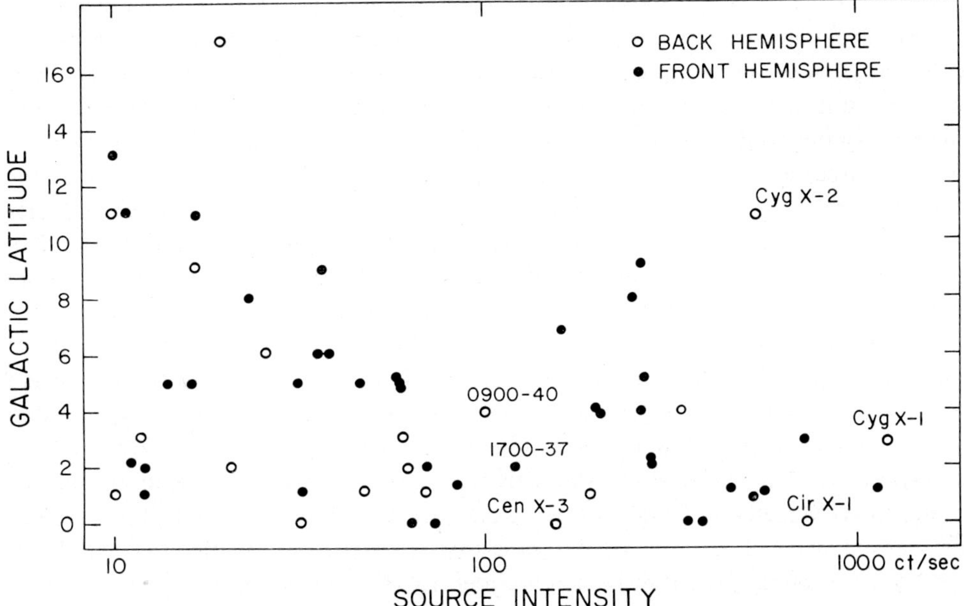

Fig. 2. The intensity-latitude distribution of X-ray sources from the 3U Catalog. Open circles are those sources appearing in the central region of the Galaxy. The closed circles are sources in the outer region of the Galaxy. Sco X-1 and Her X-1 are not plotted.

provides a qualitative answer to the question of the space density and luminosity of the X-ray sources. A more general answer can be found by looking at the distribution of all the sources. To aid in discussing these sources we have plotted their distribution in latitude-intensity space in Figure 2. The sources are divided between two regions, the central region between $320° < l < 40°$ and the outer region $40° < l < 320°$. Most of the sources in the central region may belong to the central bulge of our Galaxy. In Table III, we list the correspondence between intensity in Uhuru counts and luminosi-

TABLE III

Correspondence between Luminosity (erg s^{-1})
distance and Uhuru counting rate

Uhuru count rate (counts s^{-1})	Distance (kpc)			
	1	2	5	10
10	2×10^{34}	8×10^{34}	5×10^{35}	2×10^{36}
30	6×10^{34}	2.5×10^{35}	1.5×10^{36}	6×10^{36}
100	2×10^{35}	8×10^{35}	5×10^{36}	2×10^{37}
300	6×10^{35}	2.5×10^{36}	1.5×10^{37}	6×10^{37}
1000	2×10^{36}	8×10^{36}	5×10^{37}	2×10^{38}

ty. To a good approximation, this luminosity is the total power between 2–6 keV radiated by the source and for many of the sources is close to the total radiated power.

In the outer region, it is likely that all sources > 8 counts s^{-1} have been detected by Uhuru, with the possible exception of those lying in the Centaurus region where source confusion could still be a problem. Since there is no great increase of sources at low intensity, it is likely that these sources are distributed throughout the outer portion of the Galaxy out to distances of the order of 10 kpc. What we mean by this is simply that 8 sources are seen $I > 100$ counts s^{-1}, and 25 sources $I > 10$ counts s^{-1}. If the sources were distributed in a disc, their number would increase as $1/I$, which is clearly not the case. Also, the identified sources range in distance from 600 pc for Cyg X-2 to ~ 10 kpc for both Cen X-3 and Cyg X-3. Thus, the fainter sources in this region are likely to be a mix between more distant and lower luminosity sources.

The central region is more complex since it is likely that we fail to see a significant number of the faint sources due to confusion. It is commonly believed that the bright sources in this region (I greater than several hundred counts s^{-1}) are ~ 10 kpc distant and are similar to those found in the Magellanic Clouds (c.f. Gursky, 1973; Margon and Ostriker, 1973); however, this is likely to be a serious over-simplification. For one thing, three Magellanic Cloud sources which are at a distance of ~ 60 kpc are seen with an intensity of 20 counts s^{-1}. Translated to 9 kpc, the distance to the galactic center, these sources would be seen at an intensity of ~ 900 counts s^{-1}. Even allowing for interstellar absorption, their counting rate would be expected to be at least 500 counts s^{-1}. In fact, only five are seen with such a high counting rate, in spite of the fact that the Milky Way is believed to contain ten times the number of stars as the Magel-

lanic Clouds. The average counting rate of the bright sources in the central region is more like 300 counts s^{-1}. Another difficulty with this idea is that one of the Magellanic Cloud sources, SMC X-1, is apparently in the class, close binary with B supergiant, similar to Cyg X-1, Cen X-3, 0900−40, and 1700−37 in the Milky Way. However, three of these are in the outer region and not one is included among the brighter sources in the central region.

Whether one tries to make them correspond to outer region or bright central region sources, there are too many Magellanic Cloud sources and they are too luminous. Thus, their presence in the Magellanic Clouds may be a reflection of the difference in the average kinds of stars there compared to what is found in the Milky Way; or perhaps more simply, a chance occurrence.

Fortunately, there is a more definite method for determining the space distribution of the central region sources and that is the use of the galactic ridge. As discussed above, this is the excess counting rate, seen along the Milky Way, which may be attributed to unresolved sources. Recently, Dilworth *et al.* (1973) have attempted to determine the luminosity function of the X-ray sources using the ridge argument and the actual distribution of the sources. Their conclusion is that very few unresolved sources are needed to account for that background.

2.6. SUMMARY

The distribution of the X-ray sources around the Milky Way is compatible with a dual population, one comprising very young objects, newly born and close to the galactic plane. The other category is, on the average, more distant from the plane of the Galaxy and represent a much older class of stars.

There also appears to be a very large spread in the intrinsic X-ray luminosity of the X-ray sources from at least 10^{35} to 10^{38} erg s^{-1}.

3. Observations of Specific Galactic X-Ray Sources

We turn now from the general characteristics and distribution of the galactic sources to the properties of selected sources, specifically those which have been identified with optical objects. Of the ones to be considered, all but two are confirmed to be binaries, reflecting the observational bias of identifying an optical counterpart once an X-ray periodicity is established.

We will start with Cyg X-1, a source associated with a B0 supergiant, which is currently the most likely candidate for a black hole. Cen X-3 will then be considered a pulsing binary X-ray source, which has recently been found to also be in a system with a B0 supergiant. Her X-1, the other known pulsing binary X-ray source will then be discussed. Unlike Cen X-3, it is a low mass system, involving an A or F type star. The next source to be considered in detail will be Cyg X-3, a distant source seen in X-ray, infrared and radio which also appears to be in a low mass system. We will then consider Sco X-1 and Cyg X-2 which are representative of those galactic sources which, whether or not observed to be in binary systems, are taken to be members of

an older lower mass population of stars. Finally, we will consider the other binary sources, 3U0900 − 40, 3U1700 − 37, Cir X-1, and SMC X-1; all but Cir X-1 associated with O–B stars, and SMC X-1 particularly relevant due to its known distance of 61 kpc. A summary of the characteristics of all of these sources is given in Table I.

3.1. Cygnus x-1

In the study of compact galactic X-ray sources, Cyg X-1 has occupied a central position. It is the source with the widest observed range of time variability, the source first identified with an optical binary system, and the source most commonly considered as a possible black hole. Cyg X-1 has been observed extensively in the optical and radio frequencies as well as in X-rays, but it is clear that it was the fast-time variability of the X-ray emission which sparked the early and continuing interest in the source. Indeed, it was the discovery of this variability in the early Uhuru data which led to the successful search for other variable sources, and to the discovery of the whole class of binary X-ray sources.

The first reports of X-ray variability in Cyg X-1 occurred in the years 1966–1968 and were based on comparison of results from various balloon and rocket flights (e.g., Byram *et al.*, 1966; Overbeck and Tananbaum, 1968). It was in early 1971 that variability on time scales of less than a second was first reported from Uhuru observations (Oda *et al.*, 1971). This variability appeared at times periodic, and led the authors to hypothesize that the X-ray object was a collapsed star – possibly a neutron star or a black hole. This observation was soon followed by reports from other observers pre-

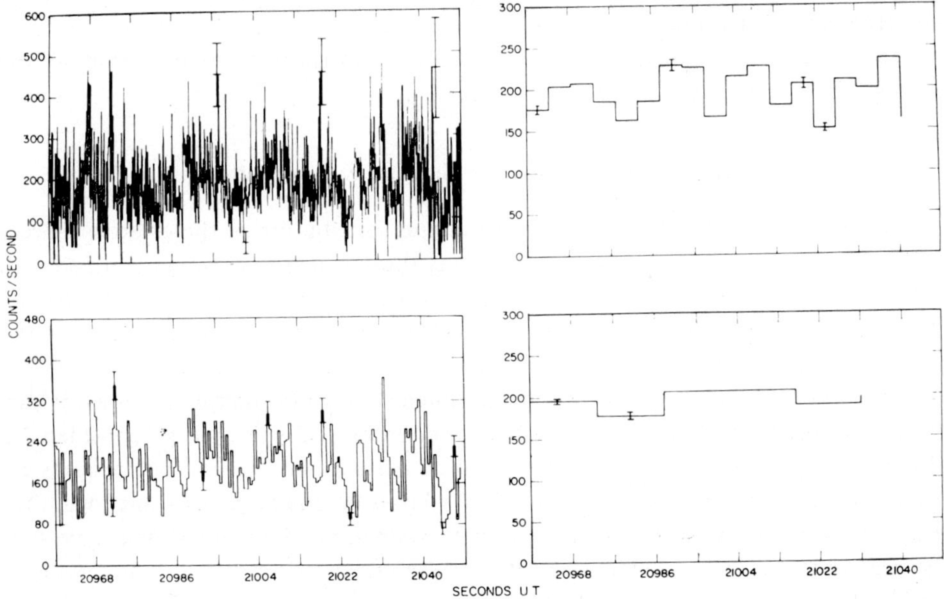

Fig. 3.   Observation of Cyg X-1 on 1971 June 10. Data have been corrected for triangular collimator response. Data are summed over 0.096 s, 0.48 s, 4.8 s, and 14.4 s intervals. Typical 1σ error bars are shown. Published by the University of Chicago Press; © 1971, University of Chicago Press.

senting varying results, including flaring on a 50 ms time scale and a 300 ms periodicity (see e.g., Rappaport *et al.*, 1971a; Holt *et al.*, 1971; Shulman *et al.*, 1971). An extensive analysis of six months of Uhuru data by Schreier *et al.* (1971), and also by Oda *et al.* (1972) led to the conclusion that pulse trains existed with periods ranging from a few tenths of a second to seconds, but only lasting to some tens of seconds. No single period was consistently present, but significant variability existed on all time scales studied, ranging from a tenth of a second to days (see e.g. Figure 3). The analysis was later extended to include a study of the spectral behavior of these time variations (Brinkman *et al.*, 1974). It was found that although the fluctuations occurred concurrently at both lower (2–5 keV) and higher (5–12 keV) energies, the higher energy pulses were typically narrower. There also appeared to be more power contained in the fluctuations at higher energies.

The range of time variability of Cyg X-1 has recently been dramatically extended by the observation of 1 ms bursts (Rothschild *et al.*, 1974). The variability observed leads us to consider a compact object smaller than about 10^7 cm as the source of the X-ray emission. This follows from the fact that the emitting region can be no larger than the distance which electromagnetic radiation can travel during the time in which the source intensity changes significantly.

The discovery of fast X-ray pulsations from Cyg X-1 first from Uhuru and then as extended by other observers has been one of the most significant achievements in X-ray astronomy in recent years. It not ony stimulated a search for similar behavior in other sources leading to the discovery of Cen X-3, Her X-1, and the other binaries, but it also led to a concentrated and mostly successful effort to identify the radio and/ or optical counterparts of both Cyg X-1 and the other sources. As was mentioned earlier, the binaries with identified optical counterparts have taken on a special status, allowing a far more detailed knowledge of their physical characteristics and a more direct means of integrating the results of X-ray observations with those of the rest of observational astronomy.

The accurate X-ray location of Cyg X-1 from Uhuru and an MIT rocket flight (Rappaport *et al.*, 1971b) led to the discovery of a radio source by Braes and Miley (1971) and Hjellming and Wade (1971). This radio source was distinguished by the fact that it was not present up to the end of March 1971. It was later found that the X-ray emission underwent a transition simultaneous with the radio source's appearance (Tananbaum *et al.*, 1972a; Hjellming, 1973). The 2–6 keV flux decreased by a factor of 4, while the 10–20 keV flux increased by 2 as shown in Figure 4. The spectrum before the transition had a low energy excess which could be fit by either a power law with energy index of 4 or an exponential with a temperature of 11×10^6 K (Schreier *et al.*, 1971; Tananbaum *et al.*, 1972a). The disappearance of this low energy component as the radio source appeared may indicate a decrease in the plasma density, reducing the X-ray emission measure and lowering the plasma frequency or the free-free self-absorption of the radio emission.

The precise error box determined for the radio source ($\lesssim 1''$) contained the star HDE226868. This star was found to be a 5.6 day spectroscopic binary (Webster and

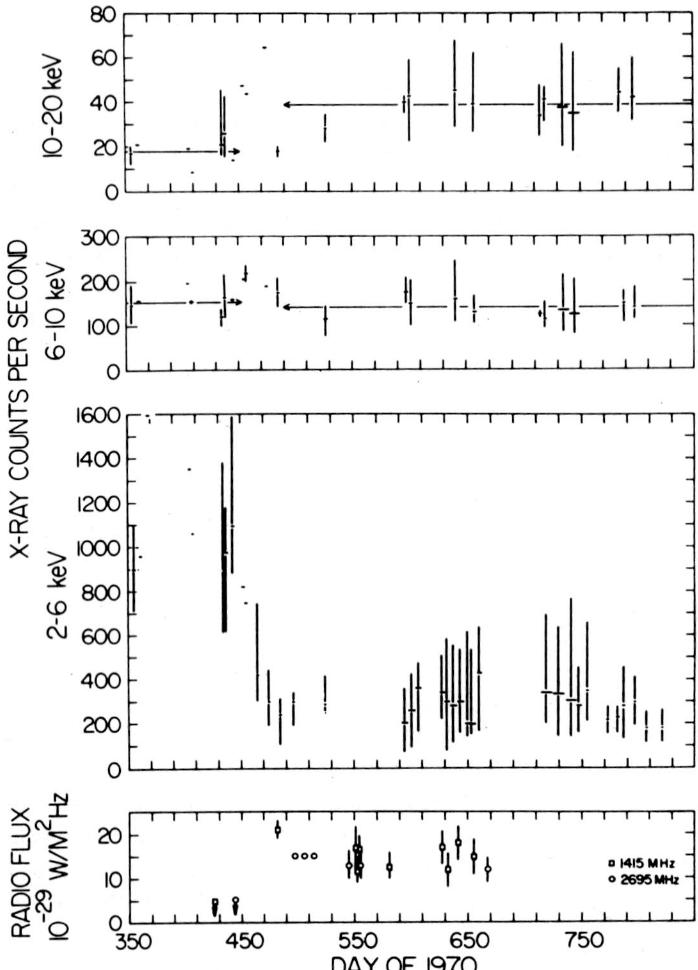

Fig. 4.   Sixteen months of observations of Cyg X-1. X-ray data are shown for three energy bands, 2–6 keV, 6–10 keV, and 10–20 keV plotted vs day of 1970. The transition discussed in the text occurred in the period near day 450. The radio data are shown at the bottom of the figure.
© 1973, International Astronomical Union.

Murdin, 1972; and Bolton, 1972). The star appears as a 9th mag. B0Ib supergiant, at a minimum distance of 2.5 kpc (Bregman *et al.*, 1973; Margon *et al.*, 1973). The primary thus has a mass of about 30 $M_\odot$, leading to a mass $\gtrsim 6$ $M_\odot$ for the X-ray source. Even if the primary consists solely of a helium burning core, the secondary mass is still $\gtrsim 3$ $M_\odot$ (Van den Heuvel and Ostriker, 1973). The lack of an observed eclipse in the hard X-rays indicates an inclination of the orbital plane to our line of sight. There is no contradiction with the possible 5.6 day effect at lower energies (Mason *et al.*, 1973) which might originate at a larger radius from the compact object. It is the high mass necessitated for the X-ray source combined with its compact nature which leads to the consideration of Cyg X-1 as a black hole.

We can summarize three main points leading to the identification of Cyg X-1 as a black hole: (1) HDE 226868 is the optical counterpart of the X-ray source, (2) the mass of HDE226868 is greater than 20 $M_\odot$, and (3) the X-ray emitting object is compact. The identification of HDE226868 with Cyg X-1 is well established via the positional coincidence of the optical and radio along with the correlated intensity variation of the X-ray and radio. Further evidence is given by the possible 5.6 day effect in soft X-rays, and also by the fact that other X-ray binaries have also been identified with B0I supergiants.

The large mass of the primary has also survived extensive critical discussion. The original conservative estimate of a primary mass of greater than 20 $M_\odot$ was based on the assumption that the star was a normal B0Iab supergiant according to its spectral characteristics. This was criticized by Paczyński (1973) and by Trimble *et al.* (1973), based on the fact that a spectrum is indicative of only the surface gravity and temperature of a star. Thus, one could create a low mass model of lower luminosity which mimics the spectrum of a B0 supergiants. This, however, was ruled out by the distance measurements already mentioned. It was also noted that the proximity of the strong X-ray source might alter the appearance of the spectrum. However, no reflection effect is observed, nor for that matter is one really predicted; the optical variations are consistent with tidal distortion, which in turn helps determine the mass function (see e.g. Mauder, 1973). Further evidence for the non-anomalous appearance of the optical spectrum comes from the normal spectrum of SMC X-1, whose distance is known (see below). Thus it appears very likely that the X-ray object does indeed have a mass of at least 6 $M_\odot$.

The X-ray variability leaves little doubt as to the compact size of the X-ray emitting region. The only alternative to a compact star is the presence of active regions on or around a normal star. However, a main sequence star of 6 $M_\odot$ should produce appreciable visible light in the system, both by emission and by reflection from the primary. No such contribution is seen. Furthermore, there is no obvious X-ray emission mechanism in this case.

In conclusion, there appears strong evidence that the X-ray object is compact with a mass significantly greater than 3 $M_\odot$. It is impossible to have a neutron star this massive (Ruffini, 1973) and it has not been demonstrated how a differentially rotating white dwarf could be stable and could emit the observed X-rays. However, as will be discussed in more detail in Section 4, an accretion disk around a black hole is consistent with all observations thus far. A modest accretion rate of $\sim 10^{-9}\ M_\odot\ \mathrm{yr}^{-1}$ can produce the X-ray luminosity, an integration over the entire disk can produce the observed power law spectrum, and irregularities in the accretion disk can produce the observed irregular variability with time scales down to a millisecond.

3.2. CENTAURUS X-3

Centaurus X-3 was found to be a pulsing binary X-ray source in the wake of the discovery of the time variability of Cyg X-1 (Giacconi *et al.*, 1971; Schreier *et al.*, 1972). Cen X-3 pulsates with a period of 4.8 s, a stable period in contrast to the erratic

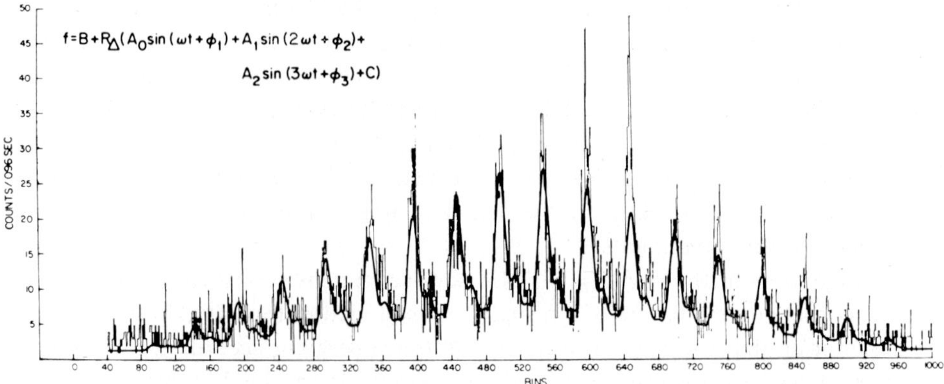

Fig. 5.   Counts accumulated in 0.096 second bins from Cen X-3 during a 100-s pass on 7 May 1971.
The functional fit obtained by minimizing χ^2 is shown as the heavier curve.
© 1973, International Astronomical Union.

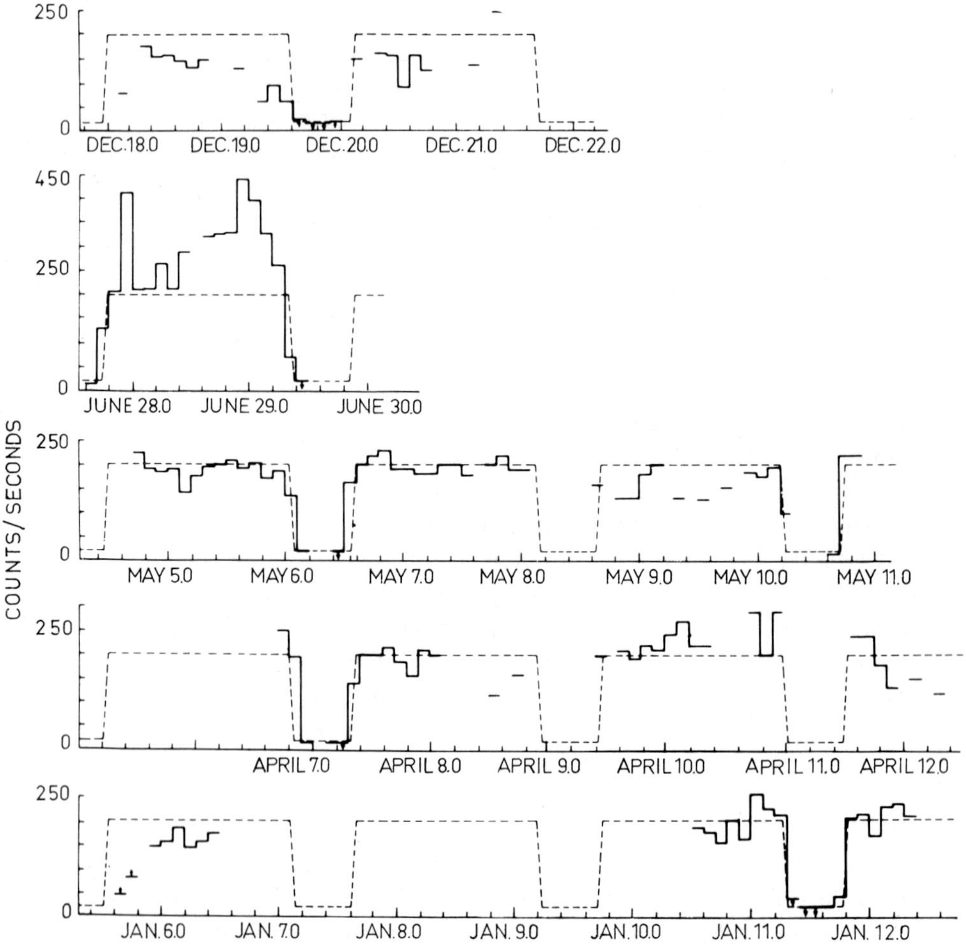

Fig. 6.   Observations of Cen X-3 by Uhuru showing regularity of eclipse. Published by the University
of Chicago Press; © 1972, University of Chicago Press.

variability of Cyg X-1, and in addition shows eclipsing behavior in the X-rays. Furthermore, the measurement of the Doppler variation of the pulsations established that the source was in a massive binary system.

The first and most obvious characteristic of Cen X-3 is its regular pulsations. Figure 5 shows 100 s of 2–6 keV counting rate data along with a minimum chi-square fit of a sinusoid with harmonics. The X-rays are seen to be over 70% pulsed with a 4.8 s period, although the shape of the pulse has been found to vary.

The binary nature of Cen X-3 is illustrated in Figure 6 where observations of the source during 1971 are shown along with an average light curve. The orbital period is 2.09 days, with an eclipse duration of 0.49 days, and with the transition into and out of eclipse taking some 0.04 days. The intensity during eclipse does not reach zero, and it can also be seen that the average non-eclipse intensity is quite variable. This variability in the average (over pulsations) intensity extends to the existence of extended low states, periods of time when the source is seen very weakly, and to times when the intensity alternates erratically between high and low states. These last may occur as the source is changing between an extended low and a 'thermal' state. No recurrent behavior has yet been observed for the extended lows, although the Uhuru data suggest time scales of 2 to 4 months for the durations of the states.

The 4.8 s period, although stable to better than 10^{-6} over time scales of at least days, has been observed to change over months. The net change in the period observed during 1971–72 was a decrease of approximately 1.5 ms yr^{-1}, i.e., $\dot{p}/p \cong -3 \times$ $\times 10^{-4}$ yr^{-1}. The changes in the period are definitely not uniform and probably not monotonic. A possible correlation between the pulsation period changes and the extended lows as determined from Uhuru data is shown in Figure 7; the period is shown on top and a characterization of the average intensity state is shown on the

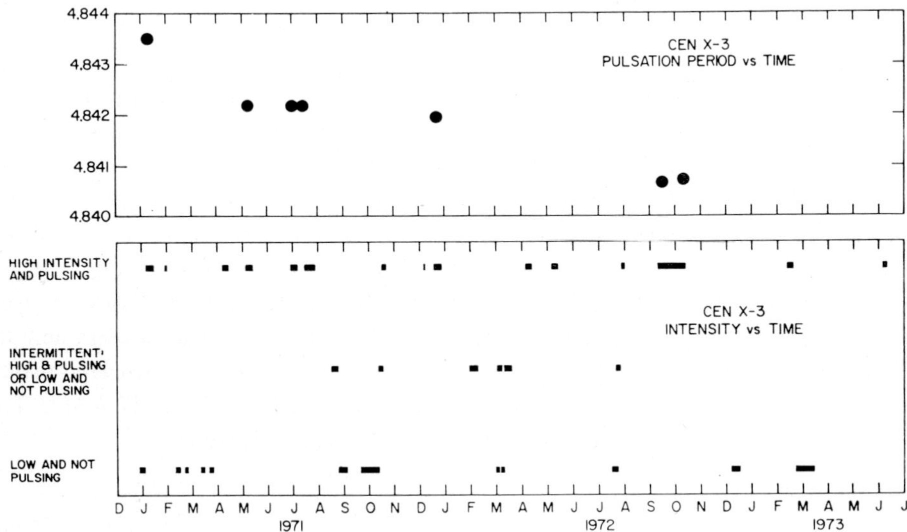

Fig. 7.   *Top*: Heliocentric period of Cen X-3. *Bottom*: Long term intensity states of Cen X-3.
© 1974, D. Reidel Publishing Co.

bottom. It can be noticed that the large changes in period occurred during intervals of time in which extended lows also occurred. If this correlation proves to be real, it may relate the spin rate changes of the X-ray source to the absorption of X-rays via an accretion model.

The orbital period of Cen X-3 has also been found to change, as shown in Figure 8. This change is seen to be non-linear and in fact to change sign. It thus cannot be explained by simple mass loss from the system, but may be due to competing mass-loss mass-exchange effects or possibly to apsidal motion.

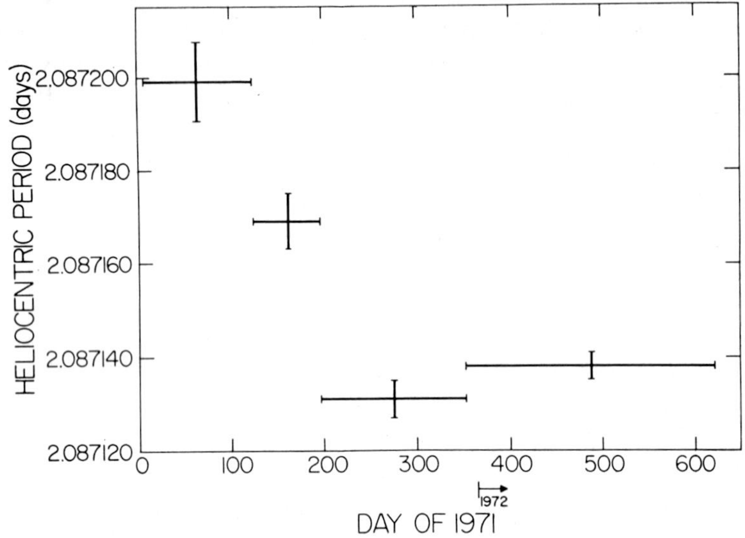

Fig. 8.   Orbital period of Cen X-3, determined over indicated baselines.
© 1974, D. Reidel Publishing Co.

The energy spectrum of Cen X-3 in the X-ray range is typically hard. There is a variable low energy cutoff ranging from 1.5 to 4 keV followed by a flat spectrum to 20 keV. Above ~ 25 keV, there is an exponential decrease (Ulmer *et al.*, 1973). There is also spectral variability across the 4.8 s pulse, with the peak intensity being harder.

Cen X-3 has recently been identified with a 13th mag. B0 supergiant by Krzeminsky (1973). The distance is thought to be about 10 kpc. The X-ray observations had previously predicted the existence of such a massive star in the system (Schreier *et al.*, 1972a); analysis of the Doppler variations of the pulsations had determined the orbital radius projected into our plane of view as 1.19×10^{12} cm, the projected orbital velocity of the X-ray source as 415 km s^{-1}, and the mass function as 15.4 $M_\odot$. This last is of course a lower limit on the mass of the optical companion.

3.3. HERCULES X-1

The X-ray source Hercules X-1 was discovered in the Uhuru data just as Cen X-3 was coming to be seen as an X-ray pulsar in a binary system (Tananbaum *et al.*, 1972b).

Since the two sources' X-ray characteristics are qualitatively the same, it was straight-forward to interpret the Her X-1 observations in the framework of a binary model. Helped by its early identification with an optically variable star of the same period, Her X-1 has become the most widely studied compact X-ray source. Many of the observations are explained in terms of a mass transfer binary system, in which a rotating neutron star is accreting matter from a companion. The amount of data compiled on this source along with the theoretical work done makes it impossible to review completely here. However, we will attempt to summarize both the main observational characteristics of the source and the basic theoretical considerations.

There are three periodicities observed in the Her X-1 X-ray data: a pulsation period of 1.24 s, an orbital period of 1.7 days, and an approximate 35 day on-off cycle. The short period pulsations, as shown in Figure 9, are similar to those of Cen X-3.

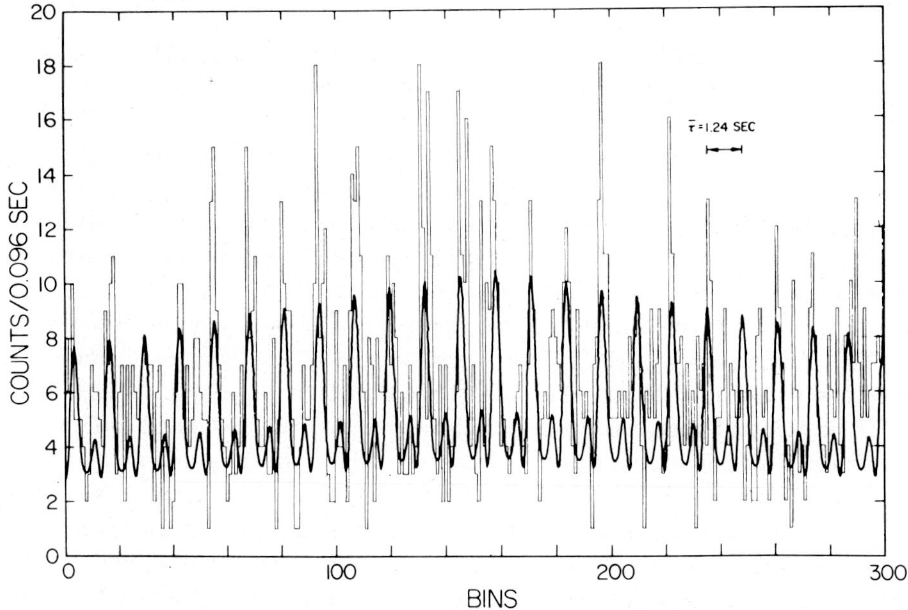

Fig. 9. Counts accumulated in 0.096-s bins from Hercules X-1 during the central 30 s of a 100-s pass on 6 November 1971. The heavier curve is a minimum χ^2 fit to the pulsations of a sine function, its first and second harmonics plus a constant, modulated by the triangular response of the collimator.
© 1974, D. Reidel Publishing Co.

The pulse shape has been studied in some detail (Doxsey *et al.*, 1973; Giacconi *et al.*, 1973; Holt *et al.*, 1974). It is variable both from pulse to pulse and on the average over minutes, sometimes appearing as a main pulse and an interpulse, and sometimes as a double peaked main pulse. Like Cen X-3, the pulsation period is stable over days, except for the binary Doppler effect, but shows changes on the order of microseconds from month to month as shown in Figure 10. A net decrease of 6 μs in 1.25 yr was observed, corresponding to $\dot{p}/p \cong -4 \times 10^{-6}\ \mathrm{yr}^{-1}$.

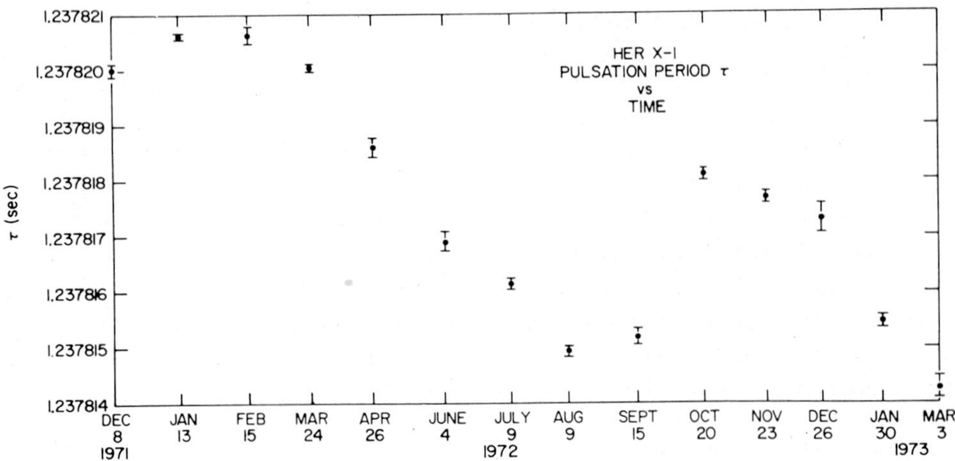

Fig. 10.   The Her X-1 heliocentric period determined for the 1.2-s pulsations during 14 high states from 1971 to 1973. © 1974, D. Reidel Publishing Co.

The 1.7 day orbital periodicity is observed both in the X-ray intensity and in the Doppler variations of the pulsation period. As shown in Figure 11, the 2–6 keV flux goes from about 100 counts s^{-1} to below a few counts per second. The transitions between high state and eclipse take less than 12 min, and the eclipse lasts for 0.24 days. No overall orbital period change has been seen in the X-ray data over more than a year to a part in 10^6. We can not exlude changes over several months of order 10^{-5} days. The orbital period of the system is also seen in visible light. Following an improved X-ray position by Clark *et al.* (1972) and the observation by Liller (1972a) that the star Hz Her had a large UV excess, Bahcall and Bahcall (1972) and Liller (1972b) discovered 1.7 day light variations of that star. The optical variations of 1.5 mag. are in phase with the X-rays, as seen in Figure 12, suggesting X-ray heating of the close side of the optical companion as the cause (Forman *et al.*, 1972). Spectroscopic observations are complicated by the heating; temperatures of 10^4 K and 7×10^3 K respectively are estimated for the two sides. The spectral type of Hz Her is late A or early F; the distance estimates range from 2 to 6 kpc. The optical light curve near minimum intensity is much narrower than the X-ray eclipse. This has been explained variously as (1) deep convective transport of the absorbed radiation (Wilson, 1973), (2) different sources of the optical light including emission from an accretion disk around Her X-1 (Boynton *et al.*, 1973; Crampton and Hutchings, 1972; Strittmatter *et al.*, 1973; Basko and Sunyaev, 1973), and (3) radiation from above the photosphere of Hz Her (Joss *et al.*, 1973).

Optical pulsations from Hz Her have been definitely detected on occasion (Lamb and Sorvari, 1972; Davidsen *et al.*, 1972; Middleditch and Nelson, 1973; Groth *et al.*, 1973). They are never more than 0.2% of the total light, and period measurements indicate that they may be emitted from gas moving in the binary system rather than from Hz Her.

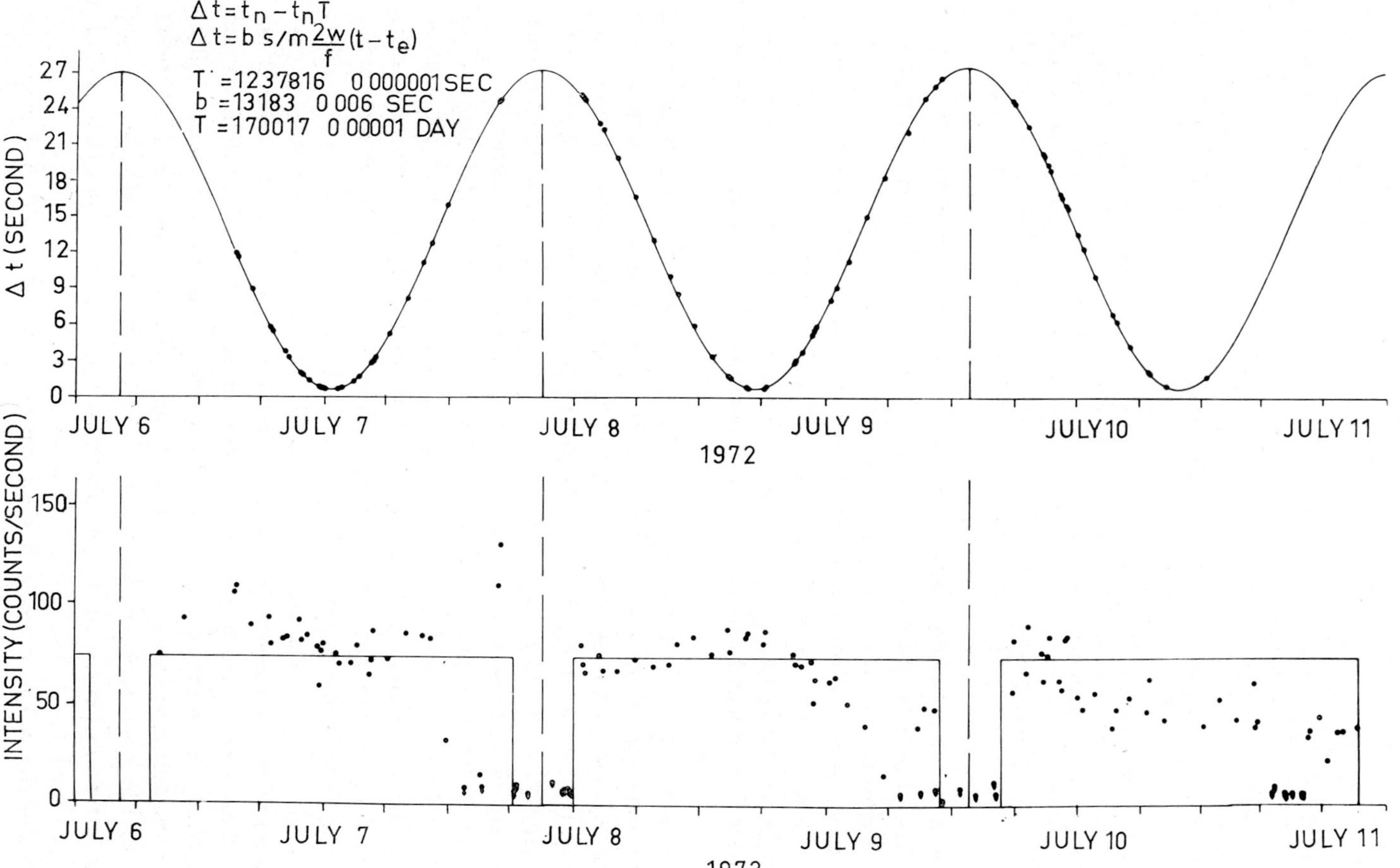

Fig. 11.   Five days of X-ray observations of Her X-1 in July 1972 are shown in the lower panel. The 2–6 keV intensity observations show three occulta-tion cycles. The data give individual sighting intensities with only upper limits observed during the eclipses. The eclipse centers are given by the vertical dashed line as determined from the Doppler variation of the 1.2-s pulse period shown in the upper panel, and the rectangular envelope is a schematic representation of the 1.7-day cycle of highs, lows and transitions. © 1974, D. Reidel Publishing Co.

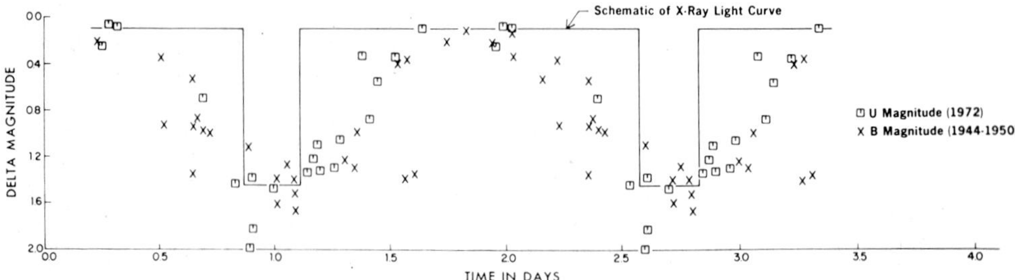

Fig. 12.   Optical observations of Hz Herculis folded with 1.700 15 days as a period. Squares indicate 1972 U magnitude observations and x's give B magnitude observations from plates taken 1944–1950. The optical curve shows much more sinusoidal variation than the rectangular 1.7-day X-ray light curve is also indicated. Observe that the phase of the optical minimum coincides with the center of the X-ray eclipse. Published by the University of Chicago Press;
© 1972, American Astronomical Society.

In addition to the pulsations and the orbital period, there is the third cycle. The source is seen to turn on about every 35 days. It then follows a regular pattern, decreasing to below detectable limits after about 11 days and remaining low for 24 days. Three such 'on-states' are shown in Figure 13. Uhuru observations have never shown any violations of this general cycle; on 67 of the 96 days in four 'off-states' as well as on scattered other 'off' days the source was not detected. However, a recent observation from the Copernicus X-ray experiment (Fabian *et al.*, 1973) reported emission during one 'off-state'.

Other features of the X-ray emission can also be seen in the figure. The turn-ons in the 35 day cycle are seen to be sharp. They all tend to occur near two orbital phases 0.2 and 0.7. Thus, within the approximate 35 day cycle, the actual turn-ons are separated by typically 34.0, 34.85 or 35.7 day intervals. Although there appears to be an underlying clock, it only keeps correct time on the average, and the turn-ons themselves are orbit phase dependent. Another important feature apparent in the figure are the intensity dips. In addition to the regular eclipses and the inherent variability of the source, additional recurring dips are seen. Although the average duration of the dips is comparable to that of an eclipse, both the intensity in a dip and the duration are quite variable. Furthermore, the dips regularly move with respect to the 1.7 day orbital phase, appearing progressively earlier before each succeeding eclipse in a given 11-day 'on-state'. Additional dips have been seen about 0.6 days after turn-on when the turn-on occurred at orbital phase 0.2. The X-ray spectrum during the dips as well as during turn-on shows a significant low energy cutoff of about 4 keV implying absorption of the X-rays. This is to be compared with a cutoff of 1.5 keV or less during the rest of the on-state. It should be noted that the 4 keV cutoff implies $\sim 10^{12}$ H atoms cm^{-3} in a region of order 10^{11} cm. Above 4 keV the spectrum is quite flat extending beyond 20 keV. At higher energies, it starts falling (Ulmer *et al.*, 1972a).

The mass of Her X-1 is of special interest, since it is thought to be an example of a neutron star in a binary system. As with Cen X-3, the Doppler variations of the pulsar

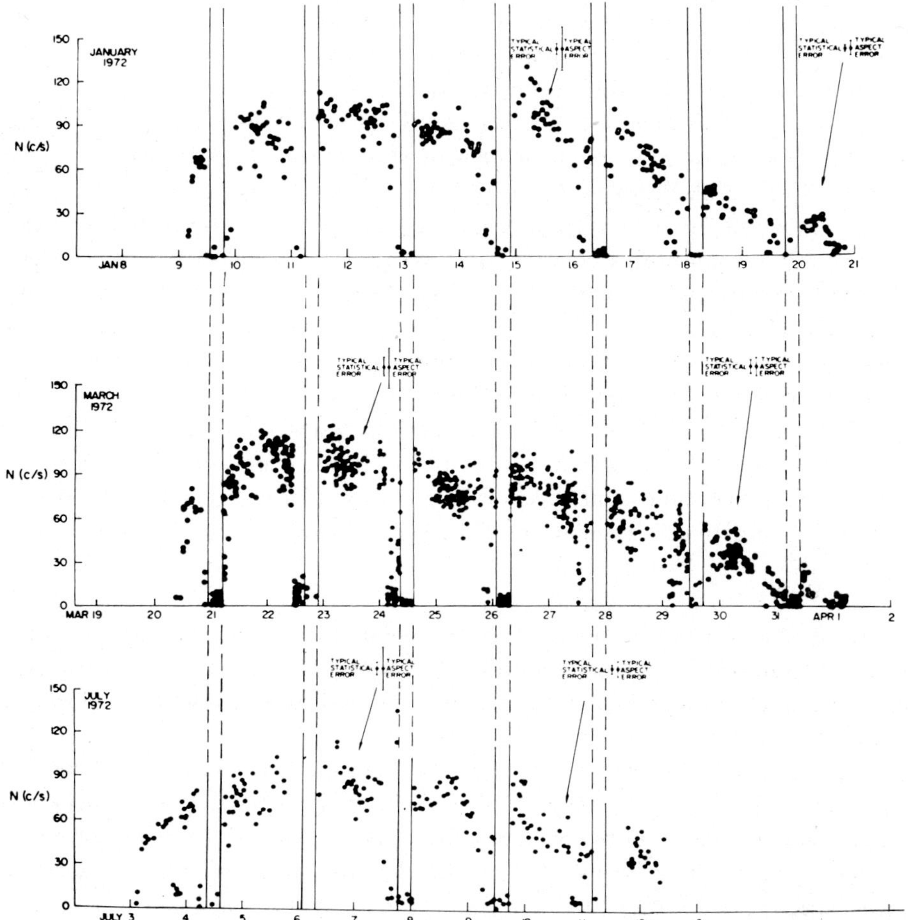

Fig. 13. Detailed 2–6 keV intensity observations of Her X-1 during January 1972, March 1972, and July 1972 high states. Each individual sighting of the source is represented by a dot with typical statistical and systematic errors indicated. The curves all show the sharp turn-on, the broad maximum, the gradual decrease over several days, and the presence of the dips discussed in the text. An example of a dip clearly separated from the eclipse can be seen on July 10.9. Published by the University of Chicago Press; © 1973, American Astronomical Society.

period allow for an accurate determination of the projected orbital radius (3.95×10^{11} cm), velocity (169.0 km s^{-1}), and mass function

$$\frac{M^3 \sin^3 i}{(M + M_x)^2} = 1.69 \times 10^{33} \text{ gm} = 0.85\, M_\odot.$$

Unfortunately, further information is needed to determine M_x itself. The determination of both the orbital velocity and the spectral type of Hz Her which would give the primary mass are complicated by the X-ray heating; current best values are 1.5–2.5 $M_\odot$. The inclination is estimated to be $>85°$ from the optical light curve. The X-ray mass is then found to be between 0.5 and 1.8 $M_\odot$. Another mass determination technique relates the mass ratio to the size of the Roche lobe and to the size of the occulting region. Using just the X-ray data then gives (for an inclination of 85°) masses of 1.04 and 1.99 $M_\odot$ for the X-ray source and Hz Her respectively.

The observations of Her X-1–Hz Her can be qualitatively explained by a model invoking an oblique rotating magnetic neutron star in orbit about Hz Her. It is generally believed that a black hole can not produce the regular pulsations. The presence of a gas stream and also an accretion disk is suggested by the presence and regular advance of the absorption dips. Accretion of matter onto the neutron star can easily power the X-ray luminosity of 10^{37} erg s^{-1}. The funneling of the matter by an oblique rotating magnetic dipole can explain the pulsations. The 35 day cycle can be caused either by cessation of mass transfer from the primary or by cessation of accretion onto the secondary, due for example to precession of the magnetic poles out of the accretion disk (see e.g., Lamb *et al.*, 1973). (Further discussion relevant to this model will be found in Chapter 4.) Although many aspects of the system are not yet explained, and there is far from universal acceptance of this model, it remains the most consistent and most detailed explanation of the Hercules system.

3.4. CYGNUS X-3

The source Cyg X-3 represents yet another mode of behavior, not yet seen in any other X-ray sources. Its intensity varies rather smoothly with a 4.8 h period, quite differently from the eclipsing binaries. It also is identified with a flaring radio source and a 4.8 h periodic infrared object.

The basic X-ray behavior is shown in Figure 14 where the 2–6 keV Uhuru observations of Cyg X-3 for 9 days in May are plotted (Parsignault *et al.*, 1973). The intensity is seen to vary by a factor of two with a period of 4.8 h. The behavior is essentially independent of energy from below 2 keV to 10 keV; thus, the light curve can not be due to photoelectric absorption which would preferentially attenuate low energies $(\sim \exp\{-(Ea/E)^{8/3}\})$. A single period of 0.199681 ± 0.000002 days is consistent with all data from December 1970 through May 1972. There is evidence for variability in the shape of the light curve, both on short time scales and on the average (Canizares *et al.*, 1973), making it difficult to ascertain period changes.

The average intensity of Cyg X-3 has also varied in the course of two years of observations. Intensity levels as averaged over the 4.8 h cycle ranging over a factor of four and persisting for possibly months have been seen, as shown in Figure 15. Although the X-ray intensity varies on times short with respect to 4.8 h, no significant fluctuations have been seen in times on the order of tenths of seconds.

The radio counterpart of Cyg X-3 is extremely variable, ranging from 20 mfu to 20 fu in several days (Braes *et al.*, 1972; Gregory *et al.*, 1972; Hjellming *et al.*, 1972). This includes the first giant radio flare of September 1972. The X-ray source was very active on the day before this radio flare was first observed; intensities a factor of two greater than previously seen were reported. The X-ray intensity observed from Uhuru at the time of the flare is shown in Figure 16. The X-ray spectrum also appeared hotter for the high intensity points.

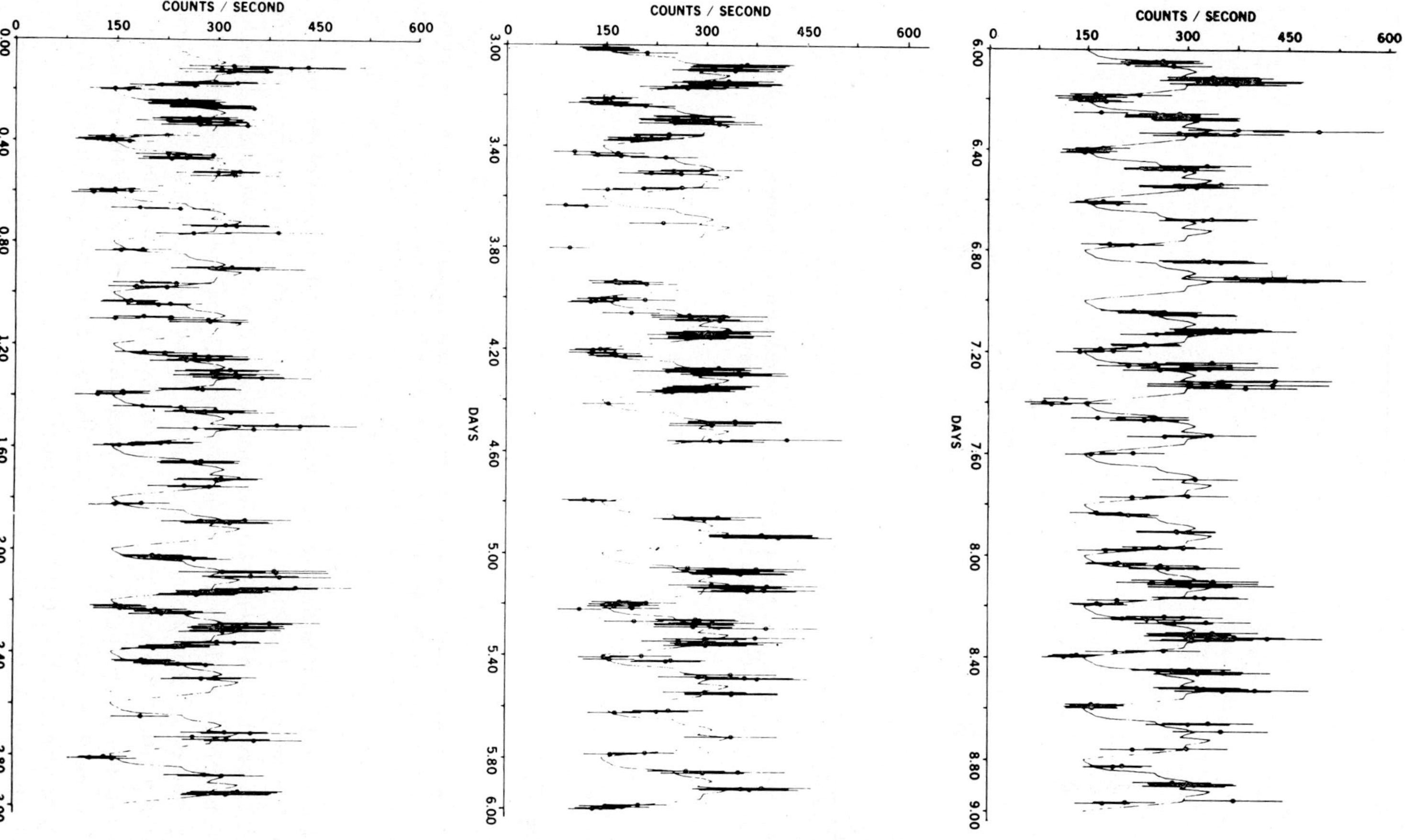

Fig. 14. Observations of Cyg X-3 from 8 May to 17 May 1972 are plotted as a function of time. The intensities are given in counts s^{-1} and have been corrected for aspect; error bars including both statistical error and systematic error due to aspect correction are shown. The time scale is in days with day 0.0 = May 8.0 (UT). Also shown is the average light curve obtained by folding all the data modulo 4.80 h, and finding the average intensity every 0.24 h.

© 1972, MacMillan Journals, Ltd.

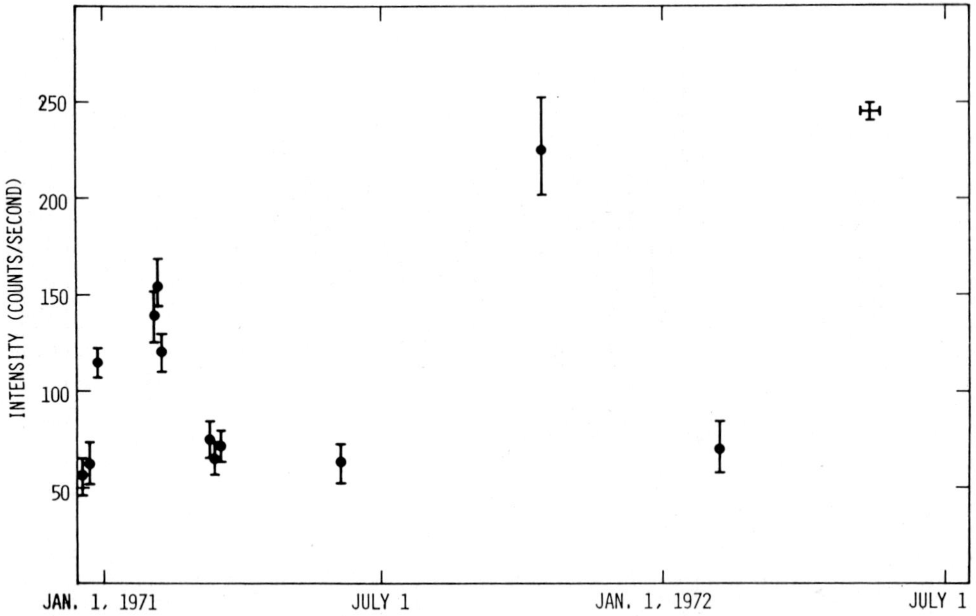

Fig. 15.   Average 2–6 keV intensity for Cyg X-3 on various days from December 1970 to July 1972 data are corrected for aspect. © 1974, D. Reidel Publishing Co.

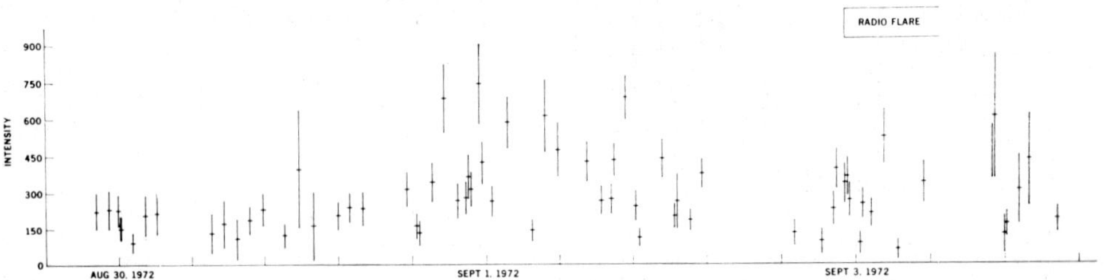

Fig. 16.   2–6 keV intensity for Cyg X-3 in late August and early September 1972. The time of the first observation of the giant radio flare is indicated. Note the difference in the X-ray behavior around September 1 compared to the preceding days. © 1974, D. Reidel Publishing Co.

Becklin *et al.* (1973) have reported coordinated X-ray and infrared observations of Cyg X-3. The X-ray and infrared fluxes vary in phase with a 4.8 h period, although there were observations when no periodic variations in the infrared were seen. In addition, there is unpredictable large amplitude flaring in the infrared flux. The agreement in 4.8 h period establishes the association of the X-ray with the infrared source and through the positional coincidence (2″) between the infrared and radio source, of the X-ray and radio source. The flux density is about 10^{-2} flux density in the infrared at 22 μm. The region is heavily obscured optically, and no visible star has been found which can be identified with the X-ray source.

The fact that Cyg X-3 displays a similar in-phase 4.8 h periodicity at wavelengths ranging from the X-rays to the infrared strongly suggests that it is an eclipsing binary system. Additional evidence in favor of this explanation is (a) the relative flatness of the infrared maxima, (b) the fact that the relative flux diffe,ences between maximum and minimum appear to be constant at different epochs despite changes in both the mean infrared flux and mean X-ray flux, and (c) the fact that the flare activity is apparently uncorrelated with the phase of the 4.8 h variation.

In a binary system with total mass M and period 4.8 h, the separation between the stellar cdnters is $1.4\,(M/M_\odot)^{1/3}\,R_\odot$. From the shape of the eclipse curve it can be concluded that the system is a contact, or near-contact, binary with each component on the order of one solar radius. Taking 10 kpc as the lower limit for the distance to the source (Braes *et al.*, 1973) and taking 6×10^{-28} W m^{-2} Hz^{-1} as the dereddened 2.2 μ flux density of the source, Becklin *et al.* (1973) obtained a 2.2 μ surface brightness temperature,

$$T_B \sim 6 \times 10^6 \left(\frac{M}{M_\odot}\right)^{-2/3} \left(\frac{R}{R_\odot}\right)^{-2} \mathrm{K}.$$

For any reasonable total mass of the Cyg X-3 binary system, the object seen at 2.2 μ has a surface temperature higher than that of the photosphere of any normal star. The high estimated value for the infrared surface brightness, and the similarity of the X-ray and infrared eclipse curve, suggests that both the X-ray and the infrared fluxes originate from the same hot object.

The phenomenological picture of Cyg X-3 as a contact system in which each component has a radius ~ 1 solar radius would appear to indicate that it is a completely different kind of object from Her X-1 and Cyg X-1 which are compact objects. However, Basko *et al.* (1973) have shown how the X-ray light curve in Cyg X-3 can be explained by the reflection of X-rays by the atmosphere of the normal companion in a close binary. If Cyg X-3 is a pulsing X-ray source in which the beam never crosses the Earth but continuously strikes the surface of the normal counterpart, then the reflected X-rays would be the only flux detected at Earth, with a 4.8 period smooth X-ray light curve. In this model, the X-ray flux should be $\lesssim 50\%$ polarized, and a jump at the K-edges of iron might be detectable in the spectrum. No attempt has been made to incorporate the flaring in the radio and infrared into a binary picture. The connection of magnetic field lines in an accretion disk (Sunyaev, 1973) in a manner analogous to solar flares could lead to the acceleration of electrons to sufficient energies for non-thermal synchrotron emission to occur (Blumenthal and Tucker, 1972). An alternative, simpler phenomenological picture just places the compact object within the atmosphere of the primary. The standard Roche model and accretion disk is then not necessarily relevant. Any short time scale variations are averaged out by the gas surrounding the source. The 4.8 h variation is then caused by Thompson scattering by the hot gas close to the system; the non-variable 2.5–3 keV low energy cutoff is caused by photoelectric absorption by the cooler gas further out. (See e.g., Davidson and Ostriker, 1973 for a model invoking a red dwarf – white dwarf binary, although they predict a periodic low energy cutoff.)

3.5. SCORPIUS X-1

Sco X-1 is the brightest source in the sky in the 1–10 keV range. It was the first X-ray source discovered and has been extensively studied, yet its nature is still a mystery. The tendency is to assume that it is a close binary source, but the evidence in support of this is only circumstantial.

The X-ray spectrum in the keV region is best fit by an exponential with an e-folding energy that varies with time but is on the average about 4 keV (Oda and Matsuoka, 1970). There is no evidence for periodic or even random pulsations of more than a few percent on a time scale ranging from a few milliseconds to about 1 min (Boldt *et al.*, 1971). However variations on the order of 50% or more have been observed on time scales of hours and days (Tananbaum, 1974). Above about 30 keV the spectrum departs from this behavior and exhibits a 'non-thermal tail' which is highly variable (Peterson, 1973). The optical and infrared continuum is consistent with an extrapolation of a flat spectrum from X-ray energies, with absorption becoming important in the infrared (Neugebauer *et al.*, 1969). The absorption leads to an estimate of 10^4–10^5 km for the emission region. Optical distance indicators such as reddening, polarization, or Ca K absorption indicate a distance greater than 200 pc; the optical luminosity of Sco X-1 at this distance is $\simeq L_\odot$. Any conventional star present in the system must be radiating at a much smaller rate.

Optically, Sco X-1 varies between 12–13 mag. The intensity varies by a few percent on a time scale of minutes and flares by a factor of two on a time scale of hours. There appear to be both active and passive states. Correlated optical/X-ray studies show that when the source is quiescent optically, the X-ray emission is also quiescent, and the variations in optical and X-ray are weakly coupled, if at all. When the source is active optically, it is also active in X-rays, and the variations are strongly correlated (Tananbaum, 1973; Price *et al.*, 1971). This suggests that some of the optical emission is generated in the hot X-ray emitting region, which dominates when the source is active, with the remainder originating in a physically distinct cooler region which dominates when the source is quiescent. The spectrum also shows a number of high excitation emission lines which must originate in a separate cooler region. There is no evidence for a periodic component in the optical intensity or for a Doppler shift of the spectral lines, although the spectral lines do vary in both intensity and position. Radio observations show three radio sources close to Sco X-1 (Wade and Hjellming, 1971). There are two weak steady sources about an arc minute on either side of Sco X-1, while the third source coincides in position with Sco X-1 and varies in an irregular manner with a typical time scale of hours. There is some evidence of a correlation between radio and optical flaring, but this is not well established. All three sources have non-thermal spectra.

Several unsuccessful or inconclusive attempts have been made to observe X-ray emission lines from sulfur ions, which should make a contribution of the order of 1% to the total intensity in the range 1–10 keV if the plasma is optically thin. The upper limits on observed lines imply that the electron scattering optical depth is about

10 for the Sco X-1 source (Stockman *et al.*, 1973; Felten *et al.*, 1972). This result, together with the total observed luminosity, implies that the characteristic size of Sco X-1 is about 10^9 cm and the density about 10^{16} cm^{-3}.

If Sco X-1 is an accreting close binary, then the orbit must have a large inclination to the line of sight. This would explain, similarly to Cyg X-1, the absence of eclipses, of Doppler variation in the optical emission lines, and the absence of a large and variable low-energy X-ray cutoff. The primary must be a low mass object because it is not bright in the optical or infrared. Basko and Sunyaev (1973) have proposed a model in which the primary is a star with 1 $M_\odot$ and the secondary is a black hole. The absence of large rapid fluctuations as in Cyg X-1 is difficult to understand, in this case, but does not rule out such a model.

The interpretation of Cyg X-3 as a low mass binary leads to the possibility of a similar model for Sco X-1. Models involving a low mass primary and a white dwarf secondary have been proposed (e.g. Prendergast and Burbidge, 1968). Davidsen and Ostriker (1973) have considered a red dwarf – white dwarf pair, with the observational differences between Cyg X-3 and Sco X-1 being caused primarily by differing stellar wind and accretion rates. A less dense wind for Sco X-1 explains its lower luminosity and the absence of binary modulation of the X-rays. However, it should be noted that some of the observations, in particular, an optical flare preceding the X-ray flare, may require more involved structure, e.g., an accretion disk.

3.6. Cygnus X-2

Cyg X-2 is believed to be similar to Sco X-1 except there is not nearly so much information available concerning it. The 1–10 keV spectrum is best fit by an exponential with $kT \simeq 4$ keV. The flux varies by as much as 25% in a second, on occasion. Cyg X-2 has a strong, variable high energy component which can account for as much as 15% of the total intensity (Matteson, 1971). An optical identification was proposed in 1967 (Giacconi *et al.*, 1967) based on the discovery of a star with characteristics similar to Sco X-1 within the area of uncertainty of the X-ray source. The X-ray position has since been refined, still including the candidate star and excluding a suggested non-variable radio source. However, simultaneous X-ray optical variations have not yet been reported, making the identification only likely and not yet positive. In both the optical and X-ray, variations in intensity of order a factor of 2 are observed within a day; however, the 2 states characteristic of Sco X-1 have not yet been reported for Cyg X-2, possibly because of much less available data.

In the optical, Cyg X-2 is actually very different from Sco X-1, although they have some common features, such as variable intensity and an ultraviolet excess. In Cyg X-2, the spectral lines show large changes in radial velocity which were at first thought to be evidence of binary motion; however, Kraft and Demoulin (1967) showed that these variations, which are of several hundred km s^{-1} were not periodic. Furthermore, they found that the bulk of the optical emission could be accounted for by the presence of a G-type sub-dwarf based on the absorption line spectrum. They estimated a distance of 500–700 pc, implying an X-ray luminosity of order 10^{36} erg s^{-1}.

With the distance, the optical emission would then be 10^{34} erg s^{-1} and is dominated by the G-type star. The optical emission from the X-ray region could then be 10^{33} erg s^{-1} or 10^{-3} of the X-ray emission as in Sco X-1. Any radio emission from Cyg X-2 might be considerably less than for Sco X-1, making it difficult to detect. In any case, the presence of a G-type sub-dwarf along with an X-ray source is evidence for the presence of two stars in the system, if the identification is correct. Wilson (1970) has given a white dwarf binary picture for Cyg X-2 that explains the observations at least qualitatively.

Several other objects appear to have similar X-ray properties to Sco X-1 and Cyg X-2. They include GX17+2 (2U1813−14), GX3+1 (2U1744−26), GX5−1 (2U1757−25), GX9+1 (2U1758−20), and GX349+2 (2U1702−36), all of which have intensity variations of order factors of 2 within a day and exponential spectra with temperatures varying from 50 to 150×10^6 K. These five additional sources all have about a 2 keV cutoff suggesting distances of order 10 kpc (if the cutoffs are caused by interstellar absorption) and thus X-ray luminosities of order 10^{38} ergs s^{-1}.

3.7. Other binaries

We return now to the remainder of the X-ray sources which have been found to be binaries.

3.7.1. *3U1700−37*

The source 3U1700−37 was discovered by Uhuru and was found to have a period of 3.412 days. The source is on for 2.3 days and off for 1.1 days, the longest occultation of any of the X-ray binaries. There is some indication of a secondary minimum at phase 0.5. The binary nature of the X-ray emission is shown in Figure 17. The intensity is very variable on time scales of hours down to tenths of a second. The fluctuations, however, are not periodic. The X-ray spectrum is flat with a variable low energy cut-off which is comparable to or greater than that of the other eclipsing X-ray binaries.

The optical identification of 3U1700−37 with the star HD153919 was suggested by Jones *et al.* (1973) and confirmed by a large number of observers. The star is an early supergiant, a 6.6 mag. O7f at a distance of 1.7 kpc. It displays a double-peaked light curve with an amplitude of about 0.1 m and a period of 3.4 days; one of the minima coincides with the X-ray eclipse. Spectroscopic observations by Hutchings *et al.* (1973) determined radial velocity variations of amplitude 29 ± 6 km s^{-1}; similar results were obtained by Hensberge *et al.* (1973) and Wolff and Morrison (1973). Strong emission lines are present, with emission and absorption lines showing different radial velocities indicating an expanding atmosphere.

If the mass of the primary is that of a typical O7f star, at least 35 $M_\odot$, then the minimum mass of the X-ray source is 1.4 $M_\odot$. Although the system as a whole bears a strong resemblance to Cyg X-1, the minimum mass and the present X-ray data cannot determine whether the X-ray source is a neutron star, a degenerate dwarf, or a black hole.

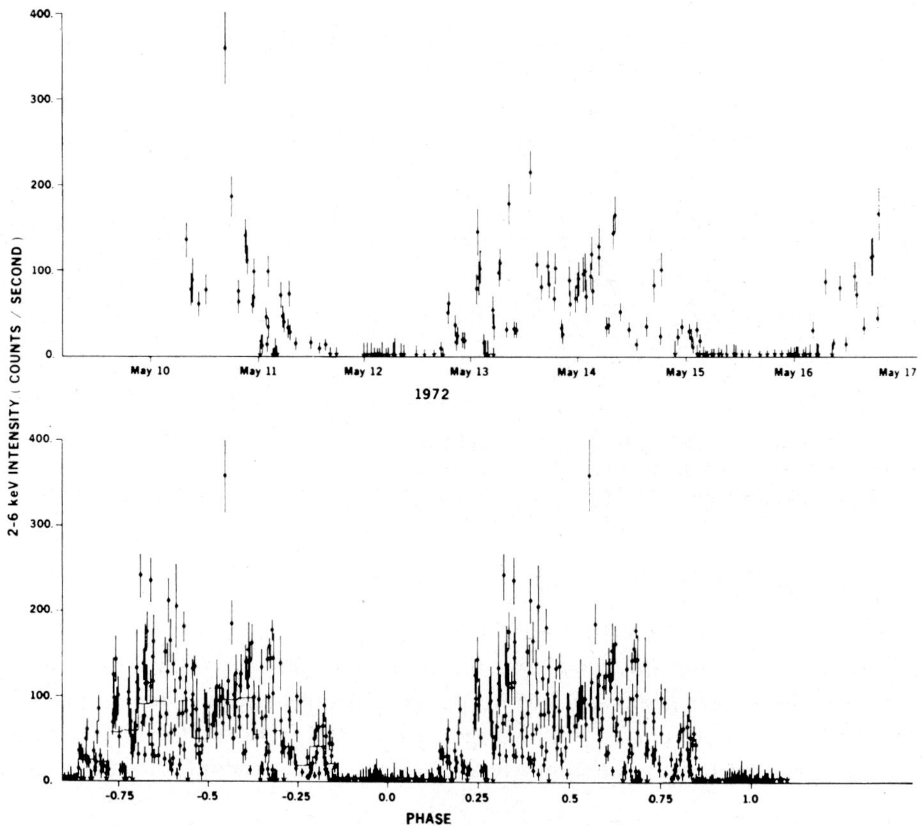

Fig. 17.   The 2–6 keV intensity of 2U1700 −37 corrected for elevation in the field of view and shown with ±1σ error bars. The upper portion shows seven days of data from May 1972. Plotted below are observations obtained between December 1970 and May 1972 folded with the 3.412-day period. A histogram of the intensity averaged in 25 equal intervals is also shown. Published by the University of Chicago Press; © 1973, American Astronomical Society.

3.7.2. *3U0900 − 40*

3U0900 − 40 is a highly variable source. It was found to be an eclipsing binary by Ulmer *et al.* (1972b) and Forman *et al.* (1973). The period is 8.95 days with an eclipse duration of 1.85 days. Its light curve (and variability) is shown by the folded Uhuru data in Figure 18. The X-ray intensity is found to be variable on time scales of seconds to days; on one occasion the intensity increased by a factor of 30 in 2 h. But no periodic variations have been seen.

The X-ray source has been identified with HD77581, a 6.9 mag. B0.5Ib star at 1.3 kpc. Both spectroscopic and photoelectric observations show agreement in period and phase between the optical and X-rays (Hiltner *et al.*, 1972; Hutchings, 1972; Jones and Liller, 1973; and others). The light curve is double peaked with an amplitude of 0.1 mag., similar to 3U1700 − 37, but exhibits significant variability. As with 3U1700 − 37, if normal mass is assumed for the primary, then spectroscopic observations imply a mass of about 1.7 $M_\odot$ for the X-ray source.

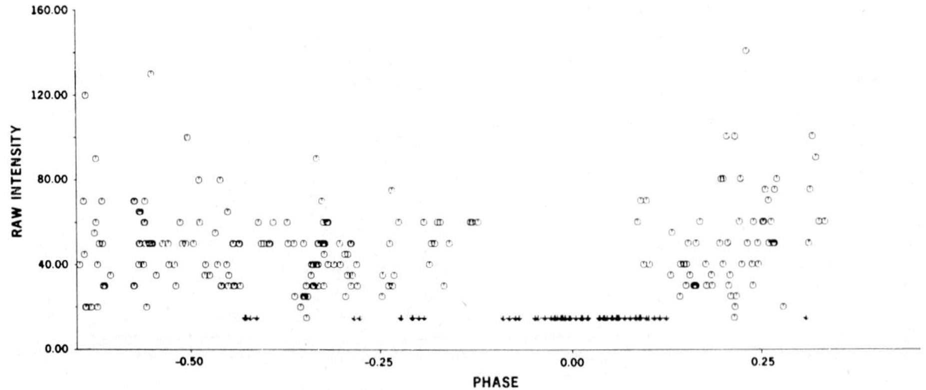

Fig. 18.   The observed 2–6 keV X-ray counting rates (uncorrected for aspect) for 2U0900 − 40 from May to July 1972 folded modulo 8.95 days. The absence of any high sightings around phase 0.0 shows the eclipsing behavior of the source. Published by the University of Chicago Press; © 1973, American Astronomical Society.

3.7.3. *Circinus X-1 (3U1516 − 56)*

The X-ray behavior of Cir X-1 is very similar to that of Cyg X-1. The source has been observed to vary significantly on all time scales observed down to 0.1 s as in Figure 19. Pulse trains have been observed, but no periodicities have been found to persist as with Cyg X-1 (Schreier *et al.*, 1971; Margon *et al.*, 1971). Different average intensity levels have been seen, with different average spectra. The variability from day to day

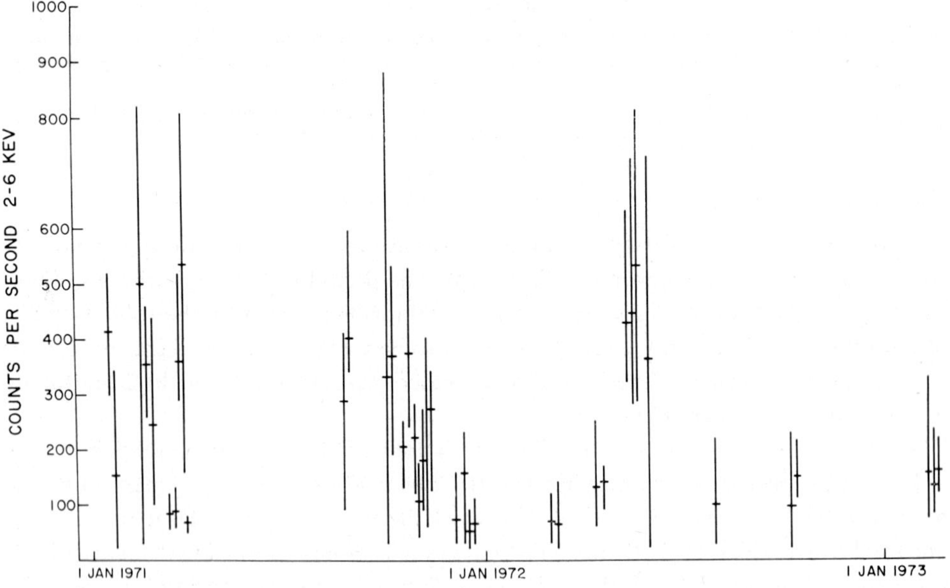

Fig. 19.   The intensity of Cir X-1 as measured over a two year period by Uhuru. Vertical extensions of data points are not uncertainties but are the observed variation in intensity on the day of observation. Published by the University of Chicago Press; © 1974, American Astronomical Society.

appears greater than that for Cyg X-1. The data is consistent with a binary period of 12.29 days, although many low points are seen outside of eclipses.

No star brighter than 14th mag. is present in the 0.0002 sq deg error box for the source. Thus, although the X-ray characteristics resemble closely those of Cyg X-1, the companion star appears quite dim. This could be due to obscuration or merely to a lower mass for the companion.

3.7.4. *SMC X-1 (3U0115−03)*

SMC X-1 is the only optically identified X-ray source in an external galaxy. The source exhibits a 3.893 day period, with an X-ray eclipse lasting 0.6 days as shown in Figure 20 (Schreier *et al.*, 1972b). Significant intensity fluctuations have been observed on

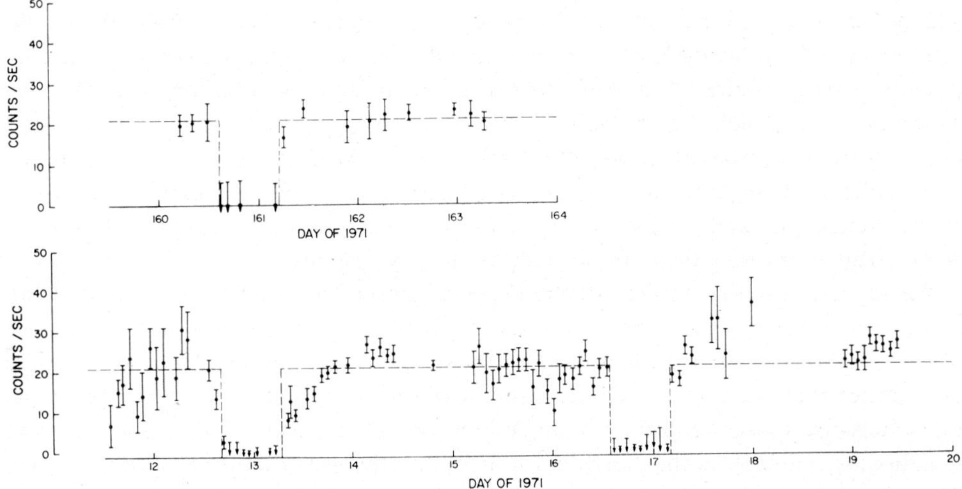

Fig. 20.   The intensity of SMC X-1 in observed counts s⁻¹ during eight days in January and three days in June 1971. Data have been corrected for elevation in the field of view; one-sigma error bars include both statistical and elevation correction errors. Published by the University of Chicago Press; © 1972, American Astronomical Society.

time scales of hours and perhaps minutes. In addition, extended lows have been observed. The source is too weak to determine whether fluctuations exist on short time scales. However, no regular short period pulsations have been observed with an upper limit of 10% of the power. The distance to the SMC of 61 kpc implies a 2–6 keV luminosity of 10^{38} erg s^{-1}. The observation of flux at 40 keV (Price *et al.*, 1971) extends the integrated X-ray luminosity to over 10^{39} erg s^{-1}, making it likely the strongest stellar X-ray source. This luminosity is comparable to or greater than the usual Eddington limit for accreting sources.

The star Sanduleak 160 was proposed as the optical counterpart for SMC by Webster *et al.* (1972) and confirmed by Liller (1973). It is a 13.2 mag. B0I supergiant. The light curve exhibits a double-peaked light curve of the correct period, with amplitude variation of about 0.1 mag. The absolute magnitude of the star is calculated as

−6.0. This is what is expected for this spectral type star, indicating that the presence of a strong X-ray source has not significantly altered the spectral appearance of the star. This is relevant in countering arguments that the spectral type of a binary companion can not be used in determining mass due to the effect of the X-ray source (i.e., for Cyg X-1). It should be noted that the ratio of X-ray to optical emission from SMC X-1 is greater than for Cyg X-1.

Spectroscopic measurements by Osmer *et al.* (1974) suggest that the mass of the X-ray source is between 1 and 4 $M_\odot$.

4. A Standard Model for the X-Ray Sources – Close Binary Systems and Accretion

The presence of a compact object in a close binary system presents a natural and efficient means of producing X-rays. As matter falls onto a compact object, gravitational potential energy is released, heating the matter and producing radiation. The efficiency of this process is such that modest accretion rates can produce the luminosities and temperatures characteristic of galactic X-ray sources. Models for X-ray sources based on this idea were suggested quite early (see e.g., Shklovsky, 1967; Cameron and Mock, 1967; Prendergast and Burbidge, 1968). Thus, it is not wholly surprising that many X-ray sources are now being found in close binary systems.

We suggest here that in fact all the presently observed galactic X-ray sources can be accounted for by such binary systems; furthermore, we suggest that the evidence favors a neutron star or a black hole as the actual X-ray source. There is a variety of evidence that leads to the conclusion. However, it is implicit in these arguments that the X-ray sources comprise a single kind of stellar system. This allows drawing together the evidence on the individual sources into the construction of a single model. There is some justification for this view. For one, there is a category of X-ray sources associated with a very specific companion; namely, a B0 supergiant. This specificity implies that the X-ray source itself is also a certain kind of stellar object, which we argue is either a neutron star or a black hole. From the point of view of making X-rays by accretion, a neutron star and a black hole are very similar since their sizes are comparable. To invoke white dwarfs, however, requires qualitatively different conditions since their size is 10^3 times greater. One consequence of this larger size is that the mass accretion rate must be 10^3 times greater than is the case for a neutron or a black hole.

The other category of X-ray sources is associated with more common stars and the evidence is not so compelling that they comprise so narrow class of stellar systems. There is at least one cross link between these two categories of X-ray sources and that is the two pulsing X-ray sources. Cen X-3 is an example of an X-ray source associated with a B0 supergiant whereas Her X-1 is associated with a low mass, halo or intermediate population star.

Another argument for a 'standard' model is the rarity of the X-ray sources. The presently observed sources comprise only 1 in 10^9 of all stars in the Galaxy, which is a very small number. The probability that nature has found a variety of means to make X-ray sources with approximately the observed space density must be exceedingly remote.

The direct evidence for the standard model is as follows. A number of the X-ray sources are positively confirmed as being in binary systems with periods varying from 0.2 days to 12 days. Considering the probability of seeing the binary character, the data are consistent with all the X-ray sources being binaries. There is direct evidence for large mass transfer rates in many of these systems. The 'dips' in Her X-1 are absorption features and are directly interpretable as due to a gas stream. In both Cen X-3 and Her X-1, the pulse period changes and possibly orbital period changes are consistent. In fact, in these two objects, as will be discussed below, the observed pulse period change is more compatible with a rotating object of neutron star size than one of white dwarf size. A more direct argument that these two objects involve neutron stars is simply that it has not been possible to find a means of making a white dwarf rotate or vibrate stably with periods as short as a second (c.f. Ostriker, 1971).

Additional direct evidence for the nature of the X-ray source is in the case of Cyg X-1 where the high mass requires a black hole.

In the remainder of this chapter, we describe what may be the inner structure of the X-ray emitting systems.

4.1. The binary system*

A standard means of describing binary star systems is via the Roche model (see e.g. Kopal, 1959). One uses a coordinate system rotating with the binary, assumes circular orbits and co-rotating stars, and defines the equipotentials of the system as in Figure 21. Close to each star, the surfaces are spherical; they then gradually increase in size

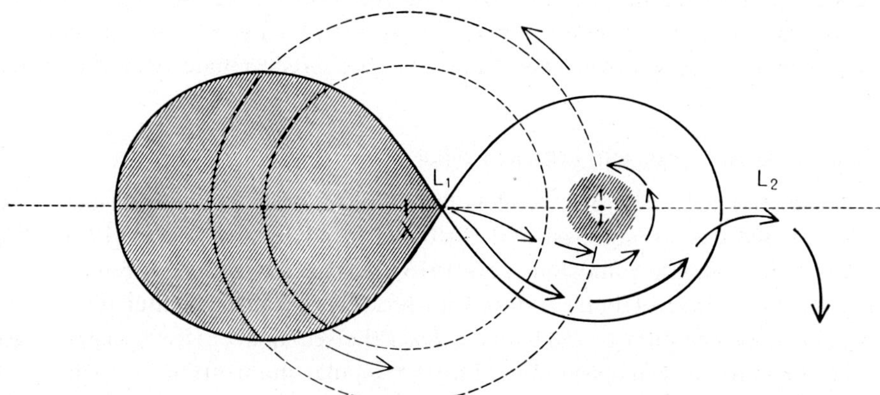

Fig. 21.   Roche equipotential surfaces (lobes) are shown by the darkest line. One star (cross-hatching) is assumed to fill its lobe and spill material through the inner Lagrange point (L_1). The flow of the material is shown by the arrows. Some may end up on the companion star and some may leave the system through the outer Lagrange point (L_2).

* See also contributions by Kraft and Ruffini.

and deform until the surfaces around each star touch at L_1, the inner Lagrangian point. This first common equipotential is called the Roche lobe or critical surface. If one star fills its Roche lobe, then matter can pass through L_1 toward the other star. In this case, the angular momentum of the accreted matter causes a disk to form around the secondary. Alternatively, mass transfer can also occur if one star has a stellar wind from which matter can accrete onto the secondary. This stellar wind can either be of a conventional type from a massive star, or can possibly be driven by the companion X-ray source which heats the outer layers of the star. Either of these means of mass transfer may be present in the observed binary X-ray sources, and it is possible that both of them are relevant to some sources.

The stellar wind models have been discussed in the literature; either of the types can yield sufficient mass to power an X-ray source (Davidsen and Ostriker, 1973). There is also an argument that the X-ray induced wind will produce X-ray luminosities close to the Eddington limit, $\sim 10^{38} (M_x/M_\odot)$ erg s^{-1} (Arons, 1973; Basko and Sunyaev, 1973). There are two features of the stellar wind model which seem particularly relevant. The first involves the fact that a very strong stellar wind may 'bury' the X-ray source, effectively absorbing the X-rays re-emitting the radiation at longer wavelengths (see e.g. Pringle, 1973). This can be used to explain the extended lows of some of the sources, especially an irregular one like that of Cen X-3. The repeating but non-periodic extended lows would then be tied to the variability of the stellar wind of the supergiant primary. (The 35-day cycle of Her X-1 can as well be explained by a precession-timed 'magnetic gate', allowing accretion only when the magnetic pole is within the accretion disk.) The second feature of a stellar wind model is the fact that radial mass loss will increase the binary period. It is interesting to note that both mass transfer (from more massive to less massive component) and mass loss from the system through L_2 in a Roche model will tend to decrease the period (see e.g. Kruszewski, 1966). Thus, in Cen X-3 where the period is seen to change in both directions, the radial stellar wind slow-down effect may at times predominate over the speed-up effects.

4.2. The accretion disk and compact object

The relevance of accretion disk models for compact sources has been discussed in some detail (see e.g. Prendergast and Burbidge, 1968; Schwartzman, 1971; Pringle and Rees, 1972; Shakura and Sunyaev, 1973). An accretion disk is required in the normal Roche picture of binary mass transfer. The accreting matter will have too much angular momentum to fall radially onto the secondary; it will instead form a disk and the material will spiral inward as the angular momentum is dissipated. It is also important to note that a black hole model must contain an accretion disk to emit X-rays. The in-fall time of matter accreting radially into a black hole is probably too short to allow sufficient radiation, whereas matter accreting radially onto a neutron star can deposit its energy on the surface which then radiates as a black body.

The properties of the accretion disk, and of the radiation from it, are dependent upon the detailed assumptions about its structure, its viscosity, etc. However, reasonable models have been constructed from which can be estimated the X-ray emissivity (see e.g. Pringle and Rees, 1972; Shakura and Sunyaev, 1973). The X-rays can be due to bremsstrahlung or black body radiation depending on the density of a gas at any given point. Integrating the radiation over the entire disk with its radially dependent temperature can then produce a non-thermal power law spectrum consistent with observations. But, in fact, the main variable in the model is probably the inner boundary condition; namely, the nature of the compact object present at the center. If the central object is a black hole, then the disk can extend inward to the closest stable orbit, corresponding to a period of about 10^{-3} $(M/M_\odot)$ seconds or a factor of a few less for a Kerr black hole.* The effective in-fall time is some 100 times this period, leading to a characteristic time scale for pulsations or fluctuations between 10^{-3} and 10^{-1} s. The radiation is also highly dependent on the mass accretion rate. Pringle and Rees (1972) point out that significant flaring is expected if the mass flux is either so large that the Eddington limit is exceeded, or so small that energy cannot be radiated efficiently.

If the central object is a neutron star, then the accretion disk is probably not the dominant influence on the radiation. The disk either extends inward to the surface of the star, in which case both the disk and the star contribute to the emission, or it extends down to the Alfvén surface. This is the point at which the stellar magnetic field dominates the accretion flow; it can be considered as the radius at which the magnetic field energy density becomes comparable to the kinetic energy of the accreting matter. The in-falling matter then follows the field lines and thus must co-rotate and flow toward the magnetic poles. The enforced co-rotation exerts a torque on the star. The sign depends on the relative sense of the orbital and spin rotations, and the time scales are on the order of 10^4 or 10^5 yr.

It may be recalled now that both Cen X-3 and Her X-1 have shown overall speed-ups of their pulsation periods, with $\dot{P}/P$ values of 3×10^{-4} and 4×10^{-6} yr^{-1} respectively. If we use a simple model to calculate the expected $\dot{P}/P$ due to transfer of angular momentum from the accreting matter to the neutron star, following Pringle and Rees (1973) (and assuming that the Alfvén radius is approximated by the co-rotation = Keplerian velocity radius, and that the X-ray luminosity is linear with the accretion rate) we determine values of 3×10^{-3} and 6×10^{-5} yr^{-1} respectively. Thus, despite the difference in absolute magnitude between predicted and observed values, the relative speed-ups for the two sources agree very well. If the central object is a white dwarf, the predicted values for $\dot{P}/P$ are substantially smaller than the observed values.

An important consideration at this point is the effect of centrifugal forces (see e.g. Lamb *et al.*, 1973). If the radius at which a particle in a Keplerian orbit co-rotates with the star is smaller than the Alfvén radius, then centrifugal forces may not allow in-fall; rather, mass may be ejected and the outer magnetosphere may be unstable. In the other case, the accreting matter will fall toward the stellar surface in the vicinity of the magnetic poles, with a free fall time of about 0.1 s. The pulsing of the X-ray

---

* See Ruffini contribution.

emission is then due to the rotation of these hot spots on the neutron star. Although the basic spectrum of the pulses is presumably thermal with a temperature of about 5×10^7 K, the details of both the spectrum and the pulse shape depend on, for example, the strength of the field, the density of the accretion column, and the possibilities of cyclotron radiation and Compton scattering (see e.g. Pringle and Rees, 1972; Davidsen and Ostriker, 1973; Lamb *et al.*, 1973; Gnedin and Sunyaev, 1973).

4.3. ORIGIN AND EVOLUTION

A final theoretical consideration is the fact that the origin and evolution of the components can not be divorced from their presence in a binary system. At any time in the life span of one of the stars when expansion is expected, the presence of the other star will short-circuit the evolution when the expansion reaches the Roche lobe. At this point, significant mass transfer may start, changing the masses and subsequent evolution of both stars. Thus, stellar evolution in close binaries was of great interest even before the X-ray binaries were discovered (for a review, see e.g. Paczyński, 1971, and contribution of Kraft).

Specific evolutionary models for X-ray binaries have been produced (e.g. Van den Heuvel and Heise, 1972). One typically starts with a massive close binary system. The more massive component evolves faster and reaches its Roche lobe after some $10^6–10^7$ yr, before the onset of helium ignition. Very rapidly ($\sim 10^4$ yr) most of the mass is transferred to the other component; the less massive component is now the more evolved star, and determines the final nature of the system. For small values of its mass, it becomes a degenerate dwarf. For larger values, it presumably undergoes a supernova after another $\sim 10^6$ yr, forming a neutron star or black hole. The more massive star now continues its evolution, and if the binary system has survived the supernova, this star will reach its Roche lobe in $10^6–10^7$ yr and start transferring matter back onto the compact companion, generating X-rays.

The crucial element in the evolution is the mass of the system. If the original mass of the system is low, one might expect to produce degenerate dwarfs or neutron stars with low mass companions, as has been discussed for Sco X-1, Cyg X-2, Cyg X-3, and Her X-1. Because of the low mass, the evolutionary times are very long, and the objects will be seen to belong to 'older' populations of stars as seems to be the case. If the original mass is high, one must end up with a high mass system as described above since the loss of more than half the original mass during a supernova explosion would disrupt a binary system. We do find high mass X-ray binary systems as represented by Cyg X-1, Cen X-3, 3U0900−40, 3U1700−37, SMC X-1 and possibly Cir X-1. Because of the high mass, these systems are expected to evolve rapidly and to belong to a 'young' population as seems to be the case. Of course, there may be alternate evolutionary schemes in which there is very substantial mass loss by other means.

An important point of current investigation is the necessary condition for a supernova not to disrupt a binary system. A relevant consideration in this respect is the eccentricity of the X-ray binaries which would be expected as the result of a supernova. Although there is some evidence for significant eccentricity in the optical light curve of, for example 3U0900−40, in the two X-ray pulsars Cen X-3 and Her X-1, upper limits for eccentricity on the order of 2×10^{-3} have been obtained. Thus, further investigation is needed to determine how binaries can survive a supernova with very little or no eccentricity, or alternatively how any eccentricity can be efficiently removed.

Finally, we wish to point out the existence of a curious coincidence which has some bearing on the standard model. It is believed that neutron stars can only be produced explosively; i.e., during a supernova event. If this is correct, then wherever we see a neutron star in an X-ray source, there had to have been a supernova. In those instances where a black hole is in evidence, we would argue that the mass of the star undergoing the supernova was too massive to form a neutron star. The coincidence is simply that the known supernova show the same duplicity in population type as do the X-ray sources. The type II supernova occur in extreme Population I objects and the type I supernova are seen primarily in halo or old disc stars. The correspondence is particularly striking in the type II supernova and the X-ray sources associated with B0 supergiants. Recent work by Moore (1973) has revealed that the supernova in spiral galaxies, most of which must be type II, not only occur in spiral arms, but at the inner edge of the spiral arms. The implication of this is that these supernovae occur within a few million years of the birth of the star; thus the stars must be ($\sim 5 - 10\ M_\odot$) massive. This is the same as saying that they occur in a very narrow spectral class of stars, which includes the class of stars associated with the high mass X-ray binaries.

Like many arguments in astronomy, this one is somewhat circular. It could be simply that both certain supernova nad certain X-ray sources require massive progenitors. However, considering that there are independent arguments for supernovae progenitors for X-ray sources, the occurrence of this coincidence is supporting evidence for X-ray sources being neutron stars and black holes.

5. Summary

When the X-ray sources were first discovered there was the general belief that some very special conditions were required for their origin. It is now possible that we have discovered those conditions – supernova explosions, close binary systems, neutron stars and black holes. Where there is unambiguous data, this view is supported, in particular, the existence of the pulsating X-ray sources, the mass estimate of Cyg X-1 and the direct evidence for the binary nature of many of the X-ray sources.

In addition, there are compelling theoretical arguments to support this view including the evolution of stars in close binary systems, the physics of neutron stars and black holes, and the great difficulty in finding alternative hypotheses to the above view.

However, one can not leave this subject without citing the problems, which are still formidable. There are details of the X-ray and optical systems, which appear to be very definite, but as yet can not be simply explained. As an example, there is the wealth of information on Her X-1 – the 35-day cycle, the optical variability – which are unexplained. The most puzzling feature is that the very large optical variability continues even when the X-rays have disappeared. Is there yet another significant source of luminosity in the system, such as cosmic rays?

Other, less dramatic observations that require explanation are the two 'states' seen in many X-ray sources. In Cyg X-1, for example, one of the states is characterized by high intensity and a soft spectrum; in the other state the intensity is about a factor of three lower. Certain sources actually seem to 'disappear' from time to time. This has been seen in X-rays in Cir X-1 and SMC X-1. The most dramatic example of this is the case of Her X-1, in which the optical variability disappears for years at a time, based on the study of historical material.

This varied and complex behavior may be attributable to the complexity of the accretion process – instabilities in the atmosphere of the companion star and structure in the immediate vicinity of the X-ray source. Right now, however, these phenomena have no satisfactory explanation.

In spite of the difficulties in understanding the details of these objects, we can not overlook the potential of the field, principally the possibility of studying the properties of neutron stars and black holes at first hand, and of learning about a branch of the evolution of stars, following their supernova phase, that has completely eluded us up to now.

6. Appendix: Galactic X-Ray Sources

A ($I > 8$ counts s^{-1}, Non-identified with extragalactic objects)

l, b	3U	I_{max}	$\dfrac{I_{max}}{I_{min}}$	Comments	
Central Region of Galaxy					
321, −6	1543 − 62	36	3		
321, −13	1626 − 67	10			
322, 0	1516 − 56	720	$\geqslant 20$	Rapid fluctuations, eclipse 12.3 days	Cir X-1
324, −11	1632 − 64	11			
324, −5	1556 − 60	17		SAO253382	B9, 8.9m (?)
327, 2	1538 − 52	11			
332, −4	1636 − 53	261			
335, 0	1624 − 49	50	5	SAO226781 Star Cluster, NGC6134	B8, 9m (?)
336, 0	1630 − 47	150	3		
339, −4	1658 − 48	344	3		
340, 0	1642 − 45	381	3		
343, −2	1705 − 44	280	3		
344, −1	1702 − 42	34			
348, 2	1700 − 37	102	> 3	Rapid fluctuations, eclipse 3.4 days	HD153919 6m 07f
346, −7	1735 − 44	210			
348, −1	1714 − 39	12			
349, 3	1702 − 36	715	2		
353, 5	1704 − 32	14			

A (Continued)

l, b	3U	I_{max}	$\dfrac{I_{max}}{I_{min}}$	Comments	
354, -5	$1746-37$	31		SAO 209 318	K1, 3.2m (?)
					Glob. Cluster NGC6441 (?)
354, 0	$1727-33$	65		SAO 208 881	B5, 6.7m (?)
				SAO 208 883	B8, 9.4m (?)
357, -11	$1822-37$	17			
357, -5	$1755-33$	47	3		
359, $+24$	$1617-15$	17,000	2.5		Sco X-1,
					12.3 mag. pec.
0, 9	$1709-23$	39	5		
2, 5	$1728-24$	60			
2, 1	$1744-26$	460	3		
3, -8	$1800-30$	250	1.5	Glob. Cluster, NGC6624?	
5, -1	$1758-25$	1127	2	SAO 186 106	K7, 8.8m?
8, 9	$1728-16$	260	1.7		
9, 1	$1758-20$	595	2		
14, 0	$1811-17$	380	3		
16, 1	$1813-14$	588	1.5		
18, 2	$1812-12$	12			

Outer Region of Galaxy

l, b	3U	I_{max}	$\dfrac{I_{max}}{I_{min}}$	Comments	
30, 6	$1822-00$	37			
31, -8	$1915-05$	23	5		
36, -4	$1908+00$	199	3		
36, 5	$1837+04$	270	2		
37, -1	$1901+03$	87	4		
43, -2	$1912+07$	21			
51, -9	$1956+11$	17			
68, 2	$1953+31$	63	5		
58, 38	$1653+35$	100	$\geqslant 6$	1.2 s pulses,	Her X-1, Hz Her
				eclipses 1.7 days	13–15 mag. pec.
71, 3	$1956+35$	1175	5	Cyg X-1	HDE226868
				Rapid fluctuations	9m, 09.71b
80, 1	$2030+40$	194	$\geqslant 3$	Cyg X-3, IR source	
				Partial Eclipse (?) 0.2 days	
87, -11	$2142+38$	540	$\geqslant 2.5$	Cyg X-2, 14m sdG	
92, -3	$2129+47$	12			
126, 1	$0115+63$	70	7		
163, -17	$0352+30$	20			X-Per (?)
201, -3	$0614+09$	60	6		
263, 4	$0900-40$	100	10	HD77581, 6m, B0.51b	
				slow flares, eclipses, 8.9 days	
263, -11	$0750-49$	9			
283, 1	$1022-55$	10			
292, 0	$1118-60$	160	$\geqslant 20$	Cen X-3, 6m, B01b	
				4.8 s pulses, eclipses 2.1 days	
294, 0	$1134-61$	9			
296, 0	$1145-61$	72	5	SAO 251 590	B0, 9.2m (?)
				SAO 251 595	B2, 8.6m (?)
300, 0	$1223-62$	32	3	SAO 251 905	A1, 9.9m (?)
303, -6	$1254-69$	26			
304, 1	$1258-61$	47	5		

A *(Continued)*

l, b	3U	I_{max}	$\dfrac{I_{max}}{I_{min}}$	Comments	
Magellanic Cloud Sources					
$0115-37$		28	$\geqslant 9$	SMC	Sanduleak 160, 13m, B01b,
$0532-66$		9.4		LMC	eclipses 3.9 days
$0539-64$		20.7		LMC	
$0540-69$		19.3		LMC	

B $(I < 8$ counts s^{-1}, $b < \pm 5°$, Non-identified$)$

l, b	3U	I_{max}	$\dfrac{I_{max}}{I_{min}}$	Comments	
$320, -1$	$1510-59$	6.4		SAO242322	G0, 10m (?)
$26, 1$	$1832-05$	6.1			
$44, 1$	$1906+09$	7.6			
$88, 2$	$2052+47$	6.2			
$101, -1$	$2208+54$	4.4			
$107, 1$	$2233+59$	4.7			
$130, -1$	$0143+61$	7.2			
$138, +2$	$0258+60$	2.9			
$143, -4$	$0305+53$	2.8			
$143, -1$	$0318+55$	4.9			
$161, 0$	$0446+44$	6.2			
$189, 5$	$0620+23$	5.0			
$244, 2$	$0757-26$	3.0			
$276, -4$	$0918-55$	6.3		SAO236863	A0, 10m (?)
				SAO236859	K0, 8.1m (?)
$294, 0$	$1134-61$	8.7			
$299, -2$	$1210-64$	6.0			

C (Other Galactic Sources)

Name	3U Designation	I (2–6 keV)	
Supernova Remnants			
Crab Nebula	$0531+21$	950	
NP0532	–	–	
Cas A	$2321-58$	53	
Tycho	$0022+63$	9.5	
Puppis A	$0821-42$	7.5	
Vela X, Y, Z	$0833-45$	9.1	
Cygnus Loop	–	–	Very Soft Spectrum
Lupus SN?	–	–	
IC443?	$0620+23$	5.0	
MSH 15-52A?	$1510-59$	6.4	
Galactic Center			
GCX	$1743-29$	40	Extended $\sim 1°$

C (Continued)

Transient Sources

		Duration	I (peak)
Cen X-2		~ 80 days	5×10^{-7} ergs cm^{-2} s
Cen X-4		~ 80 days	5×10^{-7} ergs cm^{-2} s
	1543 $-$ 47	~ 1 yr	3×10^{-8} ergs cm^{-2} s
	1735 $-$ 28	~ 1 week	10^{-8}
Cep X-4			

Acknowledgements

Although we have tried to acknowledge the most relevant references for the observations and their theoretical interpretations, we apologize for any significant omissions. In addition, we have made use of much as yet unpublished data from the Uhuru satellite, particularly in respect to Cen X-3, Cyg X-3, and Cir X-1. We thus acknowledge the efforts of the Uhuru group, and in particular, R. Giacconi, H. Tananbaum, S. Murray, W. Forman, C. Jones-Forman, R. Leach, G. Fabbiano, J. Morin, and R. Levinson. We would also like to thank D. Jarmac for assistance in preparing the manuscript.

References

Arons, J.: 1973, *Astrophys. J.* **184**, 539.

Bahcall, J. and Bahcall, N.: 1972, *Astrophys. J.* **178**, L1.

Baity, W., Ulmer, M., Wheaton, W., and Peterson, L.: 1974, *Astrophys. J.* **187**, 341.

Basko, M., Sunyaev, R., and Titarchuk, L.: 1973, preprint.

Basko, M. and Sunyaev, R.: 1973, *Astrophys. Space Sci.* **23**, 117.

Becklin, E., Hawkins, F. J., Mason, K. O., Matthews, K., Neugebauer, G., Sanford, P. W., and Wynn-Williams, C.: 1973, *Nature* **245**, 302.

Blaauw, A.: 1965, *Stars and Stellar Systems* **5**, 435.

Blaauw, A.: 1961, *Bull. Astron. Inst. Neth.* **15**, 265.

Blumenthal, G. and Tucker, W.: 1972, *Nature* **235**, 97.

Boldt, E., Holt, S., and Serlemitsos, P.: 1971, *Astrophys. J.* **164**, L9.

Bolton, C.: 1972, *Nature* **235**, 271.

Boynton, P. C., Canterna, R., Crosa, L., Deeter, J., and Gerend, D.: 1973, *Astrophys. J.* **186**, 617.

Bowyer, S., Byram, T., Chubb, T., Friedman, H.: 1964, *Science* **146**, 912.

Braes, L. and Miley, G.: 1971, *Nature* **232**, 246.

Braes, L. and Miley, G.: 1972, *Nature* **237**, 506.

Braes, L., Miley, G., Shane, W., Baars, J., and Goss, W.: 1973, *Nature Phys. Sci.* **242**, 66.

Bregman, J., Butler, D., Kemper, E., Koski, A., Kraft, R., and Stone, R.: 1973, *Astrophys. J. Letters* **185**, L117.

Brinkman, A., Parsignault, D., Schreier, E., Gursky, H., Kellogg, E., Tananbaum, H., and Giacconi, R.: 1974, *Astrophys. J.* **188**, 603.

Byram, E., Chubb, T., and Friedman, H: 1966, *Science* **132**, 66.

Cameron, A. and Mock, M.: 1967, *Nature* **215**, 464.

Canizares, C. R., McClintock, J. E., Clark, C. W., Lewin, W. H. G., Schnopper, H. W., and Sprott, G. F.: 1973, *Nature Phys. Sci.* **241**, 28.

Clark, G. W., Bradt, H. V., Lewin, W. H. G., Markert, T. H., Schnopper, H. W., and Sprott, G. F.: 1972, *Astrophys. J.* **177**, L109.

Crampton, D. and Hutchings, J. B.: 1972, *Astrophys. J.* **178**, L65.

Davidsen, A., Henry, J., Middleditch, J., and Smith, H.: 1972, *Astrophys. J.* **177**, L97.

Davidson, K. and Ostriker, J.: 1973, *Astrophys. J.* **179**, 585.

Dilworth, C., Maraschi, L., and Reina, C.: 1973, *Astron. Astrophys.* **28**, 71.

Doxsey, R., Bradt, H. V., Levine, A., Murthy, G. T., Rappaport, S., and Spada, G.: 1973, *Astrophys. J.* **182**, L25.

Fabian, A., Pringle, J., and Rees, M.: 1973, *IAU Circ.*, No. 2555.

Felten, J. E., Rees, M. J., and Adams, T. F.: 1972, *Astron. Astrophys.* **21**, 139.

Fisher, P. C., Johnson, H. M., Jordan, W. C., Meyerott, A. J., and Acton, L. W.: 1966, *Astrophys. J.* **148**, L119.

Forman, W., Jones, C., and Liller, W.: 1972, *Astrophys. J.* **177**, L103.

Forman, W., Jones, C., Tananbaum, H., Gursky, H., Kellogg, E., and Giacconi, R.: 1973, *Astrophys. J.* **182**, L103.

Friedman, H., Byram, E., and Chubb, T.: 1967, *Science* **156**, 374.

Giacconi, R., Gursky, H., Paolini, F. R., and Rossi, B. B.: 1962, *Phys. Rev. Letters* **9**, 439.

Giacconi, R., Gorenstein, P., Gursky, H., Usher, P. D., Waters, J. R., Sandage, A., Osmer, P., and Peach, J.: 1967, *Astrophys. J.* **148**, L119.

Giacconi, R., Gursky, H., Kellogg, E., Schreier, E., and Tananbaum, H.: 1971, *Astrophys. J. Letters* **167**, L67.

Giacconi, R., Murray, S., Gursky, H., Kellogg, E., Schreier, E., Matilsky, T., Koch, D., and Tananbaum, H.: 1974, *Astrophys. J. Suppl.* **237**, 27 and 37.

Giacconi, R., Gursky, H., Kellogg, E., Levinson, R., Schreier, E., and Tananbaum, H.: 1973, *Astrophys. J.* **184**, 227.

Gnedin, Y. and Sunyaev, R. 1973, *Astron. Astrophys.* **25**, 233.

Gregory, P., Kronberg, P., Seaquist, E., Hughes, V., Woodsworth, A., Viner, M., and Retallack, D.: 1972, *Nature* **239**, 440.

Groth, E., Yeung, S., Papaliolios, C., Pennypacker, R. C., Spada, G., and Middleditch, J.: 1973, *IAU Circ.*, No. 2523.

Gursky, H., Gorenstein, P., and Giacconi, R.: 1967, *Astrophys. J. Letters* **150**, L75.

Gursky, H.: 1972, *Astrophys. J. Letters* **175**, L141.

Gursky, H.: 1973, in B. DeWitt and C. DeWitt (eds.), *Black Holes*, Gordon and Breach, New York.

Hayakawa, S. and Matsuoka, M.: 1964. *Prog. Theoret. Phys. Suppl. Japan*, No. 30, 204.

Hensberge, G., Van Den Heuvel, E. P. J., and Paes de Barros, M. H.: 1973, *Astron. Astrophys.* **29**, 69.

Hiltner, W., Werner, J., and Osmer, P.: 1972, *Astrophys. J.* **175**, L19.

Hjellming, R. M. and Balick, B.: 1972, *Nature* **239**, 443.

Hjellming, R. M.: 1973, *Astrophys. J.* **182**, L29.

Hjellming, R. M. and Wade, C.: 1971, *Astrophys. J.* **168**, L21.

Holt, S., Boldt, E., Schwartz, D., Serlemitsos, P., and Bleach, R.: 1971, *Astrophys. J.* **166**, L65.

Holt, S., Boldt, E., Rothschild, R., Saba, L. R., and Serlemitsos, P.: 1974, *Astrophys. J.* **190**, L109.

Hutchings, J. P.: 1972, *IAU Circ.* No. 2537.

Hutchings, J. B., Thackeray, A. D., Webster, B. L., and Andrews, P. T.: 1973, *Monthly Notices Roy. Astron. Soc.* **163**, 13.

Jones, C. and Liller, W.: 1973, *Astrophys. J.* **184**, L65.

Jones, C., Forman, W., Tananbaum, H., Schreier, E., Gursky, H., Kellogg, E., and Giacconi, R.: 1973, *Astrophys. J.* **181**, L43.

Joss, P. C., Avni, Y., and Bahcall, J.: 1973, *Astrophys. J.* **186**, 767.

Kopal, Z.: 1959, *Close Binary Systems*, Wiley, New York.

Kraft, R. and Demoulin, M.: 1967, *Astrophys. J.* **150**, L183.

Krzeminsky, W.: 1973, *IAU Circ.*, No. 2569.

Kruszewski, A.: 1966, *Advances in Astronomy and Astrophysics*, Vol.4, Academic Press, New York, p. 233.

Lamb, D. and Sorvari, J. M.: 1972, *IAU Circ.*, No. 2422.

Lamb, F. K., Pethick, C., and Pines, D.: 1973, *Astrophys. J.* **184**, 271.

Liller, W.: 1973, *Astrophys. J.* **184**, L37.

Liller, W.: 1972a, *IAU Circ.*, No. 2415.

Liller, W.: 1972b, *IAU Circ.*, No. 2427.

Liller, W.: 1972, *Tabulation of Interesting Optical and Radio Sources near Uhuru X-ray Sources*, Harvard College Observatory.
Margon, B., Lampton, M., Bowyer, S. and Cruddace, R.: 1971, *Astrophys. J. Letters* **169**, L23.
Margon, B., Bowyer, S., and Stone, R.: 1973, *Astrophys. J.* **185**, L113.
Margon, B. and Ostriker, J.: 1973, *Astrophys. J. Letters* **186**, 91.
Mason, K., Hawkins, F., Samford, P., Murdin, P., and Savage, D.: 1974, *Astrophys. J.* **192**, L65.
Matteson, J.: 1971, Ph. D. Thesis, U.C., San Diego.
Mauder, H.: 1973, *Astron. Astrophys.* **28**, 473.
Middleditch, J. and Nelson, J.: 1973 preprint.
Moore, E.: 1973, *Publ. Astron. Soc. Pacific* **85**, 564.
Neugebauer, G., Oke, J., Becklin, E., and Garmire, G.: 1969, *Astrophys. J.* **155**, 1.
Oda, M. and Matsuoka, M.: 1970, *Prog. Elem. Part. Cosmic-Ray Phys.* **10**, 305.
Oda M., Gorenstein, P., Gursky, H., Kellogg, E., Schreier, E., Tananbaum, H., and Giacconi, R.: 1971, *Astrophys. J. Letters* **166**, L1.
Oda, M., Wada, M., Matsuoka, M., Miyamoto, S., Muranaka, N., and Ogawara, Y.: 1972, *Astrophys. J. Letters* **172**, L13.
Osmer, P. and Hiltner, W.: 1974, *Astrophys. J.* **187**, L5.
Ostriker, J.: 1971, *Ann. Rev. Astron. Astrophys.* **9**, 353.
Overbeck, J. and Tananbaum, H.: 1968, *Phys. Rev. Letters* **20**, 24.
Paczyński, B.: 1971, *Ann. Rev. Astron. Astrophys.* **9**, 183.
Paczyński, B.: 1973, *Annals N.Y. Acad. Sciences* **224**, 233.
Parsignault, D. R., Gursky, H., Kellogg, E., Matilsky, T., Murray, S., Schreier, E., Tananbaum, H., Giacconi, R., and Brinkman, A. C.: 1972, *Nature Phys. Sci.* **239**, 123.
Peterson, L.: 1973, in H. Bradt and R. Giacconi (eds.), 'X-and Gamma-Ray Astronomy', *IAU Symp.* **55**, 51.
Prendergast, K. and Burbidge, G.: 1968, *Astrophys. J.* **151**, L83.
Price, R. E., Groves, D. S., Rodrigues, R. M., Seward, F., Swift, C. D., and Toor, A.: 1971, *Astrophys. J.* **168**, L7.
Pringle, J.: 1973, *Nature Phys. Sci.* **243**, 90.
Pringle, J. and Rees, M.: 1972, *Astron. Astrophys.* **21**, 1.
Rappaport, S., Doxsey, R., and Zaumen, W.: 1971a, *Astrophys. J.* **168**, L43.
Rappaport, S., Zaumen, W., and Doxsey, R.: 1971b, *Astrophys. J.* **168**, L17.
Rothschild, R., Boldt, E., Holt, S., and Serlemitsos, P.: 1974, *Astrophys. J.* **189**, L13.
Ruffini, R.: 1973, in B. DeWitt and C. DeWitt (eds.), *Black Holes*, Gordon and Breach, New York.
Ryter, C.: 1970, *Astron. Astrophys.* **9**, 288.
Salpeter, E.: 1973, in H. Bradt, and R. Giacconi (eds.), 'X-and Gamma-Ray Astronomy', *IAU Symp.* **55**, 135.
Schreier, E., Gursky, H., Kellogg, E., Tananbaum, H., and Giacconi, R.: 1971, *Astrophys. J.* 170, L21.
Schreier, E., Levinson, R., Gursky, H., Kellogg, E., Tananbaum, H., and Giacconi, R.: 1972, *Astrophys. J.* **172**, L79.
Schreier, E., Giacconi, R., Gursky, H., Kellogg, E., and Tananbaum, H.: 1972, *Astrophys. J.* 178, L71.
Schwartzman, V. F.: 1971, *Soviet Astron. AJ* **15**, 342.
Setti, G. and Woltjer, L.: 1970, *Astrophys. Space Sci.* **9**, 185.
Seward, F. D., Burginyon, G. A., Grader, R., Hill, R., and Palmieri, T.: 1972, *Astrophys. J.* **178**, 131.
Shakura, N. I. and Sunyaev, R. A.: 1973, *Astron. Astrophys.* **24**, 337.
Shkolovsky, I.: 1967, *Astrophys. J*, **148**, L1.
Shulman, S., Fritz, G., Meekins, J., and Friedman, H.: 1971, *Astrophys. J.* **168**, L49.
Stockman, S. H., Angel, J. R. P., Novick, R., and Woodgate, B. E.: 1973, *Astrophys. J.* **183**, L63.
Strittmatter, P., Scott, J., Whelan, J., Wickramasinghe, D., and Woolf, N.: 1973, *Astron. Astrophys.* **25**, 275.
Sunyaev, R.: 1973, *Soviet Astron. AJ*, in press.
Tananbaum, H.: 1974, in F. J. Kerr (ed.), 'Galactic Radio Astronomy', *IAU Symp.* **60**, 383.
Tananbaum, H., Gursky, H., Kellogg, E., Giacconi, R., and Jones, C.: 1972a, *Astrophys. J.* **177**, L5.
Tananbaum, H., Gursky, H., Kellogg, E., Levinson, R., Schreier, E., and Giacconi, R.: 1972b, *Astrophys. J.* **174**, L143.
Trimble, V., Rose, W., and Weber, W.: 1973, *Monthly Notices Roy. Astron. Soc.* **162**, 1.
Ulmer, M., Baity, W., Wheaton, W., and Peterson, L.: 1972a, *Astrophys. J.* **178**, L61.

Ulmer, M., Baity, W., Wheaton, W., and Peterson, L.: 1972b, *Astrophys. J.* **178**, L121.
Van Den Heuvel, E. and Heise, J.: 1972, *Nature Phys. Sci.* **239**, 67.
Van Den Heuvel, E. and Ostriker, J.: 1973, *Nature Phys. Sci.* **245**, 99.
Wade, C. and Hjellming, R.: 1971, *Astrophys. J.* **170**. 523.
Webster, B. and Murdin, P.: 1972, *Nature* **235**, 37.
Webster, B., Martin, W., Feast, M., and Andrews, P.: 1972, *Nature Phys. Sci.* **240**, 183.
Wilson, R.: 1970, in L. Gratton (ed.), 'Non-Solar X-and Gamma-Ray Astronomy', *IAU Symp.* **37**, 242.
Wilson, R.: 1973, *Astrophys. J.* **181**, L75.
Wolff, S. C. and Morrison, N. D.: 1974, *Astrophys. J.* **187**, 69.

OPTICAL OBSERVATIONS OF BINARY X-RAY SOURCES

PAUL E. BOYNTON

Dept. of Astronomy, University of Washington, Seattle, Wash. 98195, U.S.A.

Compact galactic X-ray sources are undeniably an exciting discovery by the New Astronomy of X-ray observations. Some of these objects were studied briefly, and without much notice, with optical telescopes long before they were known to be sources of high-energy photons. These earlier observations gave no indication of the special properties which compel us at this time to consider the possiblity that systems of this class contain such bizarre objects as neutron stars or black holes.

The New Astronomy represents a greatly broadened vision with which we perceive the universe; one that spans the electromagnetic spectrum from low energy radio photons to multi-MeV γ-rays. Radio observations have lead to the discovery of quasars, pulsars, and the remnant radiation from the Big Bang. Infrared astronomy is just starting to reveal the secrets of the infrared sky. The New Astronomy of ultraviolet, X-ray and γ-photons, is, in spite of the difficulties of rocket and satellite instrumentation, beginning to chart and catalog a universe previously hidden from us by our opaque atmosphere. These are spectacular advances which, for the most part, have been achieved within the past decade.

What about the *old vision*, the Astronomy of visible photons that, in perspective, lie in an extremely small fractional bandwidth between the extremes of this new vision? The astronomy of the nominal three-eV-photon is alive and well, thriving in fact and needs no apologists, certainly not one like myself who might be considered a physicist-interloper. But in a time of justifiable excitement over the fascinating new discoveries which have resulted from this new vision. I would like to point out the contribution to this progress made by optical observations, particularly with reference to compact X-ray sources.

The importance of visual-light photons is neither mystery nor accident. First, because one often finds broad-band dissipation in astrophysical systems, much can be learned about their structure and dominant processes from the visual portion of the spectrum. Coupled with this is the great reserve of optical devices and techniques which may be brought to bear on a given problem. We have a whole century of optical experience to draw upon which is now being supplemented with staggering advances in electronic instrumentation. With the current capabilities of new photon counters, image tubes, digital acquisition and processing equipment, even a small observatory can be highly competitive.

With this in mind it should not seem incongruous to discuss the *optical* properties of *X-ray* sources. Although being *optical* is not their principle feat, clues to the nature of these objects are surely carried by visible-light photons.

Of the nine or so *binary X-ray* sources catalogued by the Uhuru satellite which have been identified optically, I have chosen to stress HZ Her/Her X-1 and HDE 226868/

H. Gursky and R. Ruffini (eds.), Neutron Stars, Black Holes and Binary X-Ray Sources, 221–234. All Rights Reserved.
Copyright © 1975 by D. Reidel Publishing Company, Dordrecht-Holland.

Cyg X-1. These two have become central figures in a continuing attempt to understand the physical nature of such objects. Because this area of study has opened only
within the last 2 or 3 years, one can hardly write a review of the subject. The intent here
is more to report the current observational evidence, than to record the progress we
think we have made toward explaining anything. For a summary of the optical
properties of most of the presently identified sources, the reader should refer to the
1973 Solvay conference paper of John and Neta Bahcall.

1. HZ Herculis/Hercules X-1

HZ Her, which has been established as the chief optical component of the Her X-1
system, is one of those stars referred to earlier as having been observed optically, and
without particular distinction, long before becoming famous. In the late thirties this
star was classified, understandably but perhaps hastily, as a rapid *irregular* variable
(Hoffmeister, 1941). The two exposures in Figure 1 show HZ Her (α (1950) = 16^h56^m3,
δ (1950) = $35°25'$) at maximum and minimum light. Photographically, the change in
observed flux between these two states is roughly a factor of six. Photoelectrically, in
the U band, this change can be as large as a factor of twelve. At minimum light the

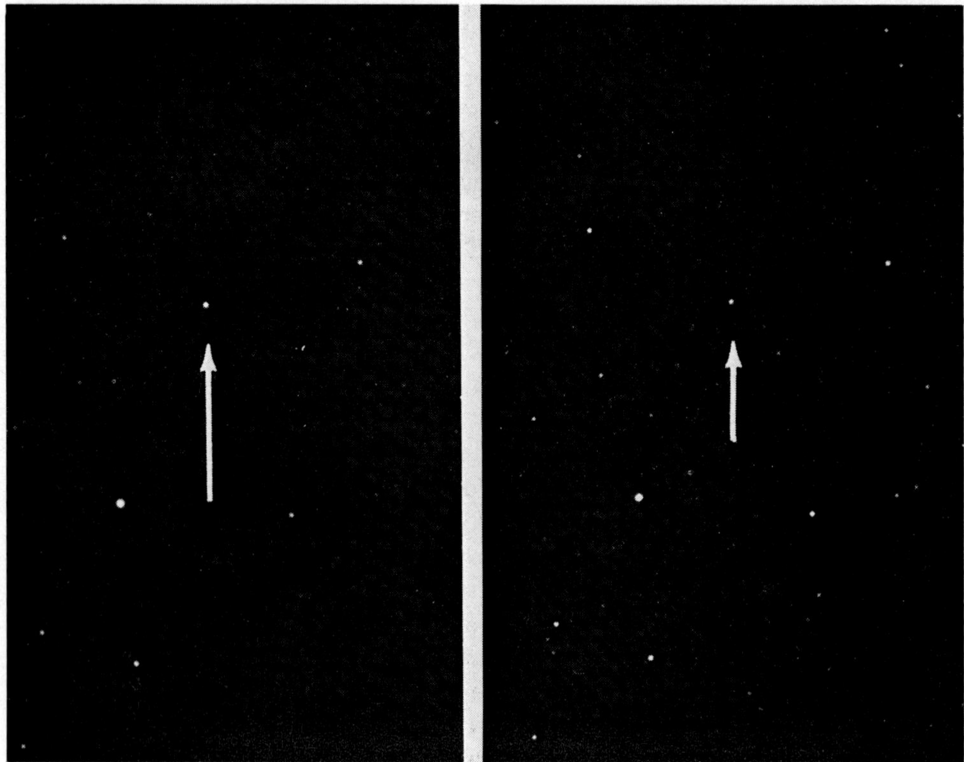

Fig. 1.   Hz Her near maximum light (left) and minimum light (right).

visual magnitude is about 14.3 with colours $U - B = 0.13$, $B - V = 0.20$, which are consistent with the spectroscopic indications of a late A star (Crampton and Hutchings, 1972). For a normal main sequence or giant star this would imply a distance greater than 2.5 kpc, and perhaps for this particular object greater than 5 kpc (Strittmatter et al., 1973), at a rather high galactic latitude, $b \simeq 37.5°$. Such distances have to be reconciled with the consequently rather high space velocity suggested by proper motion studies (Crampton, 1974).

To appreciate the significance of this optical variability it helps to be familiar with the X-ray behavior of the Her X-1 system which is reviewed in the preceding paper by Gursky and Schreier. We find that the X-ray flux changes regularly on several time scales. It varies with a ~ 35 day cycle, being observable for about ~ 11 days then absent (below detection threshold) for ~ 24 days. Next we note that when observed the X-ray flux is strongly modulated by a 1.7 day envelope, and on a still finer time scale by a 1.24 s envelope. This latter periodicity is useful for determining the motion of the pulsed X-ray source by simply interpreting the smooth change in the pulse-period with time as a doppler effect. The sinusoidal variation in period is consistent with motion in a nearly circular orbit with some unknown inclination angle, i, but with line-of-sight dimension of 8×10^6 km, and an orbital speed with a maximum component along the line-of-sight of 170 km s^{-1}. An important feature is that the disappearance of X-ray flux for 0.24 days every 1.7 days occurs at orbital phase 0.0, when the source is most distant from the observer. This clearly corresponds to an eclipse of the X-ray source by the stellar companion, Hz Her. Thus, the X-ray data alone establish Her X-1 as a member of an eclipsing binary system with orbital period of 1.7 days, which in turn implies that $\sin i$ is not far from unity.

With these facts in mind we ruturn to the optical flux variations. The most obvious periodicity to investigate is the 1.7 day eclipse cycle. In fact, it was these 1.7 day light variations which confirm the optical identification of HZ Her with the X-ray source Her X-1. (Bahcall and Bahcall, 1972a; Liller, 1972). Curiously one finds that the optical minimum coincides with X-ray minimum in phase; that is, in some sense both sources are occulted at the same time. This fact has been explained by postulating that the side of the star which faces the X-ray source is heated rather strongly by high energy photons and particles and is thus much more luminous than the unexposed side as portrayed in Figure 2 (Forman et al., 1972; Bahcall and Bahcall, 1972b). Here we see HZ Her gravitationally distorted by the proximity of the compact object, presumably a neutron star, which is responsible for X-ray emission through accretion of matter from HZ Her which spills down the gravitational potential well of the compact companion (Pines et al., 1973). This simple model is in qualitative agreement with the light curve shown in Figure 3 which was constructed by superposing on a single 1.7 day interval all the optical flux data recorded by observers in 1972 and 1973 (Lyutiy et al., 1972; Davidsen et al., 1972; Petro and Hiltner, 1973; Boynton et al., 1973; Chevalier and Ilovaisky, 1973; Grandi et al., 1973). However, there are several severe quantitative problems with this picture (Avni et al., 1973a; Henriksen et al., 1973; Wilson, 1973).

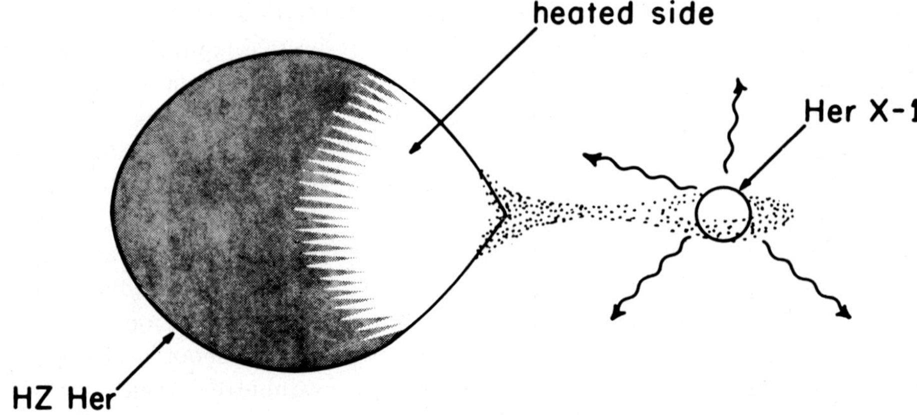

Fig. 2.   Hz Her heated on one side by its matter accreting, X-ray emitting companion, Her X-1.

(1) If the observed 2–40 keV X-ray emission, which varies in intensity by more than
an order of magnitude over the 35^d period, is responsible for heating one side
of HZ Her, why is the 35^d component of the mean optical flux variability less
than 5%?

Free precession of Her X-1 may be a viable explanation for the 35^d clock mechanism
behind the X-ray behavior, but to explain the X-ray on-off cycle in terms of a preces-
sing beam and then require that the same beamed emission to continually intersect
HZ Her constitutes a stringent geometrical limitation on the system. In any case,
because of the strongly pulsed character of the X-rays at $1^s\!24$ one would expect
persistant $1^s\!24$ optical pulsations at about the 1% level and these are not observed. A
stronger objection involves the energy budget of the system:

(2) The observed radiation in the entire 2–40 keV range does not appear adequate
to explain the extent of heating implied by the amplitude and color of the 1.7
day variation in the optical light curve.

Objections (1) and (2) have led to postulating a steady, extremely luminous
($\sim 10^{37}$ erg s^{-1} for a source distance of 5 kpc) and as yet unobserved component in
the system with maximum output in the 4 eV–1 keV range. However, there is still
another fundamental difficulty:

(3) The shape of the optical light curve, with its relatively narrow, sharp minimum
is not consistent with the simple picture of heating by a compact source of
radiation.

Clearly less than half the star should be illuminated, and in this simple model
optical eclipses would be expected to have a duration considerably longer than that
observed for X-rays. In fact, the optical eclipses appear to be more than 3 times shorter.

These fundamental problems, as important as they are, have been discussed primarily in terms of only the simplest aspects of the optical photometric data on this system.
Clarification of our understanding of HZ Her/Her X-1 may depend on more than
these basic clues; for instance, the light curve may betray system geometry and tem-

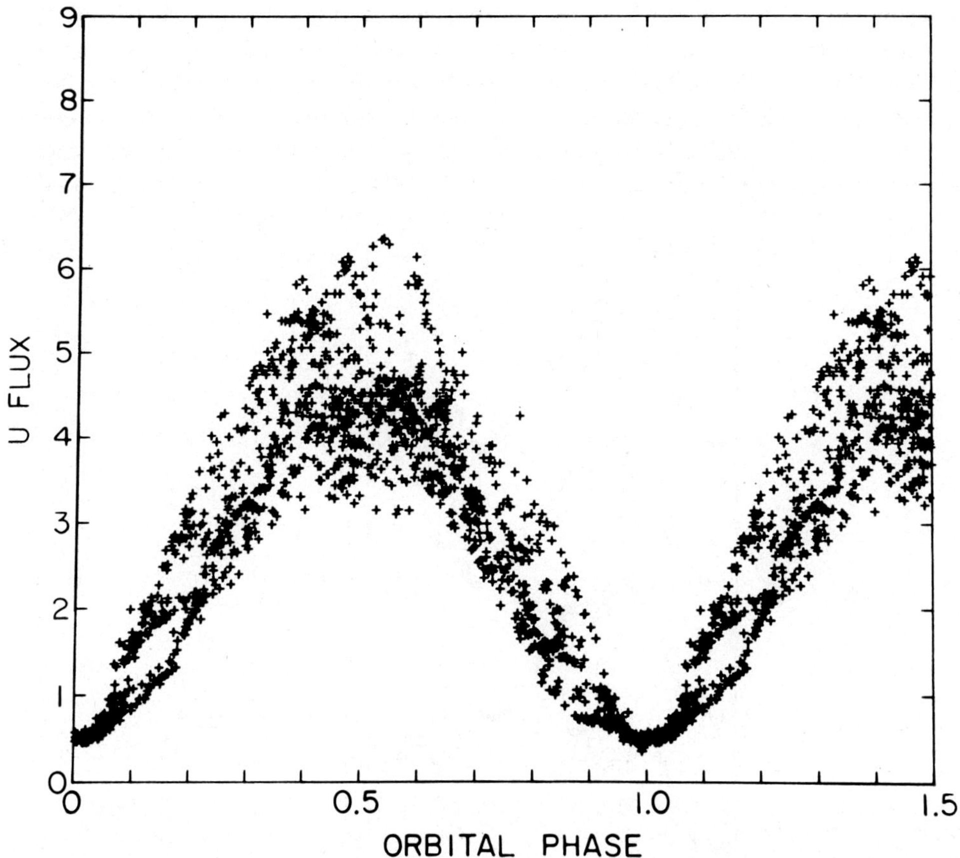

Fig. 3. A composite $1\overset{d}{.}7$ light curve of the Hercules source from the photoelectric observations of several observers in the 1972–73 observing seasons (see text). The first half cycle is repeated to aid the inspection of the primary minimum.

perature distribution through intensity and color time dependencies on time scales other than the $1\overset{d}{.}7$ orbital period. Figure 3 does not exhibit a single. well defined 1.7 day cycle. There appears to be variability or 'noise' on the light curve far beyond that attributable to measurement error, especially near maximum light. Apparently this noise does not result from random thrashings of the source, but is a remarkably regular modulation of the shape of the 1.7 day light curve during the 35-day X-ray cycle (Kurochkin, 1972; Boynton *et al.*, 1973; Chevalier and Ilovaisky, 1973) as seen in Figure 4, where the data in Figure 3 is separated into successive light curves for roughly every 3 successive orbits in the 20 to 21 orbit 35-day cycle. Each curve is composed of data sampled from several different 35-day intervals. The fact that these curves are so smooth is evidence that the features exhibited here are repeat from cycle to cycle. The most obvious 35-day modulation occurs near maximum light. During and even shortly before the ~ 7 orbit X-ray on period, the curve shows a clearly discernible secondary minimum which is replaced during the X-ray off period by a very sharp, bright feature. These two states are characterized by the fourth and seventh curves in Figure 4. The secondary minimum which develops rapidly just prior to X-ray turn-on may result from obscuration by a cloud of accreting material. This feature is observed only during the X-ray on period and has been seen also in the low energy portion of the OSO-7 data (McClintock *et al.*, 1974). The bright central peak that

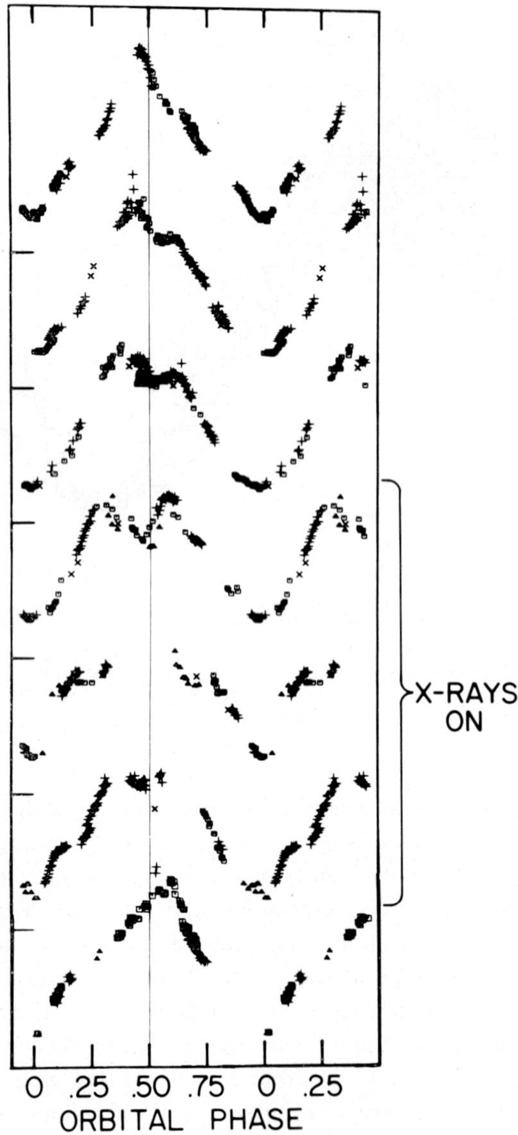

Fig. 4.   Light curves for successive 3 orbit intervals through the 35 day cycle. Data from six independent cycles has been combined. × Lyutiy *et al.* 1972, + Petro and Hiltner (1973), △ and □ Boynton *et al.* (1973).

appears near mid-orbital phase during the X-ray off period has been found to be quite blue, corresponding to a radiation temperature considerably higher than that responsible for the major portion of the 1$^\mathrm{d}$7 variability (Boynton *et al.*, 1973). These must be important clues, but as yet we do not have an adequate model to explain these optical phenomena which apparently correlate with the equally mysterious X-ray variability.

The optical light curve may have still more to say. Beyond the 'simple' 35$^\mathrm{d}$ changes at mid-orbital phase is the complex behavior which causes the scatter in flux level throughout 1$^\mathrm{d}$7 phase shown in Figure 3. The nature of this interesting and presumably significant detail is suggested by studying the progression of changes in the light curves orbit by orbit through the 35$^\mathrm{d}$ cycle. There are waves of small blue features sweeping backward with a well-defined periodicity through the dominant 1$^\mathrm{d}$7 variation (Gerend *et al.*, 1974). The 'peak' referred to earlier may be simply a part of this pattern, but the secondary minimum is apparently independent.

On much longer time scales, Harvard Observatory sky patrol plates of the Hercules region exposed as far back as 1890 have been examined for the eclipse cycle of HZ Her (Jones *et al.*, 1973). The observed intensity variations were consistent with an eclipse period which is constant to within 2 parts in 10^5 over the past 80 years. But by plotting the optical flux around maximum light ($\phi = 0.5 \pm 0.2$) against the date of observation, as shown in Figure 5, one discovers an interesting history. Since 1890, optical varia-

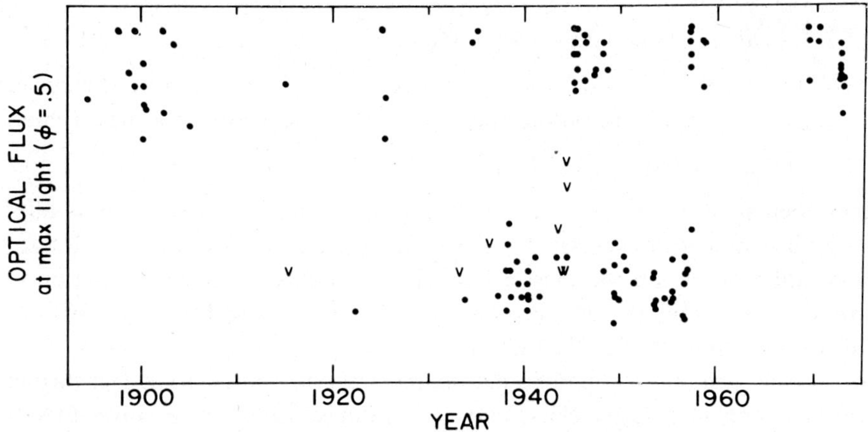

Fig. 5.   Long-term optical behavior of the Hercules source, according to Jones *et al.* (1973). 'V' denotes upper limit. Published by the University of Chicago Press; © 1973, American Astronomical Society.

tions of the kind shown in Figure 3 have been observed only about half the time. During the remaining periods, of up to 5 years duration each, HZ Her is found near minimum light level. A detailed analysis of the observations during these extended 'lows' shows a small (~ 0.2 mag.) variation at half the orbital period, as might be expected from the changing aspect and surface brightness distribution of a gravitationally distorted orbiting star.

As pointed out earlier, measurement of the doppler shifted phase of the strong 1$^\mathrm{s}$24

X-ray flux modulation allows a precise determination of the orbital period and eccentricity of the binary system, and the line-of-sight displacement and radial velocity excursions of the X-ray source. In conjunction with Kepler's third law, this knowledge of the orbital period, $P = 1\overset{d}{.}7$, and orbital radial velocity, $V_{\text{Her X-1}} \sin i = 170$ km s^{-1}, specify a relationship between the masses of binary components, the so-called mass function:

$$\frac{M^3_{\text{HZ Her}} \sin^3 i}{(M_{\text{Her X-1}} + M_{\text{HZ Her}})^2} = f\,(P,\, V_{\text{Her X-1}} \sin i) = 0.85 M_\odot\,.$$

Additional constraints allowing one to then solve for individual masses can be imposed in several ways: Three methods are discussed briefly below, they deal with the Roche lobe model, spectral classification, and the optical mass function (Avni *et al.*, 1973b).

For a close binary system with the general properties of HZ Her/Her X-1 the assumption of synchronous rotation implicit in the Roche lobe picture does not seem unreasonable (Henricksen *et al.*, 1973). In fact, the notion that HZ Her fills its critical Roche lobe and transfers matter through the inner Lagrangian point to provide accretion material to Her X-1 is a natural model for this system. By noting first that the size of the Roche lobe is determined by the mass ratio of the system (Plavec, 1968; Wilson, 1972; Tananbaum *et al.*, 1972a)

$$R_L/d \simeq 0.38 + 0.2 \log\,(M_{\text{HZ}}/M_{\text{X-1}})$$

and secondly that the actual occultation radius R is determined by the angular extent of the eclipse, b, and the orbital inclination, i, and the separation of mass centers, d:

$$(R/d)^2 = \cos^2 i + \sin^2 b \sin^2 i$$

one may then solve for the system mass ratio assuming the Roche lobe is filled, or calculate mass limits for the case $R < R_L$. Combining this information with the mass function, upper limits on the masses can be obtained for a given inclination angle. The *overall* upper limits ($i = 90°$) are very nearly $2M_\odot$ for HZ Her and $1\ M_\odot$ for Her X-1, all without the aid of optical observations.

Alternatively one might attempt a determination of the mass of HZ Her through its spectral type and luminosity class, and use the mass function to solve directly (to within a factor involving the orbital inclination) for the mass of Her X-1. As mentioned before, spectroscopic and color photometric data agree on a nominal designation of late A for the cooler side of HZ Her; however, the evolutionary status (thus luminosity) of this object is a puzzle (Strittmatter *et al.*, 1973). If this is a Population II object as suggested by its high galactic latitude (37.5°) and distance ($\lesssim 5$ kpc), then how is such massive system still evolving? Clearly the presumed presence of a neutron star companion implies a complicated history of multiple mass exchanges. If the object is sufficiently peculiar, perhaps it is grossly under luminous, much closer, and of Population I origin (Avni *et al.*, 1973b). In any case, arguments about the mass based on spectral type seem highly uncertain, but may yet be clarified by further observations.

Probably the most satisfying estimate of component masses would be to measure the orbital radial velocity of HZ Her, $V_{\mathrm{HZ\,Her}}\sin i$, from the doppler shift of *optical* absorption lines, then solve for the individual masses from the independently determined X-ray and optical mass functions. Unfortunately this is a difficult task for two reasons. First, gas streaming in the system makes the interpretation of spectral features difficult. Secondly, even without extraneous mass motion, the range of velocities due to simple rotation is weighted according to the surface brightness distribution and shape of the star (Crampton and Hutchings, 1972). The velocity curve for HZ Her can only be reconstructed by unfolding the effects of this brightness distribution which in turn is not uniquely specified by the light curve. The present best estimates are $M_{\mathrm{HZ}}=2.2+0.4\ M_{\odot}$ and $M_{\mathrm{X\text{-}1}}=1.3\pm0.4\ M_{\odot}$, which are lower limits in the sense that $i=90°$ (Crampton, 1974).

This brief discussion does not present all the observational data on Hz Her/Her X-1 but the remainder carries out a theme which is already clear: this system is remarkably regular, it exhibits clearly repeatable, periodic changes in X-ray flux on many time scales and is at least as rich in optical behavior. Furthermore, those optical data do not simply mimic the X-ray results, but provide information largely supplemental to the X-ray findings. A further example would be the 1.24 s pulsations which are usually observed optically at a very low level, 0.1 to 0.2% modulation or less (Lamb and Sorvari, 1972; Davidsen *et al.*, 1972; Middleditch and Nelson, 1973; Groth *et al.*, 1973; Groth, 1974). But the optical pulsations, when observed, do not necessarily have the doppler shifted frequency which would be appropriate to the position of the X-ray source. Often pulsed light appears to come from X-ray illuminated matter in various places in the system where one might expect streaming gas or an accretion disk as suggested in Figure 2. Optical pulsations are generally observed during the X-ray on period, but have been detected in the remaining portion of the 35^{d} cycle (Middleditch, 1974).

The regularity and richness of all these data compel one to seek out an equally ordered underlying mechanism which explains every feature in this rather complex pattern of changing optical and X-ray luminosities. In a sense the problem is that there is too much order, too many suggestive details. It is perhaps significant that HZ Her is a thousand times less luminous than the identified stellar companions of other X-ray binaries; this allows the observation of optical detail that would be swamped by the brilliance of the primary star in other sources. Perhaps we are privileged to see too much in the sense that we have to face the additional task of separating trivia from the essential.

Is the Hercules system revealing extraneous complexity; or is it representative, displaying those properties which will help us understand this class of objects that give the promise of observing high energy astrophysics in action?

2. Cygnus X-1

Conceivably optical observations may not be crucial to an understanding of HZ Her/Her X-1, but we can certainly say that the optical data appear to contain a great deal of information about the system. By contrast, the significance of optical observations to the primary issue of determining the *mass* of the X-ray source in the Cyg X-1/HDE 226868 system is quite clear.

The importance of *mass* derives from the possible existence of objects which have undergone catastrophic gravitational collapse... so called 'Black Holes' (Ruffini and Wheeler, 1971; Misner *et al.*, 1973). The fact that they may exist naturally compels us to search for them, and perhaps even to anticipate them behind yet unopened astrophysical doors. This excitement is understandable – the triumph of theory is a heady reward.

Cyg X-1 is presently the central figure in that search. It does not display the manifold properties of Her X-1 except for high X-ray luminosity and binary nature. Millisecond time-scale X-ray fluctuations are a clue to its compact nature (Holt *et al.*, 1973), but the details of how it works are in a sense secondary to a very simple question: what is the mass of the X-ray emitting component? White dwarfs, neutron stars and black holes are the kinds of objects needed to explain the X-ray luminosity in terms of matter accretion. Our present theoretical understanding of neutron stars predicts (within the context of some very general assumptions on the fluid properties of high density matter, causality, and the validity of General Relativity) an upper limit of about 3 solar masses ($3\,M_\odot$) for these objects (Rhoades and Ruffini, 1974). Since this is greater than the maximum mass of a white dwarf, an X-ray source with mass in excess of $3\,M_\odot$ must be considered a black hole candidate. Even so, some additional evidence is clearly desirable before being certain of such a discovery.

The optical counterpart to Cyg X-1, the very luminous supergiant HD 226868 shows smooth, periodic variations in the wavelength of its spectral lines indicating that it is a member of a binary system with an orbital period of 5.6 days (Webster and Murdin, 1972; Bolton, 1972a, b; Brucato and Kristian, 1973). The optical flux also shows a small variation at this period (Walker, 1972; Luytiy *et al.*, 1972; and Lester *et al.*, 1972) and it was very recently reported that the Copernicus Satellite soft X-ray flux data may show the same modulation period (Sanford *et al.*, 1974). Presently, the primary evidence for identification stems from correlated changes in X-ray flux from Cyg X-1 with radio flux from a radio source within 0.5″ of HDE 226868 (Tananbaum *et al.*, 1972b; Hjellming, 1973).

From the orbital period and the spectroscopic measure of the orbital velocity, $v_{\mathrm{orb}}\sin i$, a mass function may be calculated in the same manner as discussed in connection with Her X-1:

$$\frac{M_x \sin^3 i}{(M_x + M_s)^2} = 0.23\,M_\odot,$$

where M_x and M_s are the masses of Cyg X-1 and HDE 226868 respectively. However, this time the mass function has been determined by optical not X-ray observations, since no periodic X-ray pulsations are detected from this source. The smallest possible M_x is obtained for a given M_s if $\sin i = 1$; for example $M_x > 3\,M_\odot$ is assured if M_s is greater than $9\,M_\odot$. In this way, optical spectroscopic observations of the star HDE

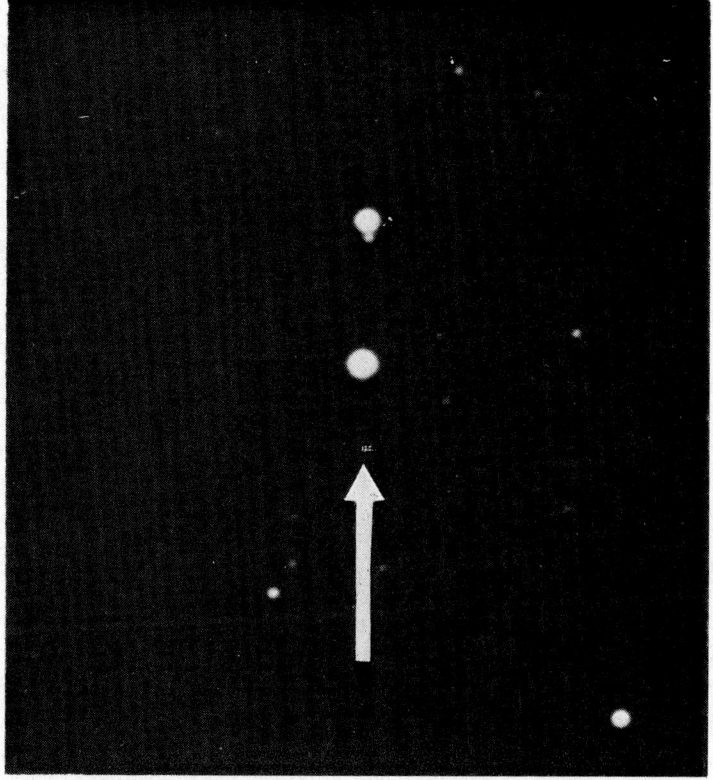

Fig. 6.   The star field containing HDE226868/Cyg X-1.

226868 can be used to set a minimum mass for Cyg X-1 if the mass of the companion star can be independently specified.

Fortunately optical spectra also provide a means for determining stellar masses if the luminosity, or equivalently the state of evolution, of the star is known. The evolutionary problem is a complicated issue, but the pessimism raised earlier regarding spectroscopic mass estimates for the Hercules system is not necessarily appropriate here. The Cygnus system is at low galactic latitude, the optical flux is not strongly affected by X-ray emission, and high dispersion spectra are easily obtained from a 9th mag. object.

In general, one argues that the spectral signatures of stars (the dominant lines, their relative strengths and the line widths) depend on the state of ionization and excitation (temperature) and the pressure (surface gravity) in the stellar photosphere. From the following simple relations

$$L = 4\pi R^2 \sigma T_e^4, \qquad g_s = \frac{MG}{R^2}$$

one can clearly specify the mass, M, of a star if the surface gravity, g_s, effective temperature, T_e, and Luminosity, L, are known. Offhand, one might say that the details of the spectrum of HDE 226868 give a surface gravity and effective temperature which are consistent with an **OB** *supergiant* thereby implying a luminosity (evolutionary state) and thus a mass in the neighborhood of 20 to 30 $M_\odot$. This value would be adequate to insure an X-ray source mass in excess of 5 $M_\odot$.

Unfortunately, there is the distinct possibility of an ambiguity in the mass determination in the absence of an independent luminosity estimate. It is possible to have a very low mass, $M_s \simeq 0.5 \ M_\odot$, star with the same surface conditions, the same high temperature and low surface gravity, which then makes it essentially indistinguishable spectroscopically from the 30 $M_\odot$ supergiant (Trimble *et al.*, 1973). Such stars are not common but a good example of this type is known (Greenstein, 1973) and one might find such an object in a binary system where mass transfer between members could strongly affect the evolution of what is now the stellar companion.

Fortunately this ambiguity can be resolved, again by optical observations. The low mass star is at least 5 times less luminous than the high mass choice. Since we know the *apparent* brightness, the intrinsic brightness (luminosity) of the star could be calculated trivially if we just knew its distance from us. Clearly a low luminosity, low mass star would have to be much closer to us (at least a factor of 2) than a high luminosity, high mass star in order to appear to have the same brightness as viewed from the Earth. How does one determine the distance to Cyg X-1? The following is a review of the recent work of two groups on this problem (Margon *et al.*, 1973; Bregman *et al.*, 1973).

We live in a spiral galaxy; a flattened, disk-shaped collection of stars, dust and gas. Just as smoke and dust in the Earth's atmosphere can make the Sun appear reddened through frequency selective scattering, interstellar dust reddens the light from distant stars. The degree of reddening is proportional to the amount of dust the light must travel through in reaching the observer. For all stars in the immediate field of view of Cyg X-1 the amount of reddening is approximately proportional to the distance to a star unless it lies beyond the dusty region. Reddening data on many stars of known distance in the Cyg X-1 field are shown in Figure 7. The trend is clear, and a star as highly reddened as HDE 226868 must be at least 2.5 kpc distant. This distance rules out the low luminosity, low mass characterization of HDE 226868 and is an important decision in favor of the black hole candidacy of Cyg X-1. But the case is not airtight, several uncertainties plague this logical chain. One is the difficulty in determining a precise orbital component of radial velocity for HDE 226868 from spectroscopically observed doppler shifts in a system where streaming gas can confuse matters. A second problem is the uncertainty in spectroscopic mass estimates even with the best spectroscopic data, which can be as large as a factor of three (van den Heuvel and Ostriker, 1973). Even so, it appears that M_s is surely greater than 10 $M_\odot$ and M_X is thus greater than 3 $M_\odot$. Finally, the neutron star mass upper limit estimates must be considered in full awareness of the assumptions involved (Leach and Ruffini, 1973).

The controversy continues. Alternatives to the black hole hypothesis suggest that

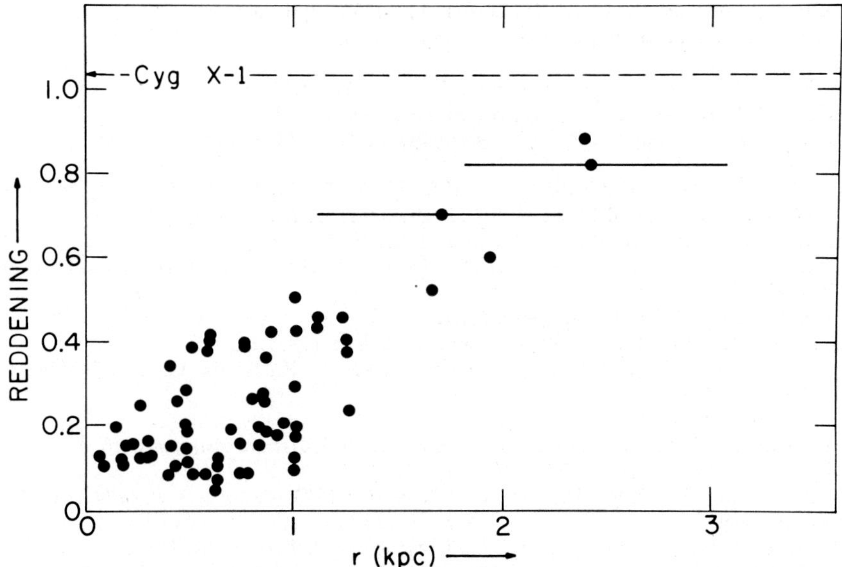

Fig. 7. Interstellar reddening data for a restricted field containing HDE226868/Cyg X-1. From Bregman *et al.* (1973). Published by the University of Chicago Press; © 1973, American Astronomical Society.

the secondary component in Cyg X-1 is a comparatively normal stellar object whose mass may well be greater than 3 $M_\odot$. Perhaps X-rays are then produced by the customarily invoked accretion of matter onto a moderate mass neutron star which is a *third* component in the system (Bahcall *et al.*, 1973a). Another view postulates a binary system of rather ordinary magnetic stars which do not corotate. The resulting buildup of magnetic energy through field twisting, which is then dissipated in plasma heating upon flux-line reconnection, is responsible for X-ray emission (Bahcall *et al.*, 1973b). But black holes are too exciting a possibility to let go easily; many of us await an unambiguous sign, perhaps a yet undiscovered relativistic signature in the data presently accumulating on this system. Whether or not a black hole lurks in the Cygnus source, optical observations, while not responsible for the direct detection of high-energy phenomena, are clearly instrumental in our continuing attempt to understand compact X-ray sources.

References

Avni, Y., Bahcall, J. N., Joss, P. C., Bahcall, N. A., Lamb, F., Pethick, C., and Pines, D.: 1973a, *Nature Phys. Sci.* **246**, 36.
Avni, Y., Bahcall, J. N., and Joss, P. C.: 1973b, preprint.
Bahcall, J. N. and Bahcall, N. A.: 1972a, *IAU Circ.*, Nos.2427 and 2428.
Bahcall, J. N. and Bahcall, N. A.: 1972b, *Astrophys. J. Letters* **178**, L1.
Bahcall, J. N., Dyson, F. J., Katz, J. I., and Paczyński, B.: 1973a, preprint.
Bahcall, J. N., Kulsrud, R. M., and Rosenbluth, M. N.: 1973b, *Nature* **243**, 27.
Bolton, C. T.: 1972a, *Nature* **235**, 271.
Bolton, C. T.: 1972b, *Nature* **240**, 124.

Boynton, P. E., Canterna, R., Crosa, L., Deeter, J., and Gerend, D.: 1973, *Astrophys. J.* **186**, 617.

Bregman, J., Butler, D., Kemper, E., Koski, A., Kraft, R. P., and Stone, R. P. S.: 1973, *Astrophys. J. Letters* **185**, L117.

Brucato, R. and Kristian, J.: 1973, *Astrophys. J. Letters* **179**, L129.

Chevalier, C. and Ilovaisky, S. A.: 1973, *Nature Phys. Sci.* **245**, 87.

Crampton, D. and Hutchings, J. B.: 1972, *Astrophys. J. Letters* **178**, L65.

Crampton, David: 1974, *Astrophys. J.* **187**, 345.

Davidsen, A., Henry, J. P. Middleditch, J., and Smith, H. E.: 1972, *Astrophys. J. Letters* **177**, L97.

Forman, W., Jones, C. A., and Liller, W.: 1972, *Astrophys. J. Letters* **177**, L103.

Gerend, D., Deeter, J., Crosa, L., and Boynton, P. E.: 1974, preprint.

Grandi, S., Hintzen, P., Jensen, E., Rydgren, A., Scott, J., Stickney, P., Whelan, J., and Worden, S.: 1973, preprint.

Greenstein, J. L.: 1973, *Astron. Astrophys.* **23**, 1.

Groth, E. J. and Nelson, M. R.: 1972, *Astrophys. J. Letters* **178**, L111.

Groth, E. J., Yeung, S. Z., Papaliolios, C., Penneypacker, C. R., Spada, G., and Middleditch, J.: 1973, *IAU Circ.*, No. 2523.

Groth, E. J.: 1974, preprint.

Henriksen, R. N., Reinhardt, M., and Aschenbach, B.: 1973, *Astron. Astrophys.* **28**, 47.

Hoffmeister, C.: 1941, *Kl. Verein. Beob. Bull.*, No. 24.

Holt, S., Rothschild, R., Boldt, E., and Serlemitsos, P.: 1973, *Comments at 1974 AAS Meeting*, Tucson, Arizona.

Jones, C. A., Forman, W., and Liller, W.: 1973, *Astrophys. J. Letters* **182**, L109.

Kurochkin, N. E.: 1972, *Variable Stars (USSR)* **18**, 425: see also *Inf. Bull. Var. Stars*, No. **753** (1973).

Lamb, D. Q. and Sorvari, J. M.: 1972, *IAU Circ.*, No. 2422.

Leach, R. and Ruffini, R.: *Astrophys. J. Letters* **180**, L15.

Lester, D. F., Nolt, I. G., and Radostitz, T. V.: 1973, *Nature* **241**, 126.

Liller, W.: 1972, *IAU Circ.*, Nos. 2415 and 2427.

Lyutiy, V. M., Sunyaev, R. A., and Cherepashchuk, A. M.: 1973, *Astron, Zh.* **50**, 3.

McClintock, J. E., Clark, G. W., Lewin, W. H. G., Schnopper, H. W., Canizares, C. R., and Sprott, G. F.: 1974, *Astrophys. J.* **188**, 159.

Margon, B., Bowyer, S., and Stone, P. S.: 1973, *Astrophys. J. Letters* **185**, L113.

Middleditch, J.: 1974, private communication.

Middleditch, J. and Nelson, J.: 1973, *Astrophys. Letters* **177**, L97.

Misner, C. W., Thorne, K. S., and Wheeler, J. A.: *Gravitation*, Walt. Freeman, San Francisco, 1973.

Petro, L. and Hiltner, W. A.: 1973, *Astrophys. J. Letters* **181**, L39.

Pines, D., Pethick, C. J., and Lamb, F. K.: 1973, *Annals of the N.Y. Academy of Sciences* **224**, 237.

Plavec, M.: 1968, *Adv. Astron. Astrophys.* **6**, 201.

Rhoades, C. E. Jr. and Ruffini, Remo: 1974, *Phys. Rev. Letters* **32**, 324.

Ruffini, R. and Wheeler, J. A.: 1971, *Relativistic Cosmology and Space Platforms*, ESRO Book SP52, Paris.

Sanford, P. W., Mason, K. O., Hawkins, F. J., Murdin, P., and Savage A.: 1974, preprint.

Strittmatter, P. A., Scott, J., Whelan, J., Wickramasinghe, D. T., and Woolf, N. J.: 1973, *Astron. Astrophys.* **25**, 275.

Tananbaum, H., Gursky, H., Kellogg, E. M., Levinson, R., Schreier, E., and Giacconi, R.: 1972a, *Astrophys. J. Letters* **174**, L143.

Tananbaum, H., Gursky, H., Kellogg, E., Giacconi, R., and Jones, C.: 1972b, *Astrophys. J. Letters* **177**, L5.

Trimble, V., Rose, W. K., and Weber, J.: 1973, *Monthly Notices Roy. Astron. Soc.* **162**, 1P.

van den Heuvel, E. P. J. and Ostriker, J. P.: 1973, *Nature* **245**, 99.

Walker, E. N.: 1972, *Monthly Notices Roy. Astron. Soc.* **160**, 9p.

Webster, B. L. and Murdin, P.: 1972, *Nature* **235**, 37.

Wilson, R. E.: 1972, *Astrophys. J. Letters* **174**, L27.

Wilson, R. E.: 1973, *Astrophys. J. Letters* **181**, L75.

BLACK HOLES AND NEUTRON STARS:
EVOLUTION OF BINARY SYSTEMS

ROBERT P. KRAFT*

Kitt Peak National Observatory,† Tucson, Ariz., U.S.A.

1. Direct Observational Evidence for the Existence of
Black Holes and Neutron Stars

The most direct observational evidence for the existence of neutron stars is, of course, the discovery of radio pulsars. A black hole, however, neither radiates electromagnetic energy nor energy in the form of gravitational radiation (Zel'dovich and Novikov, 1967, 1971). Thus, detection of such an object is limited to the influence of its static gravitational field, or to the electromagnetic radiation of gas undergoing accretion or acceleration in its vicinity. Spherically symmetrical accretion by a black hole lying in interstellar matter of typical density leads to a luminosity too small to be of much interest (Zel'dovich and Novikov, 1967, 1971; Schwarzman, 1970). But a black hole member of a binary system presents a more fruitful possibility: the companion star whose electromagnetic radiation is detectable, becomes the object of the black hole's gravitational influence and is in turn a possible rich source of accretable matter. And, of course, the same possibilities present themselves if one component of the binary is a less-exotic collapsed object – a neutron star or white dwarf. Binary systems containing a white dwarf are ubiquitous; it is a curious fact, however, that not a single radio pulsar seems to be a member of a binary system (Gott *et al.*, 1970).

The question of whether black holes or neutron stars might contribute significantly to the population of 'unseen' secondaries amongst the single-line spectroscopic binaries has been considered by several authors (Zel'dovich and Guseynov, 1965; Guseynov and Novrusova, 1971; Trimble and Thorne, 1969). The last named examined the 700-odd entries in the *Sixth General Catalogue of Spectroscopic Binaries* (Batten, 1968). By assigning masses $\mathfrak{M}_s$ appropriate to the spectral types of the observable primaries, they computed the *minimum* masses $\mathfrak{M}_u$ of the unseen companions from the mass function

$$f(\mathfrak{M}) = \frac{\mathfrak{M}_u^3 \sin^3 i}{(\mathfrak{M}_s + \mathfrak{M}_u)^2} = \text{const } PK_s^3$$

and the assignment of $i = 90°$. In the above, P is the period and K_s is the projected radial velocity-amplitude of the observable component. Forty objects were found in which $\mathfrak{M}_s \geqslant \mathfrak{M}_u \geqslant 1.4 \, \mathfrak{M}_\odot$, and an additional 10 objects had the property that $\mathfrak{M}_u \geqslant$

---

* Visiting Resident Scientist; normally at Lick Observatory, University of California, Santa Cruz.
† Operated by the Association of Universities for Research in Astronomy, Inc., under contract with the National Science Foundation.

H. Gursky and R. Ruffini (eds.), Neutron Stars, Black Holes and Binary X-Ray Sources, 235–255. All Rights Reserved.
Copyright © 1975 by D. Reidel Publishing Company, Dordrecht-Holland.

$\mathfrak{M}_s \geqslant 1.4\,\mathfrak{M}_\odot$; $\mathfrak{M} = 1.4\,\mathfrak{M}_\odot$ is the Chandrasekhar upper mass limit for (non-rotating) white dwarfs. From a comparison of the stars in these two groups with the single-line eclipsing binaries (objects obviously *not* containing a collapsed unseen component), Trimble and Thorne concluded that little evidence existed for supposing that the secondaries in the 50 systems of interest were anything other than normal stars. This conclusion was strengthened in a study by Abt and Levy (1974) who reexamined the radial velocities of half (5) the stars in Trimble and Thorne's second group and concluded that four of these were, in fact, constant, i.e. not binaries at all! All had been thought to be stars with large $f\,(\mathfrak{M})$, long period, and small velocity-amplitude: the last proved spurious.

It therefore appears that the proof of the existence of black holes cannot be established from a statistical study of the known binary stars. We are left with the possibility that ejected gas from the directly observable component may gravitationally interact with the black hole or neutron star companion. Indeed, the most plausible current explanation for the 'visibility' of X-ray binaries is based on just such a model.

2. Origins of the Accretion Hypothesis in X-Ray Binaries

The idea that X-ray emission could be generated by accretion onto a collapsed object followed from elementary considerations of the luminosity and temperature acquired on the conversion of kinetic energy into heat by the infalling matter. Thus if Φ is the gravitational potential, we have $L \sim \Phi\mathfrak{M}$, which in the Newtonian approximation becomes $L \sim G\mathfrak{M}R^{-1}$ or

$$L/L_\odot \sim 5 \times 10^7 \frac{(\mathfrak{M}/\mathfrak{M}_\odot)}{(R/R_\odot)}\,\dot{\mathfrak{M}}$$

and

$$T \sim 10^7\,\mathrm{K}\,\frac{(\mathfrak{M}/\mathfrak{M}_\odot)}{(R/R_\odot)},$$

where $\dot{\mathfrak{M}}$ is units of $\mathfrak{M}_\odot\,\mathrm{yr}^{-1}$. For typical white dwarf parameters $[(\mathfrak{M}/\mathfrak{M}_\odot)/(R/R_\odot) \sim 10]$, one obtains $T \sim 10^8\,\mathrm{K}$ and $L/L_\odot \sim 100$ when $\dot{\mathfrak{M}} \sim 2 \times 10^{-7}\,\mathfrak{M}_\odot\,\mathrm{yr}^{-1}$; these values of T and $L/L_\odot$ are reasonable fits to the luminosity and bremsstrahlung temperature found for Sco X-1, the first optically-identified compact X-ray source. The important point note is that, unless $\mathfrak{M}/R$ (in solar units) is of order 10 or more, as characteristic of collapsed objects rather than ordinary stars, the temperature of the gas will not be sufficiently high to account for the X-ray bremsstrahlung.

Detailed calculation of a model in which matter flows through the inner Lagrangian point L_1 into a disk surrounding the white dwarf, and in which radial viscous dissipation permits in the inner part of the disk ultimately to lose angular momentum and fall on the white dwarf was advanced by Prendergast and Burbidge (1968). They obtained L_X (X-ray luminosity) ~ 10 to $100\,L_\odot$ for $\dot{\mathfrak{M}} \sim 3 \times 10^{-6}\,\mathfrak{M}_\odot\,\mathrm{yr}^{-1}$; the crude order-of-magnitude estimates thus appear not too bad.

Support for such a model was derived from experience with other astronomical objects and from evolutionary calculations. It is known, for example, that old novae and U Gem stars are mass-transfer binaries in which one component, usually a main-sequence star, overflows its inner Lagrangian surface and transfers mass to its (probably) white dwarf companion – accretion heating is responsible for the anomalously high luminosities of these degenerate objects. Moreover, Kraft (1973) pointed out that the accretion rate required in the Prendergast-Burbidge model agreed with the rate at which matter is ejected through L_1 in evolving main sequence components of close binary systems, i.e., stars with masses between 2 and $10\,\mathfrak{M}_\odot$ (Figure 1) (Kippenhahn and Weigert, 1967; Kippenhahn *et al.*, 1967; Kippenhahn, 1969). This suggested that X-rays might be emitted in the later stages of evolution of rather ordinary close binaries, for example, systems with low mass subgiant secondaries, after the subgiant had become a white dwarf and its companion had begun to swell through its Roche lobe. However, in all these early considerations, it remained troublesome that no direct observational evidence was forthcoming in support of the assertion that Sco X-1 was, in fact, a binary star. This state of affairs has not changed. But the more recent discovery of a whole class of binary X-ray sources in which the collapsed component may well be a neutron star has changed the approach to the accretion picture, at least from the point of view of stellar evolution.

Accretion heating on a collapsed object in a binary system remains the most plausible model for the source of X-ray emission. In support of the general picture is the discovery of variable low energy cutoffs in the X-ray spectra of eclipsing systems, suggesting the existence of gas streams in the orbital plane, and thus insuring a 'source' of accretable material. Moreover, the rapid variability and high temperature insure the compactness of the accreting source. However, other models, which are not easily refuted at this state of our knowledge, have been advanced. One of these (Bahcall *et al.*, 1973) for example, assumes that the binary consists of two otherwise ordinary, but magnetic, stars linked by magnetic flux. The linked lines of force are twisted up by the assumed lack of synchronism between orbital motion and rotation, thus increasing the magnetic flux. Instabilities set in, as in some solar flare models and energy is released in the form of X-rays. The authors show that for quite plausible field strengths ($B \sim 10^{4,5}$ G), electron densities ($n_e \sim 10^{12}$ cm^{-3}) and rotation rates, $L_X \sim 10^{37}$ erg s^{-1} can be emitted with rapid intensity variations resulting from highly local processes of magnetic field reconnection.

The significant feature of this model is the fact that no collapsed object is required: this could have important consequences in the interpretation of Cyg X-1, in particular. Strong, variable magnetic fields have been reported in Vela X-1 and 3U 1700$-$37 (Kemp and Wolstencroft 1973) but their existence seems to be highly problematical (Angel *et al.*, 1973; Wolff and Morrison, 1974); it is also a little surprising that if this model were appropriate, X-ray sources are not found amongst binaries having Ap-star components (Preston, 1974).

Returning to the accretion picture, we now review the present state of this model for X-ray binaries.

3. Accretion onto Neutron Stars and Black Holes

The discovery of X-ray pulsars in Cen X-3 (Schreier *et al.*, 1971) and Her X-1 (Tananbaum *et al.*, 1972), with periods of 4.8 s and 1.2 s, respectively, has already been discussed and although other models have been proposed (cf. DeGregoria and Woltjer, 1973), the results suggest the presence of neutron stars in these binary systems. The orbital periods are $2^{d}09$ and $1^{d}70$, respectively. Her X-1 has been identified with the (optical) variable HZ Her, which has a primary with spectral type about F0 and an optical identification for Cen X-3 has recently been advanced (Krzeminski, 1974) in which the primary is a B-type supergiant. The four other known X-ray binaries have no apparent pulsarlike periodic X-ray variations, and contain an optical primary that is a luminous O or B-type star. The binary periods are all fairly short ($P < 10^{d}$); this means that the separation of the centers of mass exceeds the radius of the optical primary by a numerical factor only of order unity. In this sense the components can be thought of as 'close', although not necessarily 'semi-detached' (Kopal, 1959). (For a comprehensive review of the optical properties of the sources, see Bahcall and Bahcall, 1974). The minimum masses of the X-ray components have been estimated by the method outlined in Section 1; only in the case of Cyg X-1 is the minimum mass large enough to suggest the existence of a black hole (see e.g., Van den Heuvel and Ostriker, 1973). We shall return to this point later.

Also of direct interest here is the observational result that there is an upper limit to the luminosities of compact X-ray sources near 5×10^{38} erg s$^{-1} \sim 10^5 \, L/L_{\odot}$ (Tananbaum, 1973; Salpeter, 1973), most of the known objects falling in the range 10^{35}–10^{38} erg s^{-1}.

Accretion onto collapsed objects has been treated in a general way by Zel'dovich and Novikov (1967, 1971), and more specifically in binary systems for the case of neutron stars by Ostriker and Davidson (1973) and for black holes by Shakura and Sunyaev (1973). In both cases a critical mass flux is associated with the so-called Eddington luminosity

$$L_{\mathrm{cr}} = 10^{38} \, (\mathfrak{M}/\mathfrak{M}_{\odot}) \text{ erg s}^{-1},$$

corresponding to a situation in which the radiation pressure on the ionized gas is equal to the gravitational force of the collapsed object. L_{cr} corresponds to some critical value $\mathfrak{M}_{\mathrm{cr}}$ of the accretion rate. Near a black hole or on the surface of a neutron star, i.e., for $r > r_g = G\mathfrak{M}/c^2$, where r_g is the Schwarzschild radius, one can write crudely*

$$L \approx \phi\dot{\mathfrak{M}} = c^2 \left[1 - \sqrt{1 - \frac{2G\mathfrak{M}}{c^2 r}} \right] \dot{\mathfrak{M}},$$

so that, for example, on the surface of a one solar mass neutron star with radius

---

* This crude approximation takes no account either of the energy liberated by the collapsed star in absorbing mass or the gravitational 'red shift' of the emergent radiation (see Ostriker and Davidson, 1973).

$R \sim 4\, r_g,$

$$L \approx 0.1\, c^2 \mathfrak{M} = 0.9 \times 10^{20}\, \mathfrak{M}\, (\text{gm s}^{-1})$$
$$= 0.6 \times 10^{46}\, \mathfrak{M}\, (\mathfrak{M}_\odot\ \text{yr}^{-1})$$

Thus for a one-solar mass neutron star a mass inflow near $10^{-8}\, \mathfrak{M}_\odot\ \text{yr}^{-1}$ is sufficient to produce the Eddington luminosity. More generally, Shakura and Sunyaev (1973) write

$$L = \eta c^2\, \mathfrak{M},$$

where η is the efficiency of gravitational release (a number of order generally 10^{-1}), and

$$\mathfrak{M}_{\text{cr}} = 3 \times 10^{-8}\, (\mathfrak{M}/\mathfrak{M}_\odot)\, \frac{0.06}{\eta}\, (\mathfrak{M}_\odot\ \text{yr}^{-1}).$$

We are thus dealing in each case with a critical accretion rate near $10^{-8}\, \mathfrak{M}_\odot\ \text{yr}^{-1} \approx 10^{18}\ \text{gm s}^{-1}$. Accretion rates appreciably larger than this value lead to the ejection of matter by radiation pressure. In some cases, according to Shakura and Sunyaev, the object may remain an X-ray emitter with L_X near $10^{38}\ \text{erg s}^{-1}$ if viewed from certain angles; in other cases, the energy is mostly radiated in the optical and UV regions of the spectrum since the flowing matter is opaque to X-rays. In the optical spectrum, the object appears as a hot star with a rapidly expanding envelope – the black hole or neutron star is thus hidden from view by cooler overlying matter.

We are led naturally to the expectation that the upper luminosity limit observed for X-ray sources is, in fact, the Eddington limit, and that the typical X-ray sources with which we are concerned, for which $L \sim 10^{35}$ to $10^{38}\ \text{erg s}^{-1}$, lie in the domain of mass accretion rates less than $\sim 10^{-8}\, \mathfrak{M}_\odot\ \text{yr}^{-1}$.*

The consequence of all this is to ask, in the case of binary X-ray sources: what mass ejection rates are expected from the companion of the collapsed object and how much of this ejected material does the latter actually capture? To answer this, we need to consider various mechanisms of stellar mass-loss. Available processes seem to be (1) stellar winds, driven by a mechanical flux mechanism similar to that of the solar wind; (2) stellar winds driven by radiation pressure; (3) mass outflow from a binary component which fills the Roche lobe and is undergoing nuclear exhaustion in the core; (4) non-synchronous rotation and revolution in 'close' systems; (5) gravitational wave 'grinding' in close systems; (6) X-ray induced mass-ejection. If the mass-ejection rate for the solar wind of $\mathfrak{M} \sim 10^{-14}\, \mathfrak{M}_\odot\ \text{yr}^{-1}$ is any guide to the stars, (1) is too small to

---

Editors Footnote:

Kraft is here describing the 'classical' description of accretion onto a collapsed star as presented by Shakura and Sunyaev. Recent work indicates this view may be over simplified. For one, Ruffini and Wilson (*Phys. Rev. Letters* **32**, 324, 1974) have pointed out that if temperatures in the accreting matter reach 10^9 K, significant neutrino fluxes may be generated which do not contribute radiation pressure. Thus, the neutrino luminosity is not subject to the Eddington Limit.

be of interest. Mechanism (2) manifests itself in luminous stars of several kinds, and conceivably provides a sufficient outflow rate.

For example, in OB supergiants, radiation pressure driven winds eject matter at a rate of 10^{-6} to $10^{-8} \, \mathfrak{M}_\odot \, \mathrm{yr}^{-1}$ (Morton, 1967, 1969; Carruthers, 1968; Lucy and Solomon, 1970), a result based on the observation of atomic resonance lines in the rocket ultraviolet. Similar rates follow from studies of the optical spectra (Hutchings, 1968). Recently, Rosendahl (1974) surveyed the P Cyg-like Hα emission lines found in such stars, and concluded that mass outflow occurs for all B0–B1 stars with $M_V \leqslant -5.3$ and for B8–A3 stars with $M_V \leqslant -6.3$. In F-type stars, a few very luminous objects (viz., ϱ Cas and 89 Her) show evidence for ejected circumstellar material (Sargent, 1961; Bohm-Vitense, 1956; Sargent and Osmer, 1969). In luminous M-type stars, mass ejection is a widespread phenomenon (Deutsch, 1960) and takes place at a rate estimated to lie between 3×10^{-12} and $5 \times 10^{-8} \, \mathfrak{M}_\odot \, \mathrm{yr}^{-1}$, the rate increasing with advancing spectral type. In the case of ejection process (3), calculations of mass transfer rates as a function of the mass of the primary have been made by Kippenhahn and his associates (*op, cit.*) and range from $\sim 10^{-3} \, \mathfrak{M}_\odot \, \mathrm{yr}^{-1}$ for $\mathfrak{M} \sim 30 \, \mathfrak{M}_\odot$ to $\sim 10^{-7} \, \mathfrak{M}_\odot \, \mathrm{yr}^{-1}$ near $\mathfrak{M} \sim 2 \, \mathfrak{M}_\odot$ (Figure 1). Mechanism (5) has been discussed extensively by Faulkner (1973) and is probably important only in low-luminosity

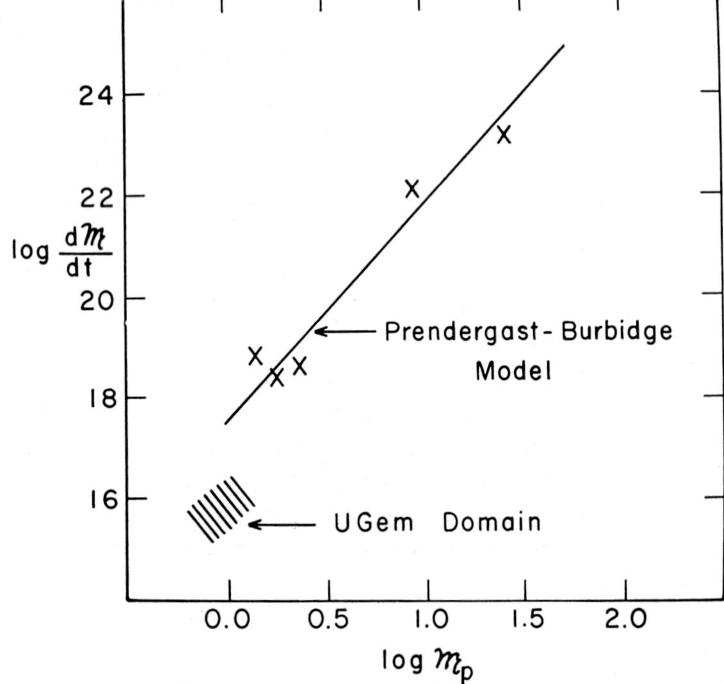

Fig. 1. *Rates of mass-transfer from Primary to Secondary as a function of the mass $\mathfrak{M}_p$ of the primary.* Examples are taken from the work of Kippenhahn, Weigert, and their associates. Transfer rate necessary to produce X-ray emission in the Prendergast-Burbidge model is indicated, as is the location of the domain of U Gem binaries (after Kraft, 1973). © 1973, International Astronomical Union.

binaries. Mechanism (4) may be of significance, but its effects have not been calculated because the problem is very difficult (Kruszewski, 1966); (6) may be applicable in Her X-1 and Cyg X-3.

We consider the location of mass-losing components of X-ray binaries in Figure 2, ignoring the accretion heating of the companion. We are concerned here with the loca-

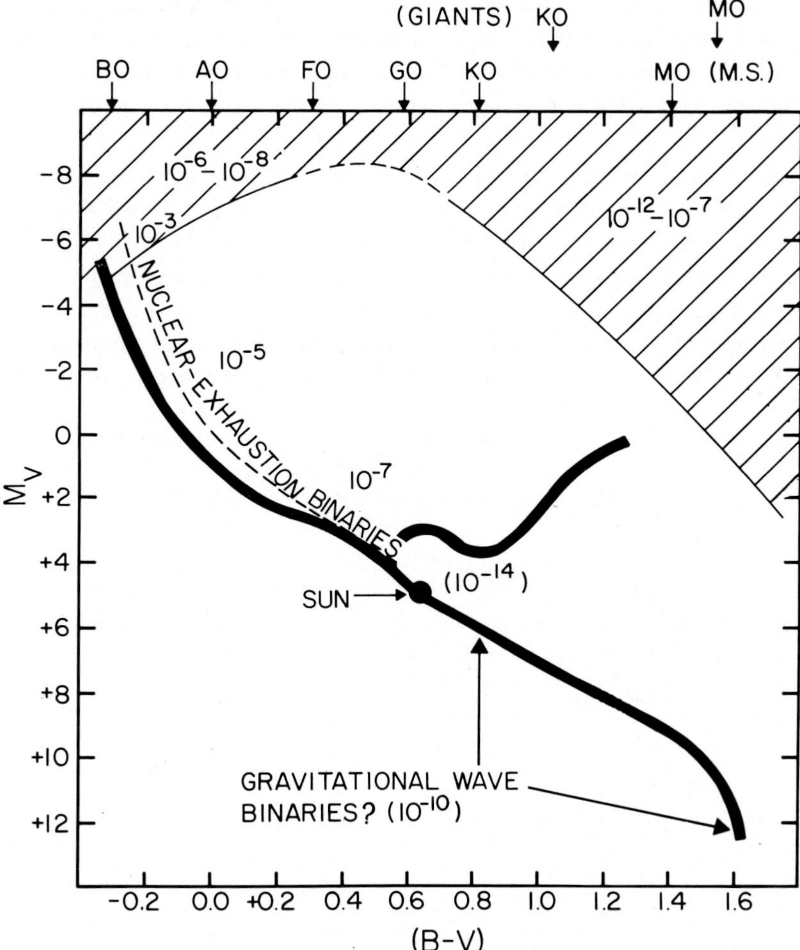

Fig. 2.   *Domains of mass-loss for stars in the HR Diagram.* Numbers given are in units of $\mathfrak{M}_\odot$ yr⁻¹.

tions of stars in connection with the theory of stellar evolution, hence with energy generation due to nuclear burning or gravitational contraction. The dashed line marked 'nuclear exhaustion binaries' is a highly schematic representation of the locations of mass-losing primaries that have left the main sequence in response to the depletion of hydrogen. The known X-ray binaries divide rather naturally into three groups. The Group I stars are located in the upper left-hand corner of the diagram near $M_V \sim -4$

to -7 and $T_e \sim 30000\,\mathrm{K}$; included are Cyg X-1, SMC X-1, Vela X-1, 3U 1700$-$37, and the X-ray pulsar Cen X-3. In all these, the ratio of the optical luminosity of the mass-losing star to the X-ray luminosity is between 10^0 and 10^2. In Group 2, we have the X-ray pulsar Her X-1 and possibly Cyg X-2, although the latter is not with certainty known to be a binary star. In this group, $M_V \sim +1$ to $+4$, $T_e \sim 7000\,\mathrm{K}$–$10000\,\mathrm{K}$, and $L_{\mathrm{opt}}/L_X \sim 10^{-2}$ to 10^{-1}. Cyg X-3 with its orbital period of 4.8 hours, may descend from the group of U Gem binaries in which mass-loss is driven by gravitational waves, as the Roche lobe gradually encroaches on the primary (Davidsen and Ostriker, 1974). It is the only representative of Group 3. It is interesting that no binary X-ray source is so far known in which the mass-losing star lies in the upper right-hand corner of the HR diagram.

The X-ray emission of Group 1 might be thought *a priori* to be driven by accretion either from a wind induced by radiation pressure or nuclear exhaustion of the companion. Models contrasting these two cases are illustrated in Figure 3, a diagram taken from the recent review article by Blumenthal and Tucker (1974). On the left, we have the case in which the non-degenerate component has swollen up, on a nuclear, followed by a Kelvin, time-scale, to fill the Roche equipotential, or inner zero-velocity surface. A particle located at the singular point L_1, experiences no net gravitational

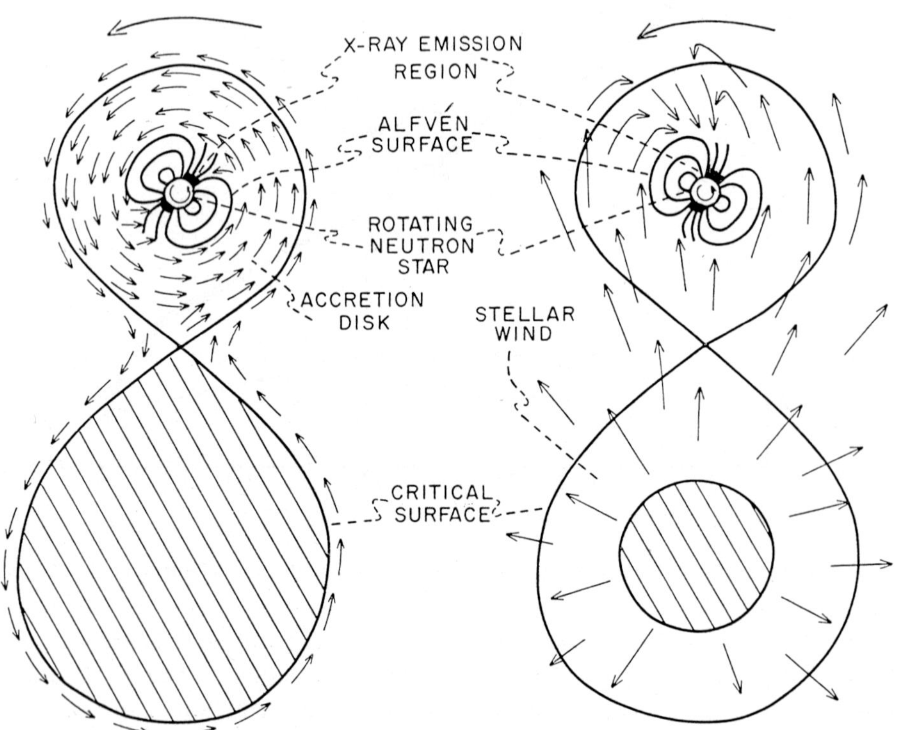

Fig. 3.   *Two Regimes of Mass-Transfer and Accretion in Binary Systems* (after Blumenthal and Tucker, 1974). The accreting object is taken to be a neutron star, but that choice is not an absolute requirement of the model. © 1974, Ann. Revs. Inc.

force; thus matter can simply 'ooze' out into the lobe surrounding the compact companion, in response to nuclear exhaustion of the primary. The ejected material takes up an orbit in a ring surrounding the secondary; radial viscous dissipation spreads the ring out and enables matter to reach the stellar surface. This is essentially the Prendergast-Burbidge model. Unless the Eddington limit is exceeded or part of the material of the injected stream or the ring is accelerated to large velocities, the collapsed star must ultimately become the sink for all the material ejected by the primary, and $\mathfrak{M}_{ejected} = k\mathfrak{M}_{acc}$, where k is of order unity. On the right-hand side of Figure 3, we have the case in which the primary is smaller than the inner zero-velocity surface but ejects matter in the form of a supersonic wind. If the material flows with high velocity, the zero-velocity surfaces are essentially irrelevant. Wind material accelerated beyond the collapsed object collides and forms a downstream accretion column, in the manner described years ago by Bondi and Hoyle (1944). The mass accretion rate is given by

$$\mathfrak{M}_{acc} \cong \pi\varrho \, \frac{(2G\mathfrak{M}_{coll})^2}{v^3},$$

where ϱ is the density of the wind near the collapsed object of mass $\mathfrak{M}_{coll}$, and v is the relative velocity.

It seems rather probable that the model pictured in the left-hand part of Figure 3 does not apply to the X-ray binaries of Group 1, since the mass-transfer rates driven by nuclear exhaustion, viz., $\sim 10^{-3}$ to $10^{-4}\,\mathfrak{M}_\odot$ yr^{-1}, are 4 or 5 order of magnitude too large for the Eddington limit, if, as seems likely, a significant amount of ejected matter is actually accreted by the collapsed object. The case of a radiation-pressure driven wind seems more promising. Ostriker and Davidson (1973) have considered a model for Cen X-3 in which accretion from such a wind onto a rotating dipolar magnetic neutron star is responsible for the X-ray luminosity. Discussion of the details of the model is beyond the scope of this paper; but for reasonable flow parameters, a one-solar-mass neutron star, and $L_X = 10^{37}$ erg s^{-1}, they require mass-loss rates between 3×10^{-7} and $9 \times 10^{-6}\,\mathfrak{M}_\odot$ yr^{-1}, depending on flow velocity. These values are a little, but not unacceptably, large, particularly since the estimates of mass-flux from luminous stars are not very reliably known (cf. Lucy and Solomon, 1970; Morton, 1967, 1969).

For Her X-1 ($=$ HZ Her), the possibility remains that mass-loss from the F-type primary could be driven by nuclear exhaustion at a rate near $10^{-8}\,\mathfrak{M}_\odot$ yr^{-1} if $\mathfrak{M} \sim 2\mathfrak{M}_\odot$ (Bahcall and Bahcall, 1973). A different model has been suggested in which the X-ray heating of the outer layers of the primary produces the wind (Pringle, 1973; Arons, 1973), since F-type stars are not known to produce stellar winds with fluxes in excess of the inadequate solar wind. The distance to Her X-1 is not well enough known to permit one to make a reliable estimate of the radius of the primary; it is thus uncertain whether the Roche lobe is actually filled, and this condition would have to be met if significant amounts of mass are to be ejected by evolutionary expansion.

We now consider several X-ray sources that illustrate the leading aspects of mass-ejection and accretion, but with emphasis not on the physics of X-ray production, but rather on the history of the compact object. We pay particular attention to the following question: *How does nature conspire to produce a binary system in which a more-or-less normal star in a state of relative 'biological' youth is locked hand-in-hand with a less massive but biologically older collapsed companion?* This state of affairs seems at first sight to violate all the basic laws of stellar evolution, according to which, everything else being equal, the evolutionary state of a star is the more advanced the larger is its mass. As we shall see, this anomaly is resolved as a consequence of mass-exchange between the components.

4. Some Evolutionary Scenarios for Mass-Transfer Binaries

The existence of evolutionary mass exchange between components of binary systems is by now a well-established astronomical phenomenon. It comes about because stars undergoing evolutionary processes in the deep interior can experience drastic changes in radius, by factors of 10 or 100, or more. If a star is a member of a binary system in which the initial separation is less than 10 to 100 $R_\odot$, it will encounter its lobe of the inner zero-velocity surface and transfer mass to its companion through the Lagrangian point at L_1. These ideas have been reviewed *in extenso* by Paczyński (1971); evolutionary calculations have been made by a number of investigators including Paczyński, and also Kippenhahn, Weigert, and their associates. All calculations assume that total mass and angular momentum are conserved. Application of these ideas to X-ray binaries has been due principally to the work of Van den Heuvel (references to follow), and the state of our knowledge up to two years ago was reviewed by the writer (Kraft, 1973). The reader is referred to this last paper for a more comprehensive treatment of the points that follow.

We are speaking of binaries whose initial periods are ~ 100 days or less, and since roughly 40% of all binaries have periods less than this value, and since further, roughly half of all stars are in binary systems, we expect mass-transfer to be a quite common stellar experience. The more massive component evolves first, in response to hydrogen exhaustion in the core, and eventually reaches the state of a collapsed object after a number of intermediate steps. What sort of collapsed object we find in the end is mostly a sensitive function of the initial mass $\mathfrak{M}_p$ of the primary star.

What happens can be understood best by considering the evolution of a single star with mass (say) $5\,\mathfrak{M}_\odot$. In Figure 4, we reproduce Iben's (1967) well-known plot of the evolutionary track in the HR diagram. Of particular interest are the events associated with points 4 and 7. After an initial period of core hydrogen burning, a hydrogen-burning shell is established at point 4, and the star rapidly moves to the right in the diagram as the exhausted core contracts. At point 7, the radius starts to decrease in response to the onset of core He-burning. Very crudely, one finds that core contraction is accompanied by an increase in radius as part of the increased energy generation goes into gravitational potential energy. On the other hand, ignition of a nuclear field in the core with corresponding expansion of the core-burning region leads to contraction of the stellar envelope.

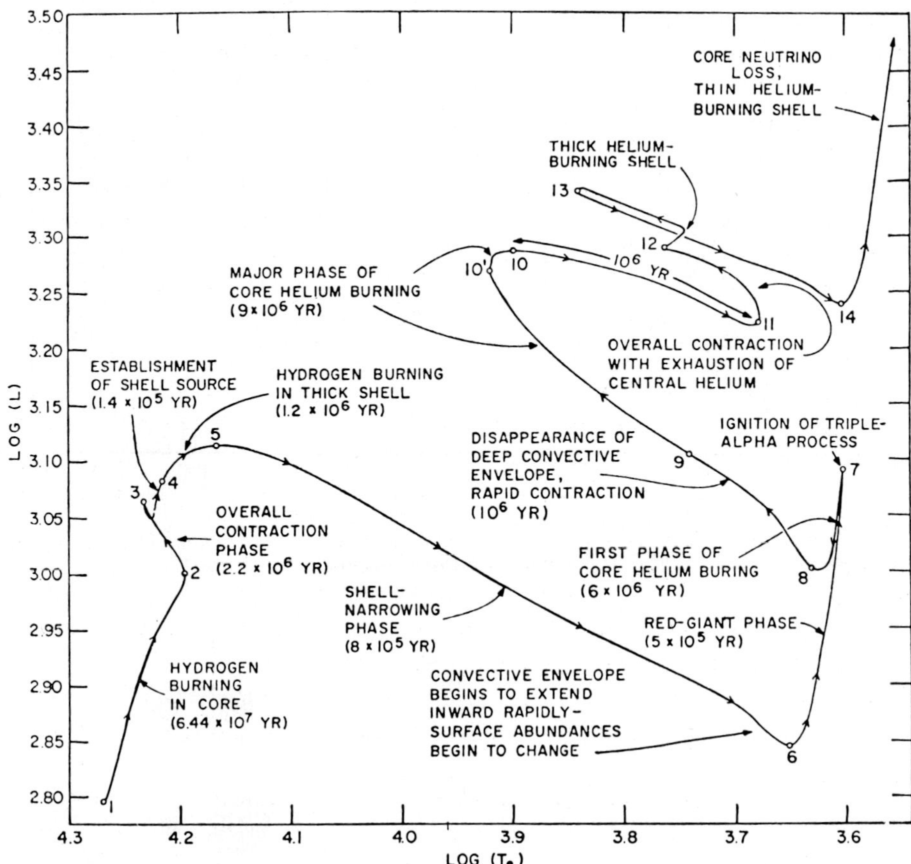

Fig. 4. Evolution of a star with $\mathfrak{M} = 5\,\mathfrak{M}_\odot$ in the HR Diagram (after Iben, 1967).
© 1967, Ann. Revs. Inc.

There are many detailed aspects to the various cases that have been considered (Paczyński, 1971; Kippenhahn and Weigert, 1967; Kippenhahn *et al.*, 1967; Kippenhahn, 1969). but the leading ideas can be summarized as follows, if it is assumed that the primary encounters the inner Lagrangian surface during the first phase of rapid expansion of the radius:

(1) $\mathfrak{M}_p \lesssim 8\,\mathfrak{M}_\odot$: The mass exchange leaves behind a helium star with mass less than the Chandrasekhar limit for white dwarfs ($< 1.4\,\mathfrak{M}_\odot$). Whether helium in the core is ever ignited depends on whether the initial mass exceeds $\sim 2.8\,\mathfrak{M}_\odot$. Only for the case of $\mathfrak{M} < 2.8\,\mathfrak{M}_\odot$ has the evolutionary calculation been carried all the way through to the white dwarf stage. This is illustrated in Figures 5 and 6 for the case $\mathfrak{M}_1 = 2\,\mathfrak{M}_\odot$, $\mathfrak{M}_2 = 1.0\,\mathfrak{M}_\odot$, and $P_{\text{initial}} = 1.15$ days (Kippenhahn *et al.*, 1967). The primary encounters the critical surface in 6×10^8 yr, and loses all but $0.26\,\mathfrak{M}_\odot$ to the secondary in only 7×10^6 yr (corresponding to an ejection rate of $2 \times 10^{-7}\,\mathfrak{M}_\odot\ \text{yr}^{-1}$). At this point, the original primary fills its critical lobe forming a semi-detached system of

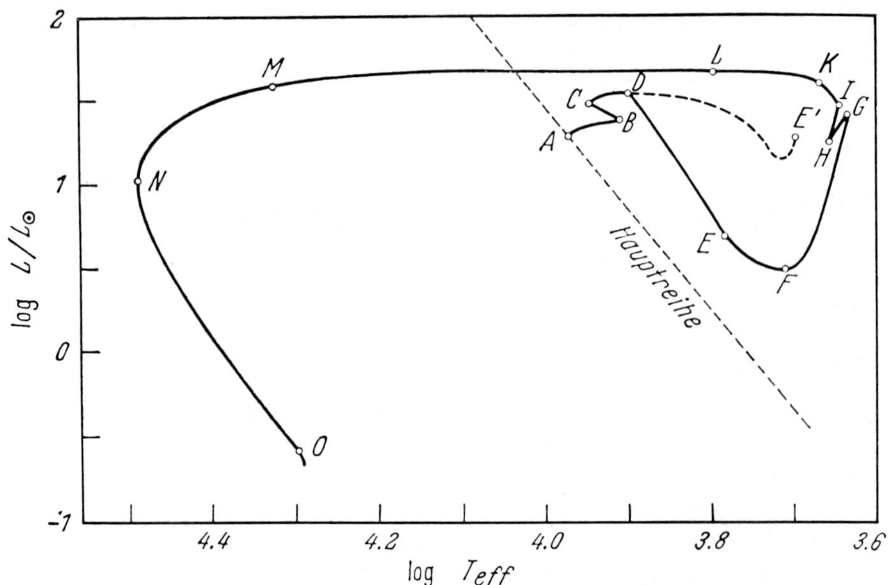

Fig. 5.   Evolutionary track of the primary component of a mass-exchange binary, with $\mathfrak{M}_1 = 2\ \mathfrak{M}_\odot$, $\mathfrak{M}_2 = 1\ \mathfrak{M}_\odot$, $P = 1.15$ days (after Kippenhahn *et al.*, 1967).
© 1967, Springer and Verlag Publ.

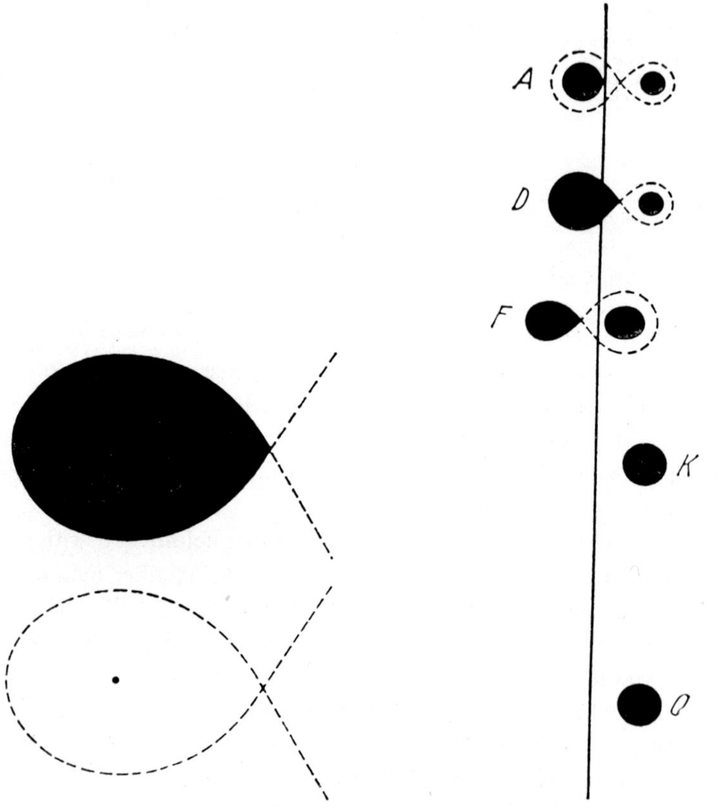

Fig. 6.   Same as Figure 5, in geometrical representation (after Kippenhahn *et al.*, 1967).
© 1967, Springer and Verlag Publ.

period 24 days. The so-called 'subgiant components of eclipsing binary systems' are of this kind (Crawford, 1955; Plavec, 1968); a typical object is DN Ori (Smak, 1964), in which $P = 13$ days and in which the (more-evolved) F5 giant and (rejuvenated) A2 main-sequence companion have masses of $0.18\,\mathfrak{M}_\odot$ and $2.65\,\mathfrak{M}_\odot$ respectively. The subgiant component develops a degenerate core which naturally stops contracting and never succeeds in igniting helium. In the end we have a main sequence A-type star accompanied by a low-mass white dwarf with the same 24-day orbital period. Subsequent mass ejection through L_1 takes place from the new primary (*née* secondary) after a nuclear time of 10^8 years, and capture of this matter by the white dwarf might then lead to an X-ray source, following Prendergast and Burbidge.

(2) $8\mathfrak{M}_\odot \lesssim \mathfrak{M}_p \lesssim 16\mathfrak{M}_\odot$: The mass exchange leaves behind a He-star with a mass lying between the white dwarf limit and $\mathfrak{M} \sim 4\,\mathfrak{M}_\odot$. According to Paczyński, such a star will have a second radial expansion during core helium or carbon burning, and thus may lose additional mass. Some of this may be captured by the companion; some may be lost to the system through the outer Lagrangian point L_2, in which case the period will probably become shorter (Kruszewski, 1966). In either case, the collapsed object may still wind up as a white dwarf, although calculation of models all the way to this state have not been made.

The possibility that the present state of Her X-1 is a result of one of the two preceding scenarios cannot be dismissed out of hand. In that case, the 1.2 s X-ray period must be interpreted as the non-radial pulsations of a white dwarf, rather than the pulsar-like flashes associated with a rotating neutron star (Lamb *et al.*, 1973). Non-radial pulsations are strongly suspected to exist, in fact, in the white dwarf components of several U Gem binaries (Warner and Robinson, 1972; Robinson, 1973), although the shortest period presently known is 17 s. This model has been discussed in connection with Her X-1 by DeGregoria and Woltjer (1973) who emphasize that the non-radial pulsations of a pure helium white dwarf with $P = 1.2$ s corresponds to a mass $\mathfrak{M} = 1.2\,\mathfrak{M}_\odot$ (Faulkner and Gribbin, 1968), a value comfortably under the white-dwarf mass limit and close to the value $\mathfrak{M} = 1.3\,\mathfrak{M}_\odot$ inferred from the spectroscopic orbit (Crampton and Hutchings, 1972).

From the same orbit, the mass of the F-type primary is inferred to be near $2\,\mathfrak{M}_\odot$ although uncertain. If the star fills its lobe of the inner Lagrangian surface, then the distance is ~ 5.8 kpc (Forman *et al.*, 1972) $L_X \sim 4 \times 10^{37}$ erg s^{-1}, and an accretion rate near $2 \times 10^{-6}\,\mathfrak{M}_\odot$ yr^{-1} is required for the white dwarf. This is 10 times the rate derived from nuclear exhaustion driving. This could mean that most or all of the X-ray emission is derived from hydrogen burning as hydrogen-rich material falls on the surface of the pure-helium white dwarf. Although the change in pulsation period induced would then be quite small, it could well fit in with the recent discovery that the pulsar period variation has long-term fluctuations and a net period decrease much smaller than previously supposed (Giacconi, 1974).

As emphasized by DeGregoria and Woltjer, the main advantage of the white dwarf, as opposed to neutron-star, model for Her X-1 is the difficulty in producing, for such a low-mass object, an evolutionary scenario in which a neutron star can be a descendant of one of the components: we examine this point in greater detail in the sequel. On the other hand, the system could conceivably have lost considerable mass through L_2 in certain evolutionary stages. One might also offer the objection to the white dwarf model that the scenario in which the descendant object has $\mathfrak{M}_p = 2.74\,\mathfrak{M}_\odot$, $\mathfrak{M}_{wd} = 0.26\,\mathfrak{M}_\odot$, $\mathfrak{M}_p/\mathfrak{M}_{wd} \sim 10/1$ and $P = 24$ days is not very much like Her X-1, in which $\mathfrak{M}_p \sim 2\,\mathfrak{M}_\odot$, $\mathfrak{M}_{wd} \sim 1\,\mathfrak{M}_\odot$, $\mathfrak{M}_p/\mathfrak{M}_{wd} \sim 2/1$, and $P = 1.7$ days. However, there exist many eclipsing binary systems with subgiant components in which the orbital elements, although frequently poorly determined, are similar to those of Her X-1 (Crawford, 1955). An example of a suitable object with quite well determined elements is UX Mon (Struve, 1947) in which the A5e type component has $\mathfrak{M} = 3.4\,\mathfrak{M}_\odot$, the more evolved, but less massive secondary, has $\mathfrak{M} = 1.5\,\mathfrak{M}_\odot$, and $P = 5.9$ days. Subsequently evolution might well produce an object not unlike Her X-1. In fact, the difficulty, as emphasized elsewhere (Kraft, 1973), may be that this scenario produces significantly more X-ray sources than are observed.

(3) $\mathfrak{M}_p \gtrsim 16\,\mathfrak{M}_\odot$: This last case we shall consider involves the production of an X-ray source as a result of the evolution of a massive binary. The transfer of mass from the primary to the secondary leaves behind a helium star with $\mathfrak{M} \gtrsim 4\,\mathfrak{M}_\odot$. According to Paczyński (1971), such a star expands very little, and cannot lose mass in a second red giant stage (unless the period is very short). It will pass rapidly through He and C-burning episodes and will presumably eventually become a neutron star or black hole after exploding as a supernova (Wheeler, 1973).* If the system remains

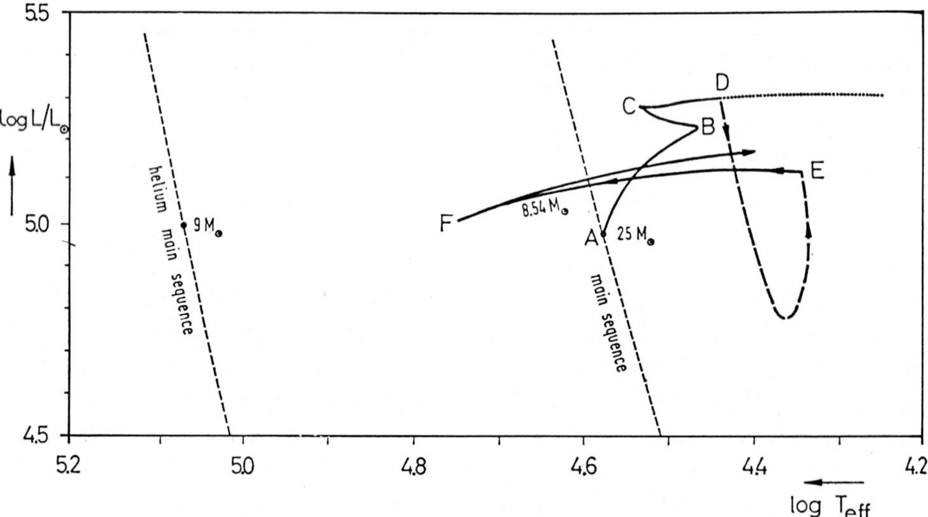

Fig. 7.   Same as Figure 5, except that $\mathfrak{M}_1 = 25\,\mathfrak{M}_\odot$, $\mathfrak{M}_2 = 15\,\mathfrak{M}_\odot$, $P = 7.0$ days (after Kippenhahn, 1969). © 1969, Springer and Verlag Publ.

* In Arnett's (1969) supernova model, carbon detonation in the degenerate core leads to ejection of matter, but the entire star disappears. Models involving a core with mass near the Chandrasekhar limit and in which photodisintegration leads to neutronization in the core and ejection of the envelope are currently under construction (Arnett, 1974).

bound in spite of the outburst, the object could eventually wind up as an X-ray source of the kind in our Group 1.

Evolutionary calculations of massive binaries have been carried through only as far as the state immediately following the episode of first mass exchange. As an example, we consider a system in which $\mathfrak{M}_1 = 25\mathfrak{M}_\odot$, $\mathfrak{M}_2 = 15\,\mathfrak{M}_\odot$, and $P_{\text{initial}} = 7.0$ days (Kippenhahn, 1969). The evolutionary tracks are shown in Figure 7. After 4.7×10^6 yr, the critical radius (point D) is reached during shell hydrogen burning,

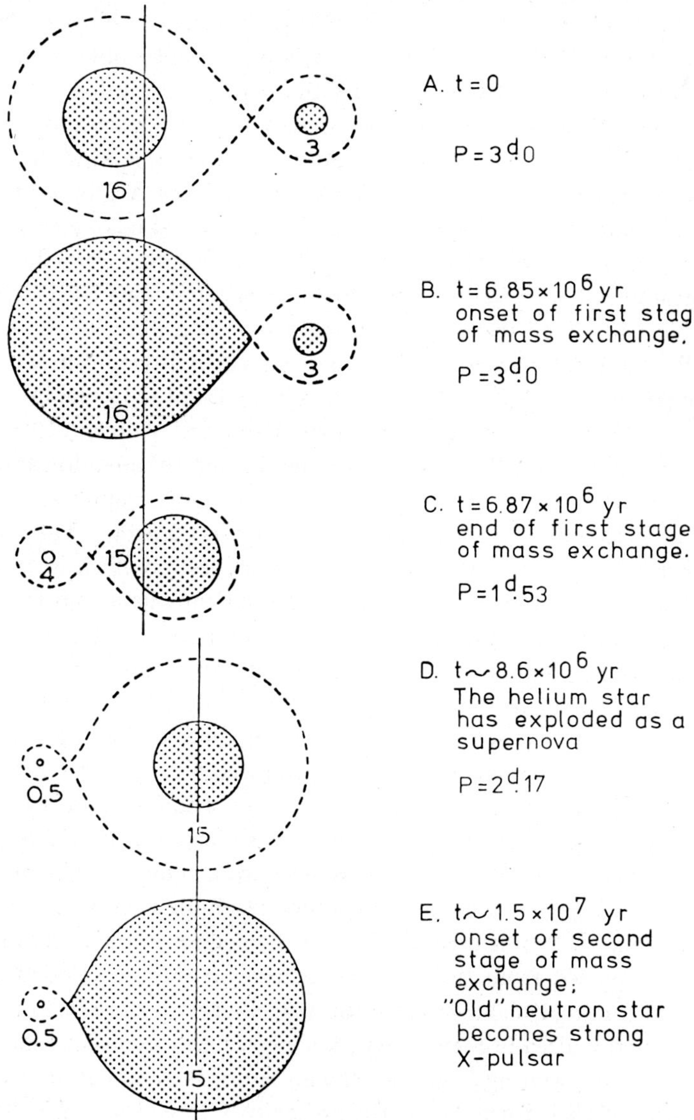

Fig. 8. An evolutionary scenario for Cen X-3 based on the evolutionary calculation illustrated in Figure 7 (after Van den Heuvel and Heise, 1972). © 1972, MacMillan Journals, Ltd.

prior to the onset of helium burning in the core. The mass exchange takes place in only 7200 years, during which time 16.5 $\mathfrak{M}_\odot$ is transferred to the secondary, and stops only when about all except the helium core has been transferred. But during the exchange, helium is ignited in the core, and it is this ignition which actually stops the expansion. The new binary, therefore consists of a helium burning core with a small mass hydrogen envelope and a hydrogen-rich companion; the masses are 8.54 $\mathfrak{M}_\odot$ and 31.46 $\mathfrak{M}_\odot$, respectively, and the period is 18 days. Note that the luminosity of the two stars is about the same, and that the less massive helium-object evolves first.

Systems of this kind are highly reminiscent of the Wolf-Rayet binaries studied some 30 years ago by Struve, Hiltner, and their associates, a point emphasized by Kippenhahn (1969). Van den Heuvel (1973) argues that further evolution of these objects could lead to binary X-ray sources, and thus that Wolf-Rayet stars and massive X-ray binaries represent two different stages of evolution of the same kind of objects. This possibility was considered in some detail for the case of Cen X-3 by Van den Heuvel and Heise (1972), and the evolutionary scenario is illustrated in Figure 8. They begin with an assumed binary in which $\mathfrak{M}_1 = 16\ \mathfrak{M}_\odot$, $\mathfrak{M}_2 = 3\ \mathfrak{M}_\odot$, and $P = 3$ days. Using the evolutionary tracks of Paczyński, they pass through the first mass exchange and find a descendant object in which the original primary is now a pure helium-burning star with $\mathfrak{M} = 4\ \mathfrak{M}_\odot$, and the companion is a rejuvenated hydrogen-rich star with $\mathfrak{M} = 15\ \mathfrak{M}_\odot$; the period is 1.53 days. The two stars have about the same luminosity, and the system resembles a Wolf-Rayet star (Kippenhahn and Weigert, 1967). The pure helium star evolves more rapidly than its hydrogen-burning companion and eventually explodes. Van den Heuvel and Heise assume, as an example, that a mass of 3.5 $\mathfrak{M}_\odot$ is ejected to infinity, leaving behind a neutron star of mass 0.5 $\mathfrak{M}_\odot$. They consider the case of fast ejection in a spherical shell and ignore the question of the interaction of the shell material with the companion star. This may be unrealistic, but the assumption enables one to apply the virial theorem and to assert that the system will remain bound since the less massive component which explodes and less than half the total mass of the system is ejected (Blaauw, 1961; Boersma, 1961; Gott, 1972). (The question of dissolution of binary systems by the explosion of supernovae has been rather nicely illustrated in Figure 1 of the paper by Gott, reproduced here as our Figure 9.) For the case of spherical ejection from one of the stars, the whole system is accelerated by 30 km s^{-1} owing to recoil, and the binary period is increased to 2.17 days (Van den Heuvel, 1968). When the 15 $\mathfrak{M}_\odot$ companion develops a sufficiently intense radiation-pressure driven wind, the entire system may then resemble Cen X-3.

A serious difficulty with this type of model is the neglect of the effect on the binary of the impulse imparted from the ejected material on the unexploded companion. McCluskey and Kondo (1971) calculated this (together with the effect on the gravitational binding energy already mentioned), but neglected to consider the decrease in effective cross-section resulting from the blowing off of the companion's outer layers. Thus, application of their formulae to the preceding Cen X-3 model leads to dissolution of the binary, principally because the companion subtends such a large solid angle as seen from the center of the exploding star. What seems rather certain, in any case,

APPENDIX I

CLASSIC PAPERS

THE HIGHLY COLLAPSED CONFIGURATIONS
OF A STELLAR MASS*

S. CHANDRASEKHAR

1

Professor Milne in his recent paper** on 'The Analysis of Stellar Structure' has put forward some essentially new considerations on the possible steady-state configurations of stellar aggregates of varying mass, luminosity, and opacity. One of the main consequences of the analysis is the explanation not only of the existence of white dwarfs – his collapsed configurations – but also of the principal physical characteristics of these configurations. The following is devoted to the development of Milne's theory of these collapsed configurations a stage further.

2

Milne's estimates for the central density and temperature of these collapsed configurations indicate that in some cases we pass beyond the range of validity of the degenerate form of the Fermi-Dirac equation of state ($p=K\varrho^{5/3}$). It can be shown that the pressure of an electron gas which is highly degenerate and which has a very highly predominant relativistic-mass variation effect, takes the limiting form [†]

$$p = \frac{n^{4/3}hc}{8}\left(\frac{3}{\pi}\right)^{1/3} \tag{1}$$

(c = velocity of light, h = Planck's constant) if the following two conditions are satisfied

$$n > \left(\frac{kT}{hc}\right)^3 \frac{4\pi G}{3} \equiv 2.86T^3 \quad (G = 2) \tag{i}$$

$$n > \frac{8\pi m^3 c^3}{3h^3} \equiv 5.88 \times 10^{29} \tag{ii}[‡]$$

(i) replaces Sommerfeld's degeneracy criterion, and (ii) ensures the predominance of the relativistic effect. As the condition that configurations are highly collapsed is precisely equivalent to the condition that in the central regions the electron assembly

---

* Reprinted from *Monthly Notices of the Royal Astronomical Society* **91**.
** *M.N.R.A.S.*, **91** (1930) 4–55, referred to hereafter as *loc. cit.* For a general exposition of his main ideas see *Nature* (1931 January 3) 'Stellar Structure and the Origin of Stellar Energy.'
[†] The corresponding expression for the energy E was obtained by E. C. Stoner, *Phil. Mag.* **9** (1930), 944. That $p = \frac{1}{3}E$ is generally true when the relativistic effect is highly predominant can easily be proved.
[‡] (ii) is given in Stoner's paper; (i) now replaces Sommerfeld's criterion. These and other inequalities are briefly discussed in my paper on 'The Dissociation Formula according to the Relativistic Statistics', *M.N.R.A.S.* **91** (1931), 446. (in course of publication).

---

is degenerate,* it would be sufficient to consider whether (ii) is satisfied in order to count the relativistic effect as highly predominant. Further, as L for these configurations is very small, T_c is normally of the order of 10^9,** and (ii) then would automatically provide for (i). (See Equation (19')).

Now the central density of a highly, collapsed configuration considered as an Emden polytrope '$n=\frac{3}{2}$' is given by †

$$\varrho_c = \frac{32\pi G^3 M^2 \beta^3}{125 K_1^3 \times 7.385} \tag{2}$$

where K_1 is the degenerate-gas constant and

$$\beta = 1 - \frac{\kappa L}{4\pi c G M}. \tag{3}$$

It is clear then that the relativistic effect will be predominant in the central regions of those collapsed configurations whose masses satisfy the inequality

$$\frac{32\pi G^3 M^2 \beta^3}{125 K_1^3 \times 7.385 \times \mu} > \frac{8\pi m^3 c^3}{3h^3}$$

or

$$M\beta^{3/2} > 5 \left[\frac{5\mu \times 7.385}{12} \right]^{1/2} \left(\frac{mcK_1}{Gh} \right)^{3/2} = 0.434 \odot \quad (\text{if } \mu = 2.5 m_H) \tag{4}$$

where μ is the mean molecular weight and $\odot$ denotes the mass of the Sun ($= 1.985 \times 10^{33}$ gms).

The purpose of this paper is to find out the consequences of introducing the equation of state $p = K_2 \varrho^{4/3}$. It will be shown that we can enumerate the complete linear sequence of steady-state configurations for an *assigned* small luminosity as the mass varies, the opacity and source-strength being constant and uniform (standard model).

3. The Equations of the Problem

We base our subsequent discussion exclusively on the standard model as it is considerably easier to work with.

We have the following set of equations for the standard model independent of the equation of state we may adopt. (The notation is identical with that used in Milne's paper.)

$$\frac{\mathrm{d}p}{\mathrm{d}r} + \frac{\mathrm{d}p'}{\mathrm{d}r} = - \frac{GM(r)}{r^2} \varrho \tag{5}$$

* Milne, *loc. cit.*, § 22.
** Milne, *loc. cit.*, p. 39.
† Equation (55), *loc. cit.*

$$\frac{\mathrm{d}p'}{\mathrm{d}r} = -\frac{\kappa L(r)}{4\pi c r^2}\,\varrho \tag{6}$$

$$\frac{L(r)}{M(r)} = \varepsilon = \frac{L}{M}\,.$$

We have also the following relation between the gas kinetic-pressure (p) and the radiation-pressure p':

$$p = p'\,\frac{\beta}{1-\beta} \tag{7}$$

where β is defined by (3).

CASE I. (The Relativistic-degenerate Case). – We have for the gas kinetic-pressure, taking into account only the electronic contribution, which is certainly by far the most important

$$p = \frac{n^{4/3} hc}{8}\left(\frac{3}{\pi}\right)^{1/3}. \tag{1'}$$

If we assume the molecular weight $\mu = 2.5\,m_H$, then

$$p = K_2 \varrho^{4/3} \tag{1''}$$

where

$$K_2 = \frac{hc}{8\,(2.5 m_H)^{4/3}}\left(\frac{3}{\pi}\right)^{1/3} = 3.619 \times 10^{14}. \tag{1'''}$$

With the equation of state given by (1″), the equation of mechanical equilibrium reduces to

$$\frac{4K_2}{3G\beta}\,r^2 \varrho^{-2/3}\,\frac{\mathrm{d}\varrho}{\mathrm{d}r} = -\,M(r). \tag{8}$$

Remembering that

$$\frac{\mathrm{d}M(r)}{\mathrm{d}r} = 4\pi r^2 \varrho,$$

we have on differentiating (8)

$$\frac{4K_2}{3G\beta}\,\frac{\mathrm{d}}{\mathrm{d}r}\left(r^2 \varrho^{-2/3}\,\frac{\mathrm{d}\varrho}{\mathrm{d}r}\right) = -\,4\pi r^2 \varrho\,. \tag{9}$$

Putting

$$\varrho = \lambda_3 \chi^3 \tag{10}$$

(9) reduces to

$$\frac{K_2\lambda_3^{-2/3}}{\pi G\beta}\cdot\frac{1}{r^2}\frac{\mathrm{d}}{\mathrm{d}r}\left(r^2\frac{\mathrm{d}\chi}{\mathrm{d}r}\right)=-\chi^3.\tag{11}$$

Changing r to the variable ζ given by

$$r=\zeta\left[\frac{K_2\lambda_3^{-2/3}}{\pi G\beta}\right]^{1/2}\tag{12}$$

we have finally

$$\frac{1}{\zeta^2}\frac{\mathrm{d}}{\mathrm{d}\zeta}\left(\zeta^2\frac{\mathrm{d}\chi}{\mathrm{d}\zeta}\right)=-\chi^3\tag{13}$$

which is Emden's polytropic equation with index 3.

From (8), using the values of ϱ and r given by (10) and (12), we have

$$M(r)=-\frac{4}{\pi^{1/2}}\left(\frac{K_2}{G\beta}\right)^{3/2}\zeta^2\frac{\mathrm{d}\chi}{\mathrm{d}\zeta}.\tag{14}$$

It may be noted that, unlike the degenerate non-relativistic case, λ_3 has disappeared from (14).

CASE II. (The Non-relativistic-degenerate Case). – We can use the equations given by Milne, *loc. cit.*, Equations (49), (48), (50), and (47) respectively:

$$\frac{1}{\eta^2}\frac{\mathrm{d}}{\mathrm{d}\eta}\left(\eta^2\frac{\mathrm{d}\psi}{\mathrm{d}\eta}\right)=-\psi^{3/2}\tag{15}$$

where

$$\varrho=\lambda_2\psi^{3/2}\tag{16}$$

$$r=\eta\left[\frac{5K_1}{8\pi G\beta}\lambda_2^{-1/3}\right]^{1/2}.\tag{17}$$

We have also

$$M(r)=-\frac{1}{4(2\pi)^{1/2}}\left(\frac{5K_1}{G\beta}\right)^{3/2}\lambda_2^{1/2}\eta^2\frac{\mathrm{d}\psi}{\mathrm{d}\eta}.\tag{18}$$

K_1 in (17) and (18) is the degenerate-gas constant.*

4. The Equations of Fit

Let $r=r'$ be the radius of the surface of demarcation, *i.e.* outside $r=r'$ the distribution of density is given by a solution of Emden's equation '$n=\frac{3}{2}$', and inside $r=r'$ by a solution of Emden's equation '$n=3$'.

* $K_1=2.138\times10^{12}$ or 3.17×10^{12}, according as $\mu=2.5m_H$ or $2m_H$. The former value will be used in the numerical work of this paper.

capable of – the degenerate fringe becomes negligible, and so we are not required to introduce the relativistic and the homogeneous core simultaneously.)

We have, since $\chi(0)=1$, if χ is Emden's solution for '$n=3$',

$$\varrho_c = \lambda_3 A^3 \tag{37}$$

or by (27)

$$\varrho_c = \left(\frac{K_2}{K_1}\right)^3 \frac{1}{[g(c_1)]^3}. \tag{38}$$

For the central temperature we have, since

$$\tfrac{1}{3}aT_c^4 = K_2\varrho_c^{4/3}\frac{1-\beta}{\beta},$$

$$T_c = \left(\frac{3K_2}{a}\right)^{1/4}\frac{K_2}{K_1}\left(\frac{1-\beta}{\beta}\right)^{1/4}\cdot\frac{1}{g(c_1)}. \tag{39}$$

We have also

$$\frac{\varrho'}{\varrho_c} = [g(c_1)]^3; \qquad \frac{T'}{T_c} = g(c_1). \tag{40}$$

The radius r_1 of the whole configuration is given by

$$\frac{r'}{r_1} = \frac{\eta'}{\eta_1} = \frac{B\eta'}{B\eta_1} = \frac{b_1}{\eta_0} \tag{41}$$

or by (25)

$$r_1 = \left[\frac{5}{8\pi G\beta K_2}\right]^{1/2} K_1\eta_0\,[\phi(b_1)]^{1/4}. \tag{42}$$

The effective temperature and the mean density are easily found to be

$$T_e = \left(\frac{L}{ac}\right)^{1/4}\left(\frac{8G\beta K_2}{5}\right)^{1/4}\cdot\frac{1}{K_1^{1/2}}\cdot\frac{1}{\eta_0^{1/2}\,[\phi(b_1)]^{1/8}} \tag{43}$$

$$\varrho_m = \tfrac{3}{4}\left[\frac{8\pi G\beta K_2}{5}\right]^{3/2}\frac{M}{\pi K_1^3}\frac{1}{\eta_0^3\,[\phi(b_1)]^{3/4}}. \tag{44}$$

It is not difficult to put the above equations in a form which makes it clear that as $b_1 \to \eta_0$ and $c_1 \to \zeta_0$ these composite configurations *continuously* pass over into the complete relativistic Emden polytrope '$n=3$' with $M=0.92\odot\beta^{-3/2}$. In making the reduction we make free use of (30), (31), and (42):

$$\varrho_c = \left(\frac{K_2}{\pi G\beta}\right)^{3/2}\frac{c_1^3}{r_1^3}\cdot\frac{\eta_0^3}{b_1^3} \tag{45}$$

$$\frac{\varrho_c}{\varrho_m} = -\tfrac{1}{3} \frac{c_1}{g'(c_1)} \frac{M(r')}{M(r_1)} \cdot \frac{\eta_0^3}{b_1^3} \tag{45'}$$

$$M(r') = -\frac{4}{\pi^{1/2}} \left(\frac{K_2}{G\beta}\right)^{3/2} \zeta_0^2 \left(\frac{d\chi}{d\zeta}\right)_{\zeta=\zeta_0} \left[\frac{c_1^2 g'(c_1)}{\zeta_0^2 g'(\zeta_0)}\right] \tag{46}$$

$$T_c = \left(\frac{K_2}{\tfrac{1}{3}a}\right)^{1/4} \left(\frac{1-\beta}{\beta}\right)^{1/4} \left(\frac{K_2}{\pi G\beta}\right)^{1/2} \frac{c_1}{r_1} \frac{\eta_0}{b_1^3}. \tag{46'}$$

But when $b_1 \to \eta_0$ and $c_1 \to \zeta_0$ it is clear from (42) and (38) that simultaneously $r_1 \to 0$ and $\varrho_c \to \infty$.* Thus the completely relativistic model considered as the limit of the composite series is a point-mass with $\varrho_c = \infty$! The theory gives this result because $p = K_2 \varrho^{4/3}$ allows any density provided the pressure be sufficiently high. We are bound to assume therefore that a stage must come beyond which the equation of state $p = K_2 \varrho^{4/3}$ is not valid, for otherwise we are led to the physically inconceivable result that for $M = 0.92 \odot \beta^{-3/2}$, $r_1 = 0$, and $\varrho = \infty$. As we do not know physically what the next equation of state is that we are to take, we assume for definiteness the equation for the homogeneous incompressible material $\varrho = \varrho_{max}$, where ϱ_{max} is the maximum density of which matter is capable. The preceding analysis would then break down when ϱ_c given by (38) exceeds ϱ_{max}. Now ϱ_{max} must at the lowest estimate be of the order of 10^{12} gm cm^{-3}, for if the "maximum density of matter is limited only by the sizes of the electrons and nuclei, densities of the order 10^{14} should not be impossible."** Our interfacial density is only 4.84×10^6 and the ratio of this to the central density $\varrho_c = \varrho_{max}$ is of the order of 10^{-6}, and a reference to the Emden tables shows that if for the moment we assume the star as *completely* relativistically degenerate, before we have proceeded $\frac{1}{1000}$th of the radius of the star into the interior, our assumption becomes valid. Hence the correction due to the degenerate 'fringe' is negligible. We can therefore to a high degree of approximation consider, as the limit of these composite series, the Emden polytrope '$n = 3$' with $\varrho_c = \varrho_{max}$ and $M = 0.92 \odot \beta^{-3/2}$, it being understood that ϱ_{max} is sufficiently high to make the correction due to the degenerate fringe negligible. (Hence for highly collapsed configurations of mass greater than $0.92 \odot \beta^{-3/2}$ we can neglect the degenerate fringe.) Thus the highly collapsed configurations for which $0.612 \odot \beta^{-3/2} < M < 0.92 \odot \beta'^{-3/2}$ are composite, consisting of a degenerate envelope surrounding a relativistic core. These composite configurations are bordered on either side by two completely determinable configurations – on the one side by an Emden polytrope '$n = \tfrac{3}{2}$,' and on the other by an Emden polytrope '$n = 3$' with $\varrho_c = \varrho_{max}$.

8. Composite Series II

So far we have considered only the Emden solutions for the relativistic core, and we have been able to specify the steady-state configurations only for $M \leqslant M_3$. Further, $M = M_3$ corresponds to a high degree of approximation to an Emden-polytrope

---

* This suggests that when g is Emden's solution, φ is of the centrally condensed type.
** R. H. Fowler, *M. N.* **87** (1926), 114, 'Dense Matter'.

'$n=3$' with $\varrho_c = \varrho_{max}$. Hence for $M > M_3$ we should expect the homogeneous core to have spread out. We should therefore consider the fit of a relativistic envelope surrounding a homogeneous core with $\varrho = \varrho_{max}$, the core and the envelope being continuous at the interface. If $r = r''$ (where $\zeta = \zeta''$ and $\chi = \chi''$) is the radius of the surface of demarcation, the equation of fit is found to reduce to

$$\tfrac{1}{3}\zeta''^3\chi''^3 = -\zeta''^2\left(\frac{d\chi}{d\zeta}\right)_{\zeta=\zeta''}. \tag{47}$$

It can easily be shown that if χ be an Emden or a centrally condensed type of solution, then (47) has no roots, except that in the former case we have the trivial solution $\zeta'' = 0$. Hence the introduction of the homogeneous core compels us to a consideration of only the collapsed-type solutions for χ.

Now by (36)

$$-\zeta_0^2\left(\frac{d\chi}{d\zeta}\right)_{\zeta=\zeta_0} = \frac{M}{\dfrac{4}{\pi^{1/2}}\left(\dfrac{K_2}{G\beta}\right)^{3/2}} = \frac{1}{C_2^{1/2}} \tag{48}$$

where C_2 can now be called the 'discriminant' of the relativistic standard model for $M > M_3$. It may be noted here that C_2 is primarily a function *only* of M, since the hypothesis that the configurations are highly collapsed provides us with $\beta \sim 1$ in any case. We can write (48) differently as

$$-\zeta_0^2\left(\frac{d\chi}{d\zeta}\right)_{\zeta=\zeta_0} = \frac{M}{M_3} \times 2.015 \tag{49}$$

where 2.015 is the corresponding boundary value for the Emden solution. Hence for $M > M_3$ we clearly see from (49) that the solutions for χ now belong to the collapsed family. Thus the condition that $M > M_3$ is precisely equivalent to the condition that there is a homogeneous core. By methods similar to Section 4 we easily obtain

$$M(r'') = \tfrac{4}{3}\pi r''^3\varrho_{max} = -\frac{4}{\pi^{1/2}}\left(\frac{K_2}{G\beta}\right)^{3/2} c_1^2 g'(c_1) \tag{50}$$

$$r_1 = \frac{r''\zeta_0}{c_1} \tag{51}$$

where $c_1 = A\zeta''$ and r_1 is the radius of the whole configuration. Comparing (50) with (49) we see that as M increases beyond M_3 the collapse proceeds further and further till finally, when $M \to \infty$, $c_1 \to \zeta_0$, and $r_1 \to r''$, and the whole configuration has completely 'collapsed' into one mass of incompressible matter at the highest density matter is capable of. We have in the limit, so to say, a *'solid star'*.

Further, if $g(c_1)$ corresponds to the Emden solution there is only one 'trivial' root for (47), namely, $c_1 = 0$, *i.e.* the central density of this completely relativistic Emden polytrope is just equal to ϱ_{max} and the radius of the star is then obviously given by

$$\frac{\frac{4}{3}\pi r_1^3 \varrho_{max}}{54.36} = 0.92 \odot \beta^{-3/2} . \tag{52}$$

Thus this Composite Series II joins continuously the Composite Series I (Section 7) and the Emden polytrope '$n=3$' with $\varrho_c = \varrho_{max}$, and $M = .92 \odot \beta^{-3/2}$ is the *common limit* of both the series. We have therefore the following complete classification of the highly collapsed configurations ($L \ll L_0$, $\beta \sim 1$) for *M considered as a variable taking the whole range of values.*

Mass	Description
Class I – $M < 0.61 \odot \beta^{-3/2}$	Emden polytropes '$n=\frac{3}{2}$'
$\quad M_{3/2} = M = .61 \odot \beta^{-3/2}$	An Emden polytrope $n=\frac{3}{2}$ with $\varrho_c = (K_2/K_1)^3$.
Class II – $0.61 \odot \beta^{-3/2} < M < 0.92 \odot \beta'^{-3/2}$	Composite I – Degenerate envelope surrounding a relativistic core.
$\quad M_3 = M = 0.92 \odot \beta'^{-3/2}$	Approximately an Emden polytrope '$n=3$' with $\varrho_c = \varrho_{max}$.
Class III – $M > .92 \odot \beta'^{-3/2}$	Composite II – relativistic envelope and homogeneous core.
$\quad M \to \infty$	Completely homogeneous ($\varrho = \varrho_{max}$).

That we are thus able to enumerate definitely the steady-state configurations for the whole range of M appears to be in complete conformity with the general scheme of Milne's ideas.

To apply the above classification to the known white dwarfs – o_2 Eridani B, Procyon B, and Van Maanen's star possibly belong to Class I. That the companion of Sirius is in Class II is also likely.But it appears that no white dwarf has yet been discovered which has a homogeneous core at the centre. This classification is made with caution, since, though they are certainly of the collapsed type, they are by no means '*highly*' collapsed, for then L and T_e would be so small that we could not see the stars.

9. Summary

In this paper Milne's theory of collapsed configurations is developed a stage further, the essential refinement being in the introduction of a relativistically degenerate core with the equation of state $p = K_2 \varrho^{4/3}$. This enables an enumeration to be made of the steady-state configurations for the whole range of M considered as a variable with $L \sim 0$. The classification arrived at is shown in the table in Section 8.

In conclusion, I wish to record my best thanks to Professor Milne for much valuable advice and criticism during the course of the work.

ON THE THEORY OF STARS*

L. LANDAU

Abstract. From the theoretical point of view the physical nature of Stellar equilibrium is considered.

The astrophysical methods usually applied in attacking the problems of stellar structure are characterised by making physical assumptions chosen only for the sake of mathematical convenience. By this is characterised, for instance, Mr Milne's proof of the impossibility of a star consisting throughout of classical ideal gas; this proof rests on the assertion that, for arbitrary L and M, the fundamental equations of a star consisting of classical ideal gas admit, in general, no regular solution. Mr Milne seems to have overlooked the fact, that this assertion results only from the assumption of opacity being constant throughout the star, which assumption is made only for mathematical purposes and has nothing to do with reality. Only in the case of this assumption the radius R disappears from the relation between L, M and R necessary for regularity of the solution. Any reasonable assumptions about the opacity would lead to a relation between L, M and R, which relation would be quite exempt from the physical criticisms put forward against Eddington's mass-luminosity-relation.

It seems reasonable to try to attack the problem of stellar structure by methods of theoretical physics, i.e. to investigate the physical nature of stellar equilibrium. For that purpose we must at first investigate the statistical equilibrium of a given mass without generation of energy, the condition for which equilibrium being the minimum of free energy F for given temperature. The part of free energy due to gravitation is negative and inversely proportional to some average linear dimensions of the system in question, or, in other words, it is proportional to $\varrho^{1/3}$ (ϱ some average density).

The remaining inner part of free energy depends on the equation of state; for the classical ideal gas it is proportional to $\log \varrho$. In view of the fact that $\log \varrho$ tends to infinity for $\varrho \to \infty$ more slowly than $\varrho^{1/3}$ we will always have a minimum of free energy at $\varrho = \infty$. That means that, in the case of classical ideal gas, we obtain no equilibrium at all. Every part of the system would tend to a point. The state of affairs becomes quite different when we consider the quantum effects. For the non-relativistic Fermi-gas the inner free energy is changing with ϱ as $\varrho^{2/3}$, that means, more rapidly than the gravitational. It would lead to the existence of a stable equilibrium. The presence of sources would produce only an additional expansion due to the radiation pressure. Mr Milne tries to escape from this conclusion by introducing a condensed inner part of the system, but he does not tell the reasons why such condensations could appear at all. The connexion of the condensed state with the normal state remains rather mysterious. It is easy to see that any equation of state leading to no discontinuous

* Reprinted from *Physikalische Zeitschrift der Sowjetunion* 1. Original article submitted 7 January, 1932.

H. Gursky and R. Ruffini (eds.), Neutron Stars, Black Holes and Binary X-Ray Sources, 271–273. All Rights Reserved.

transitions (as admitted in Milne's calculations) would never make such condensations possible.

As the velocities of electrons in the Fermi-distribution rise with the density we have to apply, for sufficiently great densities, the relativistic theory. In the extreme-relativistic case the inner free energy per unit volume varies as $\varrho^{1/3}$, i.e. the same power of density as the gravitational energy. The free energy F is therefore of the form $F=a\varrho^{1/3}$. If a is positive the system will expand in order to have F minimum, until the density becomes too small for the extreme-relativistic relation $F=a\varrho^{1/3}$ to be valid. If a is negative the system will have a tendency to collapse to a point. In order to find the criterion separating these two cases we have to investigate the solution of the general equation

$$\frac{1}{r^2}\frac{\mathrm{d}}{\mathrm{d}r}\left[r^2\frac{\mathrm{d}\mu}{\mathrm{d}r}\right]=-4\pi G\varrho. \tag{1}$$

The chemical potential μ is in the extreme-relativistic case equal to $hc\,[3\pi^2\varrho/m^4]^{1/3}$ where m is the mass per 1 electron, that means for most elements two protonic masses. The equation (1) is the $n=3$ polytropic equation of Emden. With known substitutions it can be brought to the form

$$\frac{1}{r^2}\frac{\mathrm{d}}{\mathrm{d}r}\left[r^2\frac{\mathrm{d}x}{\mathrm{d}r}\right]=-x^3$$

with the boundary condition

$$r^2\frac{\mathrm{d}x}{\mathrm{d}r}=M\left[\frac{G}{hc}\right]^{3/2}m^2\left[\frac{4}{3\pi}\right]^{1/2}$$

on the outer boundary. As Emden has shown, this equation, has a regular solution only in the case $r^2(\mathrm{d}x/\mathrm{d}r)=2.015$, in which case it admits arbitrary radii; that means, we will have an indifferent equilibrium corresponding to $a=0$ in the former rough treatment. Thus we get an equilibrium state only for masses greater than a critical mass

$$M_0=\frac{3.1}{m^2}\left[\frac{hc}{G}\right]^{3/2}=2.8\times10^{33}\text{ gr.}$$

or about $1.5\odot$ (for $m=2$ protonic masses). For $M>M_0$ there exists in the whole quantum theory no cause preventing the system from collapsing to a point (the electrostatic forces are by great densities relatively very small). As in reality such masses exist quietly as stars and do not show any such ridiculous tendencies we must conclude that all stars heavier than $1.5\odot$ certainly possess regions in which the laws of quantum mechanics (and therefore of quantum statistics) are violated. As we have no reason to believe that stars can be divided into two physically different classes according to the condition $M>\text{or}<M_0$, we may with great probability suppose that all stars possess such pathological regions. It does not contradict the above arguments,

which prove only that the condition $M > M_0$ is sufficient (but not necessary) for the existence of such regions. It is very natural to think that just the presence of these regions makes stars stars. But if it is so, we have no need to suppose that the radiation of stars is due to some mysterious process of mutual annihilation of protons and electrons, which was never observed and has no special reason to occur in stars. Indeed we have always protons and electrons in atomic nuclei very close together, and they do not annihilate themselves; and it would be very strange if the high temperature did help, only because it does something in chemistry (chain reactions!). Following a beautiful idea of Prof. Niels Bohr's we are able to believe that the stellar radiation is due simply to a violation of the law of energy, which law, as Bohr has first pointed out, is no longer valid in the relativistic quantum theory, when the laws of ordinary quantum mechanics break down (as it is experimentally proved by continuous-rays-spectra and also made probable by theoretical considerations).* We expect that this must occur when the density of matter becomes so great that atomic nuclei come in close contact, forming one gigantic nucleus.

On these general lines we can try to develop a theory of stellar structure. The central region of the star must consist of a core of highly condensed matter, surrounded by matter in ordinary state. If the transition between these two states were a continuous one, a mass $M < M_0$ would never form a star, because the normal equilibrium state (i.e. without pathological regions) would be quite stable. Because, as far as we know, it is not the fact, we must conclude that the condensed and non-condensed states are separated by some unstable states in the same manner as a liquid and its vapour are, a property which could be easily explained by some kind of nuclear attraction. This would lead to the existence of a nearly discontinuous boundary between the two states.

The theory of stellar structure founded on the above considerations is yet to be constructed, and only such a theory can show how far they are true.

---

* L. Landau and R. Peierls, *Z. Phys.* **69** (1931), 56.

SOME REMARKS ON THE STATE OF MATTER
IN THE INTERIOR OF STARS*

S. CHANDRASEKHAR

Abstract. It is shown that for *all* stars for which the radiation-pressure is greater than a tenth of the total pressure, an appeal to the Fermi-Dirac statistics to avoid the central singularity which arises in the discussions of the centrally condensed and the collapsed stars cannot be made. The bearing of this result on the possible state of matter in the interior of stars is indicated.

Since the publication of Milne's memoir on the 'Analysis of Stellar Structure' in the *Monthly Notices* for November 1930** a great deal of work has been done to consider 'composite' stellar models. But the following simple considerations seem to have escaped notice and it seems worth while to state them explicitly.

1. The Surfaces of Demarcation

As we approach the centre of a centrally-condensed or a collapsed star, we change over to the equation of state $p = K_1 \varrho^{5/3}$ if the perfect gas law breaks down. If the perfect gas-law breaks down at all, the actual transition from the perfect-gas envelope to the degenerate core must occupy a certain zone, but we could for the sake of convenience consider a definite surface of demarcation defined as the surface at which the two equations of state give the same gas-pressure.

Now in the perfect gas envelope the *total* pressure is given by

$$P = \left[\left(\frac{k}{\mu} \right)^4 \frac{3}{a} \left(\frac{1 - \beta}{\beta^4} \right) \right]^{1/3} \varrho^{4/3} , \tag{1}$$

where

$$\beta = 1 - \frac{\kappa L}{4\pi c G M} , \tag{2}$$

κ = the opacity coefficient, L = luminosity in ergs cm^{-3}, M = mass in grams, k = = Boltzmann's Constant, μ = molecular weight = αm_H (say), m_H = mass of the hydrogen atom. Since for the standard model the gas pressure p is given by

$$p = \beta P, \tag{3}$$

we have

$$p = C\varrho^{4/3} , \tag{4}$$

where

$$C = \left[\left(\frac{k}{\mu} \right)^4 \frac{3}{a} \frac{1 - \beta}{\beta} \right]^{1/3} = \frac{2.632 \times 10^{15}}{\alpha^{4/3}} \left[\frac{1 - \beta}{\beta} \right]^{1/3} . \tag{4'}$$

* Reprinted from *Zeitschrift für Astrophysik* **5**. Original article submitted 28 September, 1932.
** Refered to as l.c.

H. Gursky and R. Ruffini (eds.), Neutron Stars, Black Holes and Binary X-Ray Sources, 274–281. All Rights Reserved.

The equation of state in the degenerate zone is

$$p = K_1 \varrho^{5/3}, \tag{5}$$

where

$$K_1 = \frac{1}{20} \left(\frac{3}{\pi}\right)^{2/3} \frac{h^2}{m\mu^{5/3}} = \frac{9.890 \times 10^{12}}{\alpha^{5/3}}. \tag{6}$$

At the first surface of demarcation which we will call S_1, the density ϱ_1 is given by

$$C\varrho_1^{4/3} = K_1 \varrho_1^{5/3},$$

or

$$\varrho_1 = \left(\frac{C}{K_1}\right)^3. \tag{7}$$

Now, it is well known that the equation of state (5) changes over again into the relativistic-degenerate-equation of state

$$p = K_2 \varrho^{4/3}, \tag{8}$$

where

$$K_2 = \frac{hc}{8\mu^{4/3}} \left(\frac{3}{\pi}\right)^{1/3} = \frac{1.228 \times 10^{15}}{\alpha^{4/3}}. \tag{8'}$$

Hence if *circumstances permit* we have to consider a second surface of demarcation, S_2, where the density ϱ_2 is given by

$$\varrho_2 = \left(\frac{K_2}{K_1}\right)^3. \tag{9}$$

Hence we have *two* surfaces of demarcation if and only if

$$\varrho_2 > \varrho_1,$$

or

$$\left(\frac{K_2}{K_1}\right)^3 > \left(\frac{C}{K_1}\right)^3, \tag{10}$$

i.e. only when [cf. Equations (4'), (6), (8')]

$$\frac{1 - \beta}{\beta} < \frac{h^3 c^3 a}{512\pi k^4} = 0.1015,$$

or

$$\beta > 0.9079. \tag{11}$$

It may be remarked in passing that the above value for β is independent of the assumed molecular weight. *It depends only on the mass, luminosity and opacity in the gaseous envelope*. It is also independent of whether we consider the same opacity for the degenerate zone and the gaseous envelope, or different opacities in the two regions.

2

The meaning of the fundamental inequality (11) is made clear by the following.

In the following figure I plot $\log p$ against $\log \varrho$.

For numerical calculations I use $\alpha = 2$. The straight line ABK represents the equation of state $p = K_1 \varrho^{5/3}$ and BC the equation of state $p = K_2 \varrho^{4/3}$. These two intersect at B where the density is that which corresponds to the *second* surface of demarcation, namely ϱ_2. ABC gives roughly the equation of state of a degenerate gas.

Let us consider a star for which $\beta = 0.98$. By (4) we get

$$\log p = 14.455 + \tfrac{4}{3} \log \varrho. \tag{12}$$

DE represents this equation. It interesects the degenerate equation of state AB, C at E. The point E corresponds to the first surface of demarcation S_1. Hence for all stars for which $\beta = 0.98$, we first traverse a perfect gas envelope with an equation of state represented by DE. Then we traverse a degenerate zone corresponding to EB and finally (if we have not yet reached the centre) a relativistically degenerate zone.

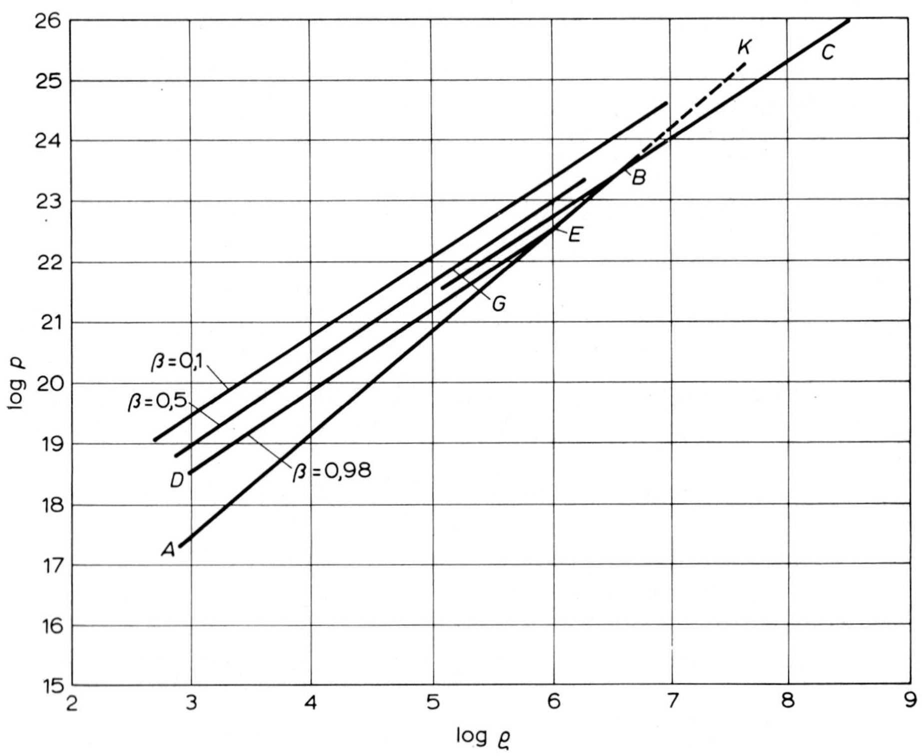

Fig. 1.

Now, if $\beta=0.9079$, then GB represents the perfect gas equation of state and the degenerate zone reduces to a single layer, and the relativistically degenerate zone is described equally well by the perfect gas equation.

Now if $\beta<0.9079$* the perfect gas equation of state has *no* intersections with ABC and this means that however high the density may become the temperature rises sufficiently rapidly to prevent the matter from becoming degenerate.

In this connection it will have to be remembered that considerations of relativity do not affect the equation of state of a perfect gas. $p = NkT$, *is true independent of relativity.*

3. Centrally-Condensed Stars

Now, for each mass M there is a unique luminosity L_0 – the 'Eddington luminosity' which makes the star a perfect gas sphere, with a polytropic index 3. This L_0 characteristizes a unique β_0 which is in fact related to M by means of Eddington's quartic equation:

$$1 - \beta = 0.003\,09 \left(\frac{M}{\odot}\right)^2 \alpha^4 \beta^4. \tag{13}$$

Now from the definition of a centrally-condensed and a collapsed star, it is clear that

$$\begin{aligned} \beta_{c \cdot c} &< \beta_0 \\ \beta_{col} &> \beta_0. \end{aligned} \tag{14}$$

Consider first the mass $\mathfrak{M}$ for which $\beta_0=0.9097$. By (13) we have

$$\mathfrak{M}/_\odot = 6.623\,\alpha^{-2}. \tag{15}$$

If we assume $\alpha=2$,

$$\mathfrak{M}/_\odot = 1.656. \tag{15'}$$

Now consider a centrally-condensed star of mass M *greater* than (or equal to) $\mathfrak{M}$. Then we obviously have

$$\begin{aligned} {}_M\beta_0 &< {}_\mathfrak{M}\beta_0 = 0.908, \\ {}_M\beta_{c \cdot c} &< {}_\mathfrak{M}\beta_0 < 0.908. \end{aligned} \tag{16}$$

Hence, we have the result that for *all centrally condensed stars of mass greater than* $\mathfrak{M}$ *the perfect gas equation of state does not break down, however high the density may become, and the matter does not become degenerate. An appeal to the Fermi-Dirac statistics to avoid the central singularity cannot be made.*

Since however we cannot allow the infinite density which the centrally condensed solution of Emden's differential equation – index 3 – allows at the centre and in the absence of our knowledge of any equation of state governing the perfect gas other than

---

* The radiation pressure is *greater* than a tenth of the total pressure if $\beta<0.9079$.

that of degenerate matter, our only way out of the singularity is to assume that there exists a maximum density $\varrho_{\max}$ which matter is capable of. We have therefore to consider the 'fit' of a gaseous envelope of the centrally condensed type on to a homogeneous core at the maximum density of matter. *If we insist on the density to be continuous at the interface* the equation of 'fit' is found to be*

$$\tfrac{1}{3}\xi'\Theta'^{3} = -\left(\frac{\mathrm{d}\Theta}{\mathrm{d}\xi}\right)_{\xi=\xi'},\tag{17}$$

where the polytropic equation describing the gaseous part of the star is

$$\frac{1}{\xi^{2}}\frac{\mathrm{d}}{\mathrm{d}\xi}\left(\xi^{2}\frac{\mathrm{d}\Theta}{\mathrm{d}\xi}\right) = -\Theta^{3},\tag{17'}$$

where ξ is the value of ξ at which $\varrho_{\max}$ begins. In $(17')$ the meaning of Θ and ξ are the following:

$$\varrho = \lambda_{3}\Theta^{3},\qquad r = \xi\left[\frac{C}{\pi G\beta}\right]^{-1/3}\lambda_{3}^{-1/3}\tag{17''}$$

(λ_{3} is a *homology constant*). But (17) has *no* solutions if Θ is of the Emden's or of the centrally-condensed type. Hence the acceptance of a $\varrho_{\max}$ does not help us out of the difficulty if we insist on the density to be continuous at the interface. The procedure then to construct an equilibrium configuration would be to proceed along the centrally condensed solution until the mean density $\varrho_{m}(r)$ of the *surviving mass* $M(r)$ equals $\varrho_{\max}$ which will occur at a determinate $r = r''$ (say) where

$$M(r'') = \tfrac{4}{3}\pi r''^{3}\varrho_{\max};\tag{18}$$

we then replace the material inside $r = r''$ by a sphere of incompressible matter at the density $\varrho_{\max}$. At r'' there will be a discontinuity of density (see Figure 2).

Now the form of Θ as $\xi \to 0$ for a centrally condensed solution is (Milne, l.c.):

$$\Theta \sim \frac{1/\sqrt{2}}{\xi\,[\log\,(D/\xi)]^{1/2}},\tag{19}$$

where D is a constant. D is *fixed* by the condition that the analytic continuation of (19) passes through $\xi = 1$ and $\Theta = 0$ and satisfies here the requisite boundary condition, namely

$$M = -\frac{4}{\pi^{1/2}}\left(\frac{C}{G\beta}\right)^{3/2}\left(\xi^{2}\frac{\mathrm{d}\Theta}{\mathrm{d}\xi}\right)_{0}.\tag{20**}$$

Hence we get the result that D is a function of L, M and κ only and hence fixed. Since

---

 * Chandrasekhar, S.: 1931, *Monthly Notices Roy. Astron. Soc.* **91**, 456, Equation (47).
** C is given by Equation $(4')$.

one compatible with a negative value of u_0, and that for equations of state of the type occurring in this problem even this possibility is excluded, so that u_0 must vanish.*

(c) A special investigation for any particular equation of state must be made to see whether solutions exist in which $0 \geqq u_0 \geqq -\infty$ and $p \to \infty$ as $r \to 0$.

3. Particular Equations of State

The above arguments show that Equations (9) and (10) together with a given equation of state completely determine the distribution of matter. The assumption $\varrho = \text{const.}$, $u_0 = 0$ makes it possible to integrate Equations (9) and (10) explicitly and leads to Schwarzschild's interior solution (Tolman, 1934, pp. 246–247). Other matter distributions corresponding to other equations of state are given by Professor Tolman in an accompanying paper.

If the matter is taken to consist of particles of rest mass μ_0 obeying Fermi statistics, and their thermal energy ** and all forces between them are neglected, then it may be shown that a parametric form for the equation of state is †

$$\varrho = K(\sinh t - t),\tag{11}$$

$$p = \tfrac{1}{3}K(\sinh t - 8\sinh \tfrac{1}{2}t + 3t),\tag{12}$$

where

$$K = \pi\mu_0^4 c^5/4h^3\tag{13}$$

and

$$t = 4\log\left(\frac{\hat{p}}{\mu_0 c} + \left[1 + \left(\frac{\hat{p}}{\mu_0 c}\right)^2\right]^{1/2}\right),\tag{14}$$

---

* This can be seen from the following argument. Having chosen some particular value of p_0 one may usually represent the equation of state in that pressure range by $\varrho = Kp^s$ with some appropriate value of s. Using this equation of state and taking the approximate form of Equation (10) near the origin for the case $u_0 < 0$, and finite p_0, one obtains:

$$\frac{dp}{dr} = \frac{p + \varrho(p)}{2r} = \frac{p + Kp^s}{2r}.$$

Integration of this equations shows that for $s < 1$ $p_0 \geqq 0$ can not be satisfied, and for $s \geqq 1$ only the value $p_0 = 0$ is possible. For the equations of state used in this problem always $s < 1$ holds. It may also be noted that the above equation together with Equation (7) show that $e^{\nu(r)} \to \infty$ as $r \to 0$.
** The condition for thermal equilibrium in a static gravitational field is given by Tolman (1934, p. 318) as $T_0(g_{44})^{1/2} = \text{const.}$ where T_0 is the proper temperature. The equilibrium state of a matter distribution which no longer radiates appreciably corresponds to a low surface temperature T_0. If g_{44} is everywhere finite, then T_0 will be small throughout the matter distribution. For those singular solutions in which g_{44} vanishes at the origin it is conceivable that the central temperature may be high. However, on the one hand from Equation (7) it is seen that the vanishing of g_{44} at the origin corresponds to infinite central pressure, and in this limit the equation of state given below reduces to $\varrho = 3p$ so that temperature introduces no radically new effects, and on the other hand zero values of g_{44} indicate the slowing down of all physical processes near the origin and thus may correspond to nonstatic solutions describing states which have not yet attained equilibrium, and which are not discussed in this paper.
† Cf. S. Chandrasekhar, *Monthly Notices Roy. Astron. Soc.* **95**, 222 (1935), but introduce *energy* density in place of his *mass* density.

where $\hat{p}$ is the maximum momentum in the Fermi distribution and is related to the proper particle density N/V by

$$\frac{N}{V} = \frac{8\pi}{3h^3}\,\hat{p}^3 .\tag{15}$$

Substituting the above expressions for p and ϱ into Equations (9) and (10) one gets:

$$\frac{du}{dr} = 4\pi r^2 K\,(\sinh t - t),\tag{16}$$

$$\frac{dt}{dr} = -\frac{4}{r\,(r - 2u)}\,\frac{\sinh t - 2\sinh\tfrac{1}{2}t}{\cosh t - 4\cosh\tfrac{1}{2}t + 3} \times$$
$$\times\,\left[(4/3)\,\pi K r^3\,(\sinh t - 8\sinh\tfrac{1}{2}t + 3t) + u\right].\tag{17}$$

These equations are to be integrated from the values $u=0$, $t=t_0$ at $r=0$ to $r=r_b$ where $t_b=0$ (which makes $p=0$), and $u=u_b$.

A note must be made of the units employed in these equations. Equations (3), (4) and (5) from which (16) and (17) are derived are stated in relativistic units (Tolman, 1934, pp. 201–2), *i.e.*, such that $c=1$, $G=1$ (c is the velocity of light, G is the gravitational constant). This determines the unit of time and the unit of mass in terms of a still arbitrary unit of length. The unit of length is now fixed by the requirement that $K=1/4\pi$. Equations (16) and (17) now become:

$$\frac{du}{dr} = r^2\,(\sinh t - t),\tag{18}$$

$$\frac{dt}{dr} = -\frac{4}{r\,(r - 2u)}\,\frac{\sinh t - 2\sinh\tfrac{1}{2}t}{\cosh t - 4\cosh\tfrac{1}{2}t + 3} \times$$
$$\times\,\left[\tfrac{1}{3}r^3\,(\sinh t + 8\sinh\tfrac{1}{2}t + 3t) + u\right].\tag{19}$$

The unit of length has been fixed to be

$$a = \frac{1}{\pi}\left(\frac{h}{\mu_0 c}\right)^{3/2}\frac{c}{(\mu_0 G)^{1/2}},$$

while the unit of mass is

$$b = \frac{c^2}{G}\,a = \frac{1}{\pi}\left(\frac{h}{\mu_0 c}\right)^{3/2}\frac{c^3}{(\mu_0 G^3)^{1/2}}.$$

For a neutron gas $a=1.36\times 10^6$ cm, $b=1.83\times 10^{34}$ g. The general character of the solution is seen to be independent of the mass of the neutron which determines only the scale of the result.

No way was found to carry out the integration analytically, so Equations (18) and (19) were integrated numerically for several finite values of t_0. For all these cases u_0

was taken to be equal to zero, since the equation of state near the origin for finite t_0 behaves like $\varrho(p)=Kp^s$, $s<1$. The first four entries in Table I were thus obtained.

For $t_0 \to \infty$ Equations (18) and (19) may be replaced by their asymptotic expressions:

$$du/dr = \tfrac{1}{2}r^2 e^t, \tag{20}$$

$$\frac{dt}{dr} = -\frac{4}{r(r-2u)}\left[\frac{r^3}{6}e^t + u\right]. \tag{21}$$

An exact solution* of these equations is:

$$e^t = 3/7r^2, \quad u = 3r/14, \tag{22}$$

which corresponds to $t_0 = \infty$, $u_0 = 0$. A careful examination of Equations (20) and (21) shows that there are no other solutions corresponding to $t_0 = \infty$, $0 \geq u_0 \geq -\infty$. The exact solution (22) of the approximate equations (20) and (21) was taken out to that value of r where $t=6$ (the approximation in the form of Equations (20) and (21) it quite good for $t \geq 6$), and then the integration of the exact Equations (18) and (19) was carried out numerically to $r=r_b$ where $t=0$. This gave the last entry in Table I.

TABLE I

Mass, radius and neutron density for various values of t_0

t_0	Mass		Radius		$\left(\dfrac{p}{\mu_0 c}\right)_{r=0}$	$\left(\dfrac{N}{V}\right)_{r=0}$ Neutrons/cm³
	In units of Equations (18), (19)	In units of ☉ for neutrons	In units of Equations (18), (19)	In kilometers, for neutrons		
1	0.033	0.30	1.55	21.1	0.25	0.062×10^{39}
2	0.066	0.60	0.98	13.3	0.52	0.56×10^{39}
3	0.078	0.71	0.70	9.5	0.82	2.2×10^{39}
4	0.070	0.64	0.50	6.8	1.17	6.4×10^{39}
∞	0.037	0.34	0.23	3.1	∞	∞

It is of interest to ask whether perhaps a finite gravitational mass might correspond to an infinite number of particles, and an infinite gravitational binding energy. It may be seen that this is not the case by the following argument. Although the proper particle density becomes infinite when the central pressure becomes infinite, still it remains integrable, so that the total number of particles always remains finite. The element of proper volume of a spherical shell is $4\pi e^{\lambda/2}r^2\,dr$. As the solution of the approximate equations shows in the neighborhood of the origin:

$$e^{\lambda/2} = \left(1 - \frac{2u}{r}\right)^{-1/2} = \left(1 - \frac{3}{7}\right)^{-1/2} = \left(\frac{7}{4}\right)^{1/2},$$

* This solution is a limiting form of the solutions V, VI given by Tolman in the accompanying paper.

$$N/V \infty \hat{p}^3 \propto e^{3t/4} \propto 1/r^{3/2} \text{ for large } t \text{ and } \hat{p};$$

$$\therefore N \propto \int_0^r \frac{r^2}{r^{2/3}} \, dr \propto r^{3/2} \text{ near the origin.}$$

A fortiori the number of particles is finite for nonsingular solutions.

For very small values of t the equation of state (11), (12) reduces to $p = K\varrho^{5/3}$ and $\hat{p} \propto t$. Using this equation of state and Newtonian gravitational theory (which is expected to give a good result for small masses and densities), one finds that $\hat{p} \propto m^{2/3}$, or that $m \propto t^{3/2}$. Figure 1 gives a schematic plot of the dependence of m on t_0 for the case that the elementary particles are neutrons. The mass m is plotted in units of sun's mass (2×10^{33} g) against $\tan^{-1} t_0$. The curve near the origin is dotted since, as has been already pointed out, a neutron core with a mass less than about $0.1 \odot$ will disintegrate into nuclei and electrons.

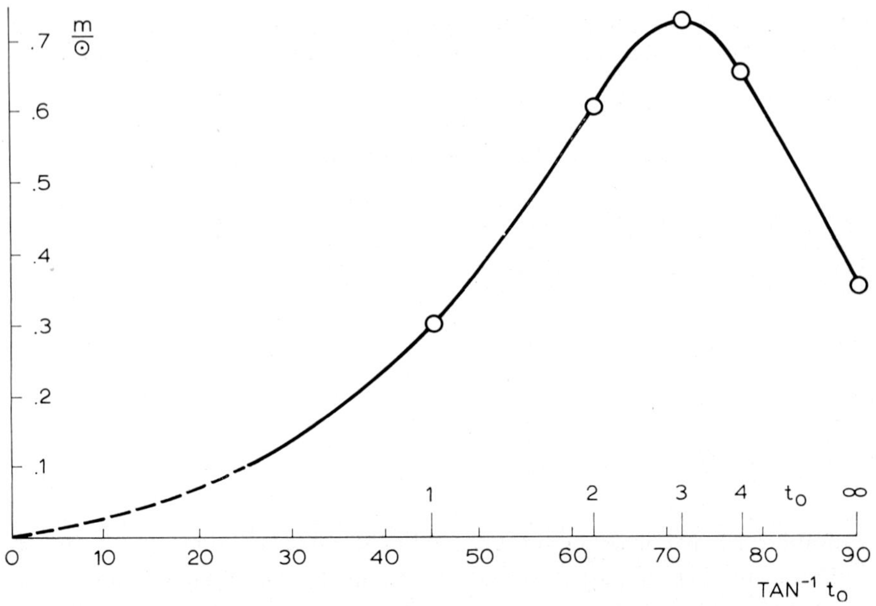

Fig. 1. Dependence of m on t_0 for neutrons.

The striking feature of the curve is that the mass increases with increasing t_0 until a maximum is reached at about $t_0 = 3$, after which the curve drops until a value roughly $\frac{1}{3} \odot$ is reached for $t_0 = \infty$. In other words no static solutions at all exist for $m > \frac{3}{4} \odot$, two solutions exist for all m in $\frac{3}{4} \odot > m > \frac{1}{3} \odot$, and one solution exists for all $m < \frac{1}{3} \odot$.

Some insight into this situation may be gained from the following considerations. In the non-relativistic polytrope solutions of Emden (1907) the equation of state was assumed to be $p = K\varrho^\gamma = K\varrho^{1+1/n}$. Solutions which at first sight seem to be quite satisfactory (i.e., giving a finite mass within a finite radius) were found for values of $n < 5$ or $\gamma > \frac{6}{5}$. But Landau (1932) pointed out that although these solutions in every case give an equilibrium configuration, they do not in every case give *stable* equilibrium. Thus, unless $\gamma \geqq \frac{4}{3}$ the equilibrium configuration is unstable. This may be seen from the following rough calculation. The gravitational part of the free energy of the system is negative and proportional to $\varrho^{1/2}$ where ϱ is an appropriate average density (Newtonian gravitational theory is used). The part of the free energy caused by compression is proportional to $\int p \, dv$, and hence to $\varrho^{\gamma-1}$ ($\gamma \neq 1$). Thus

$$F = - a\varrho^{1/3} + b\varrho^{\gamma-1}.$$

Polytrope solutions exist for both $\gamma = \frac{5}{3}$ ($> \frac{4}{3}$), i.e., for $n = \frac{3}{2}$ and for $\gamma = \frac{5}{4}$ ($< \frac{4}{3}$, but $> \frac{6}{5}$), i.e., for $n = 4$, but as may be seen from the schematic plot of the free energy curves in Figure 2, the former corresponds to stable equilibrium and the latter to unstable equilibrium.

In the present relativistic calculations the results for small masses and small central densities and pressures (small values of t_0), as was already mentioned above, may be expected to agree quite closely with nonrelativistic calculations with the equation of state $p = K\varrho^{5/3}$. Since ϱ is a monotonic function of t, the curves of free energy against t_0 for fixed total number of particles, and thus for a fixed M_0 (gravitational mass at zero density; the gravitational mass will vary somewhat along a curve of constant particle number, as the density increases), will for small masses have the same general character as the curves of free energy against some average density in the nonrela-

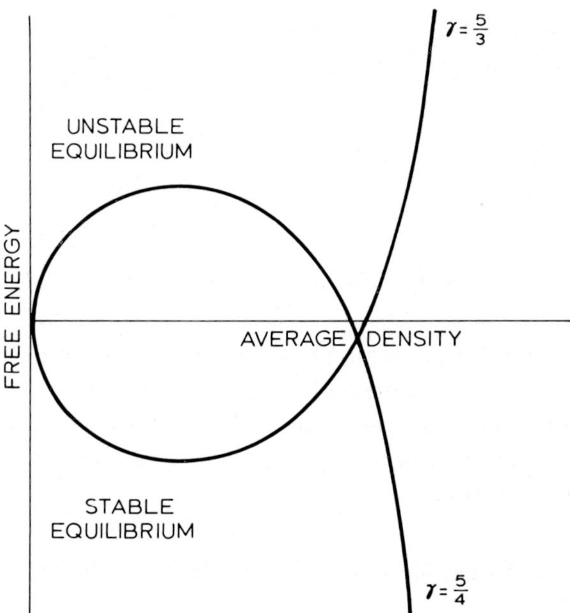

Fig. 2.   Free energy as a function of average density

tivistic case (cf. the curve for $\gamma = \frac{5}{3}$ in Figure 2). Then as the number of particles is increased the character of the free energy curves must change in order to admit the possibility of a second equilibrium position. Since the free energy must be a continuous function of t_0, and since we know from nonrelativistic calculations that for small masses (and low densities) we have a position of stable equilibrium (a minimum in the free energy curve) we can conclude that the second equilibrium position corresponds either to a maximum or to an inflection point in the free energy curve (and certainly not to a minimum). Figure 3 gives a schematic plot of free energy against t_0

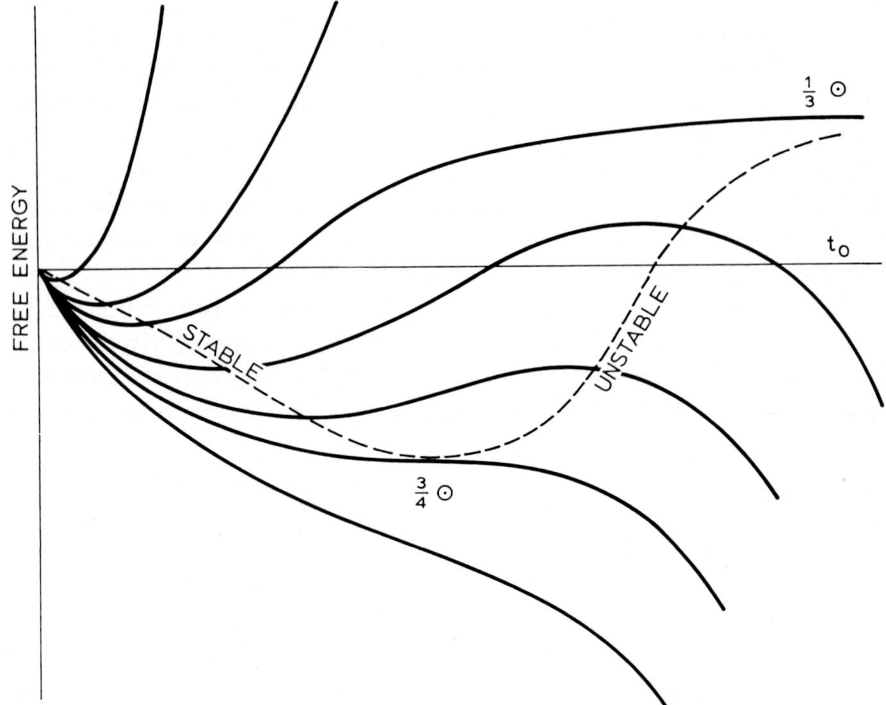

Fig. 3. Schematic plot of free energy as a function of t_0.

for different values of M_0 which would explain the existence of one equilibrium position for small masses, two for intermediate masses, and none for large masses. The masses marked on the curves are the actual gravitational masses corresponding to the equilibrium points of the critical free energy curves separating the solutions into the three types mentioned above.

4. Discussion – Relation to Tolman's Solutions

Before we study the physical implications of these results, we may try to show how their qualitative features may be obtained from the analytic solutions recently discovered by Professor Tolman.* This will also help us to understand the probable effect of alterations in the equation of state of the neutron gas at high densities.

On the one hand Tolman's solution IV, discussed in Section 6 of his paper, enables us to understand the existence of a limiting mass for static solutions and to give an estimate of its magnitude; on the other hand Tolman's solution VI, discussed in Section 8 (and less directly solution V), has for $n=\frac{1}{2}$, very much the character of our singular solution for $t_0 \to \infty$, and, with appropriately chosen constants, gives a mass of the same order of magnitude as we have found.

Tolman's solution IV is nonsingular, and corresponds to the quadratic equation of state (6.5) of his paper:

$$8\frac{(p_c - p)^2}{p_c + \varrho_c} - 5(p_c - p) + \varrho_c - \varrho = 0 \qquad \text{(Tolman, 6.5)},$$

where ϱ_c and p_c are the central density and pressure. From Equations (6.4), (6.6) and (6.9) of Tolman's paper the mass corresponding to this solution is given in terms of p_c and ϱ_c by

$$m = 4\left(\frac{p_c}{\varrho_c + 3p_c}\right)^{3/2}\left[\frac{8\pi}{3}(\varrho_c - 3p_c)\right]^{-1/2}. \qquad (23)$$

If ϱ_c and p_c are now themselves connected by the Fermi equation of state (11), (12), then $p_c \propto \varrho_c^{5/3}$ as $\varrho_c \to 0$, and $\varrho_c - 3p_c \sim 0(\varrho_c^{1/2})$ as $\varrho_c \to \infty$, and m is seen to have a maximum value. For values of ϱ_c corresponding to this maximum the equation of state (Tolman 6.5) does not differ qualitatively from the Fermi equation of state (11), (12), as may be seen by comparing for the two solutions the values of $d \ln p / d \ln \varrho$; and the maximum mass in fact turns out from (23) to be $\sim 0.4\odot$, agreeing in order of magnitude with our value of $\sim 0.7\odot$.

Tolman's solution V, with $n=\frac{1}{2}$, $R \to \infty$, and his solution VI, with $n=\frac{1}{2}$, $B/A \to 0$, are just our solution (22) corresponding to the equation of state $p=\frac{1}{3}\varrho$, a unique unstable singular solution. For solution V, with $n=\frac{1}{2}$ and finite R, the pressure differs from $\frac{1}{3}\varrho$ by terms of the order of $\varrho^{-1/6}$; however, for VI with $n=\frac{1}{2}$ and finite B/A, for large ϱ, $\varrho - 3p = \text{const.} \ \varrho^{1/2}$, which is just the behavior of a highly compressed Fermi gas. Using for the mass of this solution.

$$m = A/42B \qquad \text{(Tolman, 8.9)}$$

and adjusting the ratio B/A to make the equation of state of VI, i.e.,

$$p = \frac{1}{3}\varrho\,\frac{1 - 9(B/A)(3/56\pi)^{1/2}\,\varrho^{-1/2}}{1 - (B/A)(3/56\pi)^{1/2}\,\varrho^{-1/2}} \qquad \text{(Tolman, 8.5)}$$

agree to terms of order $\varrho^{1/2}$ with (11), (12), we get $B/A=(7/3)^{1/2}$, and $m \sim (\frac{1}{7})\odot$, to compare with the value of $\frac{1}{3}\odot$ which the Fermi equation gives.

These necessarily somewhat rough comparisons may thus serve to give an idea of the

* We are very much indebted to Professor Tolman for letting us see these results before publication, and for helpful discussions of them.

analytic character of the solutions corresponding to maximum mass and to maximum (infinite) central density which we obtained above.

5. Discussion – Application to Stellar Matter

We have seen that for a cold neutron core there are no static solutions, and thus no equilibrium, for core masses greater than $m \sim 0.7\,\odot$. The corresponding maximum mass M_0 before collapse is some ten percent greater than this. Since neutron cores can hardly be stable (with respect to formation of electrons and nuclei) for masses less than $\sim 0.1\,\odot$, and since, even after thermonuclear sources of energy are exhausted, they will not tend to form by collapse of ordinary matter for masses under $1.5\,\odot$ (Landau's limit), it seems unlikely that static neutron cores can play any great part in stellar evolution;* and the question of what happens, after energy sources are exhausted, to stars of mass greater than $1.5\,\odot$ still remains unanswered. It should be observed that for the critical solution with $m \sim 0.7\,\odot$ the potentials $g_{\mu\nu}$ are nowhere singular, and that in particular such a core does not tend to 'protect itself' from the addition of further matter by the vanishing of g_{44} at the boundary. There would then seem to be only two answers possible to the question of the 'final' behavior of very massive stars: either the equation of state we have used so far fails to describe the behavior of highly condensed matter that the conclusions reached above are qualitatively misleading, or the star will continue to contract indefinitely, never reaching equilibrium. Both alternatives require serious consideration.

The central density in the 'critical' core is even higher than nuclear density, so that our extrapolation of the Fermi equation of state can hardly rest on a very sure basis. Under these conditions the disintegration of neutrons, either into protons and electrons, or into mesotrons, will be energetically unfavorable and will not occur. And the relatively weak attractive forces which are known to act between neutrons will facilitate, and not prevent, the collapse of the core. If, however, under extreme compression, phenomena occurred which have the effect of repulsive forces, i.e., of raising the pressure for a given density above the value given by the Fermi equation of state, this could tend to prevent the collapse.

Such repulsive forces, even if they exist, will hardly make possible static solutions for arbitrarily large amounts of matter. For at low densities they cannot appreciably affect the equation of state, so that the dimensions of the core will necessarily be finite, and so will be the gravitational mass m of the core

$$m = \tfrac{1}{2} r_b (1 - e^{-\lambda_b}) \qquad \text{(Tolman, 5.5)}.$$

Now can the mass M_0 before collapse be infinite. For this to be true we should have to have a singular solution. But the effect of repulsive forces can for high density at most be to make $3p$ even more nearly equal to ϱ than for the Fermi equation of state;

---

* The mass of the shell of ordinary (but dense) matter surrounding the core must be small for cores much more massive than the lightest core stable with respect to disintegration into electrons and nuclei.

and for $\varrho = 3p$, as has been remarked above, and as is also suggested by Tolman's solutions V and VI, the *only* singular static solution is (22), for which the total particle number is finite.

We may obtain an extreme limit on the increase in the limiting mass which strong repulsive forces at high densities could give, by the following simple argument. For $\varrho < 10^{15}$ g cm^{-3} these forces can hardly be important. Let us assume that for $\varrho \geqslant 10^{15}$, they have the extreme effect of making $p = \frac{1}{3}\varrho$. Then the mass of a sphere for which this equation of state holds down to $\varrho = 10^{15}$, and for which p falls rapidly as $\varrho \to 0$, is given by our solution (22), and is of the order of $\odot$. It seems likely that our limit of $\sim 0.7 \odot$ is near the truth.

This argument is based on the requirement that even for arbitrarily high densities, $\varrho - 3p$ shall not be negative; and this is in turn closely related to the positive definite character of the (proper) energy density of neutrons and of the fields of force (apart from gravitation) associated with them. It seems probable that if p could be very much greater than $\frac{1}{3}\varrho$, static solutions of arbitrarily large mass could be found.*

From this discussion it appears probable that for an understanding of the long time behavior of actual heavy stars a consideration of nonstatic solutions must be essential. Among all (spherical) nonstatic solutions one would hope to find some for which the rate of contraction, and in general the time variation, become slower and slower, so that these solutions might be regarded, not as equilibrium solutions but as quasi-static. Some reason for this we may see in the following argument: for large enough mass the core will collapse; near the center the density and pressure will grow, and $g_{44} = e^\nu$ will be small (cf. Equation (7)); and as e^ν grows smaller, all processes will, as seen by an outside observer, slow down in the central region. Formally one sees this, in the occurrence, in Einstein's equations, of products of the form

$$e^{-\nu}\frac{d^2\lambda}{dt^2}, \qquad e^{-\nu}\left(\frac{d\lambda}{dt}\right)^2, \qquad e^{-\nu}\frac{d\lambda}{dt}\frac{d\nu}{dt}.$$

For high enough central densities it is no longer justified to neglect even a very slow time variation; and the singular solutions which presumably represent very massive neutron cores cannot be obtained unless this is taken into account. These solutions are now being investigated.

References

Edington, A.: 1926, *The Internal Constitution of the Stars*, Cambridge University Press.
Emden; 1907, *Gaskugeln* or cf. *Handbuch der Astrophys.*, Vol. 3, p. 186.
Gamow, G.: 1936, Atomic Nuclei and Nuclear Transformations, Oxford (Second ed.), p. 234.
Gamow, G.: 1938, *Phys. Rev.* **35**, 595.
Landau, L.: 1932, *Phys. Z. Sowjetunion* **1**, 285.
Landau, L.: 1938, *Nature* **141**, 333.
Oppenheimer, J. R. and Serber, R.: 1938, *Phys. Rev.* **54**, 540.
Strömgren, B.: 1937, *Ergebn. Exakt. Naturwiss.* **16**, 465.
Tolman, R. C.: 1934, *Relativity, Thermodynamics and Cosmology*, Oxford, pp. 239–241.

* Thus for $\varrho = $ const. there is a class of singular static solutions, for which $p \sim k/r^3$, and which would seem to lead, for $K \to \infty$, to infinite masses, and which one of us (G.M.V.) hopes to discuss in detail elsewhere.

ON CONTINUED GRAVITATIONAL CONTRACTION*

J. R. OPPENHEIMER and H. SNYDER

Abstract. When all thermonuclear sources of energy are exhausted a sufficiently heavy star will collapse. Unless fission due to rotation, the radiation of mass, or the blowing off of mass by radiation, reduce the star's mass to the order of that of the sun, this contraction will continue indefinitely. In the present paper we study the solutions of the gravitational field equations which describe this process. In **1**, general and qualitative arguments are given on the behavior of the metrical tensor as the contraction progresses: the radius of the star approaches asymptotically its gravitational radius; light from the surface of the star is progressively reddened, and can escape over a progressively narrower range of angles. In **2**, an analytic solution of the field equations confirming these general arguments is obtained for the case that the pressure within the star can be neglected. The total time of collapse for an observer comoving with the stellar matter is finite, and for this idealized case and typical stellar masses, of the order of a day; an external observer sees the star asymptotically shrinking to its gravitational radius.

1

Recently it has been shown (Oppenheimer and Volkoff, 1939) that the general relativistic field equations do not possess any static solutions for a spherical distribution of cold neutrons if the total mass of the neutrons is greater than $\sim 0.7 \odot$. It seems of interest to investigate the behavior of nonstatic solutions of the field equations.

In this work we will be concerned with stars which have large masses, $> 0.7 \odot$, and which have used up their nuclear sources of energy. A star under these circumstances would collapse under the influence of its gravitational field and release energy. This energy could be divided into four parts: (1) kinetic energy of motion of the particles in the star, (2) radiation, (3) potential and kinetic energy of the outer layers of the star which could be blown away by the radiation, (4) rotational energy which could divide the star into two or more parts. If the mass of the original star were sufficiently small, or if enough of the star could be blown from the surface by radiation, or lost directly in radiation, or if the angular momentum of the star were great enough to split it into small fragments, then the remaining matter could form a stable static distribution, a white dwarf star. We consider the case where this cannot happen.

If then, for the late stages of contraction, we can neglect the gravitational effect of any escaping radiation or matter, and may still neglect the deviations from spherical symmetry produced by rotation, the line element outside the boundary r_b of the stellar matter must take the form

$$\mathrm{d}s^2 = e^\nu \, \mathrm{d}t^2 - e^\lambda \, \mathrm{d}r^2 - r^2 (\mathrm{d}\theta^2 + \sin^2\theta \, \mathrm{d}\varphi^2) \tag{1}$$

with

$$e^\nu = (1 - r_0/r)$$

and

$$e^\lambda = (1 - r_0/r)^{-1}.$$

Here r_0 is the gravitational radius, connected with the gravitational mass m of the star by $r_0 = 2mg/c^2$, and constant. We should now expect that since the pressure of the

* Reprinted from *Physical Review* **56**. Original article submitted 10 July 1939.

H. Gursky and R. Ruffini (eds.), Neutron Stars, Black Holes and Binary X-Ray Sources, 296–302. All Rights Reserved.

stellar matter is insufficient to support it against its own gravitational attraction, the star will contract, and its boundary r_b will necessarily approach the gravitational radius r_0. Near the surface of the star, where the pressure must in any case be low, we should expect to have a local observer see matter falling inward with a velocity very close to that of light; to a distant observer this motion will be slowed up by a factor $(1-r_0/r_b)$. All energy emitted outward from the surface of the star will be reduced very much in escaping, by the Doppler effect from the receding source, by the large gravitational red-shift, $(1-r_0/r_b)^{1/2}$, and by the gravitational deflection of light which will prevent the escape of radiation except through a cone about the outward normal of progressively shrinking aperture as the star contracts. The star thus tends to close itself off from any communication with a distant observer; only its gravitational field persists. We shall see later that although it takes, from the point of view of a distant observer, an infinite time for this asymptotic isolation to be established, for an observer comoving with the stellar matter this time is finite and may be quite short.

Inside the star we shall still suppose that the matter is spherically distributed. We may then take the line element in the form (1). For this line element the field equations are

$$-8\pi T_1^1 = e^{-\lambda}(v'/r + 1/r^2) - 1/r^2,\tag{2}$$

$$8\pi T_4^4 = e^{-\lambda}(\lambda'/r - 1/r^2) + 1/r^2,\tag{3}$$

$$-8\pi T_2^2 = -8\pi T_3^3$$

$$= e^{-\lambda}\left(\frac{v''}{2} + \frac{v'^2}{4} - \frac{v'\lambda'}{4} + \frac{v'-\lambda'}{2r}\right)$$

$$\qquad\qquad\qquad - e^{-v}(\ddot{\lambda}/2 + \dot{\lambda}^2/4 - \dot{\lambda}\dot{v}/4),\tag{4}$$

$$8\pi T_4^1 = -8\pi e^{v-\lambda}T_1^4 = -e^{-\lambda}\dot{\lambda}/r;\tag{5}$$

in which primes represent differentiation with respect to r and dots differentiation with respect to t.

The energy-momentum tensor T_v^μ is composed of two parts: (1) a material part due to electrons, protons, neutrons and other nuclei, (2) radiation. The material part may be thought of as that of a fluid which is moving in a radial direction, and which in comoving coordinates would have a definite relation between the pressure, density, and temperature. The radiation may be considered to be in equilibrium with the matter at this temperature, except for a flow of radiation due to a temperature gradient.

We have been unable to integrate these equations except when we place the pressure equal to zero. However, one can obtain some information about the solutions from inequalities implied by the differential equations and from conditions for regularity of the solutions. From Equations (2) and (3) one can see that unless λ vanishes at least as rapidly as r^2 when $r \to 0$, T_4^4 will become singular and that either or both T_1^1 and v' will become singular. Physically such a singularity would mean that the expression used for the energy-momentum tensor does not take account of some essential physical fact which would really smooth the singularity out. Further, a star in its early stage of development would not possess a singular density or pressure; it is impossible for a singularity to develop in a finite time.

If, therefore, $\lambda(r=0)=0$, we can express λ in terms of T_4^4, for, integrating Equation (3)

$$\lambda = -\ln\left\{1 - \frac{8\pi}{r}\int_0^r T_4^4 r^2\,\mathrm{d}r\right\}. \tag{6}$$

Therefore $\lambda \geqslant 0$ for all r since $T_4^4 \geqslant 0$.

Now that we know $\lambda \geqslant 0$, it is easy to obtain some information about v' from Equation (2);

$$v' \geqslant 0, \tag{7}$$

since λ and $-T_1^1$ are equal to or greater than zero.

If we use clock time at $r=\infty$, we may take $v(r=\infty)=0$. From this boundary condition and Equation (7) we deduce

$$v \leqslant 0. \tag{8}$$

The condition that space be flat for large r is $\lambda(r=\infty)=0$. Adding Equations (2) and (3) we obtain:

$$8\pi(T_4^4 - T_1^1) = e^{-\lambda}(\lambda' + v')/r. \tag{9}$$

Since T_4^4 is greater than zero and T_1^1 is less than zero we conclude

$$\lambda' + v' \geqslant 0. \tag{10}$$

Because of the boundary conditions on λ and v we have

$$\lambda + v \leqslant 0. \tag{11}$$

For those parts of the star which are collapsing, i.e., all parts of the star except those being blown away by the radiation, Equation (5) tells us that $\dot{\lambda}$ is greater than zero. Since λ increases with time, it may (a) approach an asymptotic value uniformly as a function of r; or (b) increase indefinitely, although certainly not uniformly as a function of r, since $\lambda(r=0)=0$. If λ were to approach a limiting value the star would be approaching a stationary state. However, we are supposing that the relationships between the T_v^μ do not admit any stationary solutions, and therefore exclude this possibility. Under case (b) we might expect that for any value of r greater than zero, λ will become greater than any preassigned value if t is sufficiently large. If this were so the volume of the star

$$V = 4\pi \int_0^{r_b} e^{\lambda/2} r^2\,\mathrm{d}r \tag{12}$$

would increase indefinitely with time; since the mass is constant, the mean density in the star would tend to zero. We shall see, however, that for all values of r except r_0, λ approaches a finite limiting value; only for $r=r_0$ does it increase indefinitely.

<h1 style="text-align:center">2</h1>

To investigate this question we will solve the field equations with the limiting form of the energy-momentum tensor in which the pressure is zero. When the pressure vanishes there are no static solutions to the field equations except when all components of T_ν^μ vanish. With $p=0$ we have the free gravitational collapse of the matter. We believe that the general features of the solution obtained this way give a valid indication even for the case that the pressure is not zero, provided that the mass is great enough to cause collapse.

For the solution of this problem, we have found it convenient to follow the earlier work of Tolman (1934) and use another system of coordinates, which are comoving with the matter. After finding a solution, we will introduce a coordinate transformation to put the line element in form (1).

$$ds^2 = d\tau^2 - e^{\bar\omega}\, dR^2 - e^{\omega}(d\theta^2 + \sin^2\theta\, d\varphi^2). \tag{13}$$

Because the coordinates are comoving with the matter and the pressure is zero,

$$T_4^4 = \varrho \tag{14}$$

and all other components of the energy momentum tensor vanish.

The field equations are:

$$8\pi T_1^1 = 0 = e^{-\omega} - e^{-\bar\omega}\frac{\omega'^2}{4} + \ddot\omega + \tfrac{3}{4}\dot\omega^2 = 0, \tag{15}$$

$$8\pi T_2^2 = 8\pi T_3^3 = 0 = - e^{-\bar\omega}\left(\frac{\omega''}{2} + \frac{\omega'^2}{4} - \frac{\bar\omega'\omega'}{4}\right) +$$
$$+ \frac{\ddot{\bar\omega}}{2} + \frac{\dot{\bar\omega}^2}{4} + \frac{\ddot\omega}{2} + \frac{\dot\omega^2}{4} + \frac{\dot{\bar\omega}\dot\omega}{4}, \tag{16}$$

$$8\pi T_4^4 = 8\pi\varrho = e^{-\omega} - e^{-\bar\omega}\left(\omega'' + \tfrac{3}{4}\omega'^2 - \frac{\bar\omega'\omega'}{2}\right) + \frac{\dot\omega^2}{4} + \frac{\dot{\bar\omega}\dot\omega}{2}, \tag{17}$$

$$8\pi e^{\bar\omega} T_4^1 = - 8\pi T_1^4 = 0 = \frac{\omega'\dot\omega}{2} - \frac{\dot{\bar\omega}\omega'}{2} + \dot\omega' \tag{18}$$

with primes and dots here and in the following representing differentiation with respect to R and τ, respectively. The integral of Equation (18) is given by Tolman *:

$$e^{\bar\omega} = e^{\omega}\omega'^2/4f^2(R) \tag{19}$$

* We wish to thank Professor R. C. Tolman and Mr. G. Omer for making this portion of the development available to us, and for helpful discussions.

with $f^2(R)$ a positive but otherwise arbitrary function of R. We find a sufficiently wide class of solutions if we put $f^2(R)=1$.

Substituting (19) in (15) with $f^2(R)=1$ we obtain

$$\ddot{\omega} + \tfrac{3}{4}\dot{\omega}^2 = 0. \tag{20}$$

The solution of this equation is:

$$e^{\omega} = (F\tau + G)^{4/3}, \tag{21}$$

in which F and G are arbitrary functions of R.

The substitution of (19) in (16) gives a result equivalent to (20). Therefore the solution of the field equations is (21).

For the density we obtain from (17), (19), and (21)

$$8\pi\varrho = 4/3\,(\tau + G/F)^{-1}\,(\tau + G'/F')^{-1}. \tag{22}$$

There is less real freedom in (21) than is apparent from the two arbitrary functions F and G; for taking R a function of a new variable R^* the differential equations (15), (17) and (18) will remain of the same form. We may therefore choose

$$G = R^{3/2}. \tag{23}$$

At a particular time, say τ equal zero, we may assign the density as a function of R. Equation (22) then becomes a first-order differential equation for F.

$$FF' = 9\pi R^2 \varrho_0(R). \tag{24}$$

The solution of this equation contains only one arbitrary constant. We now see that the effect of setting $f^2(R)$ equal to one allows us to assign only a one-parameter family of functions for the initial values of $\dot{\varrho}_0$, whereas in general one should be able to assign the initial values of $\dot{\varrho}_0$ arbitrarily.

We now take, as a particular case of (24):

$$FF' = \begin{cases} \text{const.} \times R^2; & \text{const.} > 0; \quad R < R_b \\[2mm] 0 & ; \quad R > R_b. \end{cases} \tag{25}$$

A particular solution of this equation is:

$$F = \begin{cases} -\tfrac{3}{2}r_0^{1/2}\,(R/R_b)^{3/2}; & R < R_b \\[2mm] -\tfrac{3}{2}r_0^{1/2} & ; \quad R > R_b \end{cases} \tag{26}$$

in which the constant r_0 is introduced for convenience, and is the gravitational radius of the star.

We wish to find a coordinate transformation which will change the line element into form (1). It is clear, by comparison of (1) and (13), that we must take

$$e^{\omega/2} = (F\tau + G)^{2/3} = r. \tag{27}$$

A new variable t which is a function of τ and R must be introduced so that the $g_{\mu\nu}$ are of the same form as those in Equation (1). Using the contravarient form of the metric tensor, we find that:

$$g^{44} = e^{-\nu} = \dot{t}^2 - t'^2/r'^2 = \dot{t}^2 (1 - \dot{r}^2),\tag{28}$$

$$g^{11} = -e^{-\lambda} = -(1 - \dot{r}^2),\tag{29}$$

$$g^{14} = 0 = \dot{t}\dot{r} - t'/r'.\tag{30}$$

Here (30) is a first-order partial differential equation for t. Using the values of r given by (27), and the values of F and G given by (26) and (23) we find:

$$t'/\dot{t} = \dot{r}r' = \begin{cases} -(r_0 R)^{1/2} [R^{3/2} - \tfrac{3}{2}r_0^{1/2}\tau]^{-2/3}; & R > R_b \\[2mm] -r_0^{1/2}RR_b^{-3/2}[1 - \tfrac{3}{2}r_0^{1/2}\tau R_b^{-3/2}]^{1/3}; & R < R_b. \end{cases}\tag{31}$$

The general solution of (31) is:

$$t = L(x) \quad \text{for} \quad R > R_b, \quad \text{with} \quad x = \frac{2}{3r_0^{1/2}} (R^{3/2} - r^{3/2})$$

$$- 2(rr_0)^{1/2} + r_0 \ln \frac{r^{1/2} + r_0^{1/2}}{r^{1/2} - r_0^{1/2}}$$

$$t = M(y) \quad \text{for} \quad R < R_b, \quad \text{with} \quad y = \tfrac{1}{2}[(R/R_b)^2 - 1] + R_b r/r_0 R,\tag{32}$$

where L and M are completely arbitrary functions of their arguments.

Outside the star, where R is greater than R_b, we wish the line element to be of the Schwartzchild form, since we are again neglecting the gravitational effect of any escaping radiation; thus

$$e^\lambda = (1 - r_0/r)^{-1}\tag{33}$$

$$e^\nu = (1 - r_0/r).\tag{34}$$

This requirement fixes the form of L; from (28) we can show that we must take $L(x)=x$, or

$$t = x.\tag{35}$$

At the surface of the star, R equal R_b, we must have L equal to M for all τ. The form of M is determined by this condition to be:

$$t = M(y) = \tfrac{2}{3}r_0^{-1/2}(R_b^{3/2} - r_0^{3/2}y^{3/2}) - 2r_0 y^{1/2} + r_0 \ln \frac{y^{1/2} + 1}{y^{1/2} - 1}.\tag{36}$$

Equation (36), together with (27) defines the transformation from R, τ to r and t, and implicitly, from (28) and (29), the metrical tensor.

We now wish to find the asymptotic behavior of e^λ, e^ν, and τ for large values of t. When t is large we obtain the approximate relation from Equations (36) and (27):

$$t \sim -r_0 \ln \{\tfrac{1}{2}[(R/R_b)^2 - 3] + R_b/r_0 (1 - 3r_0^{1/2}\tau/2R_b^2)^{2/3}\}.\tag{37}$$

From this relation we see that for a fixed value of R as t tends toward infinity, τ tends to a finite limit, which increases with R. After this time τ_0 an observer comoving with the matter would not be able to send a light signal from the star; the cone within which a signal can escape has closed entirely. For a star which has an initial density of one gram per cubic centimeter and a mass of 10^{33} grams this time τ_0 is about a day.

Substituting (27) and (37) into (28) and (29) we find

$$e^{-\lambda} \simeq 1 - (R/R_b)^2 \{e^{-t/r_0} + \tfrac{1}{2}[3 - (R/R_b)^2]\}^{-1}, \qquad (38)$$

$$e^{\nu} \simeq e^{\lambda - 2t/r_0} \{e^{-t/r_0} + \tfrac{1}{2}[3 - (R/R_b)^2]\}. \qquad (39)$$

For R less than R_b, e^{λ} tends to a finite limit as t tends to infinity. For R equal to R_b, e^{λ} tends to infinity like e^{t/r_0} as t approaches infinity. Where R is less than R_b, e^{ν} tends to zero like e^{-2t/r_0} and where R is equal to R_b, e^{ν} tends to zero like e^{-2t/r_0}.

This quantitative account of the behavior of e^{λ} and e^{ν} can supplement the qualitative discussion given in 1. For λ tends to a finite limit for $r < r_0$ as t approaches infinity, and for $r = r_0$ tends to infinity. Also for $r \leqslant r_0$, ν tends to minus infinity. We expect that this behavior will be realized by all collapsing stars which cannot end in a stable stationary state. Of course, actual stars would collapse more slowly than the example which we studied analytically because of the effect of the pressure of matter, of radiation, and of rotation.

References

Oppenheimer, J. R. and Volkoff, G. M.: 1939, *Phys. Rev.* **55**, 374.
Tolman, R. C.: 1934, *Proc. Nat. Acad. Sci.* **20**, 3.

DETECTION AND GENERATION OF GRAVITATIONAL WAVES*[†]

J. WEBER

Abstract. Methods are proposed for measurement of the Riemann tensor and detection of gravitational waves. These make use of the fact that relative motion of mass points, or strains in a crystal, can be produced by second derivatives of the gravitational fields. The strains in a crystal may result in electric polarization in consequence of the piezoelectric effect. Measurement of voltages then enables certain components of the Riemann tensor to be determined. Mathematical analysis of the limitations is given. Arrangements are presented for search for gravitational radiation.

The generation of gravitational waves in the laboratory is discussed. New methods are proposed which employ electrically induced stresses in crystals. These give approximately a seventeen-order increase in radiation over a spinning rod of the same length as the crystal. At the same frequency the crystal gives radiation which is about thirty-nine orders greater than that of a spinning rod.

1. Introduction

The question of gravitational radiation has always been a central issue in the General Theory of Relativity. Long ago, Einstein (1916, 1918) and Eddington (1923) studied the problem and predicted that very small amounts of energy would be radiated by a spinning rod or a double star. A great deal of theoretical work on the radiation problem has appeared, during the past four decades.

Experimental work along these lines now appears possible. Two avenues of approach will be considered.** First we should like to detect the presence of gravitational radiation incident on earth from either the sun or outside the solar system. Secondly it would be highly desirable to be able to generate and detect this radiation in a small laboratory.

Devices for detection of the radiation operate essentially by measuring the Fourier transform of the Riemann tensor. These will be discussed first. This will then be followed by proposals for generation of gravitational radiation which may give an increase of many orders over the gravitational radiation from a spinning rod.

2. Detection of Gravitational Radiation

Suppose we have a system of masses which may interact with each other. We start with the action principle

$$\delta I = \delta \left[- cm \int \mathrm{d}s + W \right] = 0. \tag{1}$$

In (1) m is the rest mass and W is the part of the action function associated with

---

* Reprinted from Physical Review **117**. Original article submitted 9 February; in revised form 20 July, 1959.

† Supported by the National Science Foundation.

** A number of the results discussed here were given without proof in the author's Gravity Research Foundation Prize Essays, April 1958 and April 1959, and at the Royaumont Conference on the Relativistic Theories of Gravitation, Royaumont, France, June, 1959 (unpublished).

H. Gursky and R. Ruffini (eds.), Neutron Stars, Black Holes and Binary X-Ray Sources, 303–317. All Rights Reserved.

forces arising from the motion of the mass relative to other masses with which it interacts. The line element ds is given by

$$ds^2 = g_{\mu\nu}\, dx^\mu\, dx^\nu. \tag{2}$$

For δW we assume a function given by

$$-c\delta W = \int F_\mu \delta x^\mu\, ds; \tag{3}$$

(3) identifies F_μ as the four-force. The Euler-Lagrange equations resulting from (1) are arranged (Pauli, 1958) to obtain

$$\frac{d^2 x^\mu}{ds^2} + \Gamma^\mu_{\alpha\beta}\frac{dx^\alpha}{ds}\frac{dx^\beta}{ds} = \frac{F^\mu}{mc^2}. \tag{4}$$

$\Gamma^\mu_{\alpha\beta}$ is the Christoffel symbol of the second kind. (4) may be written in terms of the four-velocity $p^\mu = dx^\mu/ds$, as

$$\frac{\delta}{\delta s}\left(\frac{dx^\mu}{ds}\right) = \frac{\delta p^\mu}{\delta s} = \frac{F^\mu}{mc^2}. \tag{5}$$

The symbol $\delta/\delta s$ means the covariant derivative with respect to s.

Following essentially the method of Synge and Schild (1952), we now obtain from (5) an equation similar to the equation of geodesic deviation. We introduce a parameter v such that each world line corresponds to a given value of v. Taking the covariant derivative of (5) with respect to v gives

$$\frac{\delta^2 p^\mu}{\delta v\, \delta s} = \frac{\delta F^\mu/m}{c^2\, \delta v}. \tag{6}$$

Employing the commutation law for covariant differentiation enables us to express (6) in the form

$$\frac{\delta^2 p^\mu}{\delta v\, \delta s} = \frac{\delta^2 p^\mu}{\delta s\, \delta v} - R^\mu_{\alpha\beta\gamma}p^\alpha p^\beta \frac{\partial x^\gamma}{\partial v}. \tag{7}$$

In (7), $\partial x^\gamma/\partial v$ is a unit vector normal to the world lines, and the four-velocity p^μ is a unit vector tangent to the world lines. The vector n^γ, defined by

$$n^\gamma = \frac{\partial x^\gamma}{\partial v}\, dv, \tag{8}$$

is an infinitesimal vector joining points with the same value of s on neighboring world lines with values of v differing by dv. The covariant derivative of $\partial x^\gamma/\partial v$ with respect to s can be written in the forms

$$\frac{\delta}{\delta s}\left(\frac{\partial x^\gamma}{\partial v}\right) = \frac{\delta}{\delta v}\left(\frac{\partial x^\gamma}{\partial s}\right) = \frac{\delta p^\gamma}{\delta v}. \tag{9}$$

Employing (6), (7), (8), and (9) then gives

$$\frac{\delta^2 n^\mu}{\delta s^2} + R^\mu_{\alpha\beta\gamma} p^\alpha n^\beta p^\gamma = \frac{1}{c^2} \frac{\delta F^\mu/m}{\delta v}\, dv.$$ (10)

3. Mass Quadrupole Detector

In order to discuss the detector* of Figure 1, we imagine the two world lines are those of the two masses. Let n be given by

$$n^\gamma = r^\gamma + \xi^\gamma,$$ (11)

with r^γ defined by

$$\delta r^\gamma/\delta s = 0 \quad \text{for all} \quad s, r^\gamma \to n^\gamma,$$ (12)

in the limit of large internal damping and all components of $R^\mu_{\alpha\beta\delta}=0$. Equation (10) becomes

$$\frac{\delta^2 \xi^\mu}{\delta s^2} + R^\mu_{\alpha\beta\gamma} p^\alpha p^\gamma [r^\beta + \xi^\beta] = \frac{f^\mu}{mc^2}.$$ (13)

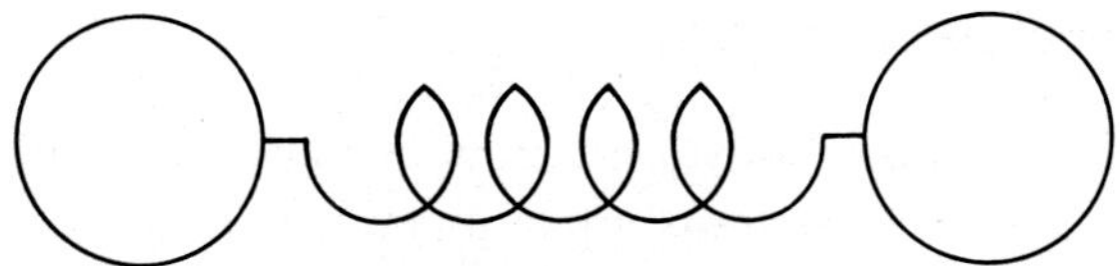

Fig. 1.   Harmonic oscillator driven by gravitational waves.

In (13) we have denoted by f^μ the differences in (nongravitational) forces at the two masses. For f^μ we assume a restoring force $-k^\mu_\alpha \xi^\alpha$ and a damping force $-cD^\mu_\alpha(\delta\xi^\alpha/\delta s)$; k^μ_α and D^μ_α are tensors associated with the spring. (13) then becomes

$$\frac{\delta^2 \xi^\mu}{\delta s^2} + \frac{D^\mu_\alpha}{cm}\frac{\delta\xi^\alpha}{\delta s} + \frac{k^\mu_\alpha \xi^\alpha}{mc^2} = -R^\mu_{\alpha\beta\gamma} p^\alpha p^\gamma [r^\beta + \xi^\beta].$$ (14)

We now let time run in the direction of the tangents to the world lines. The center of mass of the oscillator is a freely falling platform. We use coordinates in which the Christoffel symbols vanish and write (14) in the approximate form (assuming $\xi \ll r$)

$$\frac{d^2 \xi^\mu}{dt^2} + \frac{D^\mu_\alpha}{m}\frac{d\xi^\alpha}{dt} + \frac{k^\mu_\alpha \xi^\alpha}{m} = -c^2 R^\mu_{0\alpha0} r^\alpha.$$ (15)

* An arrangement somewhat similar to this was independently suggested by H. Bondi at the Royaumont Conference, Royaumont, France, June, 1959 (unpublished).

In (15) we see that the driving force for the harmonic oscillator is the Riemann tensor. Measurement of displacement amplitude or power absorbed enables one to calculate certain components of the Riemann tensor.*

Suppose now that sinusoidal (weak-field approximation) gravitational waves are incident, with angular frequency ω. An orthogonal comoving coordinate system is employed, with the oscillator oriented in the direction of the x^1 axis. k_α^μ and D_α^μ are imagined to have one component only, $k_1^1 = k$, and $D_1^1 = D$. Taking the Fourier transform of (15) leads to

$$\xi^\mu(\omega) = \frac{- mc^2 R^\mu_{0\alpha 0}(\omega)\, r^\alpha}{(-\omega^2 m + i\omega D\delta_1^\mu + k\delta_1^\mu)}. \tag{16}$$

(16) is a maximum at resonance, $-\omega^2 m + k = 0$. The total dissipation $D = D_{\text{ex}} + D_{\text{in}}$ where D_{ex} is the external dissipation and D_{in} is the internal dissipation associated with irreversible processes within the antenna. The power which can be delivered to auxiliary apparatus with D_{ex} is

$$\tfrac{1}{2}\omega^2 D_{\text{ex}}\xi^2 = \frac{m^2 c^4 (R^\mu_{0\alpha 0} r^\alpha)^2\, D_{\text{ex}}}{2(D_{\text{ex}} + D_{\text{in}})^2}. \tag{17}$$

(17) is a maximum when $D_{\text{ex}} = D_{\text{in}}$ and the maximum power P_M is given by

$$P_M = m^2 c^4 (R^\mu_{0\alpha 0} r^\alpha)^2 / (8 D_{\text{in}}). \tag{18}$$

The sinusoidal gravitational waves are now assumed to be radiated by a linear mass quadrupole oscillator. The transformation laws indicate that to a good approximation $R^\mu_{0\alpha 0}$ as seen in a frame fixed in the center of mass of the radiator is the same as that seen in a frame fixed in the center of mass of the detector, for small velocities. Using the known (Rosen and Shamir, 1957; Bonnor, 1959) solution for the linear mass quadrupole oscillator, the mean squared value of $R^\mu_{0\alpha 0} r^\alpha$ is calculated and averaged over all possible orientations of the receiving antenna. Let t_{0_r} be the radiated power per unit area averaged over a sphere, for the linear mass quadrupole oscillator. The total radiated power P is given by

$$P = 4\pi r^2 t_{0r} = G I_0^2 \omega^6 / (60\pi c^5), \tag{19}$$

where I_0 is the amplitude of the quadrupole tensor. (19) and the known expressions

* Measurement of the Riemann tensor by comparing accelerations of free test particles has been considered by F. A. E. Pirani, *Proceedings of the Chapel Hill Conference on the Role of Gravitation in Physics*, page 61, Astia Document No. AD 118180 and *Acta Phys. Polon. XV* **6** (1956) 389. While free particles are convenient for some thought experiments, interacting particles appear to be essential, in practice, at low energies. The correspondence between voltage in a piezoelectric crystal and some components of the Riemann tensor, which is discussed in the next section, may provide a basis for consideration of the measurement problem in quantized General Relativity. In principle a very small crystal may be used since the acoustic resonance vibrations have a wavelength which is about five orders smaller than that of the gravitational wave which excites them.

for the fields then give, for the mean squared value of $R^{\mu}_{0\alpha 0}r^{\alpha}$, in a direction normal to the quadrupole radiator axis.

$$[R^{\mu}_{0\alpha 0}r^{\alpha}]^{2}_{M} = [4\pi\beta^{2}|r|^{2}G/c^{5}]\,t_{0r}. \tag{20}$$

In (20), β is the propagation vector of the gravitational wave. Employing (18) and (20) gives

$$P_{M} = \pi m^{2}\beta^{2}\,|r|^{2}\,Gt_{0r}/(2cD_{in}). \tag{21}$$

The influence of the internal dissipation D_{in} will now be considered. First we assume that no irreversible processes take place within the antenna itself and that D_{in} is due entirely to radiation damping of the detector. The known solution for a linear mass quadrupole oscillator enables us to calculate the radiation resistance of the detector, D_{in}, as

$$D_{in} = 2G\omega^{4}m^{2}\,|r|^{2}/(15c^{5}). \tag{22}$$

(21) and (22) give, in terms of the wavelength λ,

$$P_{M\,(\text{radiation damping only})} = (15\lambda^{2}/16\pi)\,t_{0_{r}}. \tag{23}$$

The implication of (23) is that the average absorption cross section S_{A} for a detector which is damped only by its own reradiation is

$$S_{A} = (15/16\pi)\,\lambda^{2}. \tag{24}$$

We see from (24) that under these conditions the average absorption cross section is roughly a wavelength squared, and is independent of the constant of gravitation. Unfortunately the condition that the internal damping be only due to radiation cannot be attained in practice because other irreversible phenomena within the antenna are many orders greater than the radiation damping. In order to make this clear, we calculate the quality factor, denoted by the symbol Q, which is defined by

$$Q = \omega\,(\text{maximum stored energy})/(\text{power dissipated}).$$

The Q associated with radiation damping, denoted by Q_{R}, is

$$Q_{R} = 15c^{5}/(2G\omega^{3}m\,|r|^{2}). \tag{25}$$

For an antenna at $\omega = 2\pi \times 10^{7}$, a reasonable value of $mr^{2} = 10$ g cm^{2} and (25) gives $Q_{R} \sim 10^{34}$. A practical antenna might be expected to have a $Q \sim 10^{6}$.

We therefore must deal with systems limited by internal damping orders larger than gravitational radiation damping, and under these conditions the average absorbed power will not be independent of the kind of antenna. For an antenna orientation arranged for maximum response, we write

$$(R^{\mu}_{0\alpha 0}r^{\alpha})^{2} = 15\pi G\beta^{2}\,|r|^{2}\,c^{-5}t_{0r}. \tag{26}$$

(26) and (18) lead to power absorbed, P_{A}, given by

$$P_{A} = (15/8)\,\pi Gm^{2}\beta^{2}\,|r|^{2}\,(cD_{in})^{-1}\,t_{0r} = (15/8)\,\pi GmQ_{in}\beta^{2}\,|r|^{2}\,(\omega c)^{-1}\,t_{0r}. \tag{27}$$

In (27), Q_{in} is the Q associated with internal irreversible processes, $Q_{in} = \omega m / D_{in}$. The cross section, S, implied by (27) is

$$S = (15/8)\, \pi G m Q_{in} \beta^2 \, |r|^2 \, \omega^{-1} c^{-1}. \tag{28}$$

For a continuous spectrum the absorbed power is

$$P_A = T^{-1} \int_{-\infty}^{\infty} \int_{-\infty}^{\infty} \frac{m^2 c^4 D_{ex} \omega \omega' R_{0\alpha 0}^{\mu}(\omega) R_{0\beta 0}^{\mu}(\omega')\, r^{\alpha} r^{\beta} e^{i(\omega - \omega')t}}{2(-\omega^2 m + i\omega D + K)(-\omega'^2 m - i\omega' D + K)} \times$$
$$\times\, d\omega\, d\omega'\, dt \approx \pi^2 G m \beta^2 \, |r|^2 \, c^{-1} t_{0r}(\omega_0). \tag{29}$$

In (29), $t_{0r}(\omega_0)$ is the power spectrum of t_{0r} in the vicinity of the resonant frequency ω_0.

In order to further discuss these results we must consider the excitation of a continuous medium by a gravitational wave. This is necessary in order to be able to account for the interaction of the mass of the spring with the wave and to account for the effects of the finite velocity of propagation of the elastic forces of the spring.

4. Interaction of a Crystal with a Gravitational Wave

The starting point for our discussion is expression (10). The infinitesimal vector n^{μ} is from a reference point in the crystal to a neighboring point. The mass m is imagined to belong to an infinitesimal volume surrounding the neighboring point. On the right side of (10) we must now include both elastic forces and dissipative forces. We write n^{μ} as

$$n^{\mu} = r^{\mu} + \varepsilon_{\alpha}^{\mu} r^{\alpha}. \tag{30}$$

r^{α} is defined by the conditions

$$\delta r^{\mu}/\delta s = 0 \quad \text{for all} \quad s;\ r^{\mu} \to n^{\mu}, \tag{31}$$

in the limit of large internal damping and flat space. We may now write Equation (10) in the form

$$r^{\mu} \frac{\delta^2 \varepsilon_{\mu\nu}}{\delta s^2} + r^{\mu} B_{\nu}^{\alpha} \frac{\delta \varepsilon_{(\alpha\mu)}}{\delta s} + r^{\mu} Y^{\alpha\beta} \frac{\delta^2 \varepsilon_{(\mu\nu)}}{\delta x^{\alpha} \delta x^{\beta}} + R_{\nu\alpha\mu\beta} \left[r^{\mu} + \varepsilon_{\gamma}^{\mu} r^{\gamma} \right] p^{\alpha} p^{\beta} = 0. \tag{32}$$

In (32) the quantity $\varepsilon_{(\mu\nu)}$ is the symmetric part of $\varepsilon_{\mu\nu}$ and is therefore the strain tensor of the crystal. The second term accounts for internal damping and the third term accounts for the elastic forces. B_{ν}^{α} and $Y^{\alpha\beta}$ are normalized to unit mass density. Again p^{α} is a unit vector tangent to the world lines. Since r^{μ} can be arbitrarily specified, we may write

$$\frac{\delta^2 \varepsilon_{\mu\nu}}{\delta s^2} + B_{\nu}^{\alpha} \frac{\delta \varepsilon_{(\alpha\mu)}}{\delta s} + Y^{\alpha\beta} \frac{\delta^2 \varepsilon_{(\mu\nu)}}{\delta x^{\alpha} \delta x^{\beta}} + R_{\nu\alpha\mu\beta} p^{\alpha} p^{\beta} + R_{\nu\alpha\gamma\beta} p^{\alpha} p^{\beta} \varepsilon_{\mu}^{\gamma} = 0. \tag{33}$$

In (33) the fourth term is clearly symmetric in the indices ν and μ. The last term in

(33) may ordinarily be dropped because it is many orders smaller than the fourth one. For the strain tensor we may therefore write

$$\frac{\delta^2 \varepsilon_{(\mu\nu)}}{\delta s^2} + B^\alpha_\nu \frac{\delta \varepsilon_{(\alpha\mu)}}{\delta s} + Y^{\alpha\beta} \frac{\delta^2 \varepsilon_{(\mu\nu)}}{\delta x^\alpha \delta x^\beta} \approx - R_{\nu\alpha\mu\beta} p^\alpha p^\beta . \tag{33A}$$

We now consider a special case of (33A), namely excitation of longitudinal waves in an isotropic medium. An approximate form suitable for the present discussion, for waves in the direction x^1 of an orthogonal coordinate system (with the time direction tangent to the world lines), is

$$y \frac{\partial^2 \varepsilon}{\partial (x^1)^2} - \rho \frac{\partial^2 \varepsilon}{\partial t^2} - b \frac{\partial \varepsilon}{\partial t} = c^2 \varrho R^1_{010} . \tag{34}$$

If (34), ϱ is the density, y is an appropriate modulus, and b is a damping constant. We assume that R^1_{010} has its origin in incident sinusoidal gravitational waves so that

$$- c^2 R^1_{010} = f \exp \left[i \left(\omega t - \beta_j x^j \right) \right] . \tag{35}$$

In (35) the index j runs from 1 to 3. Let v_s be the sound velocity $(y/\varrho)^{1/2}$, λ_s be the wavelength of sound, $k_s = 2\pi/\lambda_s$, $\alpha = b/\varrho v_s$, and $\gamma = \alpha + ik_s$. Then to a good approximation the solution of (34) may be written

$$\varepsilon = \left[A\gamma \cosh \gamma x^1 - f\omega^{-2} \exp \left(-i\beta_j x^j \right) \right] e^{i\omega t} . \tag{36}$$

Making use of the boundary condition that ε vanishes at the ends leads to

$$A = - \frac{f \lambda_s \cos \beta_1 l}{2\pi\omega^2 \left(al \sin k_s l + \cos k_s l \right)} . \tag{37}$$

In (37), l is half the length of the crystal. The first term of (36) gives the contribution of the acoustic waves and the second term gives the strains which would be set up if there were no internal forces at all* (37) must be modified if the crystal is piezo-electric.

(37) has maxima when $k_s l$ is an odd multiple of $\pi/2$; however, it is clear from the denominator that the largest maximum is the first one, for which the total length is half an acoustic wavelength. The system composed of the two masses and spring (Figure 1) must be described by an equation such as (34), when the spacing of the masses approaches half an acoustic wavelength. The largest value we can expect from (28) will occur when r is half an acoustic wavelength in the spring. This is an important limitation because the velocity of acoustic waves is about five orders smaller than the velocity of light, so the cross sections implied by (28) are limited to values ten orders smaller than would be the case if the elastic forces of the spring were

* *Note added in proof.* – Integration of (36) gives relative displacements. If the result is applied to effects of gravitational waves interacting with the Earth, the contribution of the first term of (36) is found to be very small. Apparatus on the Earth's surface acts therefore as if it were in free fall insofar as the waves are concerned. This is a consequence of the fact that the velocity of sound is much smaller than the velocity of light.

propagated with the velocity of light. Such a limitation could be overcome in a number of ways. One might employ restoring forces transmitted by electric and magnetic fields, with the velocity of light. The piezoelectric effect may be employed, in which case the polarization charges in the crystal faces may give rise to some stress components which do not change sign every half acoustic wavelength.

In a piezoelectrical crystal a strain results in an electric polarization P_μ given by

$$P_\mu = \varepsilon_{\alpha\beta}\mathscr{E}^{\alpha\beta}_\mu .$$

Here $\mathscr{E}^{\alpha\beta}_\mu$ is the piezoelectric stress tensor. The electric polarization gives rise to an electric field over the crystal. Its integrated value may give a terminal voltage large enough to be observed with a low-noise radio receiver. Measurement of this voltage measures components $R_{\alpha 0\beta 0}$ of the Riemann tensor if a crystal with suitable constants is employed.

The system of stresses in the crystal is modified in a significant way if it is piezoelectric. Additional terms involving the piezoelectric constants need to be added to Equation (33). We consider a very simple example. Suppose a single longitudinal mode is excited, with sound velocity in the x^1 direction. Let the thickness in the x^2 direction be small and assume that the crystal faces normal to the x^2 direction are plated with a conductor. The piezoelectric relations (Mason, 1948) are

$$-T = \varepsilon Y_0 + DH/4\pi,$$
$$E = D/K + H\varepsilon. \tag{38}$$

In (38), T is the stress, K is the dielectric constant, ε is the strain, Y_0 is the elastic modulus, E is the electric field intensity, and D is the electric displacement. Both D and E are assumed to have components in the x^2 direction only. H is the piezoelectric constant relating open-circuit voltage to strain. A study of (38) and the equations of motion of mass elements of the crystal indicates that a wave equation similar to (34) results with

$$y = Y_0 [1 - H^2 K/4\pi] .$$

Since the crystal surface normal to the x^2 direction is plated with a conductor, $\partial E/\partial x^1 = 0$. At the free ends of the crystal, $T = 0$. If the crystal is coupled to an external impedance Z, we may write

$$-\int E \, dx^1 = Z\frac{\partial}{\partial t}\int D \, dx^1 \, dx^3 .$$

These boundary conditions and the wave equation (34) then lead to the result

$$\varepsilon = [A_1 \gamma \cosh \gamma x^1 - f\omega^{-2} \exp(-i\beta_j x^j)] e^{i\omega t}, \tag{36a}$$

where γ, β, and f are as defined earlier; this has the same form as (36), but now the constant A_1 is given in terms of the length l_1 in the x^1 direction, and lengths l_2 and

l_3 in directions x^2 and x^3 and the 'clamped' capacitance $C/4\pi$ as

$$A_1 = \frac{(f/\beta_1\omega^2)\,[2\pi\beta_1 l_2\,(1 + i\omega CZ)\,(Y_0 - H^2 K/4\pi) \times}{2\pi l_2\,(Y_0 - H^2 K/4\pi)\,[\gamma \cosh(\gamma l_1/2)]\,(1 + i\omega CZ) + i\omega Z H^2 K^2 l_3 \sinh(\gamma l_1/2)}, \qquad \begin{array}{l} \times \cos(\beta_1 l_1/2) + iH^2 K^2 l_3 \omega Z \sin(\beta_1 l_1/2)] \\ \\ \end{array}$$

$$\tag{39}$$

and the voltage which appears at the crystal terminals when coupled to an impedance Z is

$$V_Z = [2i\omega ZHKl_3/(1 + i\omega CZ)]\,[A_1 \sinh(\gamma l_1/2) - (f/\beta_1\omega^2)\sin(\beta_1 l_1/2)]. \tag{40}$$

The electrical network theorems now permit straightforward calculation of the power which can be delivered by the detector to a radio receiver. For a crystal with constants similar to polarized barium titanate on which sinusoidal gravitational waves are incident the power which can be transferred is, roughly,

$$P_A \approx 10^{-19}\,\omega^{-1}\,VQ_t t_{0r}\;\text{ergs s}^{-1} \tag{41}$$

In (41), ω is again the angular frequency and t_{0r} is the incident gravitational power flow in ergs per square centimeter per second. V is the volume of the crystal. Q_t is the Q of the crystal and associated electric circuit. A cubic meter of crystal at $\omega \sim 10^3$ gives a cross section for absorption $\sim 10^{-10}$ cm^2. While this is a small quantity it appears sufficiently large to start some experiments. For a continuous spectrum of gravitational radiation with gravitational power flow having a power spectrum function $t_{0r}(\omega)$, the power absorbed is about

$$P_A \approx 10^{-19}\,Vt_{0r}(\omega_0)\;\text{ergs s}^{-1}. \tag{42}$$

(41) and (42) provide a basis for discussion of sensitivity. In microwave spectroscopy it has been found that all spurious effects other than random fluctuations can be recognized. A similar assumption will be made here. The random fluctuations are partly thermal in origin, partly the result of spontaneous emission processes. For synchronous detection of sinusoidal waves the power output of the detector must exceed the noise (Dicke, 1946) power P_{N1} given by

$$P_{N1} = N\hbar\omega/[8\tau_A\,(e^{\hbar\omega/kT} - 1)],$$

where k is Boltzmann's constant, T is the gravitational antenna temperature, and N is the noise factor of the receiver which is expected to be less than 25 and more than 1. τ_A is the averaging time. A different expression is required if radiation with a continuous spectrum is being studied. In this case the power delivered by the detector must exceed (Dicke, 1946)

$$P_{N2} = [\pi^3\omega/(64\tau_A Q)]^{1/2}\,N\hbar\omega/[e^{\hbar\omega/kT} - 1].$$

Experiments are being planned to search for interstellar gravitational radiation* using methods described here.** For the first method the earth itself is the block of material constituting the antenna. The normal modes of the earth (about 1 cycle per hour) are excited by incident gravitational waves. This procedure is limited by the relatively low Q of the Earth and the high noise temperature of its core. The apparatus of Figure 2 is employed in the second method, in which the strains induced in the crystal

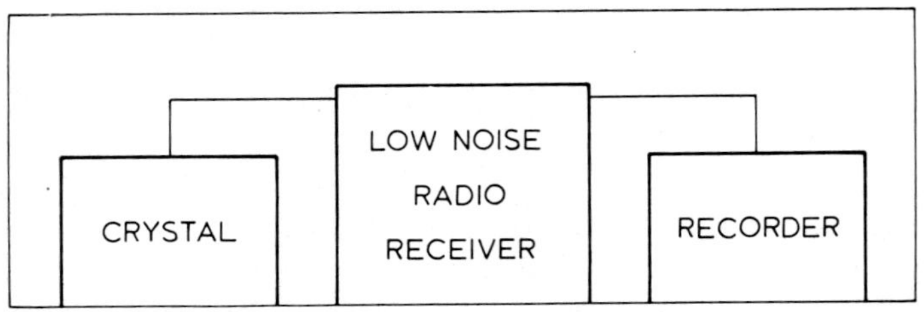

Fig. 2. Schematic diagram of piezoelectric crystal
for detection of gravitational waves.

are employed. Search at frequencies $\sim 10^3$ cycles per second is planned. The earth rotates the apparatus. If radiation is incident from some given direction it may be observed from the diurnal change in amplifier noise output. The arrangement of Figure 3 should not require rotation. If radiation is incident it will cause correlated outputs. All sources of internal fluctuations will be uncorrelated. Low-noise amplifiers such as masers (Weber, 1959) may be employed.

The discussion given here predicts that gravitational flux with a power spectrum $t_{0r}(\omega) \sim 10^{-4}$ ergs cm^{-2} s^{-1} cycle should be detectable.

* J. A. Wheeler has noted [Onzième Conseil de l'Institut International de Physique Solvay, *La Structure et l'Evolution de l'Universe* (Editions Stoops, Brussels, 1958), p. 112] that the density of gravitational radiation could be as high as 10^{-3} to 10^{-28} gm cm^{-3} ($\sim 10^3$ ergs cm^{-2} s^{-1}) and still be consistent with present information about the rate of expansion of the universe. He and M. Schwarzschild (private communication) have subsequently noted that if this radiation were set free by the same process which caused the inhomogeneous collection of matter into galaxies, it would be characterized at that time, and therefore also now, by the same scale of lengths, of the order of 10^{24} cm today (10^6 yr vibration period).

$$(\partial g_{\text{typical}}/\partial x)^2 \sim \varrho G c^{-2} \sim 0.2 \times 10^{-56} \text{ cm}^{-2},$$
$$\delta g_{\text{typical}} \sim 0.5 \times 10^{-28} \text{ cm}^{-1} \ (10^{24} \text{ cm}) \sim 10^{-4}.$$

This would appear to be not too small, but too slow to measure, by these methods.

** *Note added in proof.* – Experimental work along these lines has begun recently. It is being carried out by Dr. David M. Zipoy and Mr. Robert L. Forward, in collaboration with the author. The piezoelectric effect gives enhanced sensitivity when a mass which is many acoustic wavelengths on a side is used. At low frequencies this is not important because a mass which is one half acoustic wavelength long is already quite large and may not be obtainable as a piezoelectric crystal. Excitation of resonant acoustic vibrations in a block of metal, in accordance with expressions (36) and (37) is being considered along with the arrangements of Figures 2 and 3. Detection of the motion of the crystal ends by the change in capacitance to a nearby electrode.

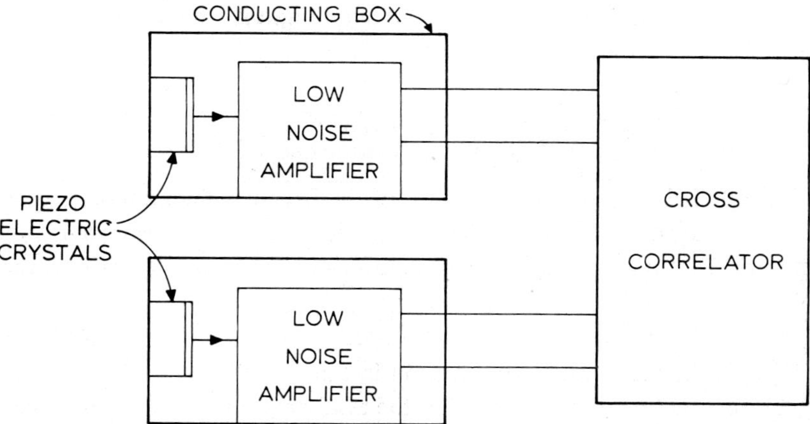

Fig. 3.   Schematic diagram of cross correlation detection of gravitational waves.

5. Rotations Induced by Gravitational Radiation

Dirac has suggested that astronomical anomalies might be correlated with effects of gravitational radiation. To discuss this and to consider detection by observing rotations we return to expression (5). Let a group of masses be situated near the space origin of a coordinate system and let the infinitesimal vector r^μ be the position vector of one of the masses. Let $v_{\mu\alpha\beta\kappa}$ be the Levi Civita tensor density. Multiply (5) by $v_{\mu\alpha\beta\kappa}r^\beta$ to obtain

$$v_{\mu\alpha\beta\kappa}r^\beta \frac{\delta}{\delta s}\frac{dx^\alpha}{ds} = \frac{d}{ds}v_{\mu\alpha\beta\kappa}r^\beta\frac{dx^\alpha}{ds} + v_{\mu\alpha\beta\kappa}r^\beta\Gamma^\alpha_{\gamma\delta}p^\gamma p^\delta = v_{\mu\alpha\beta\kappa}r^\beta F^\alpha/mc^2. \tag{43}$$

Here again p^δ is a unit vector tangent to the world lines and in the second term of (43) we have used the identity $v_{\mu\alpha\beta\kappa}p^\alpha p^\beta = 0$. Let the world line of the origin be a path for which the Christoffel symbols vanish.
Then it follows that (43) can be written

$$v_{\mu\alpha\beta\kappa}r^\beta \frac{\delta}{\delta s}\frac{dx^\alpha}{ds} = \frac{d}{ds}v_{\mu\alpha\beta\kappa}r^\beta\frac{dx^\alpha}{ds} + v_{\mu\alpha\beta\kappa}r^\beta\frac{\partial\Gamma^\alpha_{\gamma\delta}}{\partial x^\omega}p^\gamma p^\delta r^\omega. \tag{44}$$

In these coordinates $R^\alpha_{\gamma\omega\delta} = \partial\Gamma^\alpha_{\gamma\delta}/\partial x^\omega$, so (44) becomes

$$v_{\mu\alpha\beta\kappa}r^\beta \frac{\delta}{\delta s}\frac{dx^\alpha}{ds} = \frac{d}{ds}v_{\mu\alpha\beta\kappa}r^\beta\frac{dx^\alpha}{ds} - v_{\mu\alpha\beta\kappa}R^\beta_{\gamma\omega\delta}p^\gamma p^\delta r^\alpha r^\omega. \tag{45}$$

If we now use (43) and (45) and sum over all masses, we obtain

$$\sum_{\text{masses}}\frac{d}{ds}v_{\mu\alpha\beta\kappa}r^\beta\frac{dx^\alpha}{ds} = \sum_{\text{masses}}v_{\mu\alpha\beta\kappa}R^\beta_{\gamma\omega\delta}p^\gamma p^\delta r^\alpha r^\omega - \sum_{\text{masses}}v_{\mu\alpha\beta\kappa}r^\alpha F^\beta/mc^2. \tag{46}$$

(46) is a generalization of the relation between torque and change of angular momen-

tum. If there are no nongravitational forces acting and if we take the time direction tangent to the world lines (46) becomes

$$\sum_{\text{masses}} \frac{d}{ds} v_{\mu\alpha\beta 0} r^\beta \frac{dx^\alpha}{ds} = \sum_{\text{masses}} v_{\mu\alpha\beta 0} R^\beta_{0\omega 0} r^\alpha r^\omega . \tag{47}$$

We have applied (47) to the calculation of the irregular fluctuations in the period of rotation of the earth caused by incident gravitational radiation with a continuous spectrum. Under these conditions a straightforward calculation leads to the result

$$\langle I^2 \rangle_{\text{Av}}/I_\alpha^2 \approx 25\pi G t_{0r}\omega^{-2}c^{-3} . \tag{48}$$

Here $\langle I^2 \rangle_{\text{Av}}$ is the mean square fluctuation in the Earth's angular momentum; I_a is the angular momentum of rotation; t_{0r} is the total gravitational wave flux in ergs per square centimeter per second, assuming its Fourier transform to be concentrated near zero frequency; ω is the angular frequency of rotation. If we arbitrarily assume that all the earth's rotational anomalies are due to incident gravitational waves, t_{0r} is calculated to be 5×10^8 ergs per square centimeter per second. It is clear from this that the Earth's rotation is not a useful detector unless the size of the anomaly can be reduced. The other astronomical anomalies lead to larger figures.

6. Generation of Gravitational Waves

It would be very desirable to be able to generate gravitational waves with sufficient energy to be detected in the laboratory. A number of important experiments could be done.

For a spinning rod, Einstein (1916, 1918) and later Eddington (1923), gave the formula for the radiated power P_R as

$$P_R = 1.73 \times 10^{-59} I_m^2 \omega^6 \text{ ergs s}^{-1} . \tag{49}$$

Here I_m is the moment of inertia and ω is the angular frequency. ω can be increased until the rod ultimately breaks. If we write the maximum value of ω in terms of the tensile strength and express the result in terms of the elastic modulus and strain, we obtain for the length l the formula

$$l = \lambda_s (2\delta)^{1/2}/\pi . \tag{50}$$

In (50), δ is the maximum allowed strain for the material and λ_s is the wavelength of sound in the rod at the angular frequency of rupture. The implication of (50) is that the wavelength of the gravitational waves which can be radiated by a rod is at least 1 000 000 times the length of the rod. Also the moment of inertia is limited to values less than about

$$10^{-3} \varrho\lambda_s^5\delta^{5/2}/(12\pi^5) . \tag{51}$$

In (51), ϱ is again the density, and we are considering a fairly slender rod, for which

Waves one meter long could be radiated by a crystal with dimensions about fifty centimeters on a side. If it is driven just below the breaking point, each crystal would radiate $\sim 10^{-13}$ erg s^{-1}, assuming $P_{\max}$ to be its static published value. Single-crystal detectors of the type considered earlier may detect a power of about 10^{-3} erg s^{-1} at these wavelengths. A large gap therefore still exists between what can be generated and what can be detected in a small laboratory. Complex detection and generation arrays can narrow this gap. Large amounts of electrical power would have to be dissipated in crystals driven to the fracture point – perhaps 10^8 W in a crystal fifty centimeters on a side. This might well be substantially reduced if low-temperature operation can be achieved. Also one might hope that low-temperature high-frequency operation might raise the effective tensile strengths. All of these issues need careful experimental investigation. If the numbers employed earlier cannot be improved upon, it would require a crystal roughly one hundred meters on a side, and a large detection system, to generate and detect the gravitational radiation. We are not proposing that this be done. We are suggesting some investigations of crystals at low temperatures, for the purpose of exploring the possibility of improvements.

7. Conclusion

The detectors which have been proposed are sufficiently good to search for interstellar gravitational radiation. Further advances are necessary in order to generate and detect gravitational waves in the laboratory. If we compare a crystal which is excited as described above with a spinning rod of the same linear dimensions, we find that the radiation from the crystal is about seventeen orders greater and the frequency radiated by the crystal about one million times greater. If both the rod and the crystal radiate at the same frequency the crystal radiation is about thirty-nine orders greater than that of the rod. We acknowledge, with thanks, the helpful criticism of F. A. E. Pirani, P. G. Bergmann, and J. A. Wheeler. We have had very helpful discussions with R. H. Dicke.

References

Bonnor, W. B.: 1959, *Phil. Trans. Roy. Soc. London* A **251**, 233.
Dicke, R. H.: 1946, *Rev. Sci. Instr.* **17**, 268.
Dirac, P.: private communication.
Eddington, A. S.: 1923, *Proc. Roy. Soc. London* A **102**, 268.
Einstein, A.: 1916, *Sitzber. deut. Akad. Wiss. Berlin, Kl. Math. Physik und Technik* (1916), p. 688; (1918), p. 154.
Mason, W. P.: 1948, *Electromechanical Transducers and Wave Filters,* D. van Nostrand Co., Princeton, N. J., Ch. VI (sec. ed.).
Mason, W. P.: 1950, *Piezoelectric Crystals and Their Applications to Ultrasonics,* D. van Nostrand Co., Princeton, N.J., p. 64.
Pauli, W.: 1958, *Theory of Relativity,* Pergamon Press, N.Y., p. 41.
Rosen, N. and Shamir, H.: 1957, *Rev. Mod. Phys.* **29**, 429.
Synge, J. L. and Schild, A.: 1952, *Tensor Calculus,* The University of Toronto Press, Toronto, Ch. 3. (See also F. A. E. Pirani, *Helv. Phys. Acta*, Suppl. IV, p. 198.
Weber, J.: 1959, *Rev. Mod. Phys.* **31**, 681.

CONTEMPORARY PAPERS
LEADING TO THE DISCOVERY OF
GRAVITATIONALLY COLLAPSED STARS

EVIDENCE FOR X-RAYS FROM SOURCES OUTSIDE
THE SOLAR SYSTEM*†

RICCARDO GIACCONI, HERBERT GURSKY, FRANK R. PAOLINI,

and

BRUNO B. ROSSI

Data from an Aerobee rocket carrying a payload consisting of three large area Geiger counters have revealed a considerable flux of radiation in the night sky that has been identified as consisting of soft X-rays.

The entrance aperture of each Geiger counter consisted of seven individual mica windows comprising 20 cm^2 of area placed into one face of the counter. Two of the counters had windows of about 0.2-mil mica, and one counter had windows of 1.0-mil mica. The sensitivity of these detectors for X-rays was between 2 and 8 Å, falling sharply at the extremes due to the transmission of the filling gas and the opacity of the windows, respectively. The mica was coated with lampblack to prevent ultraviolet light transmission. The three detectors were disposed symmetrically around the longitudinal axis of the rocket, the normal to each detector making an angle of 55° to that axis. Thus, during flight, the normal to the detectors swept through the sky, at a rate determined by the rotation of the rocket, forming a cone of 55° with respect to the longitudinal axis. No mechanical collimation was used to limit the field of view of the detectors. Also included in the payload was an optical aspect system similar to one developed by Kupperian and Kreplin (1957). The axes of the optical sensors were normal to the longitudinal axis of the rocket. Each Geiger counter was placed in a well formed by an anticoincidence scintillation counter designed to reduce the cosmic-ray background. The experiment was intended to study fluorescence X-rays produced on the lunar surface by X-rays from the Sun and to explore the night sky for other possible sources. On the basis of the known flux of solar X-rays, we had estimated a flux from the Moon of about 0.1 to 1 photon cm^{-2} s^{-1} in the region of sensitivity of the counter.

The rocket launching took place at the White Sands Missile Range, New Mexico, at 2359 MST on June 18, 1962. The Moon was one day past full and was in the sky about 20° east of south and 35° above the horizon. The rocket reached a maximum altitude of 225 km and was above 80 km for a total of 350 s. The vehicle traveled almost due north for a distance of 120 km. Two of the Geiger counters functioned properly during the flight; the third counter apparently arced sporadically and was disregarded in the analysis. The optical aspect system functioned correctly. The rocket was spinning at 2.0 rps around the longitudinal axis. From the optical sensor data it is known that the spin axis of the rocket did not deviate from the

* Reprinted from *Physical Review Letters* 9. Original article submitted 12 October, 1962.
† The research reported in this paper was sponsored by the Air Force Cambridge Research Laboratories, Office of Aerospace Research, under Contract AF 19 (604)–8026.

H. Gursky and R. Ruffini (eds.), Neutron Stars, Black Holes and Binary X-Ray Sources, 321–328. All Rights Reserved.

vertical by more than 3°; for purposes of analysis, the spin axis is taken as pointing to zenith. The angle of rotation of the rocket corresponds with the azimuth Φ and is measured from north as zero and increasing to the east. The data were reduced by using the optical aspect information to determine the azimuth as a function of time. Each complete rotation of the rocket was divided into sixty equal intervals, and the number of counts in each of these intervals was recorded separately.

The total data accumulated in this manner during the entire flight are shown in Figure 1 for the operating Geiger counters. The observed region of the sky is shown in Figure 2. The counting rates show an altitude dependence on both the ascending and descending portions of the flight. These are shown in Figure 3, the numbers representing three-second sums. The rocket had begun tumbling during descent. The data in that portion of the flight are difficult to interpret and have not been included in the analysis.

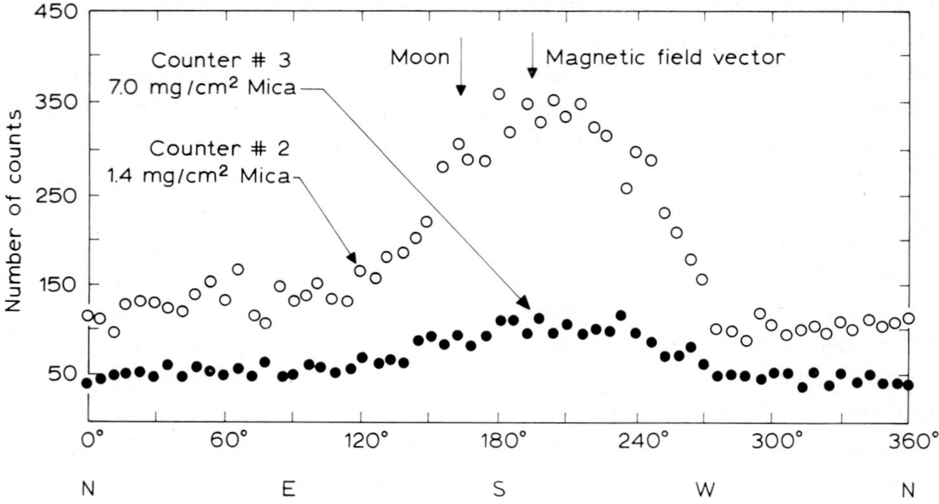

Fig. 1.   Number of counts versus azimuth angle. The numbers represent counts accumulated in 350 s in each 6° angular interval.

The residual cosmic-ray background could not be determined directly. However, the strong angular dependence of the counting rate and the large difference between the counting rates of the counters provided with windows of different thickness clearly show that most of the recorded counts are due to a strongly anisotropic and very soft radiation. Thus, the possible existence of a small cosmic-ray effect is not an essential element in the discussion of the results.

The large peak that appears at about 195° in both counters shows that part of the recorded radiation is in the form of a well collimated beam. The fact that the counting rate does not go to zero on either side of the peak shows that this beam is superimposed on a diffuse background radiation. The background radiation itself is not isotropic, but appears to have a higher intensity in directions to the east of the peak than in the direction to the west of the peak, suggesting a secondary maximum centered around 60°. The statistical significance of this conclusion may be evaluated by comparing the total number of counts recorded by counter No. 2 in an angular interval east of the maximum (from $\Phi = 102°$ to $\Phi = 18°$) with that recorded in an equivalent angular interval west of the maximum ($\Phi = 282°$ to $\Phi = 6°$). The two numbers are 2005 and 1582, respectively, yielding a difference of 423 ± 60 counts. A similar excess, although statistically less significant, appears in counter No. 3 and is 90 ± 40 counts.

At the location where the measurements were obtained, the magnetic field has an inclination of 63° and a declination of 13° east of north. Thus, the field lines are at an azimuth of 193°, which is about the same as the azimuth of the observed radiation peak. This coincidence makes one wonder whether the radiation might not consist of charged particles spiraling along the field lines. On the basis of the minimum energy necessary for the penetration of the thin- and thick-window counters, the radiation would have to consist of electrons with energies of the order of several tens of keV, or protons with energies of the order of 1 MeV. On the other hand, it would be unlikely that protons form the main component of the observed radiation, considering that they must possess a much higher energy than electrons in order to penetrate

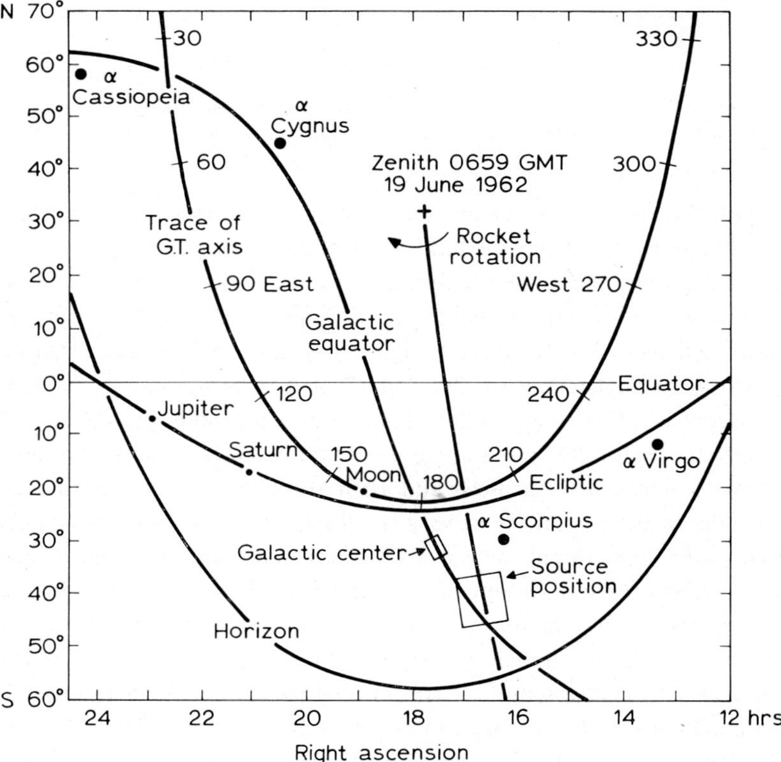

Fig. 2.   Chart showing the portion of sky explored by the counters.

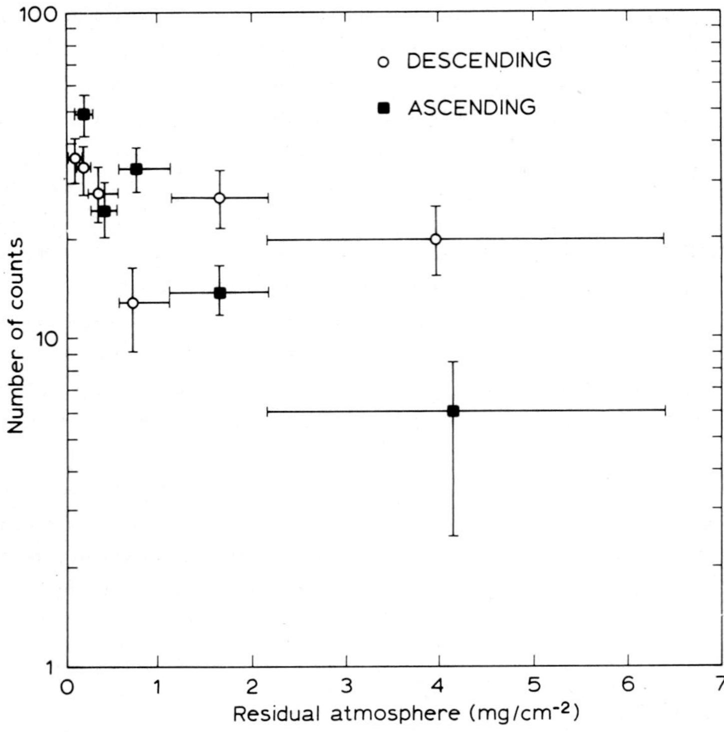

Fig. 3.   Counting rate versus residual atmosphere. The numbers represent counts accumulated in counter No. 2 in three-second intervals for azimuth angles from $102°$ to $282°$.

the windows. In any event, both protons and electrons are strongly deflected by the Earth's magnetic field and must exhibit axial symmetry with respect to the magnetic field; i.e., at a given pitch angle the flux of particles must be independent of azimuth around the field. The magnetic field makes an angle of $27°$ with the spin axis of the rocket and the detector axis makes an angle of $55°$ with the spin axis. Hence, particles moving with pitch angles between $28°$ and $82°$ would be normal to the detector axis at two different roll angles and would tend to give a double peaked distribution of counts. Particles with a $90°$ pitch angle would be detected with maximum efficiency from the north. Thus, the sharpness and azimuth of the observed peak requires that the pitch angles of these particles be very small. It is hard to find a reasonable source for particles with the required small pitch angles at the location of our measurements. In particular, particles spilling out of the inner radiation belt ought to have a broad pitch-angle distribution. One may add that it is not easy to account for the sharpness of the observed peak even under the extreme assumption that all particles had a zero pitch angle, i.e., came as a parallel beam in the direction of the field lines. The shape of the peak depends on the absorption curve of this radiation. The curve in Figure 4 represents the counting rate as a function of roll angle for a beam of particles with zero pitch angle computed under the assumption of an exponential absorption, with an absorption coefficient consistent with the relative responses of the thick-window and thin-window counters. The observed angular dependence is not consistent with the calculated curve, and it is difficult to ascribe the discrepancy to the arbitrary assumption of an exponential absorption. Moreover, it is clear that the presence of particles with finite pitch angles would broaden the predicted distribution.

It is also clear that the radiation responsible for the asymmetry of the background cannot consist of charged particles. Thus, we conclude that the bulk of the observed radiation is not corpuscular, but electromagnetic in nature.

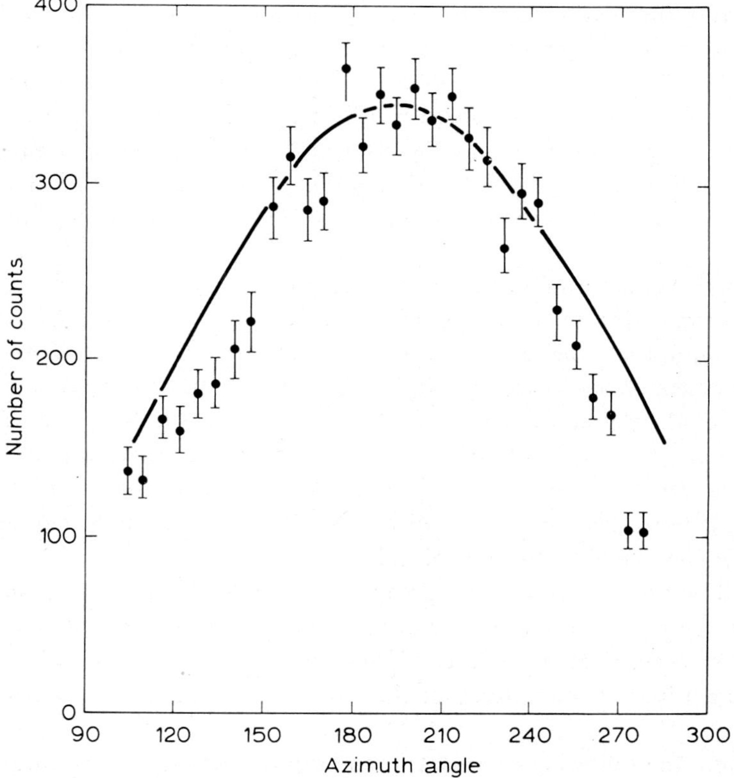

Fig. 4.   Comparison of experimental results with the computed angular dependence for a unidirectional beam of electrons exhibiting exponential absorption in the counter window.

The counters were so constructed as to be insensitive to visible or ultraviolet light. The data themselves provide a definite test on this point since a strong visible light source, the Moon, and two comparatively strong ultraviolet light sources, Virgo and presumably the Moon, went through the field of view of the counters, and yet were not detected. Thus, if the radiation is electromagnetic, it must consist of soft X-rays.

Consider first the radiation responsible for the peak and assume that it originates from a point source, and that it is monochromatic with a wavelength λ. The difference in counting rates observed in the two counters with the two different mica windows depends on the absorption coefficient $\mu_{\text{mica}}(\lambda)$ for the given radiation and on the minimum angle θ formed during the rotation of the rocket by the normal to each counter with the direction of the source. The experimental data give the product $\mu_{\text{mica}}(\lambda)\sec\theta$. The variation of the counting rate with altitude depends on the absorption coefficient $\mu_{\text{air}}(\lambda)$ for the given radiation and on the angle ψ between the horizon and the source (ψ and θ are related by a numerical constant), and yields the product $\mu_{\text{air}}(\lambda)\sec\psi$. By comparing these two pieces of information it is possible to determine values of λ and ψ for which both relations are satisfied. One obtains a λ of about 3 Å and a ψ of about $10°$.

The shape of the peak also depends on $\mu_{\text{mica}}(\lambda)$ and ψ. Use of the previously determined λ, under the assumption of a point source, allows a semi-independent determination of ψ which yields a value of $20°$. The difference between the two values of ψ can be accounted for by a source width of approximately $10°$. Thus, the peak appears to be due to a source emitting X-rays of a 3 Å wavelength whose origin is about $10°$ above the horizon. The location of this source is shown as 'source position' on the sky map in Figure 2. The measured flux from this source is 5.0 photons cm^{-2} s^{-1}.

The diffuse character of the observed background radiation does not permit a positive determination of its nature and origin. However, the apparent absorption coefficient in mica and the altitude dependence is consistent with radiation of about the same wavelength as that responsible for the peak. Assuming the source lies close to the axis of the detectors, one obtains the intensity of the X-ray background as 1.7 photons $\text{cm}^{-2}\,\text{s}^{-1}\,\text{sr}^{-1}$ and of the secondary maximum (between $102°$ and $18°$) as 0.6 photon $\text{cm}^{-2}\,\text{s}^{-1}$. In addition, there seems to be a hard component to the background of about 0.5 $\text{cm}^{-2}\,\text{s}^{-1}\,\text{sr}^{-1}$ which does not show an altitude dependence and which is not eliminated by the anticoincidence.

The question arises whether the source of the observed X radiation could be associated with the earth's atmosphere and ascribed to some form of auroral activity. The rarity of occurrence of auroras of the magnitude required to account for the observed intensities at the latitude of the measurement makes this possibility very unlikely.

In addition, the following comments are apropos. The bulk of the measurements were obtained between altitudes of 100 to 225 km and over a range of distances from the firing point of the rocket of 15 to 100 km. The variation of the measured intensity within these limits is consistent with a source at infinity. A lower limit on the position of the source places it at a distance greater than 1000 km. This number, combined with the measured elevation angle of $10°$, places a lower limit on the altitude of the source which is about 400 km above the Earth's surface. Auroral electrons of energy sufficient to produce the observed radiation would, on the other hand, penetrate down to an altitude of about 100 km and would expend the bulk of their energy at altitudes lower than 200 km. We conclude that the hypothesis of an auroral source for the observed radiation is not consistent with the data.

From Figure 2, showing the locations of the source as well as of the Moon and planets, it is clear that the observed source does not coincide with any obvious scattering body belonging to our solar system. Further, the intensity of solar X radiation at the observed wavelength is much too low in this period of the solar cycle to account for the observed intensities of the peak or of the background on the basis of backscattered solar radiation. It would thus appear that the radiation does not originate in our solar system.

From Figure 2 we see that the main apparent source is in the vicinity of the galactic center at a GT azimuthal angle of about 195°. We also see that the trace of the GT axis lies close to the galactic equator for a value of the azimuthal angle near 40°, which is the region where the background radiation is recorded with greater intensity. This apparent maximum of the background radiation is the general region of the sky where two peculiar objects – Cassiopeia A and Cygnus A – are located. It is perhaps significant that both the center of the galaxy where the main apparent source of X-rays lies, and the region of Cassiopeia A and Cygnus A where there appears to be a secondary X-ray source, are also regions of strong radio emission. Clark (1962) has pointed out that the probable mechanism for the production of the nonthermal component of the radio noise, namely, synchrotron radiation from cosmic electrons in the galactic magnetic fields, can also give rise to the X-rays we observe.

In the cosmic-ray air shower experiment presently being carried out in Bolivia,* tentative evidence has been obtained for the existence of cosmic γ rays in the energy region of 10^{14} eV at a rate of 10^{-3}–10^{-4} of the charged cosmic-ray flux at the same energy with an indication of enhanced emission in the galactic plane. Clark has shown that cosmic electrons must be produced along with γ-rays by the decay of mesons that arise in the interactions of cosmic rays with interstellar matter. Since electrons at these energies lose their energy predominantly via synchrotron radiation in the galactic magnetic field, one should observe roughly the same total energy in synchrotron radiation at the earth as in γ-ray energy. For electrons of 2×10^{14} eV in a field of 3×10^{-6} g, the peak of the synchrotron emission is at 3 Å; in a stronger field this will happen at lower electron energies. It has been shown (Strom and Strom, 1961) that X-rays in this wavelength region are not appreciably absorbed over interstellar distances.

With this one experiment it is impossible to completely define the nature and origin of the radiation we have observed. Even though the statistical precision of the measurement is high, the numerical values for the derived quantities and angles are subject to large variation depending on the choice of assumptions. However, we believe that the data can best be explained by identifying the bulk of the radiation as soft X-rays from sources outside the solar system. Synchrotron radiation by cosmic electrons is a possible mechanism for the production of these X-rays. Ordinary stellar sources could also contribute a considerable fraction of the observed radiation.

* Joint Program, MIT, University of Tokyo, University of Michigan, and University at La Paz. The MIT Cosmic Ray Group has graciously made available to us the preliminary results of the air shower experiment. We have also profited considerably from discussions with Professor M. Oda of the University of Tokyo and with Professor G. W. Clark of MIT.

We gratefully acknowledge the continued encouragement and technical assistance for this program by Dr. John W. Salisbury, Chief, Lunar and Planetary Exploration Branch of Air Force Cambridge Research Laboratories (AFCRL), and by personnel from both AFCRL and the White Sands Missile Range.

References

Clark, G. W.: 1962, unpublished memorandum.
Kupperian, J. E. and Kreplin, R. W.: 1957, *Rev. Sci. Instr.* **28**, 19.
Strom, S. E. and Strom, K. M.: 1961, *Publ. Astron. Soc. Pacific* **73**, 43.

THE FATE OF A STAR AND THE EVOLUTION OF GRAVITATIONAL ENERGY UPON ACCRETION*

YA. B. ZEL'DOVICH

It is well known that cold matter cannot resist the compressive action of gravitational attraction if its mass is greater than the mass of the sun. This result was obtained by Oppenheimer and Volkoff in 1938 in a study of the degenerate neutron gas in the Einstein theory of gravitation (the 'general theory of relativity') and is to be found in textbooks (Landau and Lifschitz, 1962). This result is not affected qualitatively by any assumptions concerning the interactions of elementary particles in the presence of high matter density.

The general theory of relativity makes a radical modification of the picture of the dynamics of condensation. In the classical theory an infinite density is attained after a finite time; one may suppose that an expansion results from this, or a shock wave arises, traveling outward from the center and ejecting a portion of the matter.

As was shown by Oppenheimer and Snyder (1939), the general theory of relativity leads to the conclusion that an infinite density is indeed reached after a finite proper time (as measured by an observer moving with any particle of the star). However, one must take the change in the time scale into account if signals are exchanged between the star and an external observer located outside the star's gravitational field. The red shift of lines emitted from the surface of a star is a special case of this change in the time variation. It turns out that for an external observer, the external surface of a star only attains the so-called gravitational (Schwarzschild) radius $r_g = 2GM/c^2$ asymptotically at $t \to \infty$. For every particle (for example, the central one) it is possible to determine the moment at which the particle must emit a signal which will arrive at the external observer as $t \to \infty$. One may speak in this sense of the gravitational self-collapse of a star (Zel'dovich, 1963a).

At the moment of emission of the signal, the density for each particle is less than the characteristic value

$$\varrho_g = \frac{3M}{4\pi r_g^3} = \frac{3}{32\pi}\frac{c^6}{M^2 G^3} = 1.8 \times 10^{16}\left(\frac{M}{M_\odot}\right)^{-2}\frac{g}{cm^3}$$

As a result, the attainment of an infinite density during the condensation is not observable, and the acquisition of information is accomplished long before this moment. Hence, the question of what happens after $\varrho = \infty$ has even less meaning.

All of these conclusions remain valid when pressure is taken into account (Podurets, 1964), even for the case of burning matter. One should regard the entropy of the matter as constant in studying the problem of the existence and stability of mechanical equilibrium in the presence of pressure and gravitational attraction. Equilibrium corresponds to a minimum of the total energy for a given entropy (and also, of course,

* Reprinted from *Soviet Physics* 9. Original article submitted 6 December, 1963.

H. Gursky and R. Ruffini (eds.), Neutron Stars, Black Holes and Binary X-Ray Sources, 329–332. All Rights Reserved.

for fixed number of conserved particles – the baryons). The entropy may be considered as constant during a rapid condensation.

For each value of entropy S there is a number of equilibrium configurations of a burning gas (the star) of different masses M; such configurations exist, however, only for masses less than the critical mass. The value of the critical mass M_c increases with increasing S; $M_c = M_c(S)$. For $S = 0$ $M_c = M_c(0) \cong M_\odot$. Stars with $M > M_\odot$ may be found in a state of mechanical equilibrium to the extent that they are burning and to the extent that, the densities being equal, the pressure of the burning matter exceeds that of the cold matter (Zel'dovich, 1963b).

As a result, the final stage in the evolution of every nonrotating star whose mass is much greater than that of the sun consists of an uncontrollable condensation, i.e., a collapse.

The brightness of a burning star drops off exponentially quite rapidly during gravitational selfcollapse, with a time constant of the order r_g/c, i.e., 10^{-4} s for $M \sim 10\ M_\odot$. As a result, condensed stars must be dark bodies which interact with the surrounding medium only via their gravitational field. Since the star's radiation, and therefore its loss in mass, during the collapse is small (Zel'dovich, 1963a; Hoyle *et al.*, 1963), the gravitational field of a condensing star at large distances is no different from the field of the same star prior to the condensation. The question has been previously raised (Zel'dovich and Smorodinskii, 1961) as to what portion of all nucleons in the universe exist at a given moment in dark condensed stars. From considerations relating to the problem of the growth of the universe, one obtains only an inequality for the over-all density $\bar{\varrho} < 2 \times 10^{-28}$ g cm^{-3}, whereas the density for normal, visible stars $\varrho \cong (0.3\text{–}1) \times 10^{-30}$ g cm^{-3}. Hoyle, Fowler, and Burbidge have decisively raised the question of whether a large number of dark stars is present. According to their estimate the mass of such stars is several times larger than the mass of the luminescent stars, which puts the mean density close to the upper limit (Zel'dovich and Smorodinskii, 1961). The interest in the catastrophic condensation of stars arose in connection with the discovery of optically intense distant radio sources (Matthews and Sandage, 1962; Schmidt, 1962, 1963), and Hoyle's hypothesis (Hoyle and Fowler, 1963) that these sources are superstars with masses of the order of $10^8\ M_\odot$. Until recently, the source of the energy of the powerful radio galaxies was not understood; ideas such as the annihilation of matter and antimatter, collisions between galaxies, and the simultaneous explosion of many supernova turned out to be untenable. How can the condensations of superstars lead to the emission of necessarily gigantic quantities of energy? According to Hoyle *et al.* (1963) the condensation is accompanied by fluctuations, whose density reaches 10^{30} g cm^{-3}; ultrarelativistic particles are emitted when the density reaches its maximum. It is impossible to agree with this point of view, merely because the break at $\varrho_m = 10^{30}$ g cm^{-3} depends on a hypothetical C-field with very strange properties (Hoyle *et al.*, 1963; Hoyle and Narlikar, 1963). For an external observer the growth in the density as a result of gravitational self-collapse stays asymptotically at a quite modest value of the order of 2–200 g cm^{-3} (for $M = 10^8\text{–}10^7$ $M_\odot$).

An alternative mechanism of energy emission is examined, in the present note, which is associated with a decrease in the surface mass in the gravitational field of a condensing star.

The velocity of a freely falling particle approaches the gravitational radius.

The relative velocity is of the order c in the collision of two particles at a distance of the order (no larger!) of the gravitational radius. The energy radiated by relativistic particles during a collision can therefore be αmc^2; the energy carried out to infinity is less because a part of the radiation and the particles falls to the star as a result of the red shift and the geometrical factor. The net amount of energy which escapes is $\alpha\beta mc^2$, where m is the mass of the particle, $\alpha<1$, $\beta<1$. This product attains a maximum which is of the order $0.1\ mc^2$ for a collision ar $r=1.5\ r_g$. It is essential that the colliding particles have different angular momenta relative to the star; the numbers given refer to the case $\mathbf{M}_1 = -\mathbf{M}_2$. If the collisions are infrequent, the energy emission is proportional to the collision cross section for a given velocity distribution of the particles far away from the star. If the cross section is large, making the mean free path small compared to the dimensions of the star, the motion of the particles is governed by collisions, making it necessary to use a hydrodynamic description of the motion. At the same time it appears that for spherically symmetric motion the gravitational energy is transformed principally into kinetic energy of motion in the radial direction; the amount of energy which can be radiated amounts to a negligibly small fraction of the rest mass of the falling matter.

However, if a flow of matter is directed against the star with a supersonic velocity at a great distance from the star, the picture of the motion is radically changed. A stationary shock wave develops on the side of the star opposite the direction of the incident matter flow. Near the star the velocity change at the front of the wave is of the order of c, and an appreciable portion of rest mass of the matter being compressed by the wave is radiated. According to Bernoulli's theorem, not even a small part of the matter can be ejected to infinity with a velocity greater than the initial velocity far from the star. But when the cloud of matter arrives at the star, surrounds it and strikes the rear side, the motion ceases to be stationary and the cumulative ejection of part of the matter with a velocity of the order of c becomes possible.

In conclusion we note that the 'particles' of which we have been speaking are not necessarily atoms and molecules, but may be localized plasma concentrations with a frozen magnetic field; the production of relativistic electrons by collisions then becomes especially probable. When examining the flow of matter from a distance one should evidently not visualize it as inter-star gas and flame with $\bar{\varrho}=10^{-25}$ g cm^{-3}, which would give a small output.

During the collapse of one member of a closely bound pair of stars, one may regard the matter of the other member as the falling material. This can be that part of the shell of the collapsing star itself which was ejected before the instant of gravitational self-collapse; in addition to the matter which acquires a hyperbolic velocity, part of the ejected matter can be stored in distant but closed orbits.

The idea of collapse in a powerful gravitational field as a source of the radiated energy of radiosources was advanced in its most general form by I. S. Shklovskii (1962).

We take this opportunity to express our sincere gratitude to I. D. Novikov and I. S. Shklovskii for numerous discussions.

References

Hoyle, F. and Fowler, W.: 1963, *Nature* **197**, 533.
Hoyle, F. and Narlikar, J. V.: 1963, *Proc. Roy. Soc.* **273**, 1352.
Hoyle, F., Fowler, W., Burbidge, G., and Burbidge, M.: 1963, 'Relativistic Astrophysics', preprint.
Landau, L. D. and Lifschitz, E. M.: 1962, *Statistical Physics* (in Russian).
Matthews, T. and Sandage, A.: 1962, *Publ. Astron. Soc. Pacific* **74**, 406.
Oppenheimer, J. and Snyder, H.: 1939, *Phys. Rev.* **56**, 455.
Oppenheimer, J. and Volkoff, G.: 1938, *Phys. Rev.* **55**, 374.
Podurets, M. A.: 1964, *Dokl. Akad. Nauk. S.S.S.R.* **154**, No. 2.
Schmidt, M.: 1962, *Astrophys. J.* **136**, 684.
Schmidt, M.: 1963, *Nature* **197**, 1040.
Shklovskii, I. S.: 1962, *Astron. Zh.* **39**, 591 (*Sov. Astron. – A.J.* **6**, 465).
Zel'dovich, Ya. B.: 1963a, *Astr. Tsikulyar*, No. 250.
Zel'dovich, Ya. B.: 1963b, *Vopr. Kosmogonii* **9**, 80.
Zel'dovich, Ya. B. and Smorodinski, Ya. A.: 1961. *ZhÉTF* **41**, 907 (*Sov. Phys. – JETP* **14**, 647).

THE NATURE OF THE X-RAY SOURCE Sco X-1*

I. S. SHKLOVSKII

Abstract. It is shown that the optical radiation of the object recently identified with the X-ray source Sco X-1 cannot represent a low-frequency extension of the radiation from the optically thin layer of hot plasma which is apparently responsible for the X-rays of this source. It is calculated that the hot plasma should be opaque in the optical frequency range, with its radiation following the Rayleigh-Jeans formula. The condition that this radiant flux be no greater than the flux of optical radiation from the object indentified with Sco X-1 yields upper and lower bounds on the size of the X-ray source of about 10^9 $(r/200)$ and 10^8 $(r/200)$ cm, respectively, where r is the distance to the source in parsecs. A 'stratified' model of the source is proposed, with the temperature rising from $\approx 10^6$ K to $\approx 5 \times 10^8$ K toward its center. It might be possible to observe the emission lines of Fe xxv and Fe xxvi in the X-ray spectrum of Sco X-1, with a substantial gravitational red shift. Evidence is given that the source is a neutron star forming a comparatively massive component of a close binary system. A stream of gas flowing out of the second component is permanently incident on the neutron star. An analysis is made of the physical conditions in the gas stream, where the H, He, N iii, and C iii emission lines originate. $N_e \approx 10^{13}$ cm^{-3} there, and the rate of mass flow through the stream is $\approx 10^{16}$–10^{17} g s^{-1}, enough to maintain the X-ray energy at the level observed. It is suggested that the optical object accompanying the X-ray source might be a cool dwarf star, with half of its surface heated by a strong flux of hard X-rays from the source.

1. Physical Characteristics of the X-Ray Source

Scorpius X-1, a very intense source of X-rays, has recently been identified with an optical object of 12–13^m (Sandage *et al.*, 1966). In all its spectroscopic, color, and photometric properties this object is very similar to a former nova. The identification has naturally attracted attention among theoreticians, and several papers have already appeared containing efforts at an interpretation (see, for example, Burbidge, 1967; Matsuoka, 1966).

One of the basic arguments for the identification suggested in (Sandage *et al.*, 1966) was considered to be the circumstance that an extrapolation to the optical frequency range of the 'flat' (for $\lambda > 10$ Å) spectrum of this source yields a radiant flux corresponding to $\approx 13^m$ (Johnson, 1966). An object of just this brightness was discovered near Sco X-1, and its color was found to be quite blue. One characteristic of the theoretical work that has been done on the source Sco X-1 is that under the assumption of bremsstrahlung from an optically thin layer of hot ($\approx 5 \times 10^7$ K) plasma, the observed X-ray flux and the estimated distance to the source (≈ 200 pc) are used to find the value of the so-called 'volume emission measure' $\frac{4}{3} R^3 N_e^2$, where R is the effective radius of the source. As for R, we only know that it is less than 10^{15} cm (corresponding to the angular size of $1''$ for the optical object). In principle, however, R could be many orders smaller than 10^{15} cm. Thus it is practically impossible to establish the physical characteristics of the source from the value found for $R^3 N_e^2$. We shall now show, however, that the belief that the radiation of the optical object identified with Sco X-1 is an extension of the bremsstrahlung from an optically thin

* Reprinted from *Soviet Astronomy* **11**. Original article submitted 28 February, 1967.

H. Gursky and R. Ruffini (eds.), Neutron Stars, Black Holes and Binary X-Ray Sources, 333–343. All Rights Reserved.

layer of hot plasma is erroneous. If this incorrect belief is repudiated, rich possibilities are opened up for understanding the real nature of Sco X-1 and, one might suppose, most other sources as well.

The following evidence argues against the theory mentioned:

(a) the flux density of the X-rays at $v_3 = 3 \times 10^{17}$ Hz ($\lambda = 10$ Å) is, according to Hayakawa *et al.* (1966), $F_{v_3} \approx 4 \times 10^{-25}$ erg cm^{-2} s^{-1} Hz^{-1}. If the Gaunt factor $gI \approx 1$ in the X-ray frequency region (see Kazachwskaya and Ivanov-Kholodnyi, 1959), then it should be taken into account in the optical frequency range. If we set

$$gI = \frac{\sqrt{3}}{\pi} \ln \left(\frac{4kT}{1.781 \cdot hv} \right)$$

(see Ginzburg, 1966), with $T = 5 \times 10^7$ K and $v_1 = 7 \times 10^{14}$ s^{-1}, we find that $gI = 5.2$. Thus if thermal X-rays had originated in an optically thin layer (even at optical frequencies!) of hot plasma, the expected value for the spectral flux density of the optical radiation would be $Fv_1 \sim 2 \times 10^{-24}$ erg cm^{-2} s^{-1} Hz^{-1}, which is ≈ 6 times as great as the observed mean value of F_{v_3}.* We note that absorption cannot be significant here. The spectrum of the optical object Sco X-1 would become flat if the color excess $CE = 0.3$, so that the total absorption would be $A \approx 0.4$, an insignificant amount.

(b) Recent observations by Friedman and his colleagues have led to the detection of fairly substantial radiation from Sco X-1 in the soft X-ray region, at $\lambda = 44$–60 Å (Byram *et al.*, 1966). According to Byram *et al.*(1966), the flux in this spectral region is $\approx 3 \times 10^{-8}$ erg cm^{-2} s^{-1}. If we make the simplest assumption, that the spectrum is flat, we find that at $v_2 = 6 \times 10^{16}$ Hz (λ 50 Å), $F_{v_2} \approx 10^{-24}$ erg cm^{-2} s^{-1} Hz^{-1}. If this radiation is regarded as thermal (there are serious difficulties with other mechanisms), then for $T \approx 10^6$ K (see below) we find, assuming that the plasma is optically thin in the visible frequency region, that the expected radiation at these frequencies would correspond to $F_{v_1} \sim 4 \times 10^{-24}$ erg cm^{-2} s^{-1} Hz^{-1}, at least 10 times as great as observed. This expected radiation would correspond to a 10^m object.

(c) The suggestion that the optical radiation of Sco X-1 amounts to the bremsstrahlung of an optically thin layer of hot plasma is contrary to the observed color variations of this object. For example, according to Sandage *et al.* (1966), in a two-week period $U - B$ varies by at least $0^m.13$. It is worth emphasizing that, as one can show, variations in the relative emission-line intensities definitely do not account for the color variations. This follows from the equivalent widths of these lines, as reported in Sandage *et al.* (1966). Thus real variations in the optical continuum are being observed, which is inconsistent with the strictly flat spectrum proposed for an optically thin layer of hot plasma.

If as previously we regard the X-rays of Sco X-1 in the range $44 < \lambda < 60$ Å as re-

---

* Since the apparent magnitudes and colors of the object identified with Sco X-1 and of 3C 273 are practically the same, we may, in accordance with Oke (1965) take the observed value to be $F_v \approx 3 \times 10^{-25}$ erg cm^{-2} s^{-1} Hz^{-1}.

presenting the bremsstrahlung of hot plasma* we could explain the absence of an optical object brighter than 10^m near this source (see point b above) by the transparency of the plasma in the optical frequency range. In this event we could write the inequality

$$\frac{2\pi k T}{\lambda_1^2} \frac{4\pi R_2^2}{4\pi r^2} < F_{v_1},\qquad(1)$$

where r is the distance from the source and R_2 is its radius. A lower and upper bound could be found for the temperature T responsible for the radiation in the frequency range v_1 to v_2 from the hot plasma. The lower bound is specified by the simple condition that F_{v_2} should be close to that observed in Byram *et al.* (1966); it follows that $T > 2 \times 10^5 \,\mathrm{K}$. On the other hand, a comparison of F_{v_2} and F_{v_3} (v_3 corresponds to $\lambda = 10$ Å) yields $T < 2 \times 10^6 \,\mathrm{K}$ as an upper bound on the plasma temperature.

We note that in the range 44–60 Å for $r > 200$ pc the absorption of X-rays by the interstellar medium should be taken into account. The total absorption in the direction toward the source Sco X-1 can be found from the profile on the 21-cm radio line as measured at this point in the sky. The profile yields the total number of hydrogen atoms in a column of unit cross section. Knowing the chemical composition of the interstellar gas and the effective cross sections for absorption by several of the most abundant atoms (H, He, O), one can with high confidence find an upper bound τ_x on the optical thickness in the range ≈ 50 Å for the source Sco X-1. The calculations made in Vainshtein *et al.* (1968) show that $\tau_x \approx 5$, whereas the optical thickness $\tau_0 \approx 1$ in the region $\lambda \approx 4000$ Å, as follows from an analysis of the color excesses of the stars in this region of the sky. An allowance for optical absorption in Equation (1) gives only a small correction. By allowing for possible absorption in the range 44–60 Å, one can lower the upper bound on T to $10^6 \,\mathrm{K}$.

Since $T > 2 \times 10^5 \,\mathrm{K}$, Equation (1) implies that for $\lambda_1 \approx 4 \times 10^5$ cm we have

$$R_2 < 10^9 \left(\frac{r}{200}\right) \mathrm{cm},\qquad(1\mathrm{a})$$

where r is expressed in parsecs.

Thus far we have discussed the radiation of Sco X-1 in the range $4000 > \lambda > 44$ Å. But what is the nature of the harder radiation of Sco X-1 in the spectral region $\lambda < 10$ Å? This radiation has repeatedly been interpreted as the bremsstrahlung of hot plasma at $T \approx 5 \times 10^7 \,\mathrm{K}$ (Burbridge, 1967; Matsuoka *et al.*, 1966). Since the optical radiation of Sco X-1 cannot be regarded as bremsstrahlung from an optically thin layer of such plasma (see above), one should acknowledge that at comparatively low frequencies

---

* If the radiation in the range $60 > \lambda > 44$ Å were synchrotron in nature, then even if there were a cutoff in the relativistic-electron energy spectrum on the low-energy side the spectrum in the frequency range v_2 to v_1 would follow the law $F_v \approx v^{1/3}$, and the flux F_{v_1} would be $\approx 3 \times 10^{-25}$ erg cm^{-2} s^{-1} Hz^{-1}. An allowance for possible interstellar absorption at v_2 (see below) would lead to even larger F_{v_1}. The assumption of a cutoff in the relativistic-electron energy spectrum clearly optimizes the situation; in all other cases F_{v_1} would be still greater.

the plasma becomes opaque. In such a case it would follow from Equation (1) that the radius of the 'hot' region $R_3 < 1.5 \times 10^8$ $(r/200)$ cm.

The idea that two small, very dense plasma clouds might exist side by side seems artificial to us. It would be far more natural to suppose that there is only one source, a hot plasma globe, whose temperature and density increase toward its center. We arrive at a model according to which the radiation of Sco X-1 in the range $\lambda < 10$ Å is due to the internal regions of the source, while the radiation in the range $\lambda > 44$ Å comes from the outer regions. By using this model one can estimate a lower bound on the source dimensions from the condition that the radiation of its inner region, having $\lambda < 10$ Å, passes without absorption through the exterior region, for which $T < 2 \times 10^6$ K. We shall write this condition as

$$\tau_2 = H_2 R_2 = \frac{1.5 \times 10^{-2}}{v^2 T^{3/2}} N_e N_i gI R_2, \tag{2}$$

where gI is the Gaunt factor, which is close to 1 for this spectral range. On the other hand, if we consider that the radiation in the range $44 < \lambda < 60$ Å arises in an optically thin plasma layer with a temperature of some hundreds of thousands of degrees, we can find the 'volume emission measure' from the value of F_{v_2}, using the standard bremsstrahlung equation:

$$\mathscr{E}_2 = N_e N_i \frac{4\pi}{3} R^3 gI \approx 10^{61} \left(\frac{r}{200}\right) \cdot e^{\tau x}. \tag{3}$$

If the plasma is opaque in the range 44–60 Å, $\mathscr{E}_2$ will be larger. It follows from Equations (2) and (3) that

$$R_2 > 3 \times 10^8 \left(\frac{r}{200}\right) \cdot e^{\tau x/2}. \tag{4}$$

A comparison of Equations (1a) and (4) indicates that the dimensions of the source Sco X-1 should be close to 5×10^8 cm if $r \approx 200$ pc, which seems to us the most probable value. With these dimensions the mean electron concentration in the outer layers of the source (where $T < 2 \times 10^6$ K) would be $N_e \approx 2 \times 10^{17}$ cm^{-3}, the total mass of plasma $\approx 3 \times 10^{20}$ g, and the content of thermal energy in the plasma $\approx 3 \times \frac{4}{3}\pi R^3 kT \approx$ $\approx 3 \times 10^{34}$ erg. This supply of thermal energy would suffice to maintain radiation of the power observed for only $\approx \frac{1}{10}$s. We therefore need a powerful and very efficient mechanism permanently heating the plasma.

The volume emission measure corresponding to the radiation in the range $\lambda < 10$ Å is $\mathscr{E}_3 \sim 2 \times 10^{59}$ cm^{-3}. It would be natural to suppose that for the hotter plasma with $T \approx 5 \times 10^7$ K, located inside the source, N_e would be greater than at the periphery. It follows that the size of this internal hot region is $< 10^8$ cm. A lower bound on the size of this region, given by the condition that at $\lambda = 10$ Å the plasma remains optically thin, would be $\approx 2 \times 10^6$ cm. Perhaps the extent of the region where $T \approx 5 \times 10^7$ K is close to 10^7 cm. In this event $N_e \approx 10^{19}$ cm^{-3}, and the total mass would be negligible, $\approx 10^{17}$ g.

We should call attention to one other circumstance. As is evident from the spectrum of Sco X-1 reproduced in Peterson and Jacobson (1966), a very flat secondary maximum occurs in the range $35 < h\nu < 50$ keV. In terms of our model for the source this spectral feature can be explained in a natural way as the thermal radiation of an even hotter and denser plasma located in the central region of the source. The temperature of this plasma would be $\approx 5 \times 10^8$ K. The volume emission measure of this innermost source might be estimated as $\approx 10^{58}$ cm^{-3}, and its diameter as $\approx 10^6$ cm.

The 'three-layer' model described above for the source Sco X-1 is of course a very crude approximation to reality. One should in fact expect a more or less continuous variation in the physical characteristics of the source with depth, with the plasma remaining transparent for sufficiently energetic X-ray quanta.

In all its properties this model, which has been obtained *solely* from an analysis of the observational data with no a priori hypotheses of any kind regarding the nature of the source, corresponds to a *neutron star in a state of accretion*. If the identification of an optical object related to a former nova with the X-ray source is correct, then a natural and moreover a highly efficient reservoir of gas for this accretion would be a stream of gas flowing permanently out of the second component of a close binary system toward the first component, a neutron star.* In this event we would evidently be observing a binary system similar to WZ Sagittae, with one of the components a neutron star. In order to investigate the physical properties of the gas stream joining the two components (as would be very important for an independent substantiation of our conclusion regarding the nature of the X-ray source), it is necessary to make an analysis of the results of the emission-line observations. This will be done in the next section. We should emphasize, however, that the model for Sco X-1 suggested above does not depend on whether the identification of this source with the optical object is valid. The only result of the optical observations of which we have made essential use is the absence of any fairly bright star near the source.

To conclude this section we wish to call attention to the following possibility, in principle, for testing our model by means of a special observing program. At a plasma temperature $T \approx 5 \times 10^7$ K, iron is the only abundant element that would not be fully ionized. The resonance 'Lα' line of Fe XXVI has a wavelength $\lambda = 1.35$ Å, and the Fe XXV line has a nearby wavelength. The radiant power in either of these lines should be

$$L_1 = \mathscr{E}_3 \cdot \alpha \beta h_{\nu_1} \int \varphi(V) Q(V) \, dV, \tag{5}$$

where $\alpha \approx 3 \times 10^{-5}$ is the cosmic abundance of iron, β is the fraction of iron ions in

---

* The hypothesis that the source Sco X-1 is a close binary system, one of whose components is a neutron star, with the hot plasma in the neighborhood of the neutron star being formed through accretion of a gas stream flowing out of the normal component, was formulated by us during discussions at the Noordwijk symposium on August 28, 1966.

the ionization state considered, $\chi_1 = h\nu_1$ is the energy of a quantum, and $\varphi(V)$ is a Maxwellian velocity distribution. The effective excitation cross section for the permitted resonance levels is, according to Peterson and Jacobson (1966), Mott and Massey (1965), Pikel'ner (1954) and Zel'dovich and Novikov (1965),

$$Q(V) = \frac{16\pi^3 l^4}{V^2 h^2} |x|^2 \ln \frac{2mV_2^2}{\chi_1}, \tag{6}$$

where $|x|^2$ is the square of the matrix element corresponding to the transition considered. Equation (6) may be rewritten by introducing in place of $|x|^2$ the oscillator strength

$$f = \frac{8\pi^2 m}{3h} \nu_1 |r|^2 = \frac{8\pi^2 m\nu_1}{h} |x|^2.$$

We will then have

$$Q = \frac{2\pi l^4}{mV^2\chi} f \ln\left(\frac{2mV^2}{\chi}\right), \tag{6a}$$

so that

$$L_1 = \alpha\mathscr{E}_3\beta \frac{4\pi l^4}{(2\pi muT)^{1/2}} f\left[l^{-\chi/kT} \ln 4 + \frac{\chi}{kT}\left\{ -Ei\left(-\frac{\chi}{kT}\right)\right\}\right]. \tag{7}$$

Taking $\mathscr{E}_3 = 10^{59}$ cm^{-3} (see above), $\beta = 0.3$, and $T = 5 \times 10^7$ K, we find

$$L = \alpha\mathscr{E}_3\beta 10^{-20} \simeq 10^{34} \text{ ergs s}^{-1}, \tag{8}$$

or several percent of the total X-ray power of the source Sco X-1. Although the optical thickness of the source for these lines may be considerable ($\approx 10^3$), the quanta will not be absorbed, but after experiencing a large number of scatterings will emerge through the wings of the profile (like ordinary Lα quanta in H II zones). The total profile width will be $\approx 6\Delta\nu_D \approx (1/500)\ \nu_0$. Thus the intensity in the line frequencies may be tens of times as great as in the adjacent continuum.

We have estimated above the dimensions of the region where $T \approx 5 \times 10^7$ K as $R_3 \approx 10^7$ cm. If the mass of the neutron star is $\approx 2M_\odot$, the gravitational red shift of the Fe XXV and Fe XXVI lines may be very substantial. In fact, we obtain

$$z = \frac{\nu - \nu_0}{\nu_0} = \frac{1}{\sqrt{1 - \dfrac{2GM}{c^2 R_3}}} - 1 = 2 \times 10^{-2}, \tag{9}$$

which is considerably more than the frequency shift due to any possible Doppler effect. Perhaps even in the near future the techniques of X-ray astronomy will have developed to the point where such a frequency displacement could be measured. This would form a decisive experiment for our model of the source Sco X-1. One should, however, keep in mind that our estimate for the expected intensity of the Fe XXV and

Fe XXVI emission lines might be too high because of a possible departure of the plasma temperature from the value 5×10^7 K we have adopted, as well as an insufficiently accurate value of $\sigma(V)$. Nevertheless, attempts to observe the Fe XXV and Fe XXVI emission lines in the spectrum of Sco X-1 would appear to be extremely interesting and promising.

2. Physical Characteristics of the Gas Stream

The presence of H, He II, N III, and C III emission lines in the spectrum of the optical object identified with Sco X-1 implies the existence there of a comparatively cool plasma whose temperature does not exceed several tens of thousands of degrees. We should furthermore emphasize that the regions where the Balmer lines are emitted might not coincide in space with the regions where the N III and C III lines are emitted. The first line corresponds to the comparatively low electron temperature of about 10^4 K, while in the emission regions of the N III and C III lines, whose excitation potentials are very high (≈ 25 eV) and which should certainly be excited by electron impact,* $T > 50000$ K. The He II emission is, of course, recombination in nature and originates in the region of the stream, where $T > 30000$–40000 K. It is readily demonstrated that at such values of T the intensity of the Balmer lines will be much smaller than that of the He II lines. According to the observations, the radiant intensity in the He II lines remains practically constant, whereas the intensity of the N III and C III lines varies very strongly. The constant intensity of the He II lines is easily explained by their recombination nature and the associated weak dependence on T. The large variability of the N III and C III lines is understandable; it arises from their high excitation potential and a variability of the ionization state of nitrogen and carbon in the stream. The variability of the H lines should be associated with variations in the degree of ionization of hydrogen in the regions where these lines are emitted.

One fundamental problem arising in the interpretation of the observational results for these lines is a determination of the conditions under which a comparatively cool plasma with a 'moderate' state of ionization can 'survive' in a field of very intense X-rays, as is found in the immediate vicinity of the source Sco X-1.

Before turning to an examination of these conditions, we shall determine from the observations the basic quantitative characteristics of the 'cool' plasma. We shall first estimate the volume emission measure $\mathscr{E}_H$ for the region emitting the Balmer lines. The observations indicate that the equivalent width of the Hβ emission line varies over the range $\Delta\lambda \approx 0.7$–$6.2$ Å (Sandage et al., 1966). Thus, knowing $\Delta\lambda$, we can find the radiant flux in the Hβ line, which for $\Delta\lambda = 2$ Å will be $F_{H\beta} = 8 \times 10^{-14}$ erg cm^{-2} s^{-1}, and the power of the source in this line becomes $L_{H\beta} = 4\pi r^2 \times F_{H\beta} = 4 \times 10^{29} (r)^2$ erg s^{-1}. On the other hand, $L_{H\beta} = 1.22 \times 10^{-25} \mathscr{E}_H$ (the plasma temperature is taken equal to 10^4 K). It follows that $\mathscr{E}_H = 3 \times 10^{54}$ cm^{-3}. In view of the variability of the

---

* These lines evidently cannot be excited by recombinations, since in view of the comparatively low abundance their intensity would be very small in comparison with the intensity of the He II lines. We note that the ionization potentials of He II, N III; and C III are similar (≈ 50 eV).

hydrogen lines we may consider that $\mathscr{E}_H \sim 10^{54}$ to 10^{55} cm^{-3}, which is millions of times smaller than $\mathscr{E}_2$ for the hot plasma in a compact source of X-rays.

For the hotter regions of the 'cool' plasma emitting the He II lines, where $T \approx 50\,000$ K, the volume emission measure $\mathscr{E}_H = 3 \times 10^{54}$ cm^{-3} [the equivalent width of the $\lambda 4686$ He II line is ≈ 3 Å, the relative abundance of hydrogen and helium is taken as 10:1, the coefficient $b_n(T)_{\text{He\,II}} = b_n(T/4)_H$, and the Einstein coefficient A_{21} for this line is 17 times as great as for Hβ]. If the temperature of the plasma emitting the He II lines is higher, say $\approx (1\text{–}3) \times 10^5$ K, then $\mathscr{E}_{He} \sim (1\text{–}3)\,10^{55}$ cm^{-5}. Finally, rough estimates yield for the region where the N III and C III lines are emitted the value $\mathscr{E}_{N,C} \sim 10^{55}$ cm^{-5}.

We shall make the natural assumption that in the region of 'cool' plasma where the N III and C III lines are emitted the corresponding ions are the most frequently encountered ionization stages of nitrogen and carbon. We shall write the condition for ionization equilibrium of these lines in the field of hard photon radiation of the source Sco X-1:

$$n_3 \int_{v_0}^{\infty} \frac{L(v)\, l^{-\tau_v} \sigma(v)}{4\pi R^2 hv} \, dv = N_e n_4 \alpha(T), \tag{10}$$

where $L(v)$ is the spectral density of the ionizing radiant power, R is the distance from the center of the source to the point in the 'cool' plasma for which the ionization equilibrium is being calculated, τ_v is the optical thickness of the stream material for the ionizing radiation, $\sigma(v)$ is the effective cross section for photoionization, $\alpha(T)$ is the recombination coefficient, and n_3 and n_4 are the concentrations of the N III, C III and N IV, C IV ions. If $n_3/n_4 \ll 1$, then since the spectrum either remains 'flat' or increases with increasing frequency, n_4/n_5 will also be $\ll 1$, and so on. In this event, the concentration of N and C in the third ionization stage would be negligibly small, and no emission lines of these ions would have been observed.

We shall take $\sigma(v) = \sigma(v_0)(v_0/v)^n$, where h$v_0$ is the ionization potential. Moreover, we shall regard the spectrum as flat, that is, $L_v = $ const. Using Equation (5), we write the condition that n_3 be the most frequently encountered ionization stage of the corresponding ion, under the assumption that the optical thickness τ_v for the ionizing radiation is small:

$$\frac{n_3}{n_4} = \frac{4\pi R^2 hn}{L_v \cdot \sigma(v_\theta)} \alpha(T)\, N_e. \tag{11}$$

For N III, with $T = 40\,000$ K, $\alpha(T) \sim 5 \times 10^{-13}$ (Pikel'ner, 1954). The exact value of $\sigma(v_0)$ is unknown. Without incurring a large error we may take it as $\approx 10^{-18}$ cm^2. If by analogy with the hydrogen atom we take $n = 3$ and introduce the value $L_v = 2 \times 10^{19}$ erg s^{-1} Hz^{-1} (which corresponds to the value of F_v at $\lambda 50$ Å, and $r \approx 200$ pc), we find

$$\frac{n_3}{n_4} \approx 2 \times 10^{39} R^2 N_e. \tag{12}$$

Equation (7) implies that for $R \approx 10^{11}$ cm, $N_e \approx 5 \times 10^{16}$ cm^{-3}. In order for the volume emission measure to be $\approx 10^{55}$ cm^{-3}, the effective size of the radiating region should be $\approx 10^7$ cm, an improbably low value. The result remains practically the same if we abandon the hypothesis $L_v = $ const and assume that for $\lambda > 60$ Å, L_v declines with decreasing frequency by the Rayleigh-Jeans law: $L_v \approx v^2$. In this event, as is readily shown, $N_e \approx 10^{16}$ cm^{-3}, and the size of the radiating region will increase to 3×10^7 cm, which is also improbable.

The situation changes if we take into account the absorption of the ionizing radiation in the stream. We shall first estimate the order of magnitude of N_e. Since the expected distance between the components of the binary system is $3 \times 10^{10} - 10^{11}$ cm and the diameter of the star from which the stream emerges is $\approx 10^{10}$ cm, its effective volume is most likely $\approx 10^{30}$ cm^3. For $N_e \approx 10^{12}$ cm^{-3}, the volume emission measure would be $\approx 10^{54}$ cm^{-3}, which is clearly inadequate, in view of the temperature dependence of the plasma radiation in individual lines. The value $N_e \approx 10^{14}$ cm^{-3} would give a volume emission measure $\approx 10^{58}$ cm^{-3}, too large a value. Hence a value $N_e \approx 10^{13}$ cm^{-3} should agree in order of magnitude with the true value of the electron concentration in the stream. Although this estimate is a rough one, it appears to be not far from reality.

With such a density and dimensions, the part of the stream located closer into the the star from which it emerges will be opaque for comparatively soft ionizing X-rays. In fact, from Equation (11), with $N_0 \approx 10^{13}$ cm^{-3}, $R \approx 3 \times 10^{10}$ cm, and $\sigma(v_0) \approx$ $\approx 3 \times 10^{-17}$ cm^2 (for He II), we find $n_{\text{HeII}}/n_{\text{HIII}} \approx 10^6$; thus, taking the relative abundance of helium and hydrogen in the stream as $1:10$, we find that the number of the He II ions in a column of unit cross section from the X-ray source to $R \approx 3 \times 10^{10}$ cm will be $0.1 \times 10^{13} \times 10^{-6} \times 3 \times 10^{10} = 3 \times 10^{16}$ cm^2, so that the optical thickness at the boundary of He II ionization will be ≈ 1. For $R \approx 5 \times 10^{10}$ cm the X-rays will already be strongly absorbed, and the ratio n_3/n_4 may become ≈ 1 for $N_e \approx 10^{13}$ cm^{-3}.

Thus, allowing for the absorption of X-rays in the stream leads naturally to the concept of a stratification of the regions where He II, N III–C III, and H are emitted. This last emission is localized in a comparatively cool part of the stream closer into the star from which it emerges. The recombination emission of He II, on the contrary, arises considerably closer to the X-ray source. One should, however, note that in the immediate vicinity of this source emission in the He II lines will already be insignificant because of the high temperature of the stream plasma. Finally, the emission in the N III–C III lines arises in a region intermediate between the hydrogen and helium emissions, overlapping partially with the latter.

We can now estimate the amount of mass flow in the stream. We shall assume that between the two components of the multiple system the velocity of the material in the stream (as determined by the attraction of the neutron star) is $\approx 10^8$ cm s^{-1}, and that the cross section is $\approx 10^{19} - 10^{20}$ cm^{-2}. If the density of the plasma in the stream is $\approx 2 \times 10^{-11}$ g cm^{-2}, the mass flow through the stream will be $10^{16} - 10^{17}$ g s^{-1}. Upon incidence on the neutron star the energy release per unit mass may attain $\approx 10^{20}$ ergs/g (Zel'dovich and Novikov, 1965). It follows that the type of mechanism we have suggested, accretion of gas on a neutron star, is fully capable of explaining the radiant power of the X-ray source Sco X-1 ($\approx 10^{36}$ ergs s^{-1}). We should point out, however, that the problem of the actual circumstances of interaction for plasma streams incident on a neutron star is a very complicated one, and will not be considered here.

Our model for the source Sco X-1 implies that it differs from other former novae in one important respect. As is well known, in these close binary systems one component is a cool star, most often a dwarf but not necessarily so (see Kraft, 1964), while the other is a small hot star, far advanced in its evolution.

In the case of Sco X-1, according to the model developed above, one component is a neutron star, while the other is a small hot star (this follows from its color).

One might, however, imagine that the second component of the Sco X-1 system is a small cool star, 'heated up' by the X-ray flux incident upon it. One can suppose that radiation with $\lambda < 10$ Å will not be absorbed appreciably in the stream. The flux of such radiation on the surface of the star turned toward the X-ray source will, for $R \approx 3 \times 10^{10}$ cm, be $\approx 5 \times 10^{13}$ ergs cm^{-2} s^{-1}, which may be four or five orders of magnitude greater than the star's own outwardly directed radiant flux. The hard X-rays, upon being absorbed in the outer layers of the star, will heat them. Thermal equilibrium, which sets in very rapidly, will be determined by the condition that the flux of intrinsic radiation of the heated outer layers be equal to the flux of incident X-rays (as in the case of planets illuminated by the Sun). The equilibrium temperature of the outer layers of a star upon which such an X-ray flux is incident will be ≈ 30000 K. If the radius of the star is $\approx 10^{10}$ cm, then for $R \approx 3 \times 10^{10}$ cm, it will 'intercept' $\approx \frac{1}{35}$ of the total flux of hard X-rays from the source. Applying the bolometric correction (3^{m} for $T = 30000$ K), with a bolometric absolute magnitude $M_{\mathrm{bol}} \approx 0$ for the source ($r = 200$ pc), we find that the star will have $M_V \approx 6$–7, as is in fact observed.

The phenomenon described above represents a variation of the 'reflection' effect in close binary systems. If it does take place, we may expect a periodic term to occur in the observed brightness and color variations of the optical object identified with Sco X-1.

We wish to point out in closing that one cannot exclude the possibility that in some former novae one of the components might be a neutron star, emitting comparatively soft X-rays through accretion of a stream. Possibly for such systems the effect described above, a heating of the surface of the second component by X-rays, might also be operative.

Note added by the editors (22 April, 1975). The soft X-Ray emission reported by Byram *et al.* (1966) from Sco X-1 in the wavelength range 44–60 Å and used in the above analysis has not been reported by other observers and was likely an incorrect result. However its absence does not materially alter the conclusion of the paper.

References

Burbidge, G. R.: 1967, in H. van Woerden (ed.), 'Radio Astronomy and the Galactic System', *IAU Symp.* **31**.

Byram, E. T., Chubb, T. A., and Friedman, H.: 1966, *Science* **153**, 1527.

Ginzburg, V. L.: 1966, *UFN* **89**, 549 (*Sov. Phys. – Uspekhi* **9**, 543).

Hayakawa, S., Matsuoka, M., and Yamashita, K.: 1966, *Rep. Ionos. Space Res., Japan* **20**, 480.

Johnson, H.: 1966, *Astrophys. J.* **142**, 635.

Kazachevskaya, T. V. and Ivanov-Kholodnyi, G. S.: 1959, *Astron. Zh.* **36**, 1022 (*Sov. Astron. – A.J.* **3**, 937).

Kraft, P. R.: 1964, *Astrophys. J.* **139**, 457.

Matsuoka, M., Oda, M., and Ogawara, Y.: 1966, *Nature* **212**, 885.

Mott, N. F. and Massey, H. S. W.: 1965, *Theory of Atomic Collisions*, Oxford Univ. Press.

Oke, J. B.: 1965, *Astrophys. J.* **141**, 6.

Peterson, L. E. and Jacobson, A. S.: 1966, *Astrophys. J.* **145**, 962.

Pikel'ner, S. B.: 1954, *Izv. Krym. Astrofiz. Obs.* **12**, 93.

Sandage, A. R. *et al.*: 1966, *Astrophys. J.* **146**, 316.

Vainshtein, L. A., Kurt, V. G., and Sheffer, E. K.: 1968, *Astron, Zh.* **45**, 237 (translation *Soviet Astron.* **12**, 189).

Zel'dovich, Ya. B. and Novikov, I. D.: 1965, *U.F.N.* **86**, 447 (*Sov. Phys. Aspekhi* **8**, 522).

OBSERVATION OF A RAPIDLY PULSATING RADIO SOURCE*

A. HEWISH, S. J. BELL, J. D. H. PILKINGTON, P. F. SCOTT,

and R. A. COLLINS

In July 1967, a large radio telescope operating at a frequency of 81.5 MHz was brought into use at the Mullard Radio Astronomy Observatory. This instrument was designed to investigate the angular structure of compact radio sources by observing the scintillation caused by the irregular structure of the interplanetary medium (Hewish *et al.*, 1964). The initial survey includes the whole sky in the declination range $-0.8° < \delta < 44°$ and this area is scanned once a week. A large fraction of the sky is thus under regular surveillance. Soon after the instrument was brought into operation it was noticed that signals which appeared at first to be weak sporadic interference were repeatedly observed at a fixed declination and right ascension; this result showed that the source could not be terrestrial in origin.

Systematic investigations were started in November and high speed records showed that the signals, when present, consisted of a series of pulses each lasting ~ 0.3 s and with a repetition period of about 1.337 s which was soon found to be maintained with extreme accuracy. Further observations have shown that the true period is constant to better than 1 part in 10^7 although there is a systematic variation which can be ascribed to the orbital motion of the Earth. The impulsive nature of the recorded signals is caused by the periodic passage of a signal of descending frequency through the 1 MHz pass band of the receiver.

The remarkable nature of these signals at first suggested an origin in terms of man-made transmissions which might arise from deep space probes, planetary radar or the reflexion of terrestrial signals from the Moon. None of these interpretations can, however, be accepted because the absence of any parallax shows that the source lies far outside the solar system. A preliminary search for further pulsating sources has already revealed the presence of three others having remarkably similar properties which suggests that this type of source may be relatively common at a low flux density. A tentative explanation of these unusual sources in terms of the stable oscillations of white dwarf or neutron stars is proposed.

1. Position and Flux Density

The aerial consists of a rectangular array containing 2048 full-wave dipoles arranged in sixteen rows of 128 elements. Each row is 470 m long in an E–W direction and the N–S extent of the array is 45 m. Phase-scanning is employed to direct the reception pattern in declination and four receivers are used so that four different declinations may be observed simultaneously. Phase-switching receivers are employed and the two halves of the aerial are combined as an E–W interferometer. Each row of dipole elements is backed by a tilted reflecting screen so that maximum sensitivity is obtained

* Reprinted from *Nature* **217**. Original article submitted 9 February, 1968.

H. Gursky and R. Ruffini (eds.), Neutron Stars, Black Holes and Binary X-Ray Sources, 344–353. All Rights Reserved.

at a declination of approximately $+30°$, the overall sensitivity being reduced by more than one-half when the beam is scanned to declinations above $+90°$ and below $-5°$. The beamwidth of the array to half intensity is about $\pm\frac{1}{2}°$ in right ascension and $\pm 3°$ in declination; the phasing arrangement is designed to produce beams at roughly $3°$ intervals in declination. The receivers have a bandwidth of 1 MHz centred at a frequency of 81.5 MHz and routine recordings are made with a time constant of 0.1 s; the r.m.s. noise fluctuations correspond to a flux density of 0.5×10^{-26} W m^{-2} Hz^{-1}. For detailed studies of the pulsating source a time constant of 0.05 s was usually employed and the signals were displayed on a multi-channel 'Rapidgraph' pen recorder with a time constant of 0.03 s. Accurate timing of the pulses was achieved by recording second pips derived from the MSF Rugby time transmissions.

A record obtained when the pulsating source was unusually strong is shown in Figure 1a. This clearly displays the regular periodicity and also the characteristic irregular variation of pulse amplitude. On this occasion the largest pulses approached a peak flux density (averaged over the 1 MHz pass band) of 20×10^{-26} W m^{-2} Hz^{-1}, although the mean flux density integrated over one minute only amounted to approximately 1.0×10^{-26} W m^{-2} Hz^{-1}. On a more typical occasion the integrated flux density would be several times smaller than this value. It is therefore not surprising that the source has not been detected in the past, for the integrated flux density falls well below the limit of previous surveys at metre wavelengths.

The position of the source in right ascension is readily obtained from an accurate measurement of the 'crossover' points of the interference pattern on those occasions when the pulses were strong throughout an interval embracing such a point. The collimation error of the instrument was determined from a similar measurement on the neighbouring source $3C$ 409 which transits about 52 min later. On the routine recordings which first revealed the source the reading accuracy was only ± 10 s and the earliest record suitable for position measurement was obtained on August 13, 1967. This and all subsequent measurements agree within the error limits. The position in declination is not so well determined and relies on the relative amplitudes of the signals obtained when the reception pattern is centred on declinations of $20°$, $23°$ and $26°$. Combining the measurements yields a position

$$\alpha_{1950} = 19\text{h } 19\text{m } 38\text{s} \pm 3\text{s}$$
$$\delta_{1950} = 22° \ 00' \pm 30'$$

As discussed here, the measurement of the Doppler shift in the observed frequency of the pulses due to the Earth's orbital motion provides an alternative estimate of the declination. Observations throughout one year should yield an accuracy of $\pm 1'$. The value currently attained from observations during December-January is $\delta = 21° 58' \pm 30'$, a figure consistent with the previous measurement.

2. Time Variations

It was mentioned earlier that the signals vary considerably in strength from day to

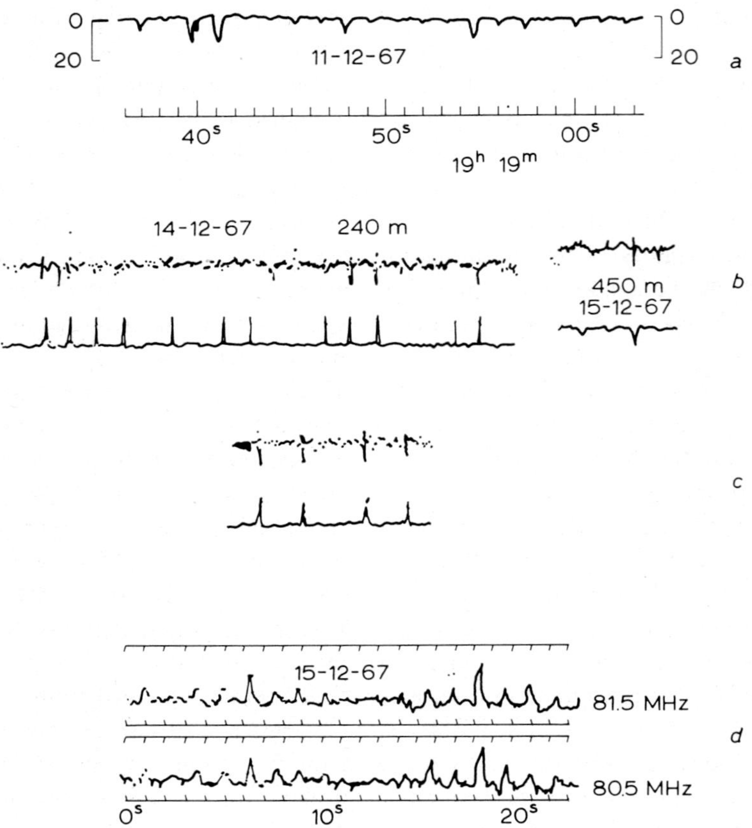

Fig. 1.   (a) A record of the pulsating radio source in strong signal conditions (receiver time constant 0.1 s). Full scale deflexion corresponds to 20×10^{-25} W m^{-2} Hz^{-1}. (b) Upper trace: records obtained with additional paths (240 m and 450 m) in one side of the interferometer. Lower trace: normal interferometer records. (The pulses are small for $l = 240$ m because they occurred near a null in the interference pattern; this modifies the phase but not the amplitude of the oscillatory response on the upper trace.) (c) Simulated pulses obtained using a signal generator. (d) Simultaneous reception of pulses using identical receivers tuned to different frequencies. Pulses at the lower frequency are delayed by about 0.2 s.

day and, typically, they are only present for about 1 min, which may occur quite randomly within the 4 min interval permitted by the reception pattern. In addition, as shown in Figure 1a, the pulse amplitude may vary considerably on a time-scale of seconds. The pulse to pulse variations may possibly be explained in terms of interplanetary scintillation (Hewish *et al.*, 1964) but this cannot account for the minute to minute variation of mean pulse amplitude. Continuous observations over periods of 30 min have been made by tracking the source with an E–W phased array in a 470 m × 20 m reflector normally used for a lunar occultation programme. The peak pulse amplitude averaged over ten successive pulses for a period of 30 min is shown in Figure 2a. This plot suggests the possibility of periodicities of a few minutes dura-

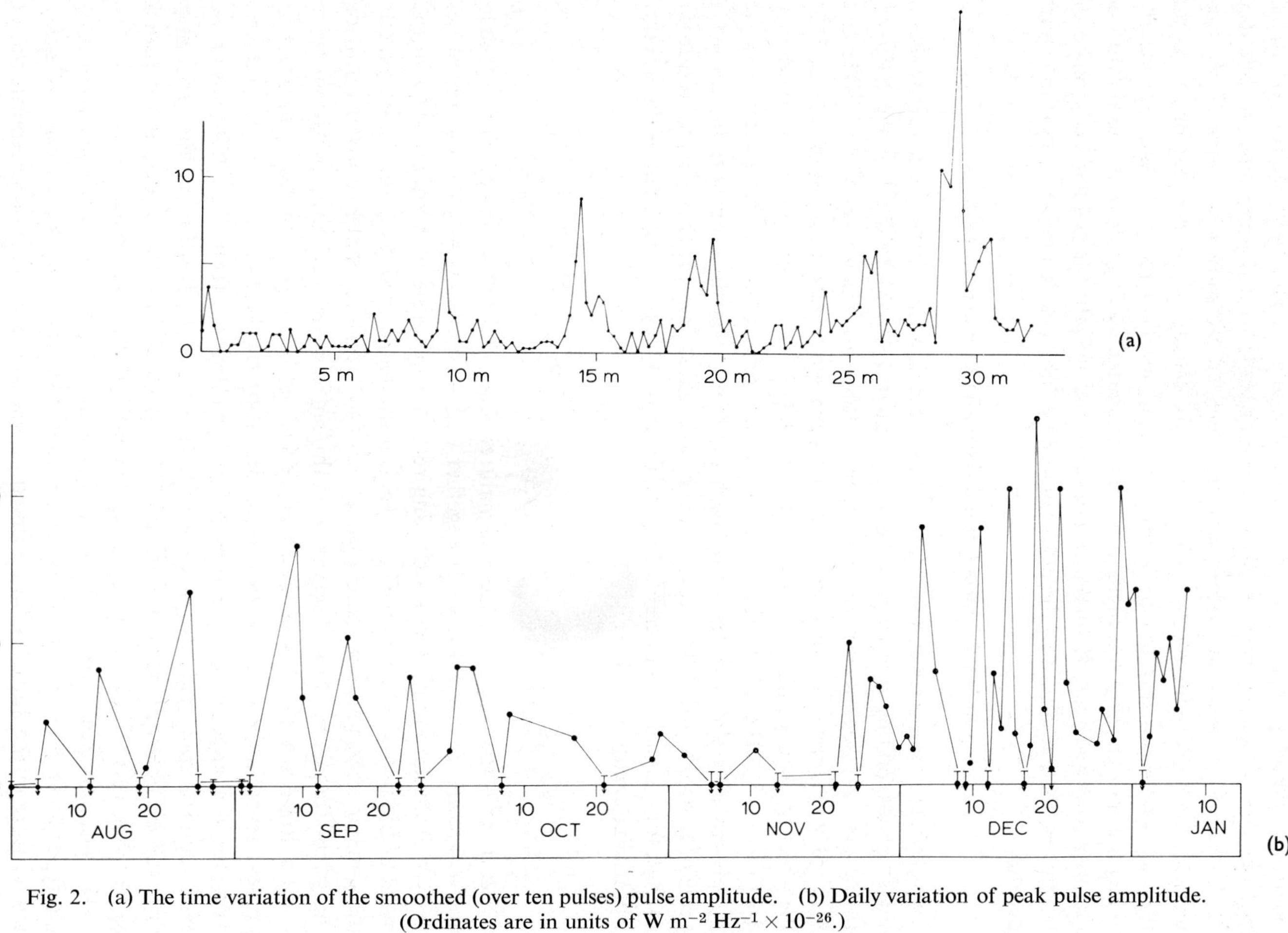

Fig. 2.   (a) The time variation of the smoothed (over ten pulses) pulse amplitude.   (b) Daily variation of peak pulse amplitude. (Ordinates are in units of $\mathrm{W\ m^{-2}\ Hz^{-1}} \times 10^{-26}$.)

tion, but a correlation analysis yields no significant result. If the signals were linearly polarized, Faraday rotation in the ionosphere might cause the random variations, but the form of the curve does not seem compatible with this mechanism. The day to day variations since the source was first detected are shown in Figure 2b. In this analysis the daily value plotted is the peak flux density of the greatest pulse. Again the variation from day to day is irregular and no systematic changes are clearly evident, although there is a suggestion that the source was significantly weaker during October to November. It therefore appears that, despite the regular occurrence of the pulses, the magnitude of the power emitted exhibits variations over long and short periods.

3. Instantaneous Bandwidth and Frequency Drift

Two different experiments have shown that the pulses are caused by a narrow-band signal of descending frequency sweeping through the 1 MHz band of the receiver. In the first, two identical receivers were used, tuned to frequencies of 80.5 MHz and 81.5 MHz. Figure 1d, which illustrates a record made with this system, shows that the lower frequency pulses are delayed by about 0.2 s. This corresponds to a frequency drift of ~ -5 MHz s^{-1}. In the second method a time delay was introduced into the signals reaching the receiver from one-half of the aerial by incorporating an extra cable of known length l. This cable introduces a phase shift proportional to frequency so that, for a signal the coherence length of which exceeds l, the output of the receiver will oscillate with period

$$t_0 = \frac{c}{l}\left(\frac{\mathrm{d}v}{\mathrm{d}t}\right)^{-1}$$

where $\mathrm{d}v/\mathrm{d}t$ is the rate of change of signal frequency. Records obtained with $l = 240$ m and 450 m are shown in Figure 1b together with a simultaneous record of the pulses derived from a separate phase-switching receiver operating with equal cables in the usual fashion. Also shown, in Figure 1c, is a simulated record obtained with exactly the same arrangement but using a signal generator, instead of the source, to provide the swept frequency. For observation with $l > 450$ m the periodic oscillations were slowed down to a low frequency by an additional phase shifting device in order to prevent severe attenuation of the output signal by the time constant of the receiver. The rate of change of signal frequency has been deduced from the additional phase shift required and is $\mathrm{d}v/\mathrm{d}t = -4.9 \pm 0.5$ MHz s^{-1}. The direction of the frequency drift can be obtained from the phase of the oscillation on the record and is found to be from high to low frequency in agreement with the first result.

The instantaneous bandwidth of the signal may also be obtained from records of the type shown in Figure 1b because the oscillatory response as a function of delay is a measure of the autocorrelation function, and hence of the Fourier transform, of the power spectrum of the radiation. The results of the measurements are displayed in Figure 3 from which the instantaneous bandwidth of the signal to exp (-1), assuming a Gaussian energy spectrum, is estimated to be 80 ± 20 kHz.

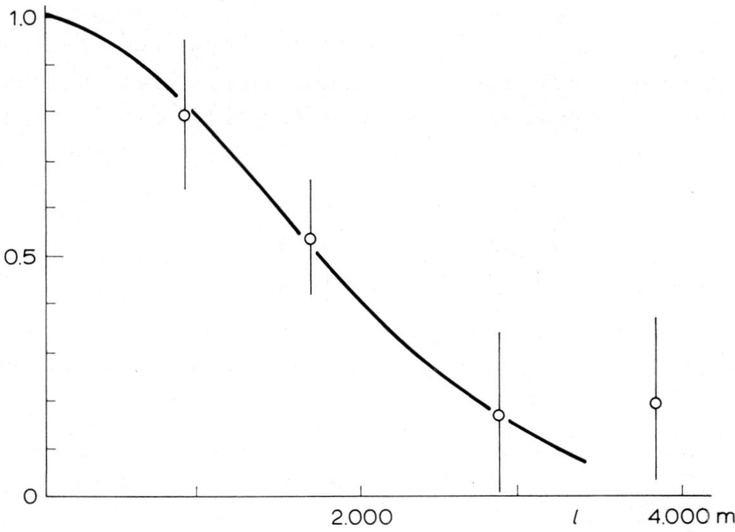

Fig. 3.   The response as a function of added path in one side of the interferometer.

4. Pulse Recurrence Frequency and Doppler Shift

By displaying the pulses and time pips from *MSF* Rugby on the same record the leading edge of a pulse of reasonable size may be timed to an accuracy of about 0.1 s. Observations over a period of 6 h taken with the tracking system mentioned earlier gave the period between pulses as $P_{obs} = 1.33733 \pm 0.00001$ s. This represents a mean value centred on December 18, 1967, at 14 h 18 m UT. A study of the systematic shift in the frequency of the pulses was obtained from daily measurements of the time interval T between a standard time and the pulse immediately following it as shown in Figure 4. The standard time was chosen to be 14 h 01 m 00 s UT on December 11 (corresponding to the centre of the reception pattern) and subsequent standard times were at intervals of 23 h 56 m 04 s (approximately one sidereal day). A plot of the variation of T from day to day is shown in Figure 4. A constant pulse recurrence frequency would show a linear increase or decrease in T if care was taken to add or subtract one period where necessary. The observations, however, show a marked curvature in the sense of a steadily increasing frequency. If we assume a Doppler shift due to the Earth alone, then the number of pulses received per day is given by

$$N = N_0\left(1 + \frac{v}{c}\cos\varphi \sin\frac{2\pi n}{366.25}\right)$$

where N_0 is the number of pulses emitted per day at the source, v the orbital velocity of the Earth, φ the ecliptic latitude of the source and n an arbitrary day number obtained by putting $n=0$ on January 17, 1968, when the Earth has zero velocity along the line of sight to the source. This relation is approximate since it assumes a circular orbit for the Earth and the origin $n=0$ is not exact, but it serves to show that the increase of N observed can be explained by the Earth's motion alone within the accuracy currently attainable. For this purpose it is convenient to estimate the values of n for which $\delta T/\delta n=0$, corresponding to an exactly integral value of N. These occur at $n_1=15.8\pm0.1$ and $n_2=28.7\pm0.1$, and since N is increased by exactly one pulse between these dates we have

$$1 = \frac{N_0 v}{c}\cos\varphi\left[\sin\frac{2\pi n_2}{366.25} - \sin\frac{2\pi n_1}{366.25}\right].$$

This yields $\varphi=43°\,36'\pm30'$ which corresponds to a declination of $21°\,58'\pm30'$, a value consistent with the declination obtained directly. The true periodicity of the

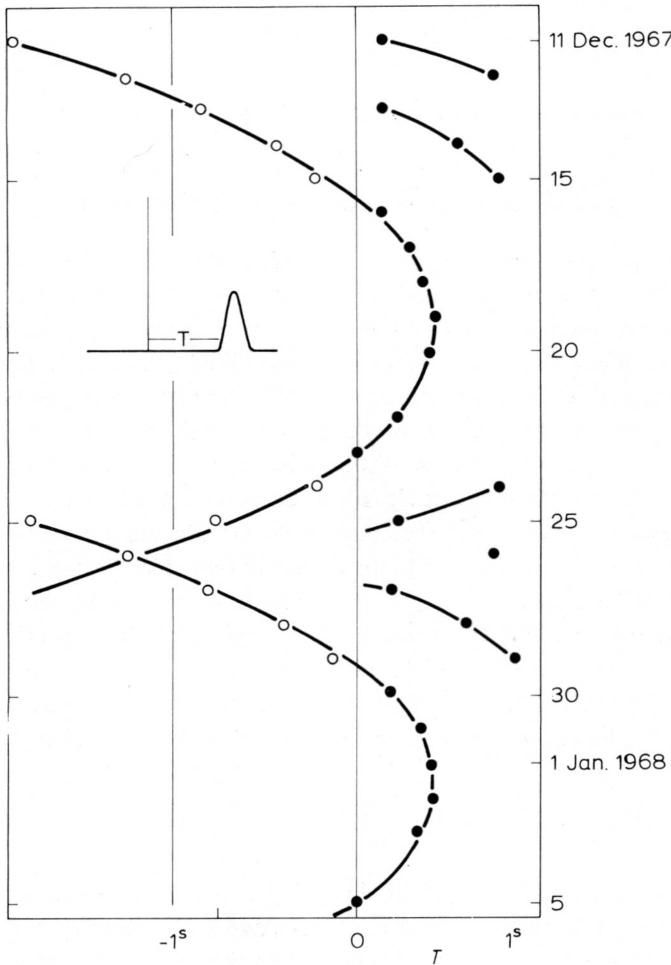

Fig. 4.   The day to day variation of pulse arrival time.

source, making allowance for the Doppler shift and using the integral condition to refine the calculation, is then

$$P_0 = 1.3372795 \pm 0.0000020 \text{ s}$$

By continuing observations of the time of occurrence of the pulses for a year it should be possible to establish the constancy of N_0 to about 1 part in 3×10^8. If N_0 is indeed constant, then the declination of the source may be estimated to an accuracy of $\pm 1'$; this result will not be affected by ionospheric refractions.

It is also interesting to note the possibility of detecting a variable Doppler shift caused by the motion of the source itself. Such an effect might arise if the source formed one component of a binary system, or if the signals were associated with a planet in orbit about some parent star. For the present, the systematic increase of N is regular to about 1 part in 2×10^7 so that there is no evidence for an additional orbital motion comparable with that of the Earth.

5. The Nature of the Radio Source

The lack of any parallax greater than about $2'$ places the source at a distance exceeding 10^3 AU. The energy emitted by the source during a single pulse, integrated over 1 MHz at 81.5 MHz, therefore reaches a value which must exceed 10^{17} erg if the source radiates isotropically. It is also possible to derive an upper limit to the physical dimension of the source. The small instantaneous bandwidth of the signal (80 kHz) and the rate of sweep (-4.9 MHz s^{-1}) show that the duration of the emission at any given frequency does not exceed 0.016 s. The source size therefore cannot exceed 4.8×10^3 km.

An upper limit to the distance of the source may be derived from the observed rate of frequency sweep since impulsive radiation, whatever its origin, will be dispersed during its passage through the ionized hydrogen in interstellar space. For a uniform plasma the frequency drift caused by dispersion is given by

$$\frac{d\nu}{dt} = - \frac{c}{L} \frac{\nu^3}{\nu_p^2}$$

where L is the path and ν_p the plasma frequency. Assuming a mean density of 0.2 electron cm^{-3} the observed frequency drift (-4.9 MHz s^{-1}) corresponds to $L \sim 65$ parsec. Some frequency dispersion may, of course, arise in the source itself; in this case the dispersion in the interstellar medium must be smaller so that the value of L is an upper limit. While the interstellar electron density in the vicinity of the Sun is not well known, this result is important in showing that the pulsating radio sources so far detected must be local objects on a galactic distance scale.

The positional accuracy so far obtained does not permit any serious attempt at optical identification. The search area, which lies close to the galactic plane, includes two twelfth magnitude stars and a large number of weaker objects. In the absence of further data, only the most tentative suggestion to account for these remarkable sources can be made.

The most significant feature to be accounted for is the extreme regularity of the pulses. This suggests an origin in terms of the pulsation of an entire star, rather than some more localized disturbance in a stellar atmosphere. In this connexion it is interesting to note that it has already been suggested[2,3] that the radial pulsation of neutron stars may play an important part in the history of supernovae and supernova remnants.

A discussion of the normal modes of radial pulsation of compact stars has recently been given by Meltzer and Thorne[4], who calculated the periods for stars with central densities in the range 10^5 to 10^{19} g cm^{-3}. Figure 4 of their paper indicates two possibilities which might account for the observed periods of the order 1 s. At a density of 10^7 g cm^{-3}, corresponding to a white dwarf star, the fundamental mode reaches a minimum period of about 8 s; at a slightly higher density the period increases again as the system tends towards gravitational collapse to a neutron star. While the fundamental period is not small enough to account for the observations the higher order modes have periods of the correct order of magnitude. If this model is adopted it is difficult to understand why the fundamental period is not dominant; such a period would have readily been detected in the present observations and its absence cannot be ascribed to observational effects. The alternative possibility occurs at a density of 10^{13} g cm^{-3}, corresponding to a neutron star; at this density the fundamental has a period of about 1 s, while for densities in excess of 10^{13} g cm^{-3} the period rapidly decreases to about 10^{-3} s.

If the radiation is to be associated with the radial pulsation of a white dwarf or neutron star there seem to be several mechanisms which could account for the radio emission. It has been suggested that radial pulsation would generate hydromagnetic shock fronts at the stellar surface which might be accompanied by bursts of X-rays and energetic electrons (Cameron, 1965; Finzi, 1965). The radiation might then be likened to radio bursts from a solar flare occurring over the entire star during each cycle of the oscillation. Such a model would be in fair agreement with the upper limit of $\sim 5 \times 10^3$ km for the dimension of the source, which compares with the mean value of 9×10^3 km quoted for white dwarf stars by Greenstein (Greenstein, 1958). The energy requirement for this model may be roughly estimated by noting that the total energy emitted in a 1 MHz band by a type III solar burst would produce a radio flux of the right order if the source were at a distance of $\sim 10^3$ AU. If it is assumed that the radio energy may be related to the total flare energy ($\sim 10^{32}$ erg) (Fichtel and McDonald, 1967) in the same manner as for a solar flare and supposing that each pulse corresponds to one flare, the required energy would be $\sim 10^{39}$ erg yr^{-1}; at a distance of 65 pc the corresponding value would be $\sim 10^{47}$ erg yr^{-1}. It has been estimated that a neutron star may contain $\sim 10^{51}$ erg in vibrational modes so the energy requirement does not appear unreasonable, although other damping mechanisms are likely to be important when considering the lifetime of the source (Meltzer and Thorne, 1966).

The swept frequency characteristic of the radiation is reminiscent of type II and type III solar bursts, but it seems unlikely that it is caused in the same way. For a white dwarf or neutron star the scale height of any atmosphere is small and a travelling disturbance would be expected to produce a much faster frequency drift than is actually observed. As has been mentioned, a more likely possibility is that the impulsive radiation suffers dispersion during its passage through the interstellar medium.

More observational evidence is clearly needed in order to gain a better understanding of this strange new class of radio source. If the suggested origin of the radiation is confirmed further study may be expected to throw valuable light on the behaviour of compact stars and also on the properties of matter at high density.

We thank Professor Sir Martin Ryle, Dr J. E. Baldwin, Dr P. A. G. Scheuer and Dr J. R. Skakeshaft for helpful discussions and the Science Research Council who financed this work. One of us (S. J. B.) thanks the Ministry of Education of Northern Ireland and another (R. A. C.) the SRC for a maintenance award; J. D. H. P. thanks ICI for a research fellowship.

References

Cameron, A. G. W.: 1965, *Nature* **205**, 787.
Fichtel, C. E. and McDonald, F. B.: 1967, *Ann. Rev. Astron. Astrophys.* **5**, 351.
Finzi, A.: 1965, *Phys. Rev. Letters* **15**, 599.
Greenstein, J. L.: 1958, in *Handbuch der Physik* 161.
Hewish, A., Scott, P. F., and Wills, D.: 1964, *Nature* **203**, 1214.
Meltzer, D. W. and Thorne, K. S.: 1966, *Astrophys. J.* **145**, 514.

ROTATING NEUTRON STARS AS THE ORIGIN OF THE PULSATING RADIO SOURCES*

T. GOLD

Abstract. The constancy of frequency in the recently discovered pulsed radio sources can be accounted for by the rotation of a neutron star. Because of the strong magnetic fields and high rotation speeds, relativistic velocities will be set up in any plasma in the surrounding magnetosphere, leading to radiation in the pattern of a rotating beacon.

The case that neutron stars are responsible for the recently discovered pulsating radio sources (Hewish *et al.*, 1968; Pilkington *et al.*, 1968; Drake *et al.*, 1968; Drake, 1968; Drake and Craft, 1968; Tanenbaum *et al.*, 1968) appears to be a strong one. No other theoretically known astronomical object would possess such short and accurate periodicities as those observed, ranging from $1 \cdot 33$ to $0 \cdot 25$ s. Higher harmonics of a lower fundamental frequency that may be possessed by a white dwarf have been mentioned; but the detailed fine structure of several short pulses repeating in each repetition cycle makes any such explanation very unlikely. Since the distances are known approximately from interstellar dispersion of the different radio frequencies, it is clear that the emission per unit emitting volume must be very high; the size of the region emitting any one pulse can, after all, not be much larger than the distance light travels in the few milliseconds that represent the lengths of the individual pulses. No such concentrations of energy can be visualized except in the presence of an intense gravitational field.

The great precision of the constancy of the intrinsic period also suggests that we are dealing with a massive object, rather than merely with some plasma physical configuration. Accuracies of one part in 10^8 belong to the realm of celestial mechanics of massive objects, rather than to that of plasma physics.

It is a consequence of the virial theorem that the lowest mode of oscillation of a star must always have a period which is of the same order of magnitude as the period of the fastest rotation it may possess without rupture. The range of $1 \cdot 5$ s to $0 \cdot 25$ s represents periods that are all longer than the periods of the lowest modes of neutron stars. They would all be periods in which a neutron star could rotate without excessive flattening. It is doubtful that the fundamental frequency of pulsation of a neutron star could ever be so long (Thorne and Ipser, 1968 and unpublished work of A. G. W. Cameron). If the rotation period dictates the repetition rate, the fine structure of the observed pulses would represent directional beams rotating like a lighthouse beacon. The different types of fine structure observed in the different sources would then have to be attributed to the particular asymmetries of each star (the 'sunspots', perhaps). In such a model, time variations in the intensity of emission will have no effect on the precise phase in the repetition period where each pulse appears; and this is indeed a striking observational fact. A fine structure of pulses could be generated within the repetition period, depending only on the distribution of emission regions around the

* Reprinted from *Nature* **218**. Original article submitted 20 May, 1968.

H. Gursky and R. Ruffini (eds.), Neutron Stars, Black Holes and Binary X-Ray Sources, 354–356. All Rights Reserved.

circumference of the star. Similarly, a fine structure in polarization may be generated, for each region may produce a different polarization or be overlaid by a different Faraday-rotating medium. A single pulsating region, on the other hand, could scarcely generate a repetitive fine structure in polarization as seems to have been observed now (Lyne and Smith, 1968).

There are as yet not really enough clues to identify the mechanism of radio emission. It could be a process deriving its energy from some source of internal energy of the star, and thus as difficult to analyse as solar activity. But there is another possibility, namely, that the emission derives its energy from the rotational energy of the star (very likely the principal remaining energy source), and is a result of relativistic effects in a co-rotating magnetosphere.

In the vicinity of a rotating star possessing a magnetic field there would normally be a co-rotating magnetosphere. Beyond some distance, external influences would dominate, and co-rotation would cease. In the case of a fast rotating neutron star with strong surface fields, the distance out to which co-rotation would be enforced may well be close to that at which co-rotation would imply motion at the speed of light. The mechanism by which the plasma will be restrained from reaching the velocity of light will be that of radiation of the relativistically moving plasma, creating a radiation reaction adequate to overcome the magnetic force. The properties of such a relativistic magnetosphere have not yet been explored, and indeed our understanding of relativistic magneto-hydrodynamics is very limited. In the present case the coupling to the electromagnetic radiation field would assume a major role in the bulk dynamical behaviour of the magnetosphere.

The evidence so far shows that pulses occupy about $\frac{1}{30}$ of the time of each repetition period. This limits the region responsible to dimensions of the order of $\frac{1}{30}$ of the circumference of the 'velocity of light circle'. In the radial direction equally, dimensions must be small; one would suspect small enough to make the pulse rise-times comparable with or larger than the flight time of light across the region that is responsible. This would imply that the radiation emanates from the plasma that is moving within 1 per cent of the velocity of light. That is the region of velocity where radiation effects would in any case be expected to become important.

The axial asymmetry that is implied needs further comment. A magnetic field of a neutron star may well have a strength of 10^{12} g at the surface of the 10 km object. At the 'velocity of light circle', the circumference of which for the observed periods would range from 4×10^{10} to 0.75×10^{10} cm, such a field will be down to values of the order of 10^3–10^4 g (decreasing with distance slower than the inverse cube law of an undisturbed dipole field. A field pulled out radially by the stress of the centrifugal force of a whirling plasma would decay as an inverse square law with radius). Asymmetries in the radiation could arise either through the field or the plasma content being non-axially symmetric A skew and non-dipole field may well result from the explosive event that gave rise to the neutron star; and the access to plasma of certain tubes of force may be dependent on surface inhomogeneities of the star where sufficiently hot or energetic plasma can be produced to lift itself away from the intense gravitational field

(10–100 MeV for protons; much less for space charge neutralized electron-positron beams).

The observed distribution of amplitudes of pulses makes it very unlikely that a modulation mechanism can be responsible for the variability (unpublished results on P. A. G. Scheuer and observations made at Cornell's Arecibo Ionospheric Observatory) but rather the effect has to be understood in a variability of the emission mechanism. In that case the observed very sharp dependence of the instantaneous intensity on frequence (1 MHz change in the observation band gives a sub-stantially different pulse amplitude) represents a very narrow-band emission mechanism, much narrower than synchrotron emission, for example. A coherent mechanism is then indicated, as is also necessary to account for the intensity of the emission per unit area that can be estimated from the lengths of the sub-pulses. Such a coherent mechanism would represent non-uniform static configurations of charges in the relativistically rotating region. Non-uniform distributions at rest in a magnetic field are more readily set up and maintained than in the case of high individual speeds of charges, and thus the configuration discussed here may be particularly favourable for the generation of a coherent radiation mechanism.

If this basic picture is the correct one it may be possible to find a slight, but steady, slowing down of the observed repetition frequencies. Also, one would then suspect that more sources exist with higher rather than lower repetition frequency, because the rotation rates of neutron stars are capable of going up to more than 100/s, and the observed periods would seem to represent the slow end of the distribution.

Work in this subject at Cornell is supported by a contract from the US Office of Naval Research.

References

Drake, F. D.: 1968, *Science* **160**, 416.
Drake, F. D. and Craft, H. D., Jr.: 1968, *Science* **160**, 758.
Drake, F. D., Gundemann, E. J., Jauncey, D. L., Comella, J. M., Zeissig, G. A., and Craft, H. D., Jr.: 1968, *Science* **160**, 503.
Hewish, A., Bell, S. J., Pilkington, J. D. H., Scott, P. F., and Collins, R. A.: 1968, *Nature* **217**, 709.
Lyne, A. G. and Smith, F. G.: 1968, *Nature* **218**, 124.
Pilkington, J. D. H., Hewish, A., Bell, S. J., and Cole, T. W.: 1968, *Nature* **218**, 126.
Tanenbaum, B. S., Zeissig, G. A., and Drake, F. D.: 1968, *Science* **170**, 760.
Thorne, K. S. and Ipser, J. R.: 1968, *Astrophys. J.* **152**, L71.

GRAVITATIONAL COLLAPSE:
THE ROLE OF GENERAL RELATIVITY*

R. PENROSE

Stars whose masses are of the same order as that of the Sun $(M_\odot)$ can find a final equilibrium state either as a white dwarf or, apparently, (after collapse and ejection of material) as a neutron star. These matters have been nicely discussed in the lectures of Hewish and Salpeter. But, as they have pointed out, for larger masses no such equilibrium state appears to be possible. Indeed, many stars are observed to have masses which are much larger than $M_\odot$ – so large that it seems exceedingly unlikely that they can ever shed sufficient material so as to be able to fall below the limit required for a stable *white dwarf* ($\sim 1.3\, M_\odot$: Chandrasekhar [1]) or neutron star ($\sim 0.7\, M_\odot$: Oppenheimer and Volkoff [2]) to develop. We are thus driven to consider the consequences of a situation in which a star collapses right down to a state in which the effects of general relativity become so important that they eventually dominate over all other forces.

I shall begin with what I think we may now call the 'classical' collapse picture as presented by general relativity. Objections and modifications to this picture will be considered afterwards. The main discussion is based on Schwarzschild's solution of the Einstein vacuum equations. This solution represents the gravitational field exterior to a spherically symmetrical body. In the original Schwarzschild co-ordinates, the metric takes the familiar form

$$\mathrm{d}s^2 = (1 - 2m/r)\,\mathrm{d}t^2 - (1 - 2m/r)^{-1}\,\mathrm{d}r^2 - r^2(\mathrm{d}\theta^2 + \sin^2\theta\,\mathrm{d}\varphi^2). \qquad (1)$$

Here θ and φ are the usual spherical polar angular co-ordinates. The radial co-ordinate r has been chosen so that each sphere $r =$ const, $t =$ const has intrinsic surface area $4\pi r^2$. The choice of time co-ordinate t is such that the metric form is invariant under $t \rightarrow t + $ const and also under $t \rightarrow -t$. The static nature of the space-time is thus made manifest in the formal expression for the metric. The quantity m is the mass of the body, where 'general-relativistic units' are chosen, so that

$$c = G = 1,$$

that is to say, we translate our units according to

$$1\,s = 3 \times 10^{10}\,\mathrm{cm} = 4 \times 10^{38}\,\mathrm{g}.$$

When $r = 2m$, the metric form (1) breaks down. The radius $r = 2m$ is referred to as the *Schwarzschild radius* of the body.

Let us imagine a situation in which the collapse of a spherically symmetrical (non-rotating) star takes place and continues until the surface of the star approaches the Schwarzschild radius. So long as the star remains spherically symmetrical, its *external*

* Reprinted from *Rivesta Nuovo Cimento*, Serie I, Vol. 1.

H. Gursky and R. Ruffini (eds.), Neutron Stars, Black Holes and Binary X-Ray Sources, 357–378. All Rights Reserved.

field remains that given by the Schwarzschild metric (1). The situation is depicted in Figure 1. Now the particles at the surface of the star must describe *timelike* lines. Thus, from the way that the 'angle' of the light cones appears to be narrowing down near $r = 2m$, it would seem that the surface of the star can never cross to within the $r = 2m$ region. However, this is misleading. For suppose an observer were to follow the surface of the star in a rocket ship, down to $r = 2m$. He would find (assuming that the collapse does not differ significantly from free fall) that the total proper time that

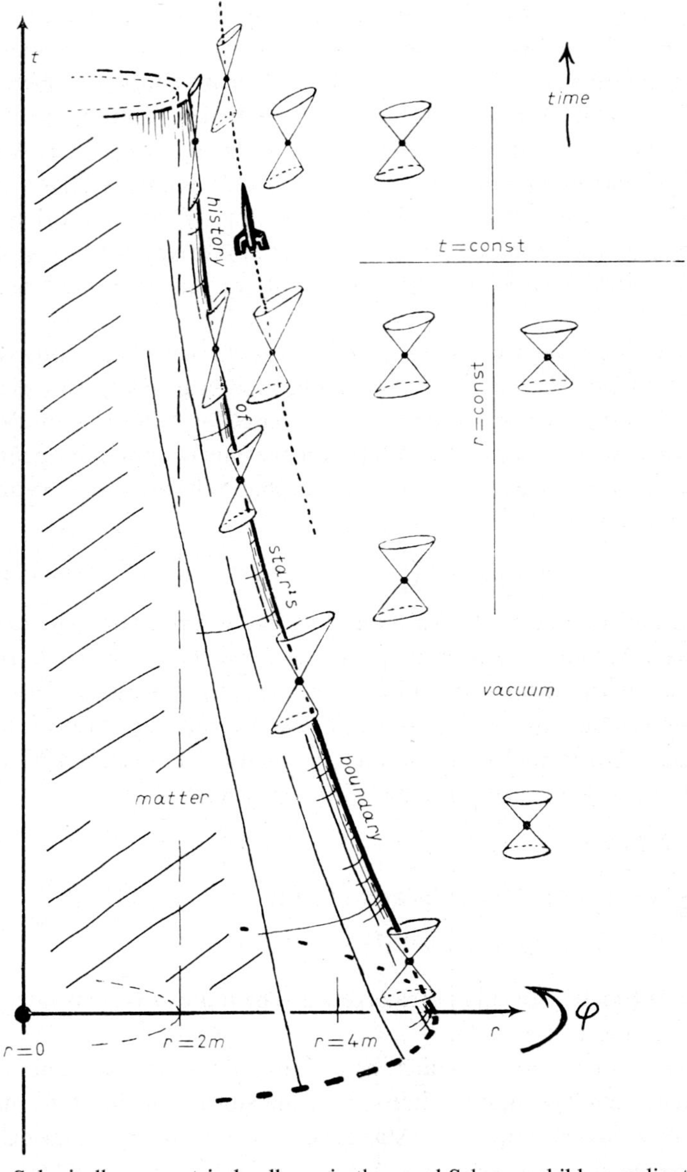

Fig. 1.   Spherically symmetrical collapse in the usual Schwarzschild co-ordinates.

he would experience as elapsing, as he finds his way down to $r = 2m$, is in fact *finite*. This is despite the fact that the world line he follows has the *appearance* of an 'infinite' line in Figure 1. But what does the observer experience after this finite proper time has elapsed? Two possibilities which suggest themselves are: (i) the observer encounters some form of space-time singularity – such as infinite tidal forces – which inevitably destroys him as he approaches $r = 2m$; (ii) the observer enters some region of space-time not covered by the (t, r, θ, φ) co-ordinate system used in (1). (It would be unreasonable to suppose that the observer's experiences could simply *cease* after some finite time, without his encountering some form of violent agency.)

In the present situation, in fact, it is possibility ii) which occurs. The easiest way to see this is to replace the co-ordinate t by an advanced time parameter v given by

$$v = t + r + 2m \log(r - 2m),$$

whereby the metric (1) is transformed to the form (Eddington [3], Finkelstein [4])

$$\mathrm{d}s^2 = (1 - 2m/r)\,\mathrm{d}v^2 - 2\,\mathrm{d}r\,\mathrm{d}v - r^2(\mathrm{d}\theta^2 + \sin^2\theta\,\mathrm{d}\varphi^2). \tag{2}$$

This form of metric has the advantage that it does not become inapplicable at $r = 2m$. The whole range $0 < t < \infty$ is encompassed in a nonsingular fashion by (2). The part $r > 2m$ agrees with the part $r > 2m$ of the original expression (1). But now the region has been extended inwards in a perfectly regular way across $r = 2m$ and right down towards $r = 0$.

The situation is as depicted in Figure 2. The light cones tip over more and more as we approach the centre. In a sense we can say that the gravitational field has become so strong, within $r = 2m$, that even light cannot escape and is dragged inwards towards the centre. The observer on the rocket ship, whom we considered above, crosses freely from the $r > 2m$ region into the $0 < r < 2m$ region. He encounters $r = 2m$ at a perfectly finite time, according to his own local clock, and he experiences nothing special at that point. The spacetime there is locally Minkowskian, just as it is everywhere else $(r > 0)$.

Let us consider another observer, however, who is situated far from the star. As we trace the light rays from his eye, back into the past towards the star, we find that they cannot cross into the $r < 2m$ region after the star has collapsed through. They can only intersect the star at a time *before* the star's surface crosses $r = 2m$. No matter how long the external observer waits, he can always (in principle) still see the surface of the star as it *was* just before it plunged through the Schwarzschild radius. In practice, however, he would soon see nothing of the star's surface – only a 'black hole' – since the observed intensity would die off exponentially, owing to an infinite red shift.

But what will be the fate of our original oberver on the rocket ship? After crossing the Schwarzschild radius, he finds that he is compelled to enter regions of smaller and smaller r. This is clear from the way the light cones tip over towards $r = 0$ in Figure 2, since the observer's world line must always remain a timelike line. As r decreases, the space-time curvature mounts (in proportion to r^{-3}), becoming theoretically infinite

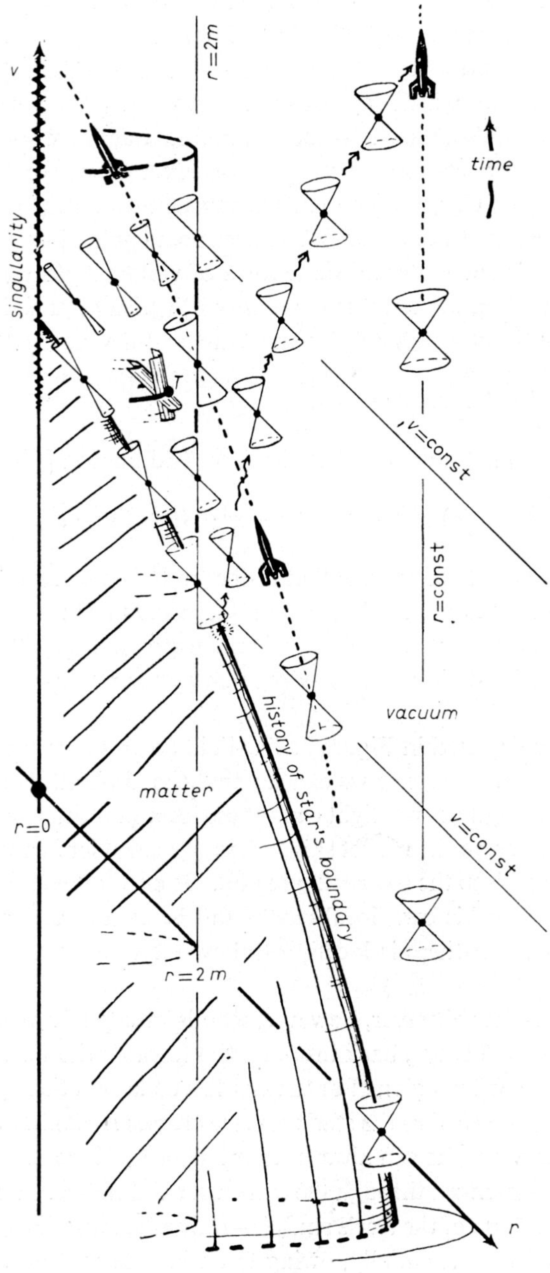

Fig. 2.   Spherically symmetrical collapse in Eddington-Finkelstein co-ordinates.

at $r=0$. The physical effect of space-time curvature is experienced as a *tidal force*: objects become squashed in one direction and stretched in another. As this tidal effect mounts to infinity, our observer must eventually* be torn to pieces – indeed, the very atoms of which he is composed must ultimately individually share this same fate!

Thus, the true *space-time singularity*, resulting from a spherically symmetrical collapse, is located not at $r=2m$, but at $r=0$. Although the hypersurface $r=2m$ has, in the past, itself been frequently referred to as the 'Schwarzschild singularity', this is really a misleading terminology since $r=2m$ is a singularity merely of the t co-ordinate used in (1) and not of the space-time geometry. More appropriate is the term 'event horizon', since $r=2m$ represents the absolute boundary of the set of all events which can be observed in principle by an external inertial observer. The term 'event horizon' is used also in cosmology for essentially the same concept (cf. Rindler [5]). In the present case the horizon is less observer-dependent than in the cosmological situations, so I shall tend to refer to the hypersurface $r=2m$ as the *absolute event horizon*** of the space-time (2).

This, then, is the standard spherically symmetrical collapse picture presented by general relativity. But do we have good reason to trust this picture? Need we believe that it necessarily accords, even in its essentials, with physical reality? Let me consider a number of possible objections:

(a) densities in excess of nuclear densities inside,
(b) exact vacuum assumed outside,
(c) zero net charge and zero magnetic field assumed,
(d) rotation excluded,
(e) asymmetries excluded,
(f) possible λ-term not allowed for,
(g) quantum effects not considered,
(h) general relativity a largely untested theory,
(i) no apparent tie-up with observations.

As regards (a), it is true that for a body whose mass is of the order of $M_\odot$, its surface would cross $r=2m$ only after nuclear densities had been somewhat exceeded. It may be argued, then, that too little is understood about the nature of matter at such densities for us to be at all sure how the star would behave while still outside $r=2m$. But this is not really a significant consideration for our general discussion. It could be of relevance only for the least massive collapsing bodies, if at all. For, the larger the mass involved, the smaller would be the density at which it would be expected to cross $r=2m$. It could be that very large masses indeed may become involved in gravitational

---

* In fact, if m is of the order of a few solar masses, the tidal forces would already be easily large enough to kill a man in free fall, even at $r=2m$. But for $m>10^8\,M_\odot$ the tidal effect at $r=2m$ would be no greater than the tidal effect on a freely falling body near the Earth's surface.
** In a general space-time with a well-defined external future infinity, the absolute event horizon would be defined as the boundary of the union of all timelike curves which escape to this external future infinity. In the terminology of Penrose [6], if $\mathcal{M}$ is a weakly asymptotically simple space-time, for example, then the absolute event horizon in $\mathcal{M}$ is $\dot{I}_-[\mathscr{I}^+]$.

collapse. For $m > 10^{11} M_\odot$ (e.g. a good-sized galaxy), the averaged density at which $r = 2m$ is crossed would be less than that of air!

The objections (b), (c), (d), (e) and, to some extent, (f) can all be partially handled if we extract, from Figure 2, only that essential qualitative piece of information which characterizes the solution (2) as describing a collapse which has passed a 'point of no return'. I shall consider this in more detail shortly. The upshot will be that if a collapse situation develops in which deviations from (2) near $r = 2m$ at one time are not too great, then two consequences are to be inferred as to the subsequent behaviour. In the first instance an absolute event horizon will arise. Anything which finds itself inside this event horizon will not be able to send signals to the outside worlds. Thus, in this respect at least, the qualitative nature of the '$r = 2m$' hypersurface in (2) will remain. Similarly, an analogue of the physical singularity at $r = 0$ in (2) will still develop in these more general situations. That is to say, we know from rigorous theorems in general-relativity theory that there must be *some* space-time singularity resulting inside the collapse region. However, we do not know anything about the detailed *nature* of this singularity. There is no reason to believe that it resembles the $r = 0$ singularity of the Schwarzschild solution very closely.

In regard to (c), (d) and (f) we can actually go further in that *exact* solutions are known which generalize the metric (2) to include angular momentum (Kerr [7]) and, in addition, charge and magnetic moment (Newman *et al.* [8]), where a cosmological constant may also be incorporated (Carter [9]). These solutions appear to be somewhat special in that, for example, the gravitational quadrupole moment is fixed in terms of the angular momentum and the mass, while the magnetic-dipole moment is fixed in terms of the angular momentum, charge and mass. However, there are some reasons for believing that these solutions may actually represent the general exterior asymptotic limit resulting from the type of collapse we are considering. Any extra gravitational multipole moments of quadrupole type, or higher, can be radiated away by gravitational radiation; similarly, extra electromagnetic multipole moments of dipole type, or higher, can be radiated away by electromagnetic radiation. (I shall discuss this a little more later.) If this supposition is correct, then (e) will to some extent also be covered by an analysis of these exact solutions. Furthermore, (b) would, in effect, be covered as well, provided we assume that all matter (with the exception of electromagnetic field – if we count that as 'matter') in the neighbourhood of the 'black hole', eventually falls into the hole. These exact solutions (for small enough angular momentum, charge and cosmological constant) have absolute event horizons similar to the $r = 2m$ horizon in (2). They also possess space-time curvature singularities, although of a rather different structure from $r = 0$ in (2). However, we would not expect the detailed structure of these singularities to have relevance for a generically perturbed solution in any case.

It should be emphasized that the above discussion is concerned only with collapse situations which do not differ too much initially from the spherically symmetrical case we originally considered. It is not known whether a gravitational collapse of a *qualitatively different character* might not be possible according to general relativity. Also, even if an absolute event horizon *does* arise, there is the question of the 'stability' of the horizon. An 'unstable' horizon might be envisaged which itself might develop into a curvature singularity. These, again, are questions I shall have to return to later.

As for the possible relevance of gravitational quantum effects, as suggested in (g), this depends, as far as I can see, on the existence of regions of space-time where there are extraordinary local conditions. If we assume the existence of an absolute event horizon along which curvatures and densities remain small, then it is very hard to believe that a classical discussion of the situation is not amply adequate. It may well be that quantum phenomena have a dominating influence on the physics of the deep interior regions. But whatever effects this might have, they would surely not be observable from the outside. We see from Figure 2 that such effects would have to propagate outwards in *spacelike* directions over 'classical' regions of space-time. However, we must again bear in mind that these remarks might not apply in some qualitatively different type of collapse situation.

We now come to (h), namely the question of the validity of general relativity in general, and its application to this type of problem in particular. The inadequacy of the observational data has long been a frustration to theorists, but it may be that the situation will change somewhat in the future. There are several very relevant experiments now being performed, or about to be performed. In addition, since it has become increasingly apparent that 'strong' gravitational fields probably play an important role in some astrophysical phenomena, there appears to be a whole new potential testing-ground for the theory.

Among the recently performed experiments, designed to test general relativity, one of the most noteworthy has been that of Dicke and Goldenberg [10], concerning the solar oblateness. Although the results have seemed to tell against the pure Einstein theory, the interpretations are not really clear-cut and the matter is still somewhat controversial. I do not wish to take sides on this issue. Probably one must wait for further observations before the matter can be settled. However, whatever the final outcome, the oblateness experiment had, for me, the importance of forcing me to examine, once more, the foundations of Einstein's theory, and to ask what parts of the theory are likely to be 'here to stay' and what parts are most susceptible to possible modification. Since I feel that the 'here to stay' parts include those which were most revolutionary when the theory was first put forward, I feel that it may be worth-while, in a moment, just to run over the reasoning as I see it. The parts of the theory I am referring to are, in fact, the geometrical interpretation of gravity, the curvature of space-time geometry and general-relativistic causality. These, rather than any particular field equations, are the aspects of the theory which give rise to what perhaps appears most immediately strange in the collapse phenomenon. They also provide the physical basis for the major part of the subsequent mathematical discussion.

To begin with, let us agree that it is legitimate to regard space-time as constituting a four-dimensional smooth manifold (or 'continuum'). I do not propose to give a justification of this, because on an ordinary macroscopic level it is normally taken as 'obvious'. (On the other hand, I think that at a deeper submicroscopic level it is almost certainly 'false', but this is not likely to affect the normal discussion of space-time structure – except perhaps *at* a space-time singularity!) Next, we must establish the existence of a *physically well-defined metric* ds which defines for our manifold a (pseudo-) Riemannian structure, with signature $(+ - - -)$. The meaning of ds is to be such that when integrated along the world line of any particle, it gives the lapse of proper times as experienced by that particle. Thus, the existence of ds depends on the existence of *accurate clocks* in nature. These clocks must behave locally according to the laws of special relativity. Also, for any two such clocks following the same world line, the time rates they register must agree with one another along the line and should not depend on, say, differing histories for the two clocks. That such clocks do seem to exist in nature, in effect, is a consequence of the fact that any mass m has associated with it a natural frequency mh^{-1}. Thus, the existence of accurate clocks comes down ultimately, via quantum mechanics, to the existence of well-defined masses in nature, whose relative values are in strict proportion throughout space-time. Of course, it might ultimately turn out that the mass ratios of particles are not constant throughout space-time. Then different particles might define slightly different (conformally related) metrics for space-time. But the evidence at present is strongly against any *appreciable* difference existing.

If two neighbouring events in space-time have a separation such that d$s^2 \geqslant 0$, then according to special relativity, it is possible for one to have a causal influence on the other; if d$s^2 < 0$, then it is not. We expect this to persist also on a global scale. Thus, it is possible, of two events, for one to influence the other causally if and only if there is a timelike or null curve connecting them.

The existence of a physically well-defined metric and causal structure for space-time, then, seems to be fairly clearly established. It is not so clear, however, that this metric, as so defined, is going to be nonflat. However, we can take the experiment of Pound and Rebka [11] as almost a direct measurement establishing the nonflat nature of space-time. (For this, strictly speaking, the experiment would have to be repeated at various points on the Earth's surface.) The measured ds near the Earth's surface and the ds further from the Earth's surface cannot both be incorporated into the same Minkowskian framework because of the 'clock slowing' effect (cf. Schild [12]). Furthermore, owing to energy balance considerations it is clear that it is with *gravitational* fields that this 'clock slowing' effect occurs (owing to the fact that it is *energy*, i.e. *mass* which responds to a gravitational field). Thus gravitation must be directly related to space-time curvature.

Since we have a (pseudo-) Riemannian manifold, we can use the standard techniques of differential geometry to investigate it. In particular, we can construct a *physically meaningful* Riemann tensor R_{abcd} and thence its Einstein tensor

$$G_{ab} = R_{ab} - \tfrac{1}{2} R g_{ab}.$$

Because of the contracted Bianchi identities we know that this satisfies the usual vanishing divergence law. But we also have a symmetric tensor T_{ab}, namely the local energy-momentum tensor (composed of all fields but gravitation), which must satisfy a similar vanishing divergence law. It does not then *necessarily* follow that

$$G_{ab} + \lambda g_{ab} = - 8\pi T_{ab} \tag{3}$$

for some constant λ, but it is worth remarking that if we do *not* postulate this equation, then we have not just one, but *two* (linearly unrelated) conserved 'energylike' quantities, namely G_{ab} and T_{ab}. In fact, this is just what happens in the theory of Brans and Dicke [13]. (Such a motivation for the choice of Einstein's field Equations (3), does not to my mind have quite the force of the earlier argument, so alternatives to (3) are certainly well worth considering.) Finally, the geodesic motion of monopole test particles may be taken as a consequence of the vanishing divergence condition on T_{ab} (Einstein and Grommer [14]).

So I want to admit the possibility that Einstein's field equations may be wrong, but *not* (that is, in the macroscopic realm, and where curvatures or densities are not fantastically large) that the general pseudo-Riemannian geometric framework may be wrong. Then the mathematical discussion of the collapse phenomenon can at least be applied. It is interesting that the general mathematical discussion of collapse actually uses very little of the details of Einstein's equations. All that is needed is a certain inequality related to positive-definiteness of energy. In fact, the adoption of the Brans-Dicke theory in place of Einstein's would make virtually no qualitative difference to the collapse discussion.

The final listed objection to the collapse picture is (h), namely the apparent lack of any tie-up with observed astronomical phenomena. Of course it could be argued that the prediction of the 'black hole' picture is simply that we will not see anything – and this is precisely consistent with observations since no 'black holes' have been observed! But the real argument is really the other way around. Quasars *are* observed. And they apparently have such large masses and such small sizes that it would seem that gravitational collapse ought to have taken over. But quasars are also long-lived objects. The light they emit does not remotely resemble the exponential cut-off in intensity, with approach to infinite red shift, that might be inferred from the spherically symmetrical discussion. This has led a number of astrophysicists to question the validity of Einstein's theory, at least in its applicability to these situations.

My personal view is that while it is certainly possible (as I have mentioned earlier) that Einstein's equations may be wrong, I feel it would be very premature indeed to dismiss these equations just on the basis of the quasar observations. For, the *theoretical* analysis of collapse, according to Einstein's theory, is still more or less in its infancy. We just do not know, with much certainty, what the consequences of the theory really are. It would be a mistake to fasten attention just on those aspects of general-relativistic collapse which *are* known and to assume that this gives us essentially the complete picture. (It is perhaps noteworthy that many general-relativity theorists have a tendency, themselves, to be a bit on the sceptical side as regards the 'classical' collapse picture!) Since it seems to me that there are a number of intriguing largely unexplored possibilities, I feel it may be worth-while to present the 'generic' general-relativistic collapse picture as I see it, not only as regards the known theorems, but also in relation to some of the more speculative and conjectural aspects of the situation.

To begin with, let us consider what the general theorems *do* tell us. In order to characterize the situation of collapse 'past a point of no return', I shall first need the concept of a *trapped surface*. Let us return to Figure 2. We ask what qualitative peculiarity of the region $r < 2m$ (after the star has collapsed through) is present. Can such peculiarities be related to the fact that everything appears to be forced inwards in the direction of the centre? It should be stressed again that *apart* from $r = 0$, the space-time at any individual point inside $r = 2m$ is perfectly regular, being as 'locally Min-kowskian' as any other point (outside $r = 0$). So the peculiarities of the $0 < r < 2m$ region must be of a partially 'global' nature. Now consider any point T in the (v, r)-plane of Figure 2 ($r < 2m$). Such a point actually represents a *spherical 2-surface* in space-time, this being traced out as the θ, φ co-ordinates vary. The surface area of this sphere is $4\pi r^2$. We imagine a flash of light emitted simultaneously over this spherical surface T. For an ordinary spacelike 2-sphere in flat space-time, this would result in an ingoing flash imploding towards the centre (surface area decreasing) together with an outgoing flash exploding outwards (surface area increasing). However, with the surface T, while we still have an ingoing flash with decreasing surface area as before, the 'out-going' flash, on the other hand, is in effect also falling inwards (though not as rapidly) and its surface area also decreases. The surface T ($v = \text{const}$, $r = \text{const} < 2m$) of metric (2) serves as the prototype of a trapped surface. If we perturb the metric (2) slightly, in the neighbourhood of an initial hypersurface, then we would still expect to get a surface T with the following property:

> T is a spacelike closed* 2-surface such that the null geodesics which meet it orthogonally all *converge* initially at T.

This convergence is taken in the sense that the local surface area of cross-section *decreases*, in the neighbourhood of each point of T, as we proceed into the future. (These null geodesics generate, near T, the boundary of the set of points lying causally to the future of the set T.) Such a T is called a *trapped surface*.

We may ask whether any connection is to be expected between the existence of a trapped surface and the presence of a physical space-time singularity such as that occurring at $r = 0$ in (2). The answer supplied by some general theorems (Penrose [6, 15], Hawking and Penrose [16]) is, in effect, that the presence of a trapped surface always *does* imply the presence of some form of space-time singularity.

There are similar theorems that can also be applied in cosmological situations. For example (Hawking [17–19], Hawking and Penrose [16]), if the universe is *spatially closed*, then (excluding exceptional limiting cases, and assuming $\lambda \leqslant 0$) the conclusion is that there must be a space-time singularity. This time we expect the singularity to reside in the past (the 'big bang'). Other theorems (Hawking [17, 19], Hawking and Penrose [16]) can be applied also to spatially open universes. For example, if there is

---

* By a 'closed' surface, hypersurface, or curve, I mean one that is 'compact without boundary'.

any point (e.g. the Earth at the present epoch) whose past light cone starts 'converging again' somewhere in the past (i.e. objects of given size start to have *larger* apparent angular diameters again when their distance from us exceeds some critical value), then, as before, the presence of space-time singularities is implied ($\lambda \leqslant 0$). According to Hawking and Ellis ([20], cf. also Hawking and Penrose [16]) the presence and isotropy of the 3 K radiation strongly indicates that the above condition on our past light cone is actually satisfied. So the problem of space-time singularities does seem to be very relevant to our universe, also on a large scale.

The main significance of theorems such as the above, is that they show that the presence of space-time singularities in exact models is not just a feature of their high symmetry, but can be expected also in generically perturbed models. This is *not* to say that *all* general-relativistic curved space-times are singular – far from it. There are many exact models known which are complete and free from singularity. But those which resemble the standard Friedmann models or the Schwarzschild collapse model sufficiently closely must be expected to be singular ($\lambda \leqslant 0$). The hope had often been expressed (cf. Lindquist and Wheeler [21], Lifshitz and Khalatnikov [22]) that the actual space-time singularity occurring in a collapsing space-time model might have been a consequence more of the fact that the matter was all hurtling simultaneously towards one central point, than of some intrinsic feature of general-relativistic space-time models. When perturbations are introduced into the collapse, so the argument could go, the particles coming from different directions might 'miss' each other, so that an effective 'bounce' might ensue. Thus, for example, one might envisage an 'oscillating' universe which on a *large* scale resembles the cycloidal singular behaviour of an 'oscillating' spatially closed Friedmann model; but the *detailed* behaviour, although perhaps involving enormous densities while at maximal contraction, might, by virtue of complicated asymmetries, contrive to avoid actual space-time singularities. However, the theorems seem to have ruled out a singularity-free 'bounce' of this kind. But the theorems do *not* say that the singularities need resemble those of the Friedmann or Schwarzschild solutions at all closely. There is some evidence (cf. Misner [23], for example) that the 'generic' singularities may be very elaborate and possess a qualitative structure very different from that of their smoothed-out counterparts. Very little is known about this, however.

It is worth mentioning the essential basic assumptions that enter into the theorems. In the first place we require an 'energy condition' which, by virtue of Einstein's Equations (3), may be stated as a negative-definiteness condition on the Ricci tensor:

$$t^a t_a = 1 \quad \text{implies} \quad R_{ab} t^a t^b \leqslant 0, \tag{4}$$

that is to say, the time-time component R_{00} of R_{ab} is nonpositive in any orthonormal frame. If we assume $\lambda = 0$ in Einstein's Equations (3), then (4) becomes

$$t^a t_a = 1 \quad \text{implies} \quad T_{ab} t^a t^b \geqslant \tfrac{1}{2} T_c^c.$$

This, when referred to an eigenframe of T_{ab}, can be stated as

$$E + p_\alpha \geqslant 0 \quad \text{and} \quad E + \sum p_\alpha \geqslant 0, \tag{5}$$

where $\alpha = 1, 2, 3$. Here E is the energy density (referred to this frame) and p_1, p_2, p_3 are the three principal pressures. If (3) holds with $\lambda < 0$, then it is still true that (4) is a *consequence* of (5). The significance of the energy condition (4) lies in the effect of Raychaudhuri [24] which states that whenever a system of timelike geodesics normal to a spacelike hypersurface starts converging, then this convergence inevitably increases along the geodesics until finally the geodesics cross over one another (assuming the geodesics are complete).

There is a corresponding focussing effect in the case of *null* geodesics. This depends on the 'weak energy condition':

$$l^a l_a = 0 \quad \text{implies} \quad R_{ab} l^a l^b \leqslant 0. \tag{6}$$

This condition (6) is a consequence of (4) (as follows by a limiting argument) but not conversely. If we assume Einstein's Equations (3) with, now, *any value of* λ, then (6) is equivalent to

$$E + p_\alpha \geqslant 0 \tag{7}$$

for $\alpha = 1, 2, 3$. The conditions (7) are, in fact, a consequence of the *nonnegative definiteness of the energy density*:

$$t^a t_a = 1 \quad \text{implies} \quad T_{ab} t^a t^b \geqslant 0$$

(that is $T_{00} \geqslant 0$ in each orthonormal frame). Thus, there is a strong physical basis for (6). The physical basis for (4) is not quite so strong, but provided $\lambda \leqslant 0$, we would certainly expect (4) to hold for all normal matter. (Note that if $E > 0 \geqslant \lambda$, only large *negative* pressures could cause trouble with (4). Usually people only worry about large positive pressures!) It is the 'strong' condition (4) that is required for the proofs of most of the theorems, but much can be said, concerning the qualitative nature of a collapse situation, even on the basis of the 'weak' condition (6) alone (cf. Penrose [5]).

A remark concerning the condition on the cosmological constant λ seems appropriate here. It is a weakness of the theorems that most of them do require $\lambda \leqslant 0$ for their strict applicability. However, it would appear that the condition $\lambda \leqslant 0$ is only really relevant to the initial setting of the global conditions on the space-time which are required for applicability of the theorem. If curvatures are to become large near a singularity, then (from dimensional considerations alone) the λ-term will become more and more insignificant. So it seems unlikely that a λ-term will really make much difference to the singularity structure in a collapse. The relevance of λ is really only at the cosmological scale.

Most of the theorems (but not all, cf. Hawking [19]) require, as an additional assumption, the nonexistence of closed timelike curves. This is a very reasonable requirement, since a space-time which possesses closed timelike curves would allow an observer to travel into his own past. This would lead to very serious interpretative difficulties! Even if it could be argued, say, that the accelerations involved might be such as to make the trip impossible in 'practice' (cf. Gödel [25]), equally serious difficulties would arise for the observer if he merely reflected some light signals into his own past! In addition closed timelike curves can lead to unreasonable consistency conditions on the solutions of hyperbolic differential equations. In any case, it seems unlikely that closed timelike curves can substitute for a space-time singularity, except in special unstable models.

Some of the theorems require an additional 'generality' condition, to the effect that every timelike or null geodesic enters some region in which the curvature is not everywhere lined up in a particular way with the geodesic. (More precisely, $t_{[a}R_{b]cd[e}t_{f]}t^c t^d \neq \neq 0$ somewhere along the geodesic, t^a being its tangent vector.) This condition plays a role in the mathematics, but from the physical point of view it is really no condition at all. We would always expect a little bit of matter or randomly oriented curvature along any geodesic in a physically realistic solution. It is only in very special limiting cases that we would expect the condition to be violated. (Curiously enough, however, practically every explicitly known solution does violate the condition!)

Finally, it should be remarked that none of the theorems *directly* establishes the existence of regions of approaching infinite curvature. Instead, all one obtains is that the space-time is not geodesically complete (in timelike or null directions) and, furthermore, cannot be *extended* to a geodesically complete space-time. ('Geodesically complete' means that geodesics can be extended indefinitely to arbitrarily large values of their length or affine parameter – so that inertially moving particles or photons do not just 'fall off the edge' of the space-time.) The most 'reasonable' explanation for why the space-time is not inextendible to a complete space-time seems to be (and I would myself believe this to be the most likely, in general) that the space-time is confronted with, in some sense, *infinite curvature* at its boundary. But the theorems do not quite say this. Other types of space-time singularity are possible, and theorems of a somewhat different nature would be required to decide which is the most likely type of singularity to occur.

We must now ask the question whether the theorems are actually likely to be relevant in the case of a collapsing star or superstar. Do we, in fact, have any reason to believe that trapped surfaces can ever arise in gravitational collapse? I think a very strong case can be made that at least *sometimes* a trapped surface must arise. I would not expect trapped surfaces necessarily always to arise in a collapse. It might depend on the details of the situation. But if we can establish that there can be nothing *in principle* against a trapped surface arising – even if in some very contrived and outlandish situation – then we must surely accept that trapped surfaces must at least occasionally arise in real collapse.

Rather than use the trapped-surface condition, however, it will actually be somewhat easier to use the alternative condition of the existence of a point whose light cone starts 'converging again'. From the point of view of the general theorems, it really makes no essential difference which of the two conditions is used. Space-time singularities are to be expected in either case. Since we are here interested in a collapse situation rather than in the 'big bang', we shall be concerned with the *future* light cone C of some point p. What we have to show is that it is possible in principle for enough matter to cross to within C, so that the divergence of the null geodesics which generate C changes sign somewhere to the future of p. Once these null geodesics start to converge, then 'weak energy condition' (6) will take over, with the implication that an absolute event horizon must develop (outside C). As a consequence of the stronger 'energy condition' (4) it will also follow that space-time singularities will occur.

Since we ask only that it be possible *in principle* to reconverge the null rays generating C, we can resort to an (admittedly far-fetched) 'gedanken experiment'. Consider an elliptical galaxy containing, say, 10^{11} stars. Suppose, then, that we contrive to alter the motion of the stars slightly by eliminating the transverse component of their velocities. The stars will then fall inwards towards the centre. We may arrange to steer them, if we like, so as to ensure that they all reach the vicinity of the centre at about the same time without colliding with other stars. We only need to get them into a volume of diameter about fifty times that of the solar system, which gives us plenty of room for all the stars. The point p is now taken near the centre at about the time the stars enter this volume (Figure 3). It is easily seen from the orders of magnitude involved, that the relativistic light deflection (an *observed* effect of general relativity) will be sufficient to cause the null rays in C to reconverge, thus achieving our purpose.

Let us take it, then, that absolute event horizons can sometimes occur in a gravitational collapse. Can we say anything more detailed about the nature of the resulting

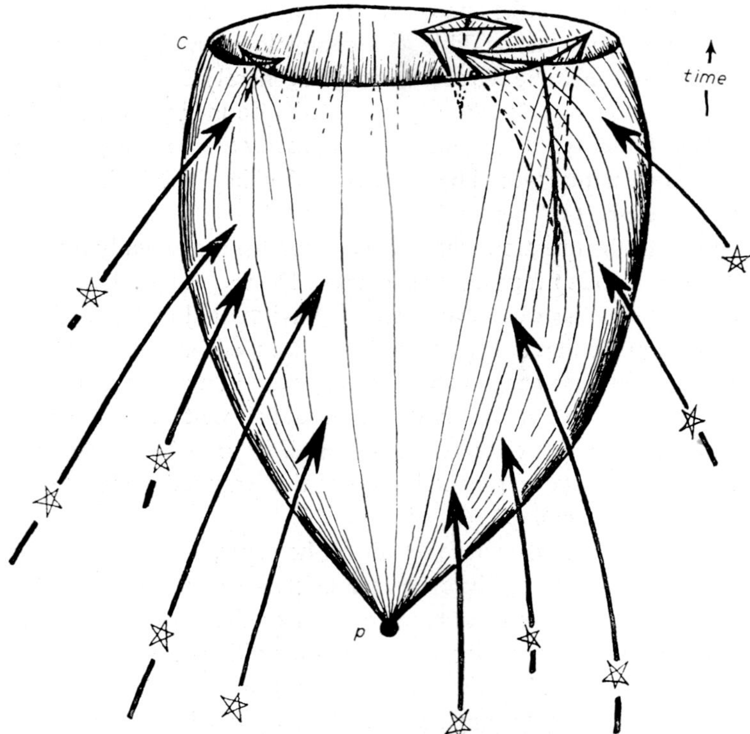

Fig. 3.   The future light cone of p is caused to reconverge by the falling stars.

situation? Hopeless as this problem may appear at first sight, I think there is actually a reasonable chance that it may find a large measure of solution in the not-too-distant future. This would depend on the validitiy of a certain result which has been independently conjectured by a number of people. I shall refer to this as the *generalized** *Israel conjecture* (abbreviated GIC). Essentially GIC would state: if an absolute event horizon develops in an asymptotically flat space-time, then the solution exterior to this horizon approaches a Kerr-Newman solution asymptotically with time.

The Kerr-Newman solutions (Kerr [7], Newman *et al.* [8]) are explicit asymptotically flat stationary solutions of the Einstein-Maxwell equation $(\lambda=0)$ involving just *three* free parameters m, a and e. As with the metric (1), the *mass*, as measured asymptotically, is the parameter m (in gravitational units). The solution also possesses angular momentum, of magnitude am. Finally, the total charge is given by e. When $a=e=0$ we get the Schwarzschild solution. Provided that

$$m^2 \geqslant a^2 + e^2$$

the solution has an absolute event horizon. Carter [9] has shown how to obtain all the geodesics and charged orbits for this solution, reducing the problem to a single quadrature. Thus, if GIC is true, then we shall have remarkably complete information as to the asymptotic state of affairs resulting from a gravitational collapse.

But what reason is there for believing that GIC has any chance of being true? One indication comes from a perturbation analysis of the Schwarzschild solution (Regge and Wheeler [28], Doroshkevich *et al.* [29]) which seems to indicate that all perturbations except rotation have a tendency to be damped out. Another indication is the theorem of Israel [26] which states, in effect, that the Schwarzschild solution is the only static asymptotically flat vacuum solution with an absolute event *horizon* (although there is a nontrivial side-condition to the theorem; cf. also Thorne [30] for the axially symmetric case). Israel [27] has also generalized his result to the Einstein-Maxwell theory, finding the spherically symmetric Reissner-Nordstrom solution to be the only asymptotically flat static solution with an absolute event horizon. Carter [31] has made some progress, in the vacuum rotating case, towards the objective of establishing the Kerr solution $(e=0)$ as the general asymptotically flat stationary solution with an absolute event horizon. In addition, there are solutions of the vacuum equations known (Robinson and Trautman [32]), which are suitably asymptotically flat and nonrotating, which apparently possess absolute event horizons, but are nonstatic. As time progresses they become more and more symmetrical, approaching the Schwarzschild solution asymptotically with time [33]. In the process, the higher multipole moments are radiated away as gravitational radiation.

The following picture then suggests itself. A body, or collection of bodies, collapses down to a size comparable to its Schwarzschild radius, after which a trapped surface can be found in the region surrounding the matter. Some way outside the trapped sur-

---

* Israel conjectured this result only in the stationary case, hence the qualification 'generalized'. In fact, Israel has expressed sentiments opposed to GIC. However, Israel's theorem [26, 27] represents an important step towards establishing of GIC, if the conjecture turns out to be true.

face region is a surface which will ultimately be the absolute event horizon. But at present, this surface is still expanding somewhat. Its exact location is a complicated affair and it depends on how much more matter (or radiation) ultimately falls in. We assume only a finite amount falls in and that GIC is true. Then the expansion of the absolute event horizon gradually slows down to stationarity. Ultimately the field settles down to becoming a Kerr solution (in the vacuum case) or, a Kerr-Newman solution (if a nonzero net charge is trapped in the 'black hole').

Doubts have frequently been expressed concerning GIC, since it is felt that a body would be unlikely to throw off all its excess multipole moments just as it crosses the Schwarzschild radius. But with the picture presented above this is not necessary. I would certainly not expect the body itself to throw off its multipole moments. On the other hand, the gravitational field *itself* has a lot of settling-down to do after the body has fallen into the 'hole'. The asymptotic measurement of the multipole moments need have very little to do with the detailed structure of the body itself; the *field* can contribute very significantly. In the process of settling down, the field radiates gravitationally – and electromagnetically too, if electromagnetic field is present. Only the mass, angular momentum and charge need survive as ultimate independent parameters. (Presumably the charge parameter e would be likely to be very small by comparison with a and m.)

But suppose GIC is not true, what then? Of course, it may be that there are just a lot more possible limiting solutions than that of Kerr-Newman. This would mean that much more work would have to be done to obtain the detailed picture, but it would not imply any qualitative change in the set-up. On the other hand there is the more alarming possibility that the absolute event horizon may be *unstable*! By this I mean that instead of settling down to become a nice smooth solution, the space-time might gradually develop larger and larger curvatures in the neighbourhood of the absolute event horizon, ultimately to become effectively singular there. My personal opinion is that GIC is more likely than this, but various authors have expressed the contrary view*).

If such instabilities are present then this would certainly have astrophysical implications. But even if GIC is true, the resulting 'black hole' may by no means be so 'dead' as has often been suggested. Let us examine the Kerr-Newman solutions, in the case $m^2 > a^2 + e^2$ in a little more detail. But before doing so let us refer back to the Schwarzschild solution (2). In Figure 4, I have drawn what is, in effect, a cross-section of the space-time, given by $v - r = \text{const}$. The circles represent the location of a flash of light which had been emitted at the nearby point a moment earlier. Thus, they indicate the orientation of the light cones in the space-time. We note that for large r the point lies inside the circle, which is consistent with the static nature of the space-time (i.e. one can 'stay in the same place' while retaining a timelike world line). On the other hand for $r < 2m$ the point lies outside the circle, indicating that all matter must be dragged inwards if it is to remain moving in a timelike direction (so, to 'stay in the

---

* Some recent work of Newman [34] on the charged Robinson-Trautman solutions suggests that new features indicating instabilities may arise when an electromagnetic field is present.

same place' one would have to exceed the local speed of light). Let us now consider the corresponding picture for the Kerr-Newman solutions with $m^2 > a^2 + e^2$ (Figure 5). I shall not be concerned, here, with the curious nature of the solution inside the absolute event horizon H, since this may not be relevant to GIC. The horizon H itself is represented as a surface which is tangential to the light cones at each of its points. Some distance outside H is the 'stationary limit' L, at which one must travel with the local light velocity in order to 'stay in the same place'.

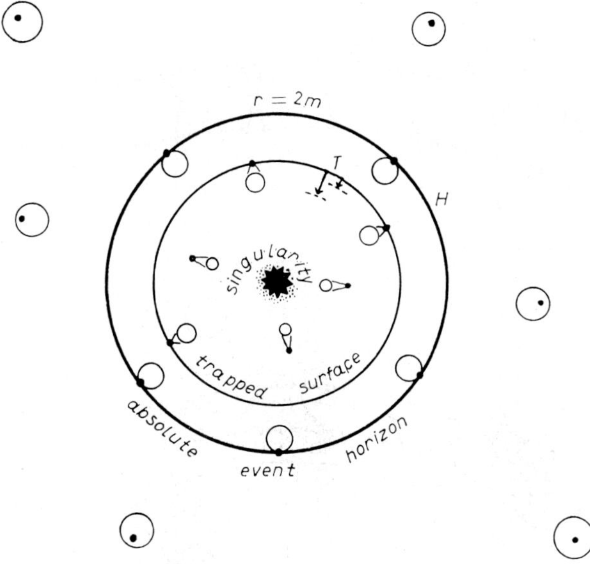

Fig. 4.   Spatial view of spherical 'black hole' (Schwarzschild solution).

I want to consider the question of whether it is possible to extract energy out of a 'black hole'. One might imagine that, since the matter which has fallen through has been lost for ever, so also is its energy content irrevocably trapped. However, it is not totally clear to me that this need be the case. There are at least two methods (neither of which is very practical) which might be construed as mechanisms for extracting energy from a 'black hole'. The first is due to Misner [35]. This requires, in fact, a whole *galaxy* of 2^N 'black holes', each of mass m. We first bring them together in pairs and allow them to spiral around one another, ultimately to swallow each other up. During the spiraling, a certain fraction K of their mass-energy content is radiated away as gravitational energy, so the mass of the resulting 'black hole' is $2m(1-K)$. The energy of the gravitational waves is collected and the process is repeated. Owing to the scale invariance of the gravitational vacuum equations, the same fraction of the mass-energy is collected in the form of gravitational waves at each stage. Finally we end up with a single 'black hole' of mass $2^N m(1-K)^N$. Now, the point is that *however small K* may in fact be, we can always choose N large enough so that $(1-K)^N$ is as small as we please. Thus, in principle, we can extract an arbitrarily large fraction of the mass-energy content of Misner's galaxy.

But anyone at all familiar with the problems of detecting gravitational radiation will be aware of certain difficulties! Let me suggest another method which actually tries to do something a little different, namely extract the 'rotational energy' of a 'rotating black hole' (Kerr solution). Consider Figure 5 again. We imagine a civilization which has built some form of stabilized structure S surrounding the 'black hole'. If

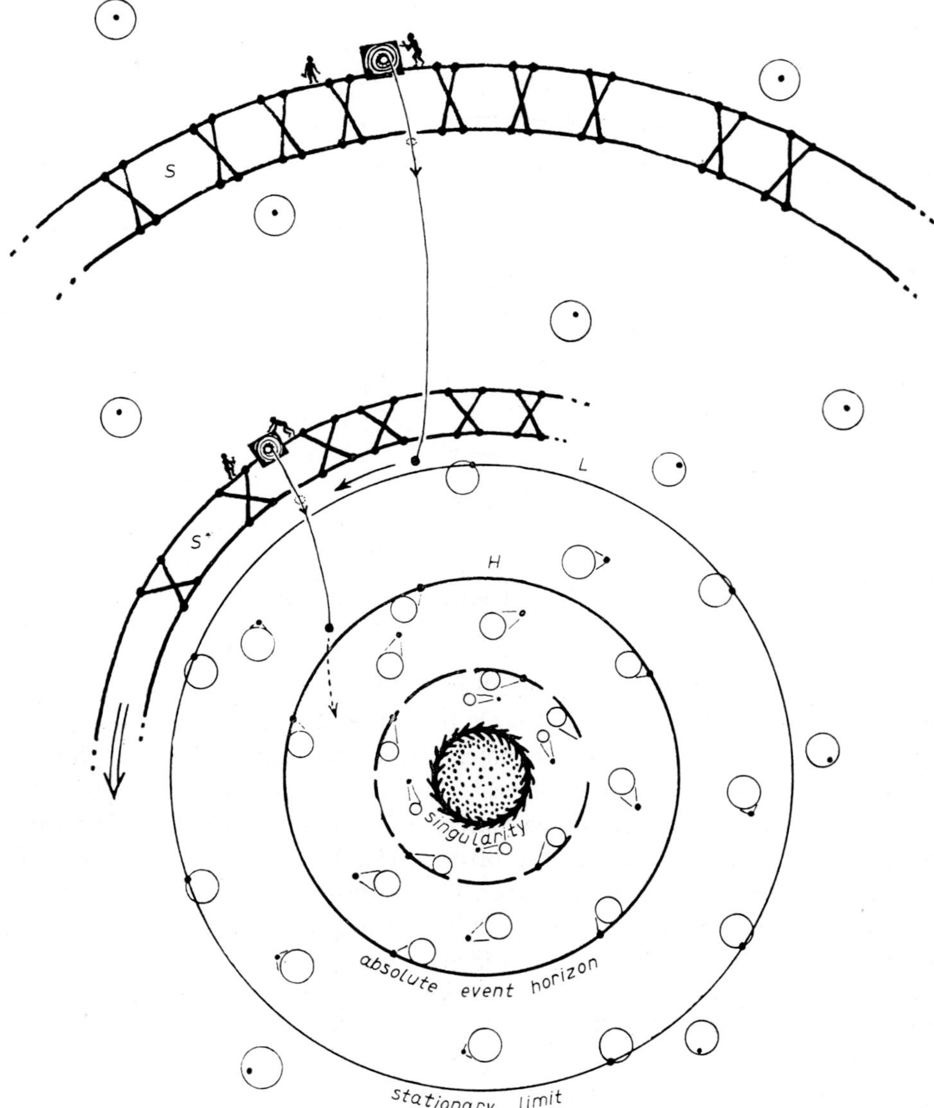

Fig. 5. Rotating 'black hole' (Kerr-Newman solution with $m^2 > a^2 + e^2$). The inhabitants of the structures S and S^* are extracting rotational energy from the 'black hole'.

they lower a mass slowly on a (light, inextensible, unbreakable) rope until it reaches L, they will be able to recover, at S, the entire energy content of the mass. If the mass is released as it reaches L then they will simply have bartered the mass for its energy content. (This is the highest-grade energy, however, namely wound-up springs!) But they can do better than this! They also build another structure S^*, which rotates, to some extent, with the 'black hole. The lowering process is continued, using S^*, to beyond L. Finally the mass is dropped through H, but in such a way that its energy content, as measured from S, is *negative*! Thus, the inhabitants of S are able, in effect, to lower masses into the 'black hole' in such a way that they obtain *more* than the energy content of the mass. Thus they extract some of the energy content of the 'black hole' itself in the process. If we examine this in detail, however, we find that the angular momentum of the 'black hole' is also reduced.

Thus, in a sense, we have found a way of extracting *rotational energy* from the 'black hole'. Of course, this is hardly a practical method! Certain improvements may be possible, e.g., using a ballistic method*. But the real significance is to find out what can and what cannot be done *in principle* since this may have some indirect relevance to astrophysical situations.

Let me conclude by making a few highly speculative remarks. In the first place, suppose we take what might be referred to, now, as the most 'conservative' point of view available to us, namely that GIC is not only true, but it also represents the *only* type of situation that can result from a gravitational collapse. Does it follow, then, that nothing of very great astrophysical interest is likely to arise out of collapse? Do we merely deduce the existence of a few additional dark 'objects' which do little else but contribute, slightly, to the overall mass density of the universe? Or might it be that such 'objects', while themselves hidden from direct observation, could play some sort of catalytic role in producing observable effects on a much larger scale. The 'seeding' of galaxies is one possibility which springs to mind. And if 'black holes' are born of violent events, might they not occasionally be ejected with high velocities when such events occur! (The one thing we can be sure about is that they *would* hold together!) I do not really want to make any very specific suggestions here. I only wish to make a plea for 'black holes' to be taken seriously and their consequences to be explored in full detail. For who is to say, without careful study, that they cannot play some important part in the shaping of observed phenomena?

But need we be so cautious as this? Even if GIC, or something like it, is true, have we any right to suggest that the *only* type of collapse which can occur is one in which the space-time singularities lie hidden, deep inside the protective shielding of an absolute event horizon? In this connection it is worth examining the Kerr-Newman solutions for which $m^2 < a^2 + e^2$. The situation is depicted in Figure 6. The absolute event horizon has now completely disappeared! A region of space-times singularity still exists in the vicinity of the centre, but now it is possible for information to escape

---

* Calculations show that this can indeed be done. A particle p_0 is thrown from S into the region between L and H, at which point the particle splits into two particles p_1 and p_2. The particle p_2 crosses H, but p_1 escapes back to S possessing *more* mass-energy content than p_0!

from the singularity to the outside world, provided it spirals around sufficiently. In short, the singularity is *visibile*, in all its nakedness, to the outside world!

However, there is an essential difference between the logical status of the singularity marked at the centre of Figure 6 and that marked at the centres of Figures 4 and 5. In the cases of Figures 4 and 5 there are trapped surfaces present, so we have a *theorem* which tells us that even with generic perturbation a singularity will still exist. In the situation of Figure 6, however, we have no trapped surfaces, no known theorem guaranteeing singularities and certainly no analogue of GIC. So it is really an open question whether a situation remotely resembling Figure 6 is ever likely to arise.

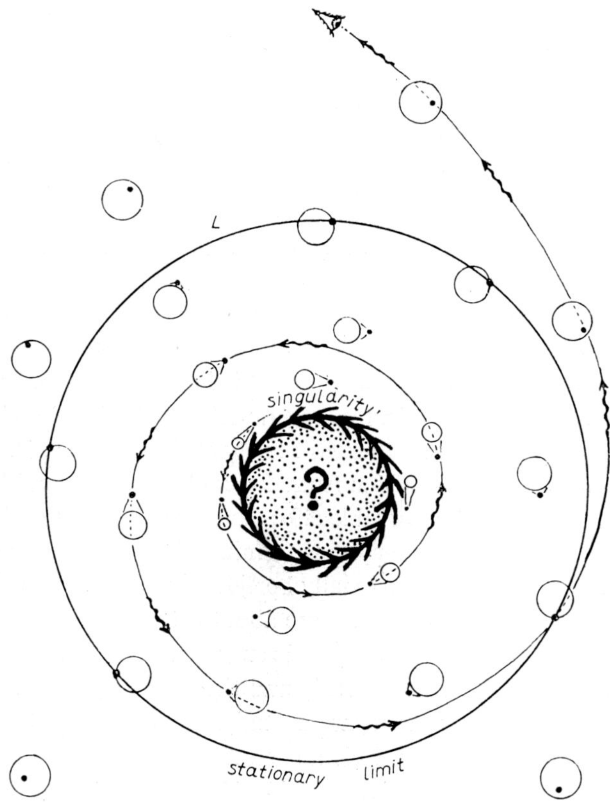

Fig. 6. A 'naked singularity' (Kerr-Newman solution with $m^2 < a^2 + e^2$).

We are thus presented with what is perhaps the most fundamental unanswered question of general-relativistic collapse theory, namely: does there exist a 'cosmic censor' who forbids the appearance of naked singularities, clothing each one in an absolute event horizon? In one sense, a 'cosmic censor' can be shown *not* to exist. For it follows from a theorem of Hawking [19] that the 'big bang' singularity is, in principle, observable. But it is not known whether singularities observable from outside will ever arise in a generic *collapse*, which starts off from a perfectly reasonable nonsingular initial state.

If in fact naked singularities do arise, then there is a whole new realm opened up for wild speculations! Let me just make a few remarks. If we envisage an isolated naked singularity as a source of new matter in the universe, then we do not *quite* have unlimited freedom in this! For although in the neighbourhood of the singularity we have no equations, we still have normal physics holding in the space-time *surrounding* the singularity. From the mass-energy flux theorem of Bondi *et al.* [36] and Sachs [37], it follows that it is *not* possible for *more* mass to be ejected from a singularity than the original total mass of the system, *unless* we are allowed to be left with a singularity of *negative* total mass. (Such a singularity would *repel* all other bodies, but would still be attracted by them!)

While in the realm of speculation concerning matter production at singularities, perhaps one further speculative remark would not be entirely out of place. This is with respect to the manifest large-scale time asymmetry in the behaviour of matter in the universe (and also the apparent large-scale asymmetry between matter and anti-matter). It is often argued that small observed violations of T (and C) invariance in fundamental interactions can have no bearing on the cosmological asymmetry problem. But it is not at all clear to me that this is necessarily so. It is a space-time singularity (i.e. presumably the 'big bang') which appears to govern the production of matter in the universe. When curvatures are fantastically large – as they surely are at a singularity – the local physics will be drastically altered. Can one be sure that the asymmetries of local interactions will not have the effect of being as drastically magnified?

When so little is known about the geometrical nature of space-time singularities and even less about the nature of the physics which takes place there, it is perhaps futile to speculate in this way about them. However, ultimately a theory will have to be found to cope with the situation. The question of the quantization of general relativity is often brought up in this connection. My own feeling is that the purpose of correctly combining quantum theory with general relativity is really somewhat different. It is simply a step in the direction of discovering how nature fits together as a whole. When eventually we have a better theory of nature, then perhaps we can try our hands, again, at understanding the extraordinary physics which must take place at a spacetime singularity.

References

[1] S. Chandrasekhar: 1935, *Monthly Notices Roy, Astron. Soc.*, **95**, 207.
[2] J. R. Oppenheimer and G. Volkoff: 1939, *Phys. Rev.* **55**, 274.
[3] A. S. Eddington: 1924, *Nature* **113**, 192.
[4] D. Finkelstein: 1956, *Phys. Rev.* **110**, 965.
[5] W. Rindler: 1958, *Monthly Notices Roy. Astron. Soc.* **116**, 6.
[6] R. Penrose: 1968, in *Battelle Rencontres* (ed. by C. M. De Witt and J. A. Wheeler) New York.
[7] R. P. Kerr: 1963, *Phys. Rev. Letters.* **11**, 237.
[8] E. T. Newman, E. Couch, K. Chinnapared, A. Exton, A. Prakash and R. Torrence: 1965, *Math. Phys.* **6**, 918.
[9] B. Carter: 1968, *Phys. Rev.* **174**, 1559.

[10] R. H. Dicke and H. M. Goldenberg: 1967, *Phys. Rev. Letters* **18**, 313.
[11] R. V. Pound and G. A. Rebka: 1960, *Phys. Rev. Letters* **4**, 337.
[12] A. Schild: 1967, in *Relativity Theory and Astrophysics*. Vol. **1**: *Relativity and Cosmology* (ed. by J. Ehlers), Providence, R. I.
[13] C. Brans and R. H. Dicke: 1961, *Phys. Rev.* **124**, 925.
[14] A. Einstein and J. Grommer: 1927, *S. B. Preuss. Akad. Wiss.* **1**, 2.
[15] R. Penrose: 1965, *Phys. Rev. Letters* **14**, 57.
[16] S. W. Hawking and R. Penrose: 1969, *Proc. Roy. Soc.*, A (in press).
[17] S. W. Hawking: 1966, *Proc. Roy. Soc.* A **294**, 511.
[18] S. W. Hawking: 1966, *Proc. Roy. Soc.* A **295**, 490.
[19] S. W. Hawking: 1967, *Proc. Roy. Soc.* A **300**, 187.
[20] S. W. Hawking and G. F. R. Ellis: 1968, *Astrophys. J.* **152**, 25.
[21] R. W. Lindquist and J. A. Wheeler: 1957, *Rev. Mod. Phys.* **29**, 432.
[22] E. M. Lifshitz and I. M. Khalatnikov: 1963, *Adv. Phys.* **12**, 185.
[23] C. W. Misner: 1969, *Phys. Rev. Letters.* **22**, 1071.
[24] A. K. Raychaudhuri: 1955, *Phys. Rev.* **98**, 1123.
[25] K. Gödel: 1959, in *Albert Einstein Philosopher Scientist* (ed. by P. A. Schilpp), New York, p. 557.
[26] W. Israel: 1967, *Phys. Rev.* **164**, 1776.
[27] W. Israel: 1968, *Comm. Math. Phys.* **8**, 245.
[28] T. Regge and J. A. Wheeler: 1957, *Phys. Rev.* **108**, 1063.
[29] A. G. Doroshkevich, Ya. B. Zel'dovich and I. D. Novikov: 1965, *Žurn. Éksp. Teor. Fiz.* **49**, 170; English trans., *Sov. Phys. JETP* **22**, 122 (1966).
[30] K. S. Thorne: 1965, Ph. D. thesis, Princeton University Princeton, N. J.
[31] B. Carter: 1969, personal communication.
[32] I. Robinson and A. Trautman: 1962, *Proc. Roy. Soc.* A **265**, 463.
[33] J. Foster and E. T. Newman: 1967, *Math. Phys.* **8**, 189.
[43] E. T. Newman: 1969, personal communication.
[35] C. W. Misner: 1968, personal communication.
[36] H. Bondi, M. G. J. van der Burg and A. W. K. Metzner: 1962, *Proc. Roy. Soc.* A **269**, 21.
[37] R. K. Sachs: 1962, *Proc. Roy. Soc.* A **270**, 103.

INTRODUCING THE BLACK HOLE*

*According to Present Cosmology, Certain Stars End Their Careers in a Total
Gravitational Collapse that Transcends the Ordinary Laws of Physics*

REMO RUFFINI and JOHN A. WHEELER

The quasistellar object, the pulsar, the neutron star have all come onto the scene of physics within the space of a few years. Is the next entrant destined to be the black hole? If so, it is difficult to think of any development that could be of greater significance. A black hole, whether of 'ordinary size' (approximately one solar mass, $1\ M_\odot$, or much larger (around $10^6\ M_\odot$ to $10^{10}\ M_\odot$, as proposed in the nuclei of some galaxies) provides our 'laboratory model' for the gravitational collapse, predicted by Einstein's theory, of the universe itself.

A black hole is what is left behind after an object has undergone complete gravitational collapse. Spacetime is so strongly curved that no light can come out, no matter can be ejected and no measuring rod can ever survive being put in. Any kind of object that falls into the black hole loses its separate identity, preserving only its mass, charge, angular momentum and linear momentum (see Figure 1). No one has yet found a way to distinguish between two black holes constructed out of the most different kinds of matter if they have the same mass, charge and angular momentum. Measurement of these three determinants is permitted by their effect on the Kepler orbits of test objects, charged and uncharged, in revolution about the black hole.

How the physics of a black hole looks depends more upon an act of choice by the observer himself than on anything else. Suppose he decides to follow the collapsing matter through its collapse down into the black hole. Then he will see it crushed to indefinitely high density, and he himself will be torn apart eventually by indefinitely increasing tidal forces. No restraining force whatsoever has the power to hold him away from this castatrophe, once he crossed a certain critical surface known as the 'horizon'. The final collapse occurs a finite time after the passage of this surface, but it is inevitable. Time and space are interchanged inside a black hole in an unusual way; the direction of increasing proper time for the observer is the direction of decreasing values of the coordinate r. The observer has no more power to return to a larger r value than he has power to turn back the hands on the clock of life itself. He can not even stay where he is, and for a simple reason: no one has the power to stop the advance of time.

Suppose the observer decides instead to observe the collapse from far away. Then, as price for his own safety, he is deprived of any chance to see more than the first steps on the way to collapse. All signals and all information from the later phases of collapse never escape; they are caught up in the collapse of the geometry itself.

That a sufficient mass of cold matter will necessarily collapse to a black hole (Oppenheimer and Snyder, 1939) is one of the most spectacular of all the predictions of

* Reprinted from *Physics Today* **24**.

H. Gursky and R. Ruffini (eds.), Neutron Stars, Black Holes and Binary X-Ray Sources, 379–393. All Rights Reserved.

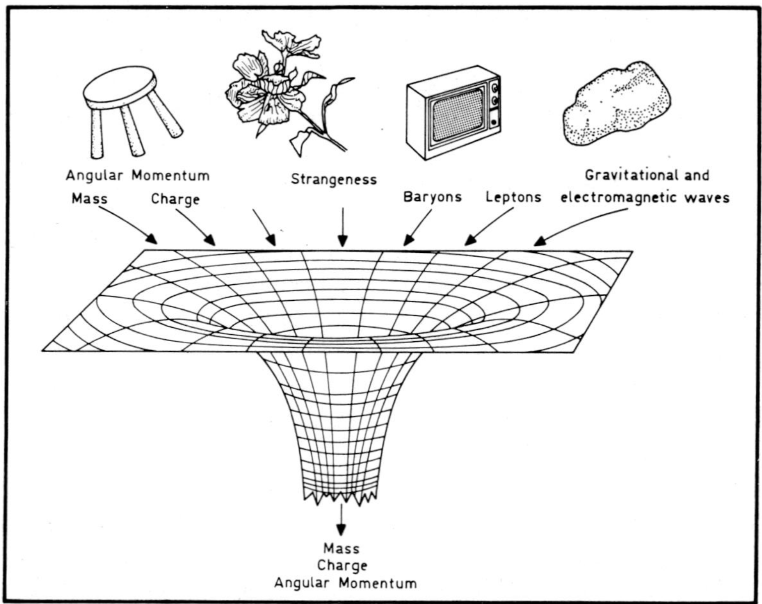

Fig. 1.   Figurative representation of a black hole in action. All details of the infalling matter are washed out. The final configuration is believed to be uniquely determined by mass, electric charge, and angular momentum.

Einstein's standard 1915 general relativity. The geometry around a collapsed object of spherical symmetry (nonrotating!) was worked out by Karl Schwarzschild of Göttingen, father of the American astrophysicist Martin Schwarzschild, as early as 1916. In 1963 Roy Kerr found the geometry associated with a rotating collapsed object. James Bardeen has recently emphasized that all stars have angular momentum and that most stars – or star cores – will have so much angular momentum that the black hole formed upon collapse will be rotating at the maximum rate, or near the maximum rate, allowed for a black hole ('surface velocity' equal to speed of light). Roger Penrose (1969) has shown that a particle coming from a distance into the immediate neighborhood of a black hole (the 'ergosphere') can extract energy from the black hole. Demetrios Christodoulou (1970) has shown that the total mass-energy of a black hole can be split into three parts,

$$E^2 = m_{\mathrm{ir}} + L^2/4m_{ir}^2 + p^2$$

The first part is 'irreducible' (left constant in 'reversible transformations'; always increased in 'irreversible transformations') and the second and third parts (arising from a rotational angular momentum L and a linear momentum p) can be added and subtracted at will.

The three most promising ways now envisaged to detect black holes are:
pulses and trains of gravitational radiation given out at the time of formation (see

Physics Today, August 1969, page 61, and August 1970, page 41, for accounts of Joseph Weber's pioneering attempts to detect gravitational radiation),

broadband electromagnetic radiation extending into the hard x-ray and gamma-ray regions emitted by matter falling into a black hole after it has been formed (this is the concept of Ya. B. Zel'dovich and I. D. Novikov. The radiation is not emitted by the individual particles as they fall in, but by the gas as a whole as it is compressed and heated to 10^{10} or 10^{11} K by the 'funnel effect' on its way towards the black hole),

jets and other activity produced in the ergosphere of rotating black holes.

1. Equilibrium Configurations

The mass of a superdense star (reached in collapse that does not go to a black hole) is determined uniquely by its central density, provided that the equation of state linking pressure and density is specified. Then, by integrating the equation for relativistic hydrostatic equilibrium (Harrison *et al.*, 1965) outwards to the point where the pressure drops to zero, we find the total mass corresponding to each value of the central density. The idea that a sufficiently massive star would contract without limit under the influence of its own gravitational field was suggested by study of the white dwarf stars. These are very dense stars in which the pressure arises primarily from a degenerate Fermi gas of electrons. No stable solution exists for a white dwarf with a mass above the Chandrasekhar limit, which is about 1.2 solar masses. What is the endpoint of stellar evolution for a star more massive than this critical mass?

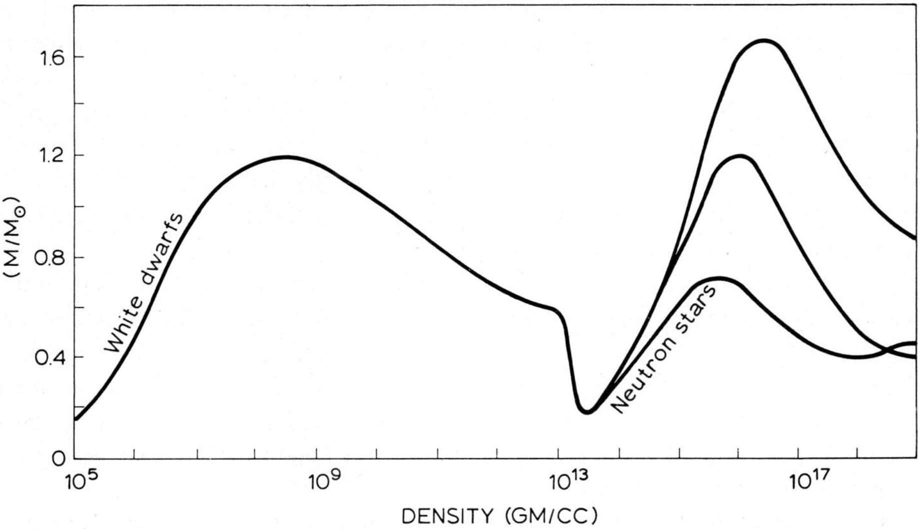

Fig. 2. Mass of a cold star calculated by numerical integration from center to surface for selected values of the central density. The upper curves assume Newtonian hydrostatic equilibrium. The upper includes only rest mass and mass-energy of compression, while the lower adds the correction for mass-energy of gravitational binding. The lower curve assumes relativistic hydrostatic equilibrium. The Harrison-Wheeler equation of state is used in all cases.

The answer depends on the 'local physics', summarized in the equation of state, and the 'global properties' determined by the gravitational field. One would hope that the different predictions of the Newton, Einstein, and Jordan-Brans-Dicke theories would provide a way to discriminate between the theories. However, the equation of state must take into account all physical phenomena, including high-energy physics. Ignorance of the equation of state at supranuclear densities blurs distinction between the contending gravitational theories. Nevertheless, a family of stable neutron stars exists for all reasonable equations of state. The minimum mass of this family is about 0.16 solar mass, but the maximum mass is uncertain by about a factor of four. In Figure 2, mass is plotted versus central density with the assumption of a particular equation of state (the Harrison-Wheeler equation) to show the difference between Newtonian gravitation and general relativity in the neutron-star region.

2. Neutron Star or Black Hole?

The physics of the formation of a neutron star or a black hole is more complicated than the physcis of either object itself. It is believed that in this process, the core of a star, possibly a late giant, collapses from its original radius of a few thousand kilometers to a compact object with a radius of a few tens of kilometers. The core has slowly evolved over thousands of years to a degree where it is unstable against gravitational collapse. This does not necessarily mean that its mass lies precisely at 1.2 solar masses, the first peak in figure 2. It may be two or five or ten times more massive and still not collapse, when inflated by sufficiently high temperatures. But cooling such a system will automatically bring it to the point of collapse. Colgate and White (1966) and May and White (1967) have made computer investigations of what happens, under the simplifying assumption of spherical symmetry. The material of the star starts moving inwards, at first slowly, then more and more rapidly, with a characteristic speedup time of less than a tenth of a second. Soon, a substantial portion of this mass, the inner part of the core, contracts sufficiently to increase greatly the strength of the gravitational fields drawing the inner core together. As a consequence, the core accelerates more rapidly than the surrounding envelope.

Two very different outcomes ensue, depending on whether the core mass and its kinetic energy of implosion do or do not suffice to drive the system on beyond nuclear densities to the point of complete gravitational collapse. Complete collapse produces a 'black hole'. On the other hand, when the mass is too small or the velocity of implosion is too low the collapse is halted at nuclear or near-nuclear density. The stopping of so large a mass implies the sudden conversion of an enormous kinetic energy into thermal energy, as if a 'charge of dynamite' had been set off at the center of the system. The high temperature (about 10^{12} K) develops high pressure. The envelope surrounding the inner core is falling more slowly and suddenly feels this pressure. The implosion is reversed. The envelope is propelled outward, producing cosmic rays and an expanding ion cloud. A famous example of such a supernova event is the Crab Nebula, with an estimated mass of the rough order of magnitude of a solar mass.

Rotation, and magnetic fields and magnetic fields coupled to rotation, can significantly change the character of the implosion, as shown by the recent work of Le-Blanc and Wilson (1969). As the center shrinks, it turns faster and faster to conserve angular momentum, winding up the magnetic lines of force like string on a spool. The Faraday-Maxwell repulsion between the lines of force causes the spool to elongate. The lines of force carry matter with them, shooting jets out from the two poles. It will be interesting to see how these effects will be modified when the calculation is expanded to include all the physical details of the Colgate-May-White analysis and nuclear reactions as well.

3. Continuing Collapse

When the core of the collapsing star is too massive or imploding with too much kinetic energy, the implosion may still slow down as nuclear densities are encountered, but nuclear forces will not stop the implosion. Gravitational forces become overwhelming, the system zooms through the neutron-star stage, and complete collapse follows. The resulting system has been variously termed 'continuing collapse', a 'frozen star', and a 'black hole'. Each name emphasizes a different aspect of the collapsing system. The collapse is continuing because even after and infinite time, as measured by a distant observer, the collapse is still not complete. Rather, the departure from a static configuration of Schwarzschild radius $r=2m$ as seen by a distant observer diminishes exponentially in time, with a chararcteristic time of the order of $2m$, or about 10 microseconds for an object of one solar mass. Table I explains the purely geometrical system of units employed in general relativety. In this sense, the system is a 'frozen star'.

In another sense, the system is not frozen at all. On the contrary, the dimensions shrink to indefinitely small values in a finite and very short proper time for an observer moving with the collapsing matter (see Figure 3). Moreover, a spherical system

TABLE I

Geometrical units

Einstein's account of gravitation is purely geometrical, and every quantity that arises is expressed in units of length. From this point of view the distinction between grams and meters, or between seconds and meters, is as artificial as the distinction between miles and feet.

Thus, in geometrical units:

1 cm of time (that is, 1 cm of light travel time) is $1 \text{ cm}/(3 \times 10^{10} \text{ cm s}^{-1} = 3.3 \times 10^{-11} \text{ s} = 1/30 \text{ ns}$.

1 cm of mass is $1 \text{ cm}/(G/c^2) = 1 \text{ cm}/(0.742 \times 10^{-28} \text{ cm gm}^{-1} = 1.4 \times 10^{28}$ gm, which is comparable to the mass of the Earth. The mass of the Sun, 1.987×10^{33} gm in conventional units, is 1.47 km in geometrical units. The deflection of light passing an object of mass m in geometrical units at a distance of closest approach b is $\theta = 4m/b$.

1 cm^2 of angular momentum is $1 \text{ cm}^2/(G/c^3) = 1 \text{ cm}^2/(2.47 \times 10^{-30} \text{ s gm}^{-1}) = 4.05 \times 10^{38} \text{ gm cm}^2 \text{ s}^{-1}$. The maximum angular momentum for a black hole of 1 km mass is $(1 \text{ km})^2$.

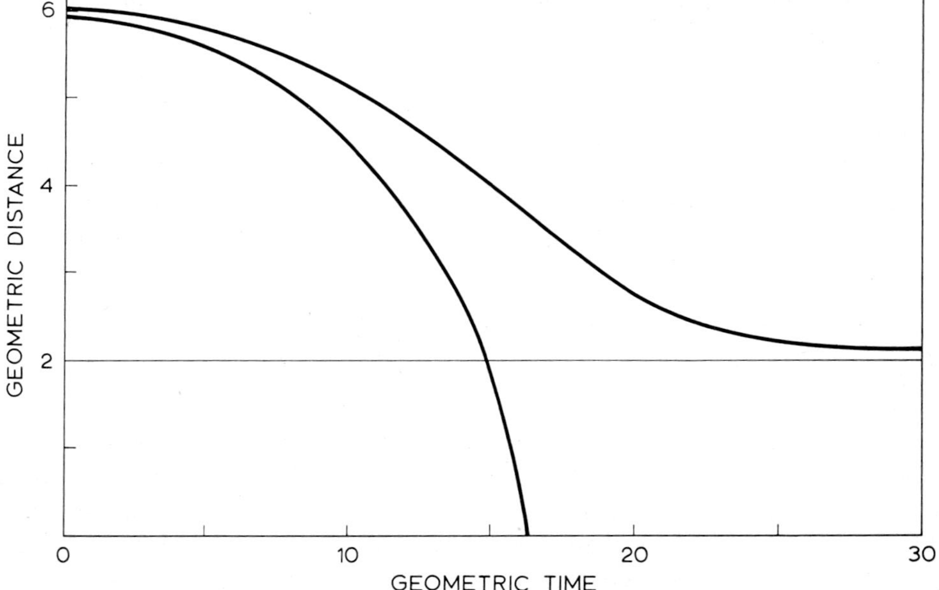

Fig. 3.　Fall towards a Schwarzschild black hole as seen by a comoving observer (upper) and a distant observer (lower). The proper time for the free fall to the center is finite, although the approach to the Schwarzschild radius as seen by a distant observer is asymptotic in time. 'Relative distance' and 'relative time' are measured in units of the mass of the black hole.

appears black from outside; no light can escape. Light shot at it falls in. A particle shot at it falls in. A 'meter stick' would be let down in vain to measure the dimensions of the object. The stick is pulled to pieces by tidal forces, and the broken-off pieces fall in without a trace. In these senses, the system is a black hole.

4. Process of Formation

At least three processes suggest themselves for the formation of a black hole:

Direct catastrophic collapse of a star with a white-dwarf core, a collapse going through neutron-star densities without a stop.

A two-step process: the collapse of a star with a white-dwarf core to a hot neutron star followed by cooling and collapse to a black hole.

A multistep process, with first the formation of a stable neutron star and then the slow accretion of enough matter to raise the mass above the critical value for collapse.

What happens in the collapse has been well analyzed in the case of a system of spherical symmetry, and for small departures from spherical symmetry that lend themselves to analysis by perturbation methods. However, in the general and very important case of large departures from spherical symmetry, only a few highly simplified situations have so far been treated. This fascinating field is largely unexplored. The central question is easily stated: Does every system after complete gravitational collapse go to a 'standard final state', uniquely fixed by its mass, charge and angular momentum and by no other adjustable parameter?

5. A Dust Cloud

Start with a cloud of dust of specific density 10^{-16} and radius 1.7×10^{10} cm. Let the cloud be imagined to draw itself together by its own gravitational attraction until its radius falls to 10^{-5} of its original value, or 1.7×10^{14} cm. The dust is still dust. No pressure will arise to prevent the continuing collapse. However, despite the everyday nature of the local dynamics, the global dynamics has clearly reached extreme relativistic conditions. How then does one properly describe what is going on?

A variety of treatments of this problem have been given, from the original analysis of Oppenheimer and Snyder (1939) to treatments of O. Klein (1961) and others. The simplest analysis for our purposes is that of Beckedorff and Misner (1962) in which the geometry interior to the cloud of dust is identical with that of a Friedmann universe, that is, a three-sphere of uniform curvature.

The geometry within the three-sphere is

$$ds^2 = a^2(\eta)[-d\eta^2 + d\chi^2 + \sin^2\chi\,(d\theta^2 + \sin^2\theta\,d\phi^2)]$$

where $a(\eta)$ is the radius of curvature, and the hyperspherical angle χ would go from 0 to π if the sphere were complete. It is not; it extends only from the center to the surface of the cloud.

The density of the cloud at the starting instant is related to the initial curvature a_0 by the standard formula for the Friedmann universe $\varrho_0 = 3/8\pi a_0^2$. As the collapse proceeds, an increasing fraction of the gravitational energy of the dust cloud is converted into kinetic energy. However, the total mass-energy remains constant.

Outside the dust cloud, the geometry remains the static geometry of Schwarzschild (Birkhoff theorem)

$$ds^2 = -(1 - 2m/r)\,dt^2 + (1 - 2m/r)^{-1}\,dr^2 + r^2(d\theta^2 + \sin^2\theta\,d\phi^2).$$

The Friedmann geometry and the Schwarzschild geometry match at the boundary of the dust cloud. A particle located at this boundary falls according to different laws as calculated from the Friedmann and Schwarzschild solution. But the results must agree, and they do.

6. Light Cone

Light given off from a particle at the periphery of the dust cloud, before arrival at the Schwarzschild radius, will always escape if emitted radially outward. However, if it makes an angle to the radial direction in its own local Lorentz frame, it will make a still larger angle to the radial direction in a local Lorentz frame that happens to have zero velocity at the moment in question. The photon will be trapped unless emitted in an allowed cone around the outward direction. The allowed cone shrinks to extinction when the dust cloud contracts to the Schwarzschild radius. Light that emerges radially 'outwards' after the cloud has contracted within the Schwarzschild radius never escapes to a faraway observer. It is caught, not in the matter but in the collapse of the geometry surrounding the matter.

7. The Kruskal Diagram

According to Figure 3, the fall of a test particle towards a black hole ends at $r=2m$ as seen by a distant observer. The fall ends at $r=0$ according to someone falling with the test mass itself. How can two such different versions of the truth be compatible? For an answer, it is enough to focus attention on the Schwarzschild geometry itself, and on a test particle falling in this geometry.

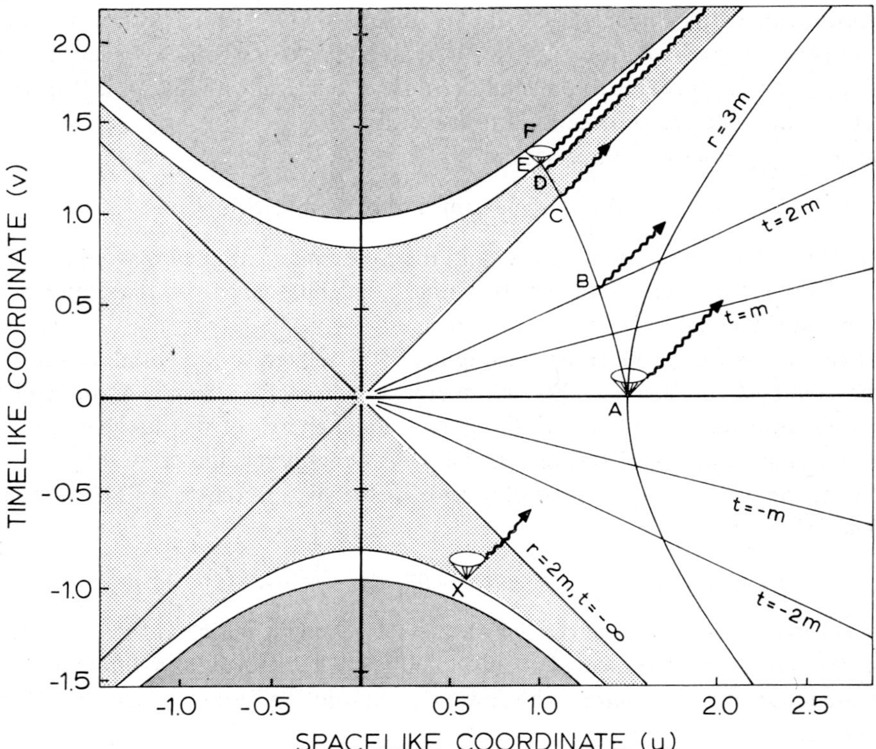

Fig. 4.   Kruskal coordinates for Schwarzschild spacetime, showing the relation to the usual coordinate (r, t). Radial light rays are straight lines with slope $= \pm 1$. Only the unshaded region is covered by the usual range of coordinates: $2m < r < \infty$, $-\infty < t < +\infty$. The colored line is the world line of a particle that starts at A and falls straight towards the black hole. A distant observer receives the signal it gives out at A and B. The ray emitted at C is the last ray that can escape to infinity, and it only gets to a distant observer after an infinite Schwarzschild time. Rays D and E are caught in the collapse of the geometry and never reach a distant observer. The curvature is nonsingular at C but rises toward infinity as $F(r=0)$ is approached. Point F is reached in a finite *proper* time. There is as little reason to expect a photon to escape from inside a black hole at X and to cross the inner boundary of everyday space, $r=2m$, as there is to expect advanced electromagnetic waves to travel inward from infinity.

The central point is simple. The range of coordinates $2m \leqslant r \leqslant \infty$, $-\infty < t < +\infty$ fails to cover all the Schwarzschild space-time. Time 'goes beyond infinity' just as Achilles goes beyond the tortoise in the famous paradox of Zeno. In no way can one see the incompleteness of the usual coordinate range more clearly by reference to the Kruskal coordinates (Kruskal, 1960; Fronsdal, 1959) as shown in Figure 4.

In this diagram, u is spacelike and v is timelike. Points of the same t-value lie on the straight line $v/u = $ constant. Points of the same r-value lie on the hyperbola $u^2 - v^2 = $ = constant, with asymptote $u = \pm v$. A light ray travelling radially outward is always represented by a straight line of slope $dv/du = +1$; one travelling radially inward, by a line of slope $dv/du = -1$.

One sees that r is a reasonable 'position coordinate' for values of r greater than $2m$; but for values of r less than $2m$ this coordinate changes character; it becomes a time coordinate rather than a space coordinate. The reverse happens to t; it changes from a time coordinate to a position coordinate. One can maintain oneself at a fixed value of r, with r greater than $2m$, by means of a rocket lift or otherwise. However, one can not maintain oneself at a fixed r less than $2m$ any more than one can make time stand still. The evolution of time forces such a person from $r = 1.9m$ to $r = 1.8m$ and so on, all the way to $r = 0$. No escape is possible; he is hemmed in by the light cone.

8. Departures from Symmetry

A spherical cloud of dust falls into a Schwarzschild 'black hole'. What happens if the cloud departs in a minor way from sphericity? If it is not endowed with angular momentum, it still collapses to a Schwarszchild black hole. If it has less than a critical angular momentum, it ends up as a uniquely defined but distorted black hole, given by the Kerr geometry, which is appropriate to a rotating system.

The 'standard solution' for a black hole of given mass and angular momentum has certain well defined quadrupole and higher moments. One finds (Israel, 1967; Doroshkevich et al., 1965, 1966) that any perturbation from the standard Kerr solution decreases exponentially with time. To the outside observer, all details of the gravitational field get washed out except mass and angular momentum, provided that the original perturbation was not too large.

In a similar way, all distributions of charge near a black hole appear to a distant observer to have spherical symmetry. The extreme gravitational field near a black hole greatly distorts the lines of force from the normal pattern. Far from the black hole, the lines appear to diverge from a point much closer to the center of the sphere than the actual location of the charge. The dipole moment goes to zero as the charge approaches $2m$. Nothing in the final pattern reveals the true location of the charge. We see in the black hole simply mass plus charge, and no other details. The law for the disappearance of the dipole, p, as given by Price, is

$$p \propto \frac{\log t}{t^4}.$$

This disappearance of the dipole takes place according to the same kind of law as the fadeout of perturbations of the quadrupole and higher moments of the mass distribution.

The collapse leads to a black hole endowed with mass and charge and angular momentum but, so far as we can now judge, no other adjustable parameters: 'a black hole has no hair'. Make one black hole out of matter; another, of the same mass, angular momentum and charge, out of antimatter. No one has ever been able to propose a workable way to tell which is which. Nor is any way known to distinguish

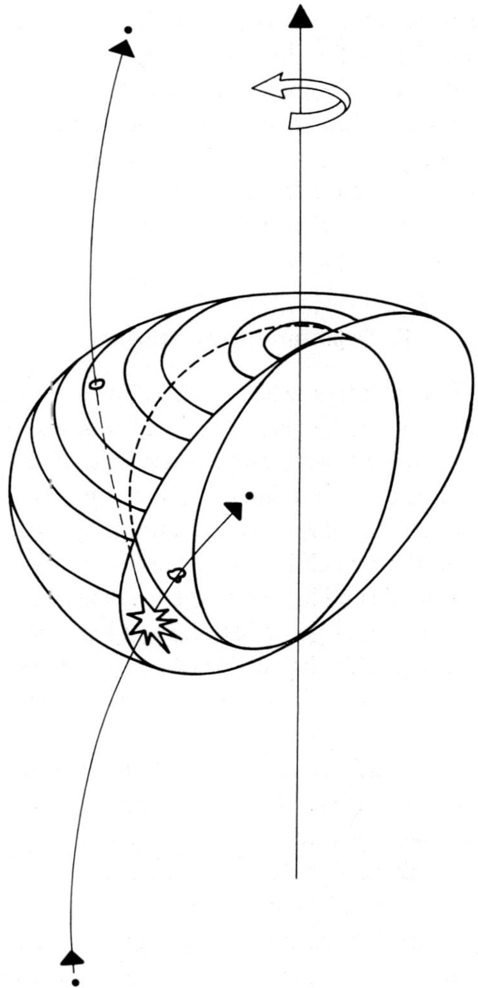

Fig. 5.   Ergosphere of a rotating black hole. The region between the surface of infinite redshift (outer) and the event horizon (inner), here shown in a cutaway view, is called the "ergosphere." When a particle disintegrates in this region and one of the fragments falls into the black hole, the other fragment can escape to infinity with more rest plus kinetic energy than the original particle.

either from a third black hole, formed by collapse of a much smaller amount of matter, and then built up to the specified mass and angular momentum by firing in enough photons, or neutrinos, or gravitons. And on an equal footing is a fourth black hole, developed by collapse of a cloud of radiation altogether free from any 'matter'.

Electric charge is a distinguishable quantity because it carries a long-range force (conservation of flux; Gauss's law). Baryon number and strangeness carry no such long-range force. They have no Gauss's law. It is true that no attempt to observe a change in baryon number has ever succeeded. Nor has anyone ever been able to give a convincing reason to expect a direct and spontaneous violation of the principle of conservation of baryon number. In gravitational collapse, however, that principle is not directly violated; it is transcended. It is transcended because in collapse one loses the possibility of measuring baryon number, and therefore this quantity can not be well defined for a collapsed object. Similarly, strangeness is no longer conserved.

9. Angular Momentum

The third property of a black hole is angular momentum. When it is nonzero, the geometry becomes more complicated. One deals with the Kerr (1963) solution to the field equations instead of the Schwarzschild solution. There are two interesting surfaces associated with the Kerr geometry, the 'surface of infinite red shift' and inside it, the 'event horizon'. An object at or within the event horizon can send no photons to a distant observer, independent of the object's state of motion or the direction of photon emission. For this reason, the event horizon is also called the 'one-way membrane'.

The Schwarzschild geometry represents the degenerate case of the Kerr geometry, in which the surface of infinite red shift and the event horizon coincide. In the general case, the two surfaces are separated everywhere except at the poles, as shown in Figure 6. The very interesting region between these surfaces is called the 'ergosphere'. A particle that comes within the ergosphere can still, if properly powered, escape again to infinity. However, its life in this region has an unusual feature; there is no way for it to remain at rest, rocket powered or not.

Energy can be extracted from the ergosphere by a mechanism that may occasionally have significance for a cosmic ray. Consider a particle that enters the ergosphere and disintegrates, one fragment falling into the hole and the other escaping to infinity (see Figure 6). Penrose (1969) has shown that the process can be so arranged that the emerging fragment has more energy at infinity than the original particle.

The extra energy is effectively extracted from the rotational energy of the black hole. If a particle can dip through the ergosphere and escape with some of the energy and angular momentum of the black hole, it is also true that a particle that is captured can increase the energy and angular momentum of the black hole. Capture is possible when the particle passes by sufficiently close to the black hole. The critical impact is smaller for a capture that increases the angular momentum of the system. Therefore, random accretion of particles leads to a gradual decrease of the angular momentum of the system. Selective accretion of particles with the maximum positive impact parameter is

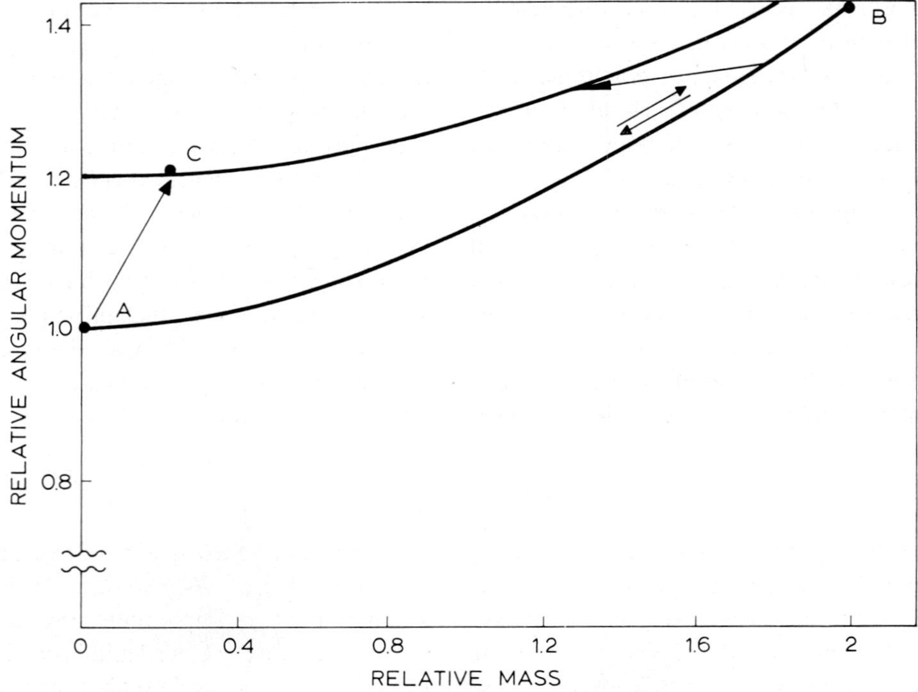

Fig. 6. Transformations of a black hole between the static Schwarzschild case (A) and the extreme Kerr case of maximum angular momentum (B) are accomplished by accretion of particles. The 'irreducible mass' (mass in the absence of rotation) remains constant in reversible transformations $(A \rightarrow B)$, which involve capture at grazing incidence to the black hole. In the irreversible transformation $A \rightarrow C$, the irreducible mass increases from 1.0 m_ir to 1.2 m_ir. The 'relative mass' is an abbreviation for 'mass in units of the original value of the irreducible mass'. The 'relative angular momentum' is in units of $m_\mathrm{ir}{}^2$, as explained below.

more appropriate to a black hole immersed in the debris from the collapse of a rotating star. Such favorable accretion can increase the angular momentum of the black hole to a critical value $L = m^2 = 2m_\mathrm{ir}^2$, at which 29% the total energy of the black hole is rotational energy. When this value is reached, further speedup is impossible.

The possibility of increasing and decreasing the angular momentum of the black hole leads to a 'phase diagram' somewhat similar to those in thermodynamics. In Figure 6, due to Demitrios Christodoulou, reversible changes in angular momentum are made by the accretion of a particle out of the most favorable orbit, at grazing incidence to the black hole. If the angular momentum is changed by the capture of a particle out of a less favorable orbit, the 'irreducible mass' (the mass of the black hole in the absence of rotation) must increase. There is no process whereby the irreducible mass can be caused to decrease. Therefore transformations accomplished by the accretion of particles from unfavorable orbits (for example, from head-on impact) are irreversible and lead to a steady movement upwards on the 'phase diagram' of mass versus angular momentum.

10. Gravitational Radiation

The discovery of quasars, objects with enormous energy release, led many workers to investigate gravitational collapse as a mechanism superior to fission or fusion for converting mass to energy. A closer look caused discouragement.

The difficulty can be understood by a study of Figure 7, which shows the radii of orbits of different angular momentum in the Kerr, Schwarzschild and Newtonian cases. In the Newtonian case, there are stable circular orbits of all radii down to zero. The one-dimensional potential for the radial motion has a minimum for all values of the angular momentum. This is not the case for general relativity. There is a lowest value for the angular momentum and a corresponding radius for which there is a minimum in the potential. This is the stable orbit closest to the black hole.

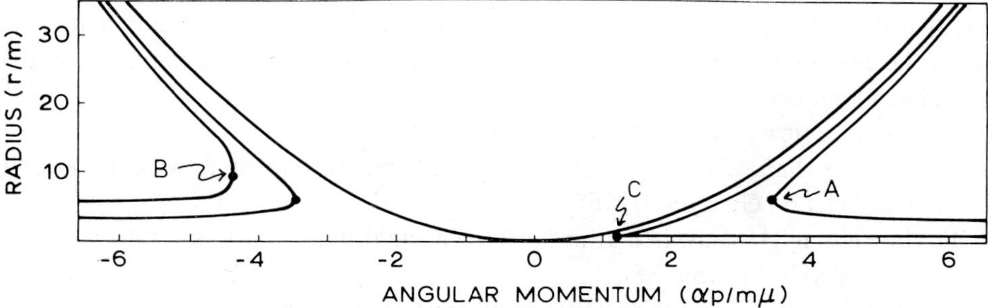

Fig. 7. Closest stable circular orbits for the Schwarzschild and Kerr black holes. For Newtonian gravity there are stable orbits of all radii down to zero. The parabola gives the radius of each orbit as a function of angular momentum. For the curved geometries, there are both a minimum and a maximum in the effective potential for each value of the angular momentum down to a critical value below which there is only a point of inflection–hence no stable orbits. A is the minimum Schwarzschild stable orbit; B and C are the minimum stable Kerr orbits for counterrotating and corotating particles respectively. These results have great significance for the amount of gravitational radiation a particle can emit before falling into a black hole.

When a particle emits gravitational radiation, it spirals into the black hole, moving to lower and lower orbits as it loses energy. When the energy of the particle decreases below the value in the last stable orbit, the particle is captured directly without further radiation. For the Schwarzschild black hole, the closest stable orbit is quite far from the center, and the particle can emit only about 5.7% of its mass as gravitational radiation before it precipitously falls into the black hole (point A in Figure 7). If the particle is orbiting contrary to the rotation of the Kerr black hole, it leaps in from an even greater distance where it has lost only 3.8% of its energy (point B in Figure 7). But if the particle is *corotating* with the black hole, it remains in stable orbit until it radiates 42.3% of its mass as gravitational energy (point C in Figure 7). These results, obtained by James Bardeen, give a great incentive for reexamining other energy-release mechanisms in the context of the Kerr geometry.

11. Search for Black Holes

The existence of black holes has been predicted for over thirty years. No one who accepts general relativity has seen any way to escape their existence. Moreover, a black hole is a characteristic geometrodynamical entity. A neutron star could still exist in Newtonian theory; not so a black hole.

It took 34 years from the prediction of a neutron star by F. Zwicky (Baade and Zwicky, 1934) in 1934 to its discovery as a pulsar by Anthony Hewish and others in 1968. If the prediction of black holes by Oppenheimer and Snyder in 1939 is also followed after 34 years by their discovery, what will be the technique by which they are detected in 1973?

Of all objects that one can conceive to be travelling through empty space, few offer poorer prospects of detection than a solitary black hole of solar mass. No light comes directly from it. It can not be seen by its lens action or other effect on a more distant star. It is difficult enough to see Venus, 12000 km in diameter, swimming across the disc of the sun; looking for a 15-km object moving across a far-off stellar light source would be unimaginably difficult.

Therefore, we turn to a black hole that is not isolated:

A black hole that affects a companion normal star only through its gravitational pull (Zel'dovich and Guseynov, 1965)

One close enough to draw in matter from the normal star (Shklovsky, 1967)

One embedded in a normal star

A black hole moving through a cloud of dispersed matter.

The possibility of capitalizing on a double-star system is most favorable when the black hole is so near to a normal star that it draws in matter from its companion. Such a flow from one star to another is well known in close binary systems (Gaposchkin, 1958; Struve, 1952) but no unusual radiation emerges. When one of the components is a neutron star or a black hole, a strong emission in the x-ray region is expected.

Gas being funneled down into a black hole undergoes heating by compression (Zel'dovich and Novikov, 1964). The temperature is extremely high (10^{10}–10^{12} K), but only a fraction of the radiation escapes, because it comes from a region of high red shift close to the Schwarzschild radius of the black hole. Zel'dovich, Novikov, and Schwarzmann conclude that the bulk of the radiation emerges in the visible part of the spectrum or in the X-ray and gamma-ray region, depending on the mass of the black hole.

References

Baade, W. and Zwicky, F.: 1934, *Proc. Nat. Acad. Sci. U.S.* **20**, 254.
Beckedorff, D. L. and Misner, C. W.: 1962, Unpublished communication.
Christodouloau, D.: 1970, *Phys. Rev. Letters*, to be published.
Colgate, S. and White, R. H.: 1966, *Astrophys. J.* **142**, 626.
Doroshkevich, A. G., Zel'dovich, Ya. B., and Novikov, I. D.: 1965, *Zh. Eksp. Teor. Fiz.* **49**, 170.
Doroshkevich, A. G., Zel'dovich, Ya. B., and Novikov, I. D.: 1966, *Sov. Phys. J, ETP* **22**, 122.
Fronsdal, C.: 1959, *Phys. Rev.* **116**, 778.
Gaposchkin, V. F.: 1958, *Handbuch der Physik* **L-225**, Springer Verlag, Berlin.

Harrison, B. K., Thorne, K. S., Wakano, M., and Wheeler, J. A.: 1965, *Gravitational Theory and Gravitational Collapse*, Univ. of Chicago Press, Chicago.
Hewish, A. *et al.*: 1968, *Nature* **217**, 709.
Israel, W.: 1967, *Phys. Rev.* **164**, 1776.
Kerr, R. P.: 1963, *Phys. Rev. Letters* **11**, 237.
Klein, O.: 1961, in *Werner Heisenberg und die Physik unserer Zeit*, Vieweg, Braunschweig.
Kruskal, M. D.: 1960, *Phys. Rev.* **119**, 1743.
Leblanc, J. M. and Wilson, J. R.: 1969, Lawrence Radiation Laboratory publication, UCRL-71873.
May, M. M. and White, R. H.: 1967, *Relativity Theory and Astrophysics*: Vol. 3, *Stellar Structure* (ed. by J. Ehlers), American Mathematical Society (see also *Phys. Rev.* **141** (1966) 1232).
Oppenheimer, J. R. and Snyder, H.: 1939, *Phys. Rev.* **56**, 455.
Penrose, R.: 1969, *Riv. Nuovo Cim.* **252**, special issue.
Price, R.: unpublished communication.
Shklovsky, I. S.: 1967, *Astrophys. J.* **148**, L1.
Struve, O.: 1952, *Stellar Evolution*, Princeton U.P.
Zel'dovich, Ya. B. and Guseynev, O. Kh.: 1965, *Dokl. Akad. Nauk U.S.S.R.* **162**, 791.
Zel'dovich, Ya. B. and Novikov, I. D.: 1964, *Sov. Phys.–Dokl.* **9**, 246.

HALOS AROUND 'BLACK HOLES'*

V. F. SHVARTSMAN

Abstract. 'Black holes' – bodies confined within their gravitational radius – will necessarily attract interstellar gas. If $M_{\text{hole}} > 0.03\ M_\odot$, at least 0.01 mc^2 of the infalling material should be converted into radiation. The corresponding lumonosity would be of order $10^{32} (M/10\ M_\odot)^{3/2} (\varrho_{\text{gas}}/10^{-24}\ \text{g cm}^{-3})^{1/2}$ erg s^{-1}, and would result from synchrotron radiation by magnetized plasma that would be heated to $T \approx 10^{12}$ K during the infall process. The spectrum would have a very mild slope extending from optical to radio wavelengths. Black holes might be observable as faint optical stars with no lines; they could be distinguished by intensity fluctuations on a time scale $\Delta t \approx 10^{-5}$–10^{-2} s, with no periodic component whatever. In many cases accretion by massive holes ($10 \gtrsim M \gtrsim 10^4\ M_\odot$) should engender hard radiation (X and/or γ rays) exhibiting a flare behavior on a time scale ranging from a few months to tens of years; the peak intensity should be 10–100 times the synchrotron intensity. This phenomenon is associated with the turbulence of the interstellar medium: the angular momentum of the gas would halt the infall near the gravitational radius of the hole, and the momentum would be 'annihilated' when the object moves into the adjacent turbulence cell. Only if $M > 10^6\ M_\odot$ would the angular momentum of the gas be capable of diminishing the mass falling into an isolated hole (that is, its luminosity). During infall toward a hole of $M \approx 10^5\ M_\odot$, the gas will initially remain cool ($T \approx 5000$ K) and will display an emission spectrum similar to the optical spectra of quasars. Because of accretion, a hole in a binary-star system might be observable as a visible secondary component. Accretion by a hole can be distinguished observationally from accretion by a neutron star. Possible candidates for black holes that may actually have been detected include certain type-Dc white dwarfs, the γ-ray star Sgr γ-1, the X-ray flare stars Cen X-2 and Cen X-4, and such objects as Sco X-1 and Cyg X-2.

1. Introduction

Perhaps the most interesting implication of general relativity theory is the prediction that the universe may contain masses that are confined within their gravitational radius $r_g = 2GM/c^2$. To an external observer such objects should appear rather like 'black holes', drawing matter and radiation into themselves. We will recall that although a collapsed body will of itself radiate nothing at all – neither light, neutrinos, nor gravitational waves – it will nevertheless possess a static gravitational field which will influence its surroundings. Matter that is drawn in will reach r_g only asymptotically, after an infinite time; the region $r < r_g$ could not be observed at all by an external observer, and would thereby 'drop out' of our space [1].

How might 'black holes' be formed? There are at least five ways:

(1) holes may have been present in the universe 'from the beginning', that is, left over from the epoch of singularity;

(2) holes may have developed from density fluctuations during the prestellar stage;

(3) holes may have formed from supermassive first-generation stars;

(4) holes may have resulted from the relativistic evolution of close clusters, galaxies, and the like;

(5) finally, holes may have developed from ordinary but sufficiently massive stars.

It is highly probable that this last possibility has been realized. The discovery of pulsars, as is now well recognized, has confirmed the old theoretical prediction that massive stars are unstable and should suffer a catastrophic collapse at the end of their

* Reprinted from *Soviet Astronomy* **15**. Original article submitted 20 July 1970.

H. Gursky and R. Ruffini (eds.), Neutron Stars, Black Holes and Binary X-Ray Sources, 394–405. All Rights Reserved.

evolution. But theory predicts two final states after collapse: a neutron star in the event that the mass of the object is less than 1.5 $M_\odot$; and an uninterrupted infall, or a 'collapsed' star, if $M > 1.5\, M_\odot$.

Neutron stars have been discovered because of their activity: the generation of relativistic particles and radio waves. Collapsed stars, however, are passive by their very nature. Consequently, it is usually considered that 'black holes' might most likely be discovered through their influence on radiating matter: on the motion of the normal component in a binary star [2–4], on stars in globular clusters [5] or galaxies [6], on the motion of the galaxies themselves [7] and so on.

However, the method of the 'excluded third' is risky in astronomy. And furthermore, when configurations with an invisible component are selected the objects sought might be left out of the list altogether. As we shall demonstrate in this paper, holes whose mass exceeds the solar mass ought to be surrounded by highly luminous halos.

The question of the energy released through accretion by collapsed bodies was first posed by Zel'dovich [8] and Salpeter [9]. The accretion of gas by 'superholes' ($M > 10^7\, M_\odot$) located at the centers of galaxies has been discussed by Lynden-Bell [10]. We shall be interested primarily in 'stellar' masses, with $10^{-1}\, M_\odot \gtrsim M \gtrsim 10^6\, M_\odot$. Their luminosity will turn out to be given by the single equation $L \approx 0.1\, c^2\, \mathrm{d}M/\mathrm{d}t$, while their spectra can be highly diversified.

2. Intensity of the Accretion and Flow Regimes

Let us imagine a massive object moving at a velocity u through a gas possessing no angular momentum. On the 'back' side of such an object a conically shaped shock wave will be formed in which the gas will lose the component of its velocity perpendicular to the shock front. After compression in the shock wave, particles having sufficiently small impact parameters will fall into the star [1]. One can show that the infall of gas will begin at the characteristic distance

$$r_c = \alpha GM/v_c^2 = \alpha 10^{14}\, M v_{c(10)}^{-2}\ \mathrm{cm}, \tag{1}$$

the distance at which the potential energy of the particles becomes comparable to the kinetic energy; and the flux of mass at the star will be

$$\frac{\mathrm{d}M}{\mathrm{d}t} = \beta 10^{11}\, M^2 v_{c(10)}^{-3} n_c\ \mathrm{g\ s}^{-1}. \tag{2}$$

In Equations (1) and (2) the mass M is expressed is solar masses, the velocity $v_c = = (u^2 + a_c^2)^{1/2}$ is expressed in units of 10 km s^{-1}, and the sound velocity a_c in the gas and the density n_c [atoms/cm^3] of the material drawn in refer to distances $r > r_c$. The dimensionless coefficients α and β are functions of the adiabatic index of the gas and the ratio u/a. We may take the rough approximations $\alpha \approx 1$, $\beta \approx 1$. Numerical values for various γ and u/a have been given by Salpeter [9]. Historically, Equation (2) was first obtained by Hoyle, Lyttleton, and Bondi [11–13].

Note that for any reasonable value of the adiabatic index γ the diameter of the tail will be $d \approx r_c \gg r_g$, and the pressure will be finite. It therefore seems to us beyond doubt that a symmetrization of the flow should take place during infall, so that for $r \ll r_c$ the motion of the plasma may be considered radial (see Figure 1).

In the spherically symmetric case, with back pressure absent, a supersonic flow of the gas will be established at $r \ll r_c$, that is, practically free fall [1]. To find the temperature variation, we shall write the second law of thermodynamics, $dE = -p\, dV + dQ$, in the form

$$\tfrac{3}{2} R^* \frac{dT}{dt} = R^* \frac{T}{\varrho} \frac{d\varrho}{dt} - 5 \times 10^{20}\, T^{1/2} \varrho\kappa + \frac{dQ'}{dt}. \tag{3}$$

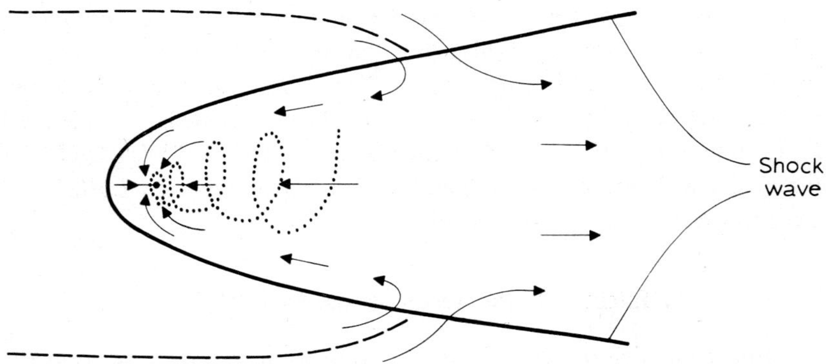

Fig. 1. The pattern of hydrodynamic accretion in the case where the star's own velocity u is much greater than the sound velocity a in the gas. The dashed curves correspond to the critical impact parameter. The dotted curve denotes the trajectory (helical) that the particles describe if an angular momentum relative to the star is present. In the figure the momentum vector is oriented parallel to the direction of motion of the object.

Here R^* is the gas constant; the second term on the right-hand side describes the losses of 1 g of plasma to radiation ($\kappa = 1$ corresponds to bremsstrahlung from a fully ionized plasma), and the third term represents the energy variation due to other non-adiabatic processes. Since $\varrho \propto r^{-3/2}$, we have

$$\frac{dT}{dr} = -\frac{T}{r} + [2 \times 10^{-4}\, M v_{c\,(10)}^{-3/2} n_c] \frac{\sqrt{T}}{r} \kappa + \frac{dQ'/dt}{v \cdot \tfrac{3}{2} R^*}. \tag{4}$$

As it falls in toward a massive object the gas will be efficiently cooled through radiation ($T = [A \ln(r/r_0) + T_0^{1/2}]^2$); when $T \approx 5000\,\mathrm{K}$ is reached, recombination will begin and a temperature $T \approx \mathrm{const}$ will be established. In the case of infall to an object of low mass, radiation will play a minor role; if the last term is neglected, the temperature variation will approach a $T \propto r^{-1}$ law, and the adiabatic index of the gas $\gamma \to \tfrac{5}{3}$. The physical explanation of this difference is clear: the smaller the object, the smaller its gravitational radius, that is, its characteristic scale and the time for infall, and the radiation processes will become slow compared to the contraction process. Taking $dQ'/dt = 0$ (see the Appendix), we obtain for the critical mass, by Equation (4),

$$M_{cr} \sim 10^4 \, M_\odot \, [T'_{(4)}]^{1/2} \, v^3_{c\,(10)} n_c^{-1} \kappa_{(2)}^{-1} \, . \tag{5}$$

Here $T'_{(4)}$ denotes the temperature at the distance of interest to us in units of $10^4 \, \mathrm{K}$, and $\kappa\,(2) \equiv \kappa/100$.

3. The Gas-Dynamics Approximation

The high temperatures developed during infall toward a mass $M < M_{cr}$ cast some doubt on the applicability of the gas-dynamical approach. At $r < r_c$, the mean free path with respect to Coulomb collisions,

$$l = kT^2/ne^4 \cdot L_{\mathrm{Coul}} \approx 10^{12} \, T_{(4)} n^{-1} \, \mathrm{cm}\,, \tag{6}$$

will in this case far exceed the characteristic size r of the region. For $M \approx M_\odot$, the free path $l \approx r$ even at the critical radius. We recall that the mass flux at the star in the approximation of noninteracting particles is smaller than the gas-dynamical flux (2) by a factor $(c/a_{\mathrm{sound}})^2 \approx 10^9$ [1].

However, the material drawn in will contain magnetic fields. The Larmor radius of protons moving at thermal velocities will be smaller than the dimensions of the region of motion even at field strengths $H > 1.3 \, T^{1/2} \, r^{-1}$ g. Inasmuch as $T \leqslant T_c (r_c/r)$, we have the condition for capture

$$H\,(r) \geqslant 10^{-12} \, T_{c\,(4)}^{3/2} M^{-1} \, (r_c/r)^{3/2} \, \mathrm{G}\,. \tag{7}$$

The field at infinity is of order 10^{-6} g, so that in order for the condition (7) to be satisfied out to $r = r_g$ it is necessary that H increase with depth at least as $r^{-0.8}$. We shall see in Section 5 that there are grounds for expecting a far steeper rise in $H(r)$, so that the gas-dynamics approach would be applicable.

The rise in the magnetic field would prevent any heat exchange between the different layers (see the Appendix). It would also affect the radiation and motion of the plasma. But first let us digress to consider one other topic.

4. Luminosity and Efficiency in the Idealized Problem

For $M > M_{cr}$ the gas will be practically isothermal, so that the luminosity

$$L = \frac{dM}{dt} \, R^* \int_{r_1}^{r_2} T \, \frac{dn}{n} \approx 10^{34} \, M_{(5)}^2 v_{c\,(10)}^{-3} n_c \, \mathrm{erg} \, \mathrm{s}^{-1}\,. \tag{8}$$

Here $M_{(5)} \equiv M/10^5 \, M_\odot$, $T \approx 5000 \, \mathrm{K}$, and r_1 is the radius below which radiation will rapidly be quenched by general-relativity effects (gravitational plus the Doppler red shift) [1]. Roughly speaking, $r_1 \approx 2r_g$. The luminosity (8) will fall mainly in the optical range and will result from line and recombination emission. If $M < M_{cr}$, the plasma temperature will increase inward along an adiabatic curve, and the luminosity of the spherical layer bounded by r and $r/2$ will be

$$L(r) \approx r^3 \kappa 10^{-27} n^2 T^{1/2}(r) \approx 10^{21} M^3 v_{c\,(10)}^{-6} n_c^2 [T_{12}(r)]^{1/2} \text{ erg s}^{-1}. \quad (9)$$

The value $T(r) \approx 10^{12} \text{ K}$ may be regarded as an upper limit. The optical depth

$$\tau(r) = \int_r^{r_c} \sigma n_c (r_c/r)^{3/2} \, dr \approx 10^{-6} (\sigma/\sigma_k) M (r_g/r)^{1/2} n_c T_{c\,(4)}^{-3/2} \quad (10)$$

(where σ_k is the Compton cross section) will always be much less than unity if $M < M_{cr}$, but if $M > M_{cr}$ the interior regions may become opaque, despite the Doppler shift.

In the approximation considered here, the efficiency of black holes would be extremely low. If accretion by an object with $M > M_{cr}$ takes place, only $\approx 10^{-8}$ of the mass of the infalling material will be converted into radiation; if $M < M_{cr}$ the efficiency will be even lower. The actual situation, however, is more favorable than this idealized one.

5. The Influence of Magnetic Fields *

Any plasma that is drawn in will necessarily contain magnetic fields. The observable effects will depend in an essential way on the law for the growth of the fields during the fall process. In our opinion there is only one real possibility: the magnetic energy ε_m and the gravitational energy ε_{gr} per unit volume should be of the same order not only at the critical radius (where all four energies – gravitational, kinetic, thermal, and magnetic – would be of the same order) but also in the zone $r < r_c$. We can prove this statement by assuming the contrary. For suppose that in some region $\varepsilon_m \ll \varepsilon_{gr}$; then the field would not affect the infall, and because of strict freezing-in (it is readily seen that during infall the magnetic viscosity will always be far smaller than the kinematic viscosity) its radial component H_r would increase according to the law $H_r \propto r^{-2}$, so that $\varepsilon_m = H^2/8\pi \propto r^{-4}$. However, $\varepsilon_{gr} \propto r^{-5/2}$, and an equality $\varepsilon_m \approx \varepsilon_{gr}$ would therefore be established very rapidly. On the other hand, the inequality $\varepsilon_m \gg \varepsilon_{gr}$ also would not be possible, because the field energy (due to freezing-in) would be derived from the energy of contraction, that is, from the kinetic energy, which would be smaller than ε_{gr}. We are therefore left with the condition $\varepsilon_m \approx \varepsilon_{gr}$. The behavior of the infall will be highly complicated in this situation: the field will have a predominantly radial character, and will be annihilated at the boundaries of sectors in which it is oppositely directed; the plasma will be inhomogeneous; the rate of infall will be nonuniform; very strong gutter instabilities will develop periodically in regions where $H \perp r$; and so on.

We shall nevertheless assume that despite the magnetic field, on the average the infall of gas does not depart seriously from free fall; that is, we shall suppose that $\bar{v} \approx (2GM/r)^{1/2}$ and $\bar{n} \propto r^{-3/2}$. Admittedly, both our idealizations may seem far-

---

* See the note added in proof on p. 403.

reaching, but it is natural to make them in our first steps toward a solution of the problem. We shall, then, let $H^2/8\pi = anGMm_p/r$, with $a \approx 1$.

In the case of infall toward a mass $M < M_{cr} \approx 10^4\, M_\odot$, the plasma will be strongly heated, and the synchrotron losses will rapidly become decisive. In the range $T > 10^{10}\,\mathrm{K}$ Equation (4) will take the form $(dQ'/dt = 0)$

$$\frac{dT}{dr} = -\frac{T}{r} + [3 \times 10^{-8} M^2 v_{c(10)}^{-3/2} n_c] \frac{T^2}{r^2}\, a. \tag{11}$$

Its solution is $T = (Cr + A/2r)^{-1}$; after the value

$$T_{\max} = 3 \times 10^{12}\, M^{-1/2} v_{c(10)}^{3/4} n_c^{-1/2} a^{-1/2}\,\mathrm{K} \tag{12}$$

is reached the 'temperature' will begin to fall. The efficiency of radiant energy release will be

$$\eta = 0.2\, mc^2 T_{\max(12.5)}, \quad \text{if} \quad T_{\max} < 3 \times 10^{12}, \tag{13a}$$

$$\eta = 0.2\, mc^2 T_{\max(12.5)}^{-3}, \quad \text{if} \quad T_{\max} > 3 \times 10^{12}. \tag{13b}$$

Of course, since equilibrium will not have been established during the accretion process [see Equation (6)], the 'temperature' T should be interpreted not as the Maxwellian parameter but as the mean energy of motion of the electrons across the magnetic field. Equations (11–13) as well as (17–19) below are, to some extent, merely illustrative in character. (In a separate paper [17] we have given a more detailed discussion of the circumctances of magnetic accretion, a theorem regarding the 'equipartition of energy', and a demonstration that in the case of accretion by 'black holes' the synchrotron mechanism should convert about $0.1\, mc^2$ of the infalling material into radiation.) We shall defer to Section 8 a description of the corresponding spectrum; we first wish to point out one possible factor that might raise the efficiency.

6. The Influence of Angular Momentum

Interstellar gas, generally speaking, would possess a certain angular momentum K relative to the line of motion of a hole. It is well recognized [1] that, in general relativity, particle capture is possible if $K < K_g = 2mcr_g$. After it approaches a gravitating center an isolated particle with $K > K_g$ will again recede to infinity. According to Section 3, however, plasma at any distance from a collapsed object may be regarded as a continuous medium. Its compression at the tail of the flow, in the shock wave, will be accompanied by radiant energy release; if this energy becomes negative, then matter can never leave the neighborhood of the hole. The corresponding criterion is evident: it is necessary that the momentum K be much smaller than the quantity $K_c = mr_c v_c$. Thus if the momentum of the material drawn in satisfied the inequality

$$2mr_g c < K \ll mr_c v_c, \tag{14}$$

then before they fall into the hole the particles will go into a stationary orbit. The minimum radius of a stable stationary orbit would be $r_{\min} = 3r_c$ [18]. In order for a particle in Keplerian motion to reach $r_{\min}$, it should emit radiation of 0.07 mc^2; if the role of viscosity is appreciable and nearly solidbody rotation prevails, the efficiency would be twice as great.

Might we expect the condition (14) to be staisfied under actual conditions? Denoting the tangential velocity component by v_t, let us write the inequality in the form

$$1.2 \times 10^{-4} v^2_{c\,(10)} < v_{t\,(10)}\,(r = r_c) \ll v_{c\,(10)}. \tag{14a}$$

First let us consider the left-hand inequality. Estimates indicate that the internal scale of interstellar turbulence, $l_0 = Re_0 v'/v_t$, will always be less than or of the order of the radius r_c (Re_0 is the critical Reynolds number, and v' is the magnetic viscosity). Hence $v_t(r_c) = v_t(L_t) \cdot (r_c/L_t)^x$, where $L_t \approx 100$ pc and $v_t(L_t) \approx 10$ km s^{-1} [19]. In the interstellar medium, because of suppression of turbulence by the magnetic field and the formation of shock waves, the spectrum $v(l)$ should be steeper than a Kolmogorov spectrum ($x = \frac{1}{3}$); evidently $\frac{1}{2} \gtrsim x < 1$ [19, 20]. Let $x \equiv 2/(3+y)$. It is then easily seen than the first of the inequalities (14) will be satisfied for masses

$$M > M^{\mathrm{I}}_t \approx 3.5\, M_{\odot} v^5_{c\,(10)}\,[1.2 \times 10^{-2} v_{c\,(10)}]^v. \tag{15}$$

Thus for objects with stellar masses, spherically symmetric accretion will evidently be realized in many cases, while in some cases the plasma will be halted not far from r_g.

In the case of accretion of gas by very massive objects, rotation will undoubtedly play a role, and the radiant efficiency $\eta \approx 0.1\, mc^2$. It is now appropriate to inquire as to the influence of the momentum on $\mathrm{d}M/\mathrm{d}t$, that is, the right-hand inequality (14). For $L_t = 100\, pc$ and $v_t(L_t) = 10$ km s^{-1} this inequality will become

$$M \ll M^{\mathrm{II}}_t \simeq 3 \times 10^6\, M_{\odot} v^2_{c\,(10)}. \tag{16a}$$

The criterion (16a) refers to the case where $v_c \approx v_t \approx 10$ km s^{-1};* if $v_c \gg 10$ km s^{-1}, only differential galactic rotation would be capable of preventing accretion. For a body 10 kpc away from the galactic center we would have in place of the criterion (16a):

$$M \ll M^{\mathrm{II}}_t \simeq 3 \times 10^9\, M_{\odot} v^3_{c\,(50)}. \tag{16b}$$

The role of the momentum of the gas in accretion by objects with $M > 10^6\, M_{\odot}$ has been pointed out qualitatively by Salpeter [9]. He has also indicated a fundamental mechanism for momentum loss: the gas in adjacent turbulence cells would have opposite directions of rotation. We shall return to this topic presently, but we first wish to call the reader's attention to an important case in which the sign of the momentum might not change.

7. Accretion in Binary Systems

Our motive in discussing this situation is clear: in addition to a high efficiency, binary systems would ensure an enormous loss rate $\mathrm{d}M/\mathrm{d}t$. Thus, variables of the β Lyrae

---

* Note that under the conditions prevailing in the Galaxy, supermassive objects necessarily would rapidly diminish their velocity u [but not $v = (u^2 + a^2_{\mathrm{sound}})^{\frac{1}{2}}$] to about 10 km s^{-1} [21].

type lose as much as 10^{-5} $M_\odot$ yr^{-1}; under favorable conditions about one-half of this mass could fall into the second component. Hence the luminosity of holes in binary systems would in principle be capable of reaching 10^{38}–10^{39} erg s^{-1}.

Around the dense component gas streams should form a disk with an approximately Keplerian velocity distribution. The mechanism for momentum loss would be viscosity, which would tend to produce solid-body rotation. The inner regions of the disk would be retarded, and the outer regions accelerated. Some of the gas would leave the binary system, while the rest would sink in toward the dense component. Very high temperatures would evidently be attained during this settling process (see the model calculation by Prendergast and Burbidge [22]).

Accretion by holes, neutron stars, and white dwarfs in close binary systems has been proposed on several occasions [8, 22–24] as a source of energy. An important question arises here: would it be possible to distinguish accretion by a hole from accretion by a star? The answer will be clear from the considerations above. In the first case the radiation would all come from a hot, thin disk; in the second case half the radiation would be due to the disk and half to emission from the surface of the star, in which an appreciable contribution should be present from equilibrium radiation [25] and/or plasma oscillations [26]. Furthermore, in the case of accretion by a neutron star the radiation ought to contain a strictly periodic term (with $p > 1$ s) arising from the rotation of the star. In the case of accretion by a hole, only random luminosity variations would be expected, with $\Delta t \approx 10^{-5}$–10^{-2} s (see Section 9 below). Also, the emission spectrum of the disk itself probably would, in general, be quite flat. Thus a search for a hole in a binary system, based on the absence of an optically visible component [2–4], might actually exclude some of the objects sought.

8. Gamma-Ray, X-Ray, and Optical Stars?

We shall now subdivide accretion into spherical, helical, and disk types, depending on the character of particle motion. The first two types would be realized around isolated objects; the last type, in binary systems. In the case of disk acceretion, most of the energy should be released in the form of X-ray photons. Helical accretion would lead to the appearance of γ-ray stars as well as X-ray stars. Indeed, as we have been in Section 6, the spectrum of interstellar turbulence is such that in many instances the plasma would be halted (transferred from a helical to a circular orbit) in the vicinity of r_g. When a hole moves into the adjacent turbulence cell, where the gas is rotating in the opposite direction, particle collisions will begin to take place near r_g, and unlike the case of settling onto a disk, the process of momentum loss would here be very rapid

It is worth noting that in the case of accretion by an isolated neutron star the role of the momentum of the gas would probably be small because of the small value of r_c (see Section 6). On the other hand, even if a disk were to develop, it would be located near the surface of the neutron star and would rapidly sink into the star. The integrated luminosity of the objects in this situation would be very low, of order 10^{30} ergs s^{-1} [see Equation (2)]; and most of the radiation would fall in the ultraviolet [27]. If the neutron star has a magnetic field, the infall of gas would be stopped, in general, far beyond r_g, and would be accompanied by Langmuir oscillations at radio frequencies [26].* Evidently, then, only 'black holes' could be 'pure' γ-stars.

In this connection it is interesting to note the recent discovery [28] of a discrete source of γ rays ($E_\gamma > 50$ MeV) which has not yet been identified with an X-ray object. Yet, the sensitivity of the X-ray equipment was one or two orders higher than the sensitivity of the γ-ray counters.

The intensity of disk-type accretion should remain unchanged over tens of thousands of years, apart from fluctuations associated with the inhomogeneity of the gas flow. (Possible eclipses of the disk by the normal component would be of special interest.) The radiation of isolated objects due to spiral-type accretion would undoubtedly exhibit a flare behavior. The minimum characteristic time for a flare would be of order r_c/v_c, that is, a few months. Curiously enough, among the known sources of hard x rays there are definitely two classes of objects: those whole luminosity has remained constant, on the average, throughout the entire period of observation (for example, Sco X-1 and Cyg X-2); and those whose luminosity has varied by tens of times within a year, either appearing or disappearing from the field of view (such as Cen X-2 and Cen X-4 [29, 30]). Observers have often suggested that Sco X-1 and Cyg X-2 might belong to binary systems [31–34].

The accretion of gas by an isolated hole of mass $M \approx 1$–$100\ M_\odot$ should, as mentioned in previous sections, be primarily spherical. Magnetic fields should play a definite role in the radiation. According to Equations (2) and (13a), the corresponding luminosity should be

$$L \sim 0.1\ T_{\max\,(12.3)} c^2\ \mathrm{d}M/\mathrm{d}t \approx 2 \times 10^{33}\ M_{100}^{3/2} v_{10}^{-9/4} n^{1/2} a^{-1/2}\ \mathrm{erg\ s}^{-1}\,;$$

$$(17)$$

here $M_{100} = M/100\ M_\odot$. The spectrum of the radiation should have a distinctive form: the intensity should remain nearly constant over a wide frequency range near

$$v_{\max} \sim 10^{14}\ M_{100}^{-3/8} v_{10}^{-9/16} n^{-9/3} a^{-13/8}\ \mathrm{Hz}\,, \qquad (18)$$

an exponential decline should set in at $v \gg v_{\max}$, and at $v \ll v_{\max}$ there should be an extremely slow decline to frequencies at which absorption of the radiation becomes appreciable.

$$L(v) = \int_{v/2}^{v} (\mathrm{d}L/\mathrm{d}v)\ \mathrm{d}v \propto v^{4/13}\,, \qquad \mathrm{d}L/\mathrm{d}v \propto v^{-9/13}\,. \qquad (19)$$

Coherent mechanisms and negative self-absorption might be operative.

In cases where γ rays are generated, they would be in addition to synchrotron radiation. However, because of the 'accumulation' of material in circular orbits, the

---

* From the observational standpoint such objects might appear as 'second generation' pulsars. Further details have been given elsewhere [26].

'peak' γ-ray luminosity should be one or two orders higher than given by Equation (17).

The radiation of a 'black hole' would of course heat the gas at $r > r_c$, thereby influencing the intensity of the accretion [see Equation (2)]. One can show that for most cases of interest this effect would be insignificant, but in certain cases it could be decisive. A more thorough discussion of this topic, together with a solution of the self-consistent problem as exemplified by a neutron star, has been given elsewhere [27].

Conceivably, then, individual 'black holes' might already be observable today, as faint optical stars with no lines but with a nonthermal spectrum extending far into the low-frequency range (as far as radio frequencies). Perhaps some of the objects heretofore regarded as type DC white dwarfs are actually 'black holes'.

9. A Critical Experiment

What properties of the radiation would allow 'black holes' of stellar mass to be distinguished reliably from other objects? In our view, such properties would be the exceptionally small size of a hole together with the absence of any rotation, in other words, because of the development of instability in a magnetized plasma the luminosity of a hole would fluctuate on a time scale $\Delta t \approx 10^{-5}$–10^{-2} s, but a periodic term should be entirely absent.

10 Quasar-Like Stars?

Equations (5) and (16) imply that in the case of accretion by a hole with $M \approx 10^5\, M_\odot$ the infalling gas will be cooled up to the onset of the 'spiral' mode (somewhere in the zone $r \approx 10^{-4}$–$10^{-3}\, r_c$); $T \approx 5000\,\mathrm{K}$. The corresponding luminosities are given by Equation (8), and are small compared to the integrated luminosity, but nevertheless they fall wholly in the optical range. The spectrum of the radiation – broadened emission lines, well-developed recombation bands, the presence of a nonthermal continuum and absorption lines – should to a large extent resemble the optical spectrum of quasars. On the other hand, holes with masses of order $10^5\, M_\odot$ are interesting in that they might have come from 'first generation' stars that developed from entropy perturbations in the pregalactic medium [35] (the remainder of these perturbations would presumably have served as the origin of the globular clusters [36]).

In the case accretion by 'superholes' with $M \gg 10^5\, M_\odot$, the momentum of the gas would be expected to play a role from the very outset; but here too it would seem that in many cases a regime of 'cold accretion' would develop, accompanied by a spectrum similar to that observed for quasars. We intend to examine this possibility in a separate paper.

Appendix

The Absence of Heat Conductivity Between Different Layers

Why have we consistently neglected the term $\mathrm{d}Q'/\mathrm{d}t$, representing the heat conductivity? For radiative conductivity, we have done so because the optical depth of the plasma is negligible [see Equation (10)]; for ion conductivity, because the infall of the gas is supersonic; and finally, for electron conductivity, because the magnetic fields grow rapidly during the infall process. Let us consider this last point more carefully. According to Equation (6), the time required for exchange of energy between different electrons will far exceed the time required for infall toward the hole. Thus heat conductivity could only arise from the migration of electrons. However, such migration will be severely limited by the small Larmor radius of the electrons $(l_L/r < 10^{-6})$, by the strictness with which the freezing-in conduction is satisfied, and by the tangling of the lines of force during infall (see Section 5). In the interior, where the temperatures and fields are larger, there will in addition be a small 'mean free path' relative to synchrotron losses, as compared to the characteristic scale of the motion.

The author is indebted to Ya. B. Zel'dovich for suggesting the topic and for much valuable counsel. He is also grateful to G. S. Bisnovatyi-Kogan and L. M. Ozernoi for their comments.

Note added in proof: Kardashev [37] was the first to point out the circumstance that in accretion by a 'black hole' a substantial fraction of the rest mass of the infalling material might be liberated because of the presence of a magnetic field; in this connection see also the remark on page 360 of the Russian edition of Zel'dovich and Novikov's book [1]. The synchrotron radiation emitted upon accretion by the magnetosphere of a neutron star has been considered in recent papers [14, 15]. Bisnovatyi-Kogan and Sunyaev, in discussing the problem of infrared sources [16], consider that in accretion by collapsed stars the annihilation of the magnetic field may lead to the formation near r_g of a shock wave at whose front a substantial part of the energy will be transformed into plasma oscillations with the emission of radiation. Our views concerning magnetic accretion have been set forth more fully in a separate paper [17].

References

[1] Ya. B. Zel'dovich and i. D. Novikov: 1971, *Relativistic Astrophysics* [in Russian], Nauka, Moscow; 1971, English translation, Vol. 1, Univ. Chicago Press.

[2] O. Kh. Guseinov and Ya. B. Zel'dovich: 1966, *Astron. Zh.* **43**, 313 [*Sov. Astron.–AJ* **10**, 251 (1966)]; Ya. B. Zel'dovich and O. Kh. Guseinov: 1965, *Astrophys. J.* **144**, 840.

[4] O. Kh. Guseinov and Kh. I. Novruzova: 1970, *Astron. Tsirk.*, No. 560.

[5] A. A. Wyller: *Astrophys. J.* **160**, 443.

[6] A. M. Wolfe and G. R. Burbidge: 1970, *Astrophys. J.* **161**, 419.

[7] S. van den Bergh: 1969, *Nature* **224**, 891.

[8] Ya. B. Zel'dovich: 1964, *Dokl. Akad. Nauk SSSR*, **155**, 67 [*Sov. Phys.–Dokl.* **9**, 195 (1964)].

[9] E. E. Salpeter: 1964, *Astrophys. J.* **140**, 796.

[10] D. Lynden-Bell: 1969, *Nature* **223**, 690.

[11] F. Hoyle and R. A. Lyttleton: 1939, *Proc. Cambridge Phil. Soc.* **35**, 405.

[12] H. Bondi and F. Hoyle: 1944, *Monthly Notices Roy. Astron. Soc.* **104**, 273.

[13] H. Bondi: 1952, *Monthly Notices Roy. Astron. Soc.* **112**, 195.

[14] P. R. Amnuél' and O. Kh. Guseinov: 1968, *Izv. Akad. Nauk Azerbaidzhan SSR, Ser. Fiz. Tekh. Mat.*, No. 3, 70.

[15] G. S. Bisnovatyi-Kogan and A. M. Fridman: 1969, *Astron. Zh.* **46**, 721. [*Sov. Astron.–AJ*, **13**, 566 (1970)].

[16] G. S. Bisnovatyi-Kogan and R. A. Sunyaev: 1970, *Preprinty Inst. Priklad. Matem. Akad. Nauk SSSR*, No. 31; *Astron. Zh.* [*Sov. Astron.–AJ* (in press)].

[17] V. F. Shvartsman: 1970, *Preprinty Inst. Priklad. Matem. Akad. Nauk SSSR*, No. 42.

[18] S. A. Kaplan, Zh. Éksp. Teor. Fiz., *19*, 951 (1949).

[19] S. A. Kaplan and S. B. Pikel'ner: 1970, *The Interstellar Medium*, Harvard Univ. Press.

[20] S. A. Kaplan: 1954, *Dokl. Akad. Nauk SSSR* **94**, 33.

[21] W. H. McCrea: 1954, *Monthly Notices Roy. Astron. Soc.* **113**, 162.

[22] K. H. Prendergast and G. R. Burbidge; 1968, *Astrophys. J., J.* **151**, L83.

[23] I. S. Shklovskii: 1967, *Astron. Zh.* **44**, 930 [*Sov. Astron.–AJ*, **11**, 749 (1968)]; *Astrophys. J.* **148**, L1 (1967).

[24] A. G. W. Cameron and M. Mock: 1967, *Nature* **215**, 464.

[25] Ya. B. Zel'dovich and N. I. Shakura: 1969, *Astron. Zh.* **46**, 225 [*Sov. Astron.–AJ* **13**, 175 (1969)].

[26] V. F. Shvartsman: 1970 *Radiofizika* **13**, 1852.

[27] V. F. Shvartsman: 1969, *Preprint Inst. Priklad. Matem. Akad. Nauk SSSR*, No. 57; *Astron. Zh.* **47**, 824 (1970) [*Sov. Astron.–AJ* **14**, 662 (1971)].

[28] G. M. Frye, J. A. Staib, A. D. Zych, V. D. Hopper, W. R. Rawlinson, and J. A. Thomas: 1969, *Nature* **223**, 1320.

[29] W. H. G. Lewin, J. E. McClintock, and W. B. Smith: 1970, *Astrophys. J.* **159**, L193.

[30] W. D. Evans, R. D. Belian, and J. P. Conner: 1970, *Astrophys. J.* **159**, L57.

[31] Yu. N. Efremov: 1967, *Astron. Tsirk.*, No. 401.

[32] E. M. Burbidge, C. R. Lynds, and A. N. Stockton: 1967, *Astrophys. J.* **150**, L95.

[33] R. P. Kraft and M.-H. Demoulin: 1967, *Astrophys. J.* **150**, L183.

[34] J. Kristian, A. R. Sandage, and J. A. Westphal: 1967, *Astrophys. J.* **150**, L99.

[35] A. G. Doroshkevich, Ya. B. Zel'dovich, and I. D. Novikov: 1967, *Astron. Zh.* **44**, 295 [*Sov. Astron.–AJ* **11**, 233 (1967)].

[36] P. J. E. Peebles and R. H. Dicke: 1968, *Astrophys. J.* **154**, 891.

[37] N. S. Kardashev: 1964, *Astron. Zh.* **41**, 807 [*Sov. Astron.–AJ* **8**, 643 (1965)].

DISK MODEL OF GAS ACCRETION ON A RELATIVISTIC STAR IN A CLOSE BINARY SYSTEM*

N. I. SHAKURA

Abstract. A model of a close binary system incorporating a relativistic (neutron or collapsed) star is considered. The formation of the disk around the relativistic star and the energy flux radiated by the disk are analyzed. It is shown that the model can be used to explain the properties of galactic X-ray sources with a thermal emission spectrum.

1. Introduction

The general theory of relativity necessarily predicts the absence of a stable state for a cold star with a mass in excess of the Oppenheimer-Volkoff limit [1]. A real star has a store of energy which, it instantaneously released, can completely scatter the material of the star into space. If in the process of catastrophic contraction after the loss of stability there is no release of energy which is sufficient to terminate the contraction process, and even if there is a partial explosion such that the remaining mass is less than the critical value, the central parts of the star will collapse and will reach the gravitational radius $r_g = 2GM/c^2$ in a short period of time. An object of this kind is frequently referred to in the literature as a frozen or collapsed star, a collapsar, gravitational tomb, or a black hole. The collapsed star does not emit electromagnetic, gravitational, or any other forms of energy. The collapsar can be detected only through the interaction of its stationary gravitational field with the surrounding objects, including the gas cloud in which it may be located. In the spherically symmetric case of accretion of a cold gas, the kinetic energy of directed motion does not exceed forms of energy and is not radiated in the outward direction [2]. In the case of supersonic motion of a collapsed star through a gas cloud, a shock wave is produced behind the star along its line of motion in which 10–20% of the rest energy $m_0 c^2$ of the incident particles may be released [3]. Shvartsman [4] has shown, taking into account the magnetic fields frozen into the interstellar gas, that even in the case of spherical accretion an energy $E \sim 0.1\, m_0 c^2$ should be released. An amount of energy of the same order of magnitude is released during the accretion of gas onto a neutron star. The total luminosity $L \sim 0.1 c^2 \cdot (\mathrm{d}M/\mathrm{d}t)$ cannot exceed the critical value $L_{\mathrm{crit}} = 10^{38} (M/M_\odot)$ erg s^{-1} When $M = M_\odot$ and $L = 0.1 L_{\mathrm{crit}}$, the mass flow becomes $\mathrm{d}M/\mathrm{d}t \sim 10^{17}$ g s^{-1}.

The possibility of the liberation of gravitational energy during the accretion of gas onto neutron and frozen stars was first discussed in [3, 5–7]. Shakura and Zel'dovich [8] have calculated the radiated energy spectrum during accretion onto a neutron star without an intrinsic magnetic field from a sufficiently dense gas cloud. The effect on the radiation spectrum of a magnetic field $H \sim 10^9 – 10^{10}$ was considered in [9, 10] for a neutron star without rotation. The result of this work was the prediction of an X-ray flux at $E \sim 1$–50 keV with a nontrivial spectral distribution. On this basis it was suggested that, possibly, some X-ray sources are neutron stars radiating as a result

* Reprinted from *Soviet Astronomy* **16**. Original article submitted 16 June, 1971.

H. Gursky and R. Ruffini (eds.), Neutron Stars, Black Holes and Binary X-Ray Sources, 406–416. All Rights Reserved.

of accretion. Possible γ-ray and relativistic-particle intensities from a neutron star in the course of accretion were calculated in [11]. The discovery of pulsars, their identification with neutron stars, and their observed properties suggests that, in the presence of a strong magnetic field, rotation may act as a very effective source of activity for these stars. It is still not clear how long the pulsar stage of a neutron star can continue, and various estimates give the time for the attenuation of the magnetic field between $t \sim 10^6$ yr and $t > 10^{10}$ yr. The repeated 'firing' of a neutron star as a result of accretion, when the time of the pulsar stage in much shorter than the cosmologic time, was considered in [12, 13].

In addition to the papers listed above on the accretion by cold stars, there is also the paper by Lynden-Bell [14], who considered the possibility of disk accretion of gas by the nucleus of a galaxy which, in his opinion, is a massive superstar which has attained a radius smaller than the gravitational radius: a dead quasar.

Previously [18] and subsequently [19–21] attempts were made to find invisible cold stars in binary systems through their influence on the motions of the observed components. This possibility would appear to be realistic in sufficiently separated pairs, where the gas density in the neighborhood of the frozen star is negligible. Unless this is so, accretion around the frozen star should result in a 'hot halo'.

In this paper we shall consider the various phenomena which are a consequence of the existence in a close binary system of a relativistic star, either a neutron star or a collapsed star. The second component will be a star with an extended atmosphere filling its entire Roche surface. It turns out that the system becomes a strong source of X rays while retaining some of the properties of ordinary close binary systems [15], i.e., high-intensity gas flows, presence of a disk, irregular luminosity variations, and complicated Doppler shifts which are occasionally difficult to use to deduce the periodic variation in the velocity due to rotation of the system.

It was reported in [16, 17] that, under particular conditions, white dwarfs can also become strong sources as a result of accretion. Burbidge and Prendergast [17] pointed out that in a pair in which the star participating in accretion is small, the accretion itself should be of the disk type rather than spherically symmetric. The necessary mass flow for a white dwarf with $M \sim M_\odot$ and $L = 0.1 L_{crit}$ is about 10^{21} g s^{-1}, which is higher by four orders of magnitude than for a neutron star of the same mass. The equations derived here enable us to confirm the validity of the conclusions reported in [17].

2. Physical Picture of Disk Accretion; Basic Equations

When the star is small, the material ejected from the region near the internal Lagrangian point in a binary system has a substantial angular momentum relative to the small star, and an efficient mechanism is necessary to dispose of the angular momentum before the gas reaches the surface of the star or, in the case of the collapsed star, its gravitational radius. The most probable mechanism is the formation of a disk in which angular momentum transfer occurs between neighboring layers through turbulent friction. The presence of a disk around one of the stars in a close binary system is confirmed by both observations [15] and theoretical calculations [22]. In the absence of a supply of mass, the disk disperses, and the nonstationary problem is described by a diffusion-type partial differential equation. For a constant mass flow, the motion reaches a certain stationary state. Owing to the inhomogeneous nature of the gas flows, the outer boundary exhibits nonperiodic oscillations which are gradually damped out in the regions of the disk near the star. On the average, a particle at each point within the disk rotates in a circular orbit:

$$v_\varphi = \sqrt{\frac{GM}{r}}. \tag{2.1}$$

In the direction perpendicular to the plane of rotation the component of the gravitational force of the star is balanced by the gas pressure gradient.* From the equations of hydrostatic equilibrium we can then estimate the thickness of the disk:

$$z_0 = 2r \frac{v_s}{v_\varphi}, \tag{2.2}$$

where v_s is the thermal velocity of the particles. The radial motion due to momentum transfer through turbulent friction is superimposed on the Keplerian rotation:

$$\frac{\partial (\omega r^2)}{\partial t} + v_r \frac{\partial}{\partial r} (\omega r^2) = \frac{1}{y_0 r} \frac{\partial}{\partial r} (\sigma_{r\varphi} r^2), \tag{2.3}$$

where $\sigma_{r\varphi}$ (dyn cm) is the turbulent friction between neighboring layers, integrated with respect to z, and $y_0 = \int \varrho(z)\, dz$ is the surface density of the disk. In a stationary state $\partial/\partial t = 0$, and we have

$$\frac{dM}{dt} = 2\pi I = 2\pi y_0 v_r r = \text{const}. \tag{2.4}$$

Substituting (2.4) into (2.3) and integrating, we obtain

$$I\omega r^2 = \sigma_{r\varphi} r^2 + A. \tag{2.5}$$

Near a neutron star or a white dwarf without a strong intrinsic magnetic field, the disk is present right up to the surface of the star, and the angular velocity reaches its maximum values $\omega_m = \sqrt{GM/R^3}$ in a boundary layer of the order of the thickness of the homogeneous atmosphere. It then rapidly falls down to the angular velocity of the star $\omega_s \ll \omega_m$. At the point where $d\omega/dr = 0$, we have $\sigma_{r\varphi} = 0$, and $A = I\omega_m R^2$. Since $I\omega r^2 \sim r^{1/2}$, the constant A is small in comparison with the other two terms, with the exception of the boundary regions, and we can write good accuracy

$$I\omega = \sigma_{r\varphi}. \tag{2.6}$$

---

* Self-gravitation in the disk is negligible.

For a collapsed star, the stable circular orbits are absent for $r < 3r_g$, and for $r \to r_g$, general relativity effects predominate and the particles move independently of the mechanism of friction. In the first approximation, we shall assume that beyond $r = 3r_g$ matter is collapsed and leaves the field of view. The maximum possible energy release is $\approx 0.1 m_0 c^2$.

In the case of the finite perturbations which occur in a binary system, the assumption of developed turbulent motion in the disk presents no difficulty. The maximum turbulence scale is not greater than the disk thickness $z_0(r)$, and since $l \sim z_0(r) \ll r$, the turbulence turns out to be small-scale and isotropic. To describe local isotropic turbulent motion, it is convenient to have the equations derived for laminar flow in which the laminar viscosity is replaced with the turbulent viscosity $\eta_t = \varrho v_t l$:

$$\sigma_{r\varphi} = y l v_t r \frac{d\omega}{dr}. \tag{2.7}$$

In this expression $v_t = \alpha v_s$ is the velocity of turbulent motion and α is a parameter which describes the degree of excitation of the turbulence. Since $l = z_0 = 2\ v_s/\omega$, we have

$$\sigma_{r\varphi} = -3\alpha y v_s^2. \tag{2.8}$$

It is important to note that an analytic solution can be obtained only in the neighborhood of the star on which accretion takes place. Near the Lagrangian point, and on the boundary of the disk where flow separation takes place, the motion of matter is much more complicated. On the other hand, near a neutron star or a collapsed star there is an observable flux of radiation, and this is the most interesting region.

The presence of differential rotation and turbulent motion can lead to an enhancement of weak magnetic fields in the gas [23]. The energy of the enhanced field is probably not greater than the turbulent energy, so that the stresses due to the magnetic fields are small in comparison with (2.8), and the field does not influence the dynamics of the problem but may affect the acceleration of particle groups and the nonthermal radiation mechanism [14].

3. The Density and Temperature Distribution over the Disk

The following assumptions were introduced in the determination of the energy balance in the disk given in Appendix 1: (a) When $v_r \ll v_\varphi$ and $z_0(r) \ll r$, energy transfer along the disk by the average radial motion is negligible and the diffusion of photons occurs only along the z-axis, so that beyond a certain value of r the luminosity is given by

$$L(r) = \frac{1}{2} \frac{dM}{dt} \left(\frac{GM}{r} - \frac{GM}{R_1} \right); \tag{3.1}$$

(b) at each point within the disk we have local thermodynamic equilibrium $\varepsilon = aT^4$, and the surface of the disk radiates like a blackbody, i.e., the energy radiated per unit area is given by $Q = b T_e^4$, $b = 5.75$ erg deg^{-4} cm^{-2}.

Assumption a presents no difficulties, whereas in the presence of strong scattering by free electrons assumption b may not be satisfied. We shall discuss this below.

From the condition

$$L(r) = 4\pi \int_r^{R_1} Q(r)\, r\, \mathrm{d}r = \frac{1}{2}\frac{\mathrm{d}M}{\mathrm{d}t}\frac{GM}{r}, \qquad R_1 \gg r \tag{3.2}$$

we find that

$$T_e(r) = T_e(R)\left(\frac{R}{r}\right)^{3/4} \tag{3.3}$$

where $T_e(R) = (L/4\pi b R^2)^{1/4}$ is the temperature on the inner boundary of the disk R, where Keplerian orbits are still maintained, and $L = (\frac{1}{2})(\mathrm{d}M/\mathrm{d}t)(GM/R)$ is the total luminosity of the disk.

Energy-balance calculations show that when $L \sim 0.1 L_{\mathrm{crit}}$ or more, there is a region in which the radiation pressure $p_r = \varepsilon/3$ is substantially greater than the gas pressure $p_g = \varrho \Re T$ in the case of a neutron star and a collapsed star. Table I gives the surface temperature T_e, the internal temperature T_c, surface density y_0, and the thickness of the disk z_0 at the points R and r_1, where $p_g = p_r$. In the numerical calculations it was assumed that $\alpha \approx 1$, i.e., we assumed near-sonic excitation of turbulence. Table I also gives the values of the optical thickness for scattering $(\tau_s = \sigma_s y_0 = 0.38 y_0)$ and absorption $(\tau_f = \sigma_f y_0)$. For a fully ionized plasma, only the free-free transitions participate in absorption, and $\sigma_f = 3.7 \times 10^{22} \varrho/T^{7/2}$. For all models $\tau_s \gg 1$, $\tau_f < 1$. Scattering increases the time spent by the photon in the medium, and thus facilitates additional absorption, so that since $\tau_s \gg \tau_f$ the true thickness for absorption is $\tau_{\mathrm{true}} = \sqrt{\tau_s \tau_f}$. Therefore, only for the models with a neutron star and a collapsed star $L \sim 0.1 L_{\mathrm{crit}}$, $\tau_{\mathrm{true}} < 1$ (although for practically all models $\tau_f \ll 1$), and the leading process which

TABLE 1

Type of star	Frozen star $M=10\,M_\odot$		Neutron star $M=M_\odot$ $R=10^6$ cm		White dwarf $M=M_\odot$ $R=10^{10}$ cm	
Luminosity, erg s^{-1}	10^{38}	10^{36}	10^{37}	10^{35}	10^{37}	10^{35}
$T_e(R)$, K	6.1×10^6	2×10^6	1.1×10^7	3.5×10^6	1.1×10^5	3.5×10^4
$T_c(R)$, K	1.4×10^7	4×10^6	2.6×10^7	6.3×10^6	4×10^5	6.3×10^4
$y_0(R)$, g/cm^2	4.2×10^2	1.3×10^4	4.2×10^2	2×10^3	3.2×10^4	2×10^3
$z_0(R)$, cm	10^6	9×10^3	10^5	4.5×10^3	10^9	4.5×10^8
r_1/R	10	–	50	–	–	–
$T(r_1)$, K	1.1×10^6	–	5.7×10^5	–	–	–
$T_c(r_1)$, K	5.9×10^6	–	6×10^6	–	–	–
$y_0(r_1)$, g cm^{-2}	1.3×10^4		1.5×10^5	–	–	–
τ_s	1.6×10^2	5×10^3	1.6×10^2	7.6×10^2	1.2×10^4	7.6×10^2
τ_s	1.5×10^{-6}	0.42	1.7×10^{-6}	0.027	0.027	2.6
$\tau = \sqrt{\tau_j\tau_f}$	1.5×10^{-2}	46	1.6×10^{-2}	4.5	20	45
$L(\nu < 8\times10^{14}$ Hz$)$, erg s^{-1}	2×10^{34}	10^{33}	8×10^{32}	4×10^{31}	4×10^{35}	2×10^{34}

equalizes the temperature of matter and of radiation is Compton scattering [24]. For the other models the assumption of blackbody radiation is fully justified.

4. The Spectrum Radiated by the Disk

For a model with Planck's law of radiation, the luminosity of the whole disk at a particular frequency v is given by

$$L(v) = 4\pi^2 \int_R^{R_1} B_v(T)\, r\, dr = \frac{8\pi^2 hv^3}{c^2} \int_R^{R_1} \frac{r\, dr}{e^{hv/kT} - 1}, \tag{4.1}$$

where R_1 is the external radius of the disk.

We shall now substitute $x = (hv/kT)(r/R)^{3/4}$ and evaluate the above integral for the various regions of the spectrum. The results are

$$
\left.
\begin{array}{lll}
\text{(a)} & hv \ll kT(R_1), & L(v) \sim R_1^2 T(R_1)\, v^2 ; \\[4pt]
\text{(b)} & kT(R_1) \ll hv \ll kT(R), & L(v) \sim R^2 T^{8/3}(R)\, v^{1/3} ; \\[4pt]
\text{(c)} & hv \gg kT(R), & L(v) \sim R^2 v^2 e^{-hv/kT(R)} .
\end{array}
\right\} \tag{4.2}
$$

The above spectral distribution is close to the spectrum of an optically transparent plasma with low-frequency self-absorption. It satisfactorily approximates the observed radiation from some X-ray sources. The almost flat dependence $v^{1/3}$ in the intermediate frequency range is a result of the fact that the maximum contribution to the radiation at the frequency v is provided by the region of the disk for which $hv \sim 3kT(r)$. This enables us to estimate the maximum possible luminosity in the optical region if we know the total luminosity of the disk (see the last line in Table I).

5. Discusssion

The most reliable of our conclusions is that normal white dwarfs in close binary systems cannot be X-ray sources of the type of ScoXR-1, CygXR-2, CenXR-2, etc. This is in conflict with the predictions given in [17]. Only a white dwarf with a strong magnetic field in a close pair can act as an X-ray source as a result of accretion. In this case, the gas flows along the lines of force and the formation of hot spots and of a shock wave is possible (these raise the observed temperature). However, for luminosities $L \sim 0.1 l_{\mathrm{crit}}$, we have $dM/dt \sim 10^{21}$ g s$^{-1} \sim 10^{-5}\, M_\odot$ yr^{-1}, and the frequency of nuclear-reaction flares due to the accumulation of hydrogen in the surface layers of the white dwarfs [24–25], observed as explosions of novas and supernovas, would be 0.1–1 per annum. Thus, if the white dwarf is, in fact, an X-ray source, then this is true for purely internal reasons such as rotation, magnetic fields, and so on.

Attempts to detect pulsars in binary systems have so far been unsuccessful. It is possible that such pairs decay at the time of formation of the neutron star as a result of the rejection of a substantial fraction of the mass. Another possibility put forward in [26] is that the ambient plasma, whose density in the close pair can be substantial, prevents the neutron star from exhibiting the properties of a pulsar. Analysis of this kind of system is very complicated and the first steps in this direction are reported in [27]. It was suggested in [26] that such objects might be found by examining strictly periodic X-ray emission. On the other hand, in the case of a neutron star without a strong field, the disk exists right up to the surface.

It was noted in Section 4 that the spectral distributions resulting from the present model are similar to those for known X-ray sources. It is also possible to explain other observed properties. For example, let us consider the possible reasons for irregular and quasiregular activity: (a) small fluctuations due to the inhomogeneity of the gas flow and the associated inhomogeneity of the outer regions of the disk; (b) presence of turbulent magnetic fields in the disk may lead to the formation of current layers and flares similar to solar flares but with much greater energy release in regions with opposite directions of the magnetic fields which approach one another under the action of turbulent pulsations; (c) global flares due to the fact that the companion star may lose an amount of matter exceeding the flow necessary for the critical luminosity of the accreting star: Here we might have ejection of plasma clumps under the action of radiation pressure; (d) when the rotation of the companion and of the whole system is asynchronous, and the angular momenta are not parallel, the disk may rotate in a plane which is not parallel to the plane of rotation of the system, and its orientation may vary with the precession period of the companion. It is then very difficult to detect the periodic variations due to orbital motion (which also depend on the orientation of the system relative to the observer) against the background of these oscillations.

It would be naive to use existing data as a basis for concluding that the above model is the only one possible for a source such as ScoXR-1. There are perhaps 10 other models, both with internal and external (accretion) energy sources, which would completely describe the observed properties. Our aim was merely to draw attention to the consequences to which the presence of a relativistic star in a close binary system would lead.

The author is indebted to Ya. B. Zel'dovich for his constant attention and to the staff and research students in the Department of Astrophysics of the Institute of Applied Mathematics for valuable discussions.

Appendix 1

Energy Balance in the Disk

Practically throughout the disk, with the exception of the boundaries, we have $v_r \ll v_\varphi$. It can then be assumed that half the gravitational energy, which has been converted into turbulence and then into heat, is completely radiated at the same radius in the form of electromagnetic radiation, and the transfer of thermal energy over the disk is negligible. For values of r greater than a given value, the luminosity is then given by

$$L(r) = \frac{1}{2}\frac{GM}{r}\frac{dM}{dt}.$$

(A1.1)

It is clear that the total luminosity is given by

$$L = \frac{1}{2}\frac{GM}{R}\frac{dM}{dt}.$$

(A1.2)

Let $Q(r)$ be the total flux of radiant energy per unit area:

$$Q(r) = \frac{1}{4\pi r}\frac{dL}{dr}.$$

(A1.3)

Using (A1.1) and (A1.2) we can rewrite this in the form

$$Q(r) = \frac{L}{4\pi R^2}\left(\frac{R}{r}\right)^3.$$

(A1.4)

In the first approximation we shall assume that, at a given radius r, the amount of energy released per unit mass is constant along the z-axis, while the flow along the z-axis is given by

$$q(z) = Q_0(r)\, 2\,\frac{y(z)}{y_0(r)}.$$

(A1.5)

In this expression y_0 is the surface density of the disk at given r and $y(z)$ is the amount of material in a unit column with height z. Let us substitute (A1.5) into transfer equation:

$$q = -\frac{c}{3\sigma}\frac{d\varepsilon}{dy}$$

(A1.6)

and integrate

$$\varepsilon(y) = \varepsilon_c(r) - \frac{3Q(r)\,\sigma y^2}{cy_0},$$

(A1.7)

where $\varepsilon_c(r)$ is the radiant energy density at the center of the disk for $z=0$. On the surface of the disk $\varepsilon(y_0/2) = \sqrt{3Q(r)}/c$. Using this relation, we can rewrite (A1.7) in the form

$$\varepsilon_c(r) = \frac{3}{4}\frac{Q(r)}{c}\left(\sigma y_0 + \frac{4}{\sqrt{3}}\right).$$

(A1.8)

For an optically thick disk $\sigma y_0 \gg 1$, and we have, finally.

$$\varepsilon_c(r) = \frac{3}{4}\frac{Q(r)}{c}\sigma y_0,$$

(A1.9)

where $\varepsilon_c = aT^4c(r)$ and, therefore, (A1.9) gives the relation between T_c and y_0 along the disk. Solving this simultaneously with the equation of motion given by (2.2), we obtain the explicit form of the functions $y_0(r)$ and $T_c(r)$. The variation of z_0 with radial distance is given by (1.4).

We shall omit the intermediate steps and merely write out the final expressions for the various parts of the disk.

(a) In the region where $p_r = \varepsilon/3 = aT^4/3 \gg p_g = \varrho T$ and the scattering by free electrons $\sigma_s = 0.38$ is greater than the true absorption σ_f, we have $v_s^2 = \varepsilon_c/3\varrho = aT_c^4/3\varrho$ and

$$\varepsilon_c(r) = aT_c^4(r) = \frac{4}{\alpha} \frac{v_\varphi(R)\,c}{R\sigma} \left(\frac{R}{r}\right)^{3/2}, \tag{A1.10}$$

$$y_0(r) = \frac{16}{3\alpha\sigma} \frac{c}{v_\varphi(R)} \left(\frac{L_k}{L}\right)\left(\frac{r}{R}\right)^{3/2}, \tag{A1.11}$$

$$z_0(r) = R\,\frac{L}{L_k}, \tag{A1.12}$$

$$v_r(r) = -\frac{3}{4}\,\alpha v_\varphi(R)\left(\frac{L}{L_k}\right)^2\left(\frac{R}{r}\right)^{5/2}; \tag{A1.13}$$

(b) In the region of the disk where $p_r \ll p_g$ but $\sigma_s \gg \sigma_f$, $v_s^2 = \varrho \Re T$, $\Re = 8.3 \times 10^7$ erg deg^{-4}:

$$T_c(r) = \left(\frac{L^2\sigma}{16\alpha\pi^2 R^4\omega(R)\,c\Re a}\right)^{1/5}\left(\frac{R}{r}\right)^{9/10}$$
$$= \frac{10^8}{\alpha^{1/5}}\left(\frac{L}{10^{38}}\right)^{2/5}\left(\frac{10^6}{R}\right)^{1/2}\left(\frac{M_\odot}{M}\right)^{1/10}\left(\frac{R}{r}\right)^{9/10}, \tag{A1.14}$$

$$y_0(r) = \frac{1.3\times 10^5}{\alpha^{4/5}}\left(\frac{L}{10^{38}}\right)^{3/5}\left(\frac{M_\odot}{M}\right)^{4/5}\left(\frac{R}{r}\right)^{3/5}, \tag{A1.15}$$

$$z_0(r) = \frac{1.8\times 10^4}{\alpha^{1/10}}\left(\frac{M_\odot}{M}\right)^{11/20}\left(\frac{R}{10^6}\right)^{5/4}\left(\frac{L}{10^{38}}\right)^{1/5}\left(\frac{r}{R}\right)^{21/20}, \tag{A1.16}$$

$$v_r(r) = -1.7\times 10^6\,\alpha^{4/5}\left(\frac{L}{10^{38}}\right)^{2/5}\left(\frac{M_\odot}{M}\right)^{3/5}\left(\frac{R}{r}\right)^{2/5}. \tag{A1.17}$$

The radius r_1 at which $p_g = p_r$ is found by equating (A1.16) and (A1.12):

$$\frac{r_1}{R} = 10^{-4}\,\alpha^{2/21}\left(\frac{L}{L_\kappa}R\right)^{20/21}\left(\frac{10^{38}}{L}\right)^{4/21}\left(\frac{M}{M_\odot}\right)^{11/21}\left(\frac{10^6}{R}\right)^{25/21}. \tag{A1.18}$$

Therefore, for $R < r < r_1$ we must use Equations (A1.10)–(A1.13) and for $r > r_1$ we use (A1.14)–(A1.17). If for given L, M, and R we find that $r_i < R$, this means that the region with $p_r > p_g$ is totally absent.

In the above equations $L_{\mathrm{crit}} = 10^{38}\,(M/M_{\odot})$ is the critical luminosity of the star and α is a parameter describing the degree of excitation of turbulence.

Appendix 2

The Poynting–Robertson Effect

During absorption and subsequent isotropic reradiation by a particle rotating in a Keplerian orbit, the particle loses a proportion of its angular momentum and approaches the surface of the star. Shvartsman has pointed out that, in the case of a neutron star with luminosity close to the critical value, this effect, which is known in the literature as the Poynting-Robertson effect, can ensure the necessary flow of matter which, in turn, produces the required luminosity as a result of deceleration in the atmosphere. For an optically semitransparent plasma (the photon undergoes only one scatter), the equation for the rate of change of the angular momentum (since the system is electrically neutral and electrons and protons move along r in the same way) is given by

$$\frac{d(\omega r^2)}{dt} = -\frac{L}{4\pi r^2}\frac{\sigma}{c}r\left(\frac{v_c}{c}\right). \tag{A2.1}$$

In this expression $\sigma = \sigma_T/\mu_e m_p$ is the scattering coefficient per unit mass and $\sigma_T = 6.65\times 10^{-25}\ \mathrm{cm}^2$ is the Thomson scattering cross section. Since at the critical luminosity $\sigma L_{\mathrm{crit}}/4\pi c = GM$, we have from Equation (2.1) the following expression for the radial velocity of the particles:

$$v_r = -2v_\varphi(R)\frac{v_\varphi(R)}{c}\frac{L}{L_\kappa}\frac{R}{r}. \tag{A2.2}$$

Comparison of (A2.2) and (A1.13) shows that as r increases, the Poynting-Robertson effect becomes predominant. However, in point of fact, the disk is optically thick: It is only the material in the surface layers of the disk with $\tau \leqslant 1$ which absorbs and reradiates the radiation arriving directly from the star and the main mass of the disk is screened off. In view of this, the equation giving the rate of change of the angular momentum of a column of matter y_0 in the disk can be written in the form

$$y_0\frac{d(\omega r^2)}{dt} = -\frac{2}{c}\frac{L}{4\pi r^2}\frac{R}{r}\frac{v_\varphi(r)}{c}r. \tag{A2.3}$$

Substituting (A1.11) into (A2.3), we obtain the following expressions for the radial velocity:

$$v_r = -\tfrac{3}{4}\alpha v_\varphi(R)\left(\frac{v_\varphi(R)}{c}\right)^2\left(\frac{L}{L_\kappa}\right)^2\left(\frac{R}{r}\right)^{7/2}, \tag{A2.4}$$

i.e., the effect is negligible during disk accretion.

References

[1] J. R. Oppenheimer and G. M. Volkoff: 1938, *Phys. Rev.* **55**, 374.

[2] Ya. B. Zel'dovich and I. D. Novikov: 1967, *Relativistic Astrophysics* [in Russian], Nauka.

[3] E. E. Salpeter: 1964, *Astrophys. J.* **140**, 796.

[4] V. F. Shvartsman: 1970, Preprint IPM, No. 42; *Astron. Zh.* **48**, 479 (1971) [*Sov. Astron.–AJ*, **15**, 377 (1971)].

[5] Ya. B. Zel'dovich: 1964, *Dokl. Akad. Nauk SSSR* **155**, 67 [*Sov. Phys.–Dokl.* **9**, 195 (1969)].

[6] Ya. B. Zel'dovich and I. D. Novikov: 1964, *Dokl. Akad. Nauk SSSR* **158**, 811 [*Sov. Phys.–Dokl.*, **9**, 834 (1965)].

[7] Ya. B. Zel'dovich and I. D. Novikov: 1965, *Usp. Fiz. Nauk* **86**, 447 [*Sov. Phys.–Usp.* **8**, 522 (1966)].

[8] Ya. B. Zel'dovich and N. I. Shakura: 1969, *Astron. Zh.* **46**, 225 [*Sov. Astron.–AJ* **13**, 175 (1969)].

[9] P. R. Amnuel' and O. Kh. Guseinov: 1968, *Izv. Akad. Nauk AzerbSSR, Ser. Fiz.-Tekh. Mat. Nauk*, No. 3, 70.

[10] G. S. Bisnovatyi-Kogan and A. M. Fridman: 1969, *Astron. Zh.* **46**, 721 [*Sov. Astron.–AJ*, **13**, 566 (1970)].

[11] V. F. Shvartsman: 1970, *Astrofizika* **6**, 123; **6**, 309 (1970).

[12] V. F. Shvartsman: 1970 *Radiofizika* **13**, 1852.

[13] J. P. Ostriker, M. J. Rees, and J. Silk: 1970, *Astrophys. J. Letter* **6**, 179.

[14] D. Lynden-Bell: 1968, *Nature* **223**, 690.

[15] R. P. Kraft: 1963, *Adv. Astron. Astrophys.* **2**, 43.

[16] A. G. Cameron and M. Mock: 1967, *Nature* **215**, 464.

[17] G. R. Burbidge and K. H. Prendergast: 1968, *Astrophys. J.* **151**, 183.

[18] O. Kh. Guseinov and Ya. B. Zel'dovich: 1966, *Astron. Zh.* **43**, 313 [*Sov. Astron.–AJ* **10**, 251 (1966)].

[19] V. L. Trimble and K. H. Thorne: 1969, *Astrophys. J.* **156**, 1013.

[20] A. G. W. Cameron: 1971, *Nature* **229**, 178.

[21] R. Stothers: 1971, *Nature* **229**, 180.

[22] K. H. Prendergast: 1960, *Astrophys. J.* **132**, 162.

[23] S. I. Vainstein and A. A. Ruzmaikin: 1971, *Astron. Zh.* **48**, 902 [*Sov. Astron.–AJ* **15**, 714 (1972)].

[24] A. A. Ilarionov and R. A. Syunyaev: 1972, *Astron. Zh.* **49**, 58 [*Sov. Astron.–AJ* **16**, 45 (1972)].

[25] L. Mestel: 1952, *Monthly Notices Roy. Astron. Soc.* **112**, 598.

[26] V. F. Shvartsman: 1971, *Astron. Zh.* **48**, 438 [*Sov. Astron.–AJ* **15**, 342 (1971)].

[27] P. R. Amnuél and O. Kh. Guseinov: 1970, *Astrofizika* **6**, 397.

[28] J. M. Cohen and A. G. W. Cameron: 1971, *Astrophys. Space Sci.* **10**, 227.

EVIDENCE FOR THE BINARY NATURE OF CENTAURUS X-3 FROM *UHURU* X-RAY OBSERVATIONS*

E. SCHREIER, R. LEVINSON, H. GURSKY, E. KELLOGG,
H. TANANBAUM, and R. GIACCONI

Abstract. Analysis of data spanning a year of observations of the pulsating X-ray source Cen X-3 from *Uhuru* has revealed the existence of periodic variations intensity of the source and correlated sinusoidal variations in the period of the $4^{s}_{.}8$ pulsations. We interpret this effect as due to an occulting binary system. The changes in intensity are then due to occultation of the X-ray source by a large massive companion, and the sinusoidal variations in the period of the $4^{s}_{.}8$ pulsations are due to Doppler effect. The orbital period of the X-ray source is $2^{d}_{.}08712 \pm 0^{d}_{.}00004$. The time of occurrence of the center of occultation is JD $(2,441,132.081 \pm 0.001) \pm n(2.08712 \pm 0.00004)$. Physical parameters for the system are derived. Evidence for the existence and nature of an extended atmosphere surrounding the massive occulting object is discussed.

1. Introduction

In a previous letter (Giacconi *et al.*, 1971) we have reported the discovery of periodic X-ray pulsations from Cen X-3. Over 70 percent of the X-ray flux was found to be pulsed with a period of approximately $4^{s}_{.}8$. The average intensity was observed to change by a factor of 10 within 1 hour during a day-long observation. The regular nature of the pulsations set the source apart from the variable pulsating sources, such as Cyg X-1 and the source in Circinus, 2U 1516–56 (Schreier *et al.*, 1971a; Margon *et al.*, 1971)**

Since the discovery of its pulsing properties, we have continued analyzing observations of Cen X-3 from *Uhuru* taken between 1971 January and December. The study of these additional data has led to the discovery that Cen X-3 undergoes regular changes in intensity between two distinct levels with a period of 2 days. Furthermore, the period of the short pulsations ($4^{s}_{.}8$) undergoes continuous variations with the same 2-day cycle, indicating the existence of a physical connection between these two effects.

We have adopted the working hypothesis that we are dealing with an eclipsing binary system consisting of a compact object and a large massive companion. We find that all our present observational data are consistent with this interpretation. We can then derive detailed physical parameters describing the mass of the system, the orbital radius and velocity, and also some information about the atmosphere of the large companion.

2. Intensity Variations

In examining the entire set of observations from 1971 January to 1971 December we were struck by the existence of two distinct levels of intensity at which the source would appear. The transitions between high and low levels, when they could actually be observed, were always found to occur within about 1 hour.

---

* Reprinted from *Astrophysical Journal* **172**. Original article submitted 21 January, 1972.
** The 2U notation for sources refers to *The Uhuru Catalog of X-Ray Sources* (Giacconi *et al.*, 1972).

H. Gursky and R. Ruffini (eds.), Neutron Stars, Black Holes and Binary X-Ray Sources, 417–426. All Rights Reserved.

In 1971 May we obtained a continuous and detailed set of observations of Cen X-3 spanning an entire week. The data are shown in Figure 1. The intensities shown represent the observed counting rates above background. Since the source pulsates, we have defined the intensity as the average counting rate over each transit containing several $4^s\!8$ periods. The observed high-level intensity corresponds to an absolute counting rate of 0.25–0.35 counts cm^{-2} s^{-1} in the 2–6 keV energy range. Three low-intensity states

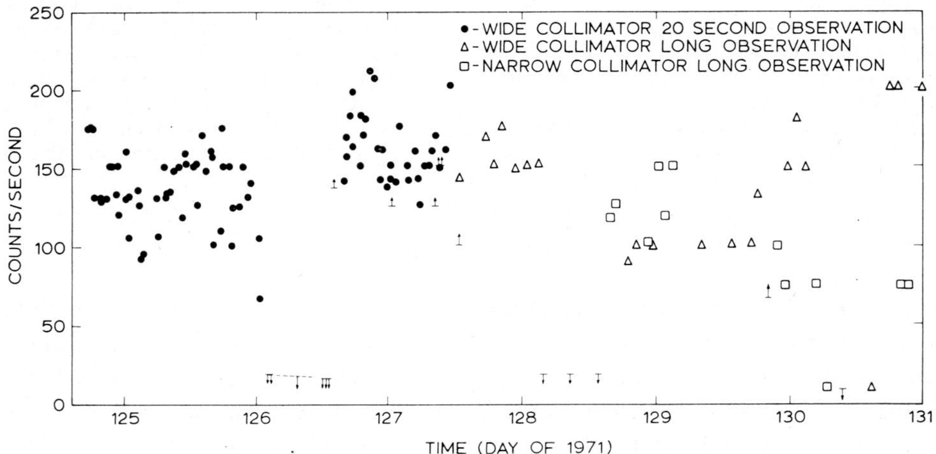

Fig. 1. The observed intensity from Cen X-3 averaged over several $4^s\!8$ cycles is shown as a function of time. The different symbols refer to sightings of different duration as noted. The data have not been corrected for elevation in the field of view of the detector.

are observable starting at days 126.02, 128.12 and 130.2, respectively. The low-level intensity is approximately 0.02–0.035 counts cm^{-2} s^{-1} (2–6 keV). The regular nature of the intensity variations clearly suggests the presence of a periodic phenomenon, with the low-intensity state comprising about one-quarter of the period.

When we examine the bulk of the data obtained between 1971 January and 1971 December, we find this behavior to be always present, as shown in Figure 2. We can fit all the observations with a periodic function as shown. The unique period obtained for this function is $2^d\!08710 \pm 0^d\!00015$. The low, high, and transition states are of $0^d\!488 \pm 0^d\!012$, $1^d\!529 \pm 0^d\!012$, and $0^d\!035 \pm 0^d\!007$ duration, respectively. The probability that the observed transitions are due to randomly occurring events is negligible.

We wish to note, however, that the variations of the average intensity from Cen X-3 are more complex than outlined above. First, the average intensity during any given 'high' state can vary by as much as 30 percent, in times shorter than 1 minute. Also, there appear to be times at which the source remains in its low emission state for days. In particular, Cen X-3 was scanned for a day or more on 1970 December 30–1971 January 1, February 10–13, February 22–23, March 12–13, and 1971 March 19–22. During all the above observations, the intensity from Cen X-3 appeared to be less than

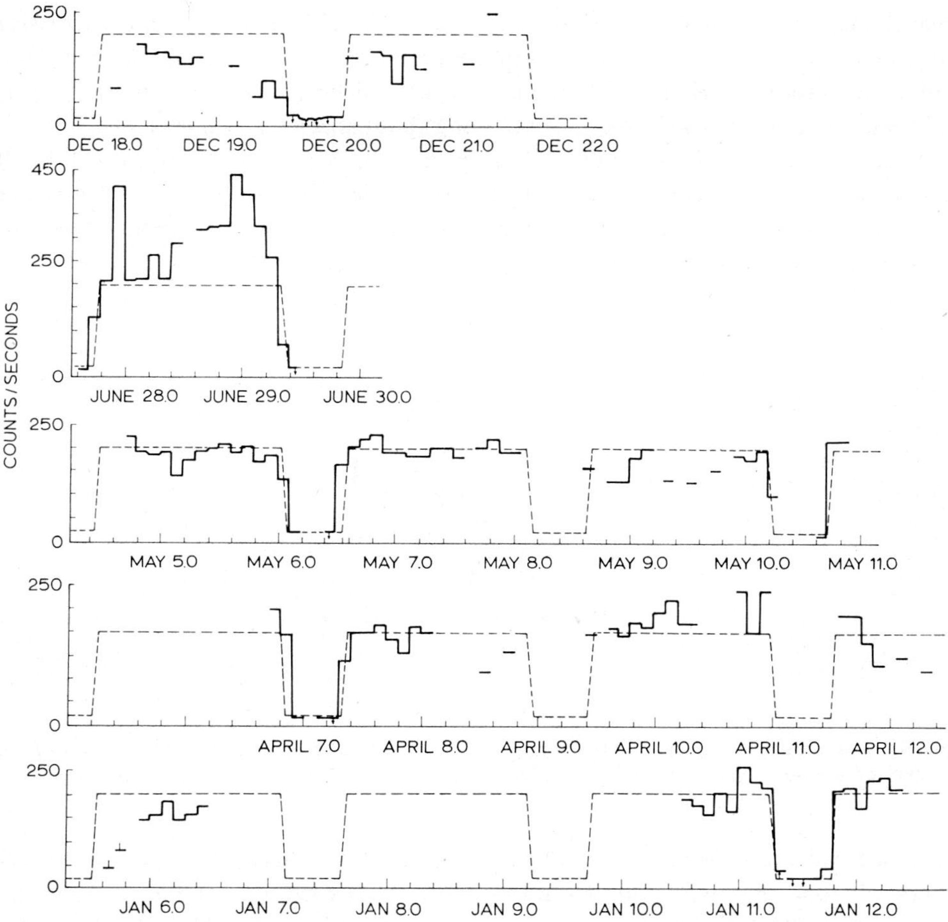

Fig. 2.   The observed intensity from Cen X-3 averaged over $0^{d}1$ intervals and corrected for elevation in the field of view is plotted as a function of time. Statistical errors are not shown and are typically of the order of 1 percent. Also plotted is the average intensity predicted by the light curve.

0.03 counts $cm^{-2} s^{-1}$ in the 2–6 keV range. We found, however, no instance in all of our data in which the source was observed at its high emission level except in accordance with the predictions of the periodic fit function discussed above.

3. Pulsation-Period Analysis

An absolute determination of the period of the $4^{s}8$ pulsation was obtained during long scans of the source, by fitting the data with a sinusoidal function via a minimum-χ^{2} technique. A typical fit to the data is shown in Figure 3. We have available 27 long-duration ($\geqslant 100$ s) sightings of Cen X-3 during 1971 April, May, and June, to each of which this technique was applied. The resulting values for the pulsation periods are shown in Table I. The scatter of these values is not consistent with statistics, indicating real changes in the values of the period of about $0\overset{s}{.}02$. The changes occur in successive days and do not indicate a long-term trend. With the exception of an anomalously high value found for the period in 1971 January 11–12, which we now believe to be in error, all values range from $4\overset{s}{.}832$ to $4\overset{s}{.}853$, oscillating about an average value of $4\overset{s}{.}842$.

It can be seen from Figure 3 that a large part of the X-ray emission is pulsed. In all 27 long passes, with the source at its high-intensity level, 80–90 percent of the power was pulsed. In the low-intensity level, we can place an upper limit of about 20 percent on the amount of power pulsed.

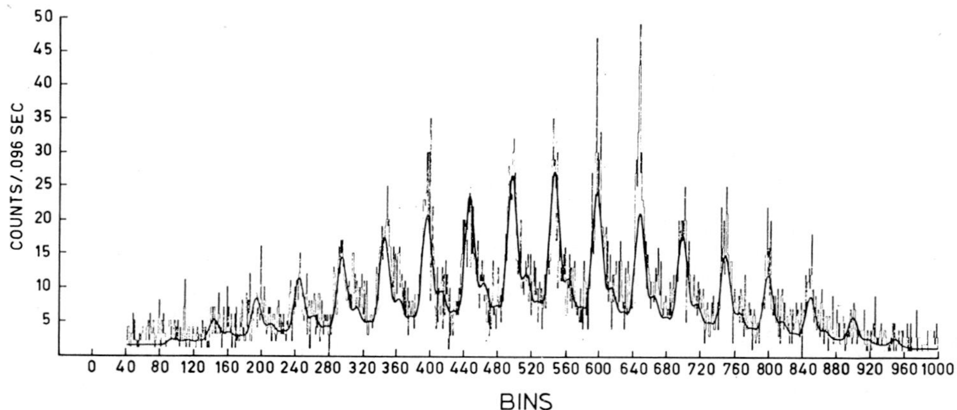

Fig. 3.   The counts accumulated in $0\overset{s}{.}096$ bins from Cen X-3 during a 100 s pass on 1971 May 7 are plotted as a function of bin number. The functional fit obtained by minimizing χ^2 is also shown.

In order to study the period changes in more detail, the phase of the pulsations was followed from sighting to sighting during those times when the satellite scanned the source every 720 s. The method was decribed in our earlier paper. This phasing analysis has been applied to three sets of data in January, May, and June. In all cases where 20000 s or more of data were analyzed, changes of period were noted. In the longer observations, it became clear that the transitions were more continuous than had previously been thought, and that the discrete-period values determined in the earlier paper were the result of straight-line approximations to short sections of a continuous curve. In fact, all the data currently available are consistent with a 2-day sinusoidal variation in the pulsational period. In Figure 4a we show a phasing analysis of two days of data obtained in May. Also shown is the best fit function of the form

$$\Delta t \equiv t_n - j_n \tau^* = at + b\sin(2\pi/T)(t - t_0), \tag{1}$$

where Δt is the difference in seconds between the time t_n at which a peak of a pulse is observed and the predicted time of occurrence $j_n\tau^*$ assuming an integral number of trial periods τ^*, t and t_0 are expressed in days, T is the occultation period in days, b is the amplitude of the oscillation in seconds, and a is a quantity which measures the fractional deviation of the average period of the pulsations from the trial period.

TABLE I

Pulsation period of Centaurus X-3

Date (1971) (1)	Long sightings τ (sec) (2)	Phasing analysis $\tau = \tau_0 + A \cos(2\pi/T)(t < t_0)$ τ_0 (sec) (3)	A (sec) (4)	t_0 (day of '71) (5)	T (days) (6)
Jan 5–6*	$4.8368 \pm 0.0004 \rightarrow 4.8449 \pm 0.0005$				
Jan 11–12†	$4.8703 \pm 0.0002 \rightarrow 4.8726 \pm 0.0005$				
Jan 5–12		4.843255 ± 0.000007 (4.843496	0.00668 ± 0.00015	11.0022 ± 0.0077	2.08712 ± 0.00011
April 8.16	4.832 ± 0.010				
8.29	4.845 ± 0.006				
8.59	4.834 ± 0.011				
8.74	4.838 ± 0.006				
8.79	4.848 ± 0.007				
9.64	4.839 ± 0.004				
9.99	4.837 ± 0.003				
10.22	4.839 ± 0.004				
10.27	4.844 ± 0.003				
10.34	4.847 ± 0.004				
10.51	4.853 ± 0.004				
11.86	4.838 ± 0.003				
12.11	4.838 ± 0.003				
May 4–7		4.842398 ± 0.000001 (4.842185)	0.006717 ± 0.000005	125.7923 ± 0.0002	2.08707 ± 0.00025
May 7.53	4.848 ± 0.005				
7.73	4.850 ± 0.004				
7.80	4.838 ± 0.005				
7.85	4.843 ± 0.004				
8.06	4.848 ± 0.004				
8.80	4.836 ± 0.004				
8.86	4.834 ± 0.003				
8.98	4.839 ± 0.003				
9.35	4.843 ± 0.003				
9.57	4.842 ± 0.004				
9.71	4.851 ± 0.005				
9.77	4.852 ± 0.003				
June 28–29		4.842435 ± 0.000012 (4.842176)	0.006699 ± 0.000029	180.0605 ± 0.0009	2.0885 ± 0.0035
June 30, 72	4.822 ± 0.002				
30, 86	4.844 ± 0.004				

The values of the period determined from each of the 100 s sightings at the times shown in column (1) are given. Note that the periods corresponding to the January 5–6 (marked with an asterisk) and 11–12 (marked with a dagger) observations have been obtained by a different technique. Since long passes were not available, a χ^2 fit to the entire set of 20 s observations allowed us a criterion to choose between discrete possible periods. The high values of the period shown for January 11–12 are due to a choice from the χ^2 fits with 95 percent confidence. The alternative choice results in a period consistent with all others shown. In column (3) the periods obtained from the phasing analysis are shown. The numbers in parentheses are corrected for the Earth's orbital motion. In the remaining columns we give the best-fit values of the parameters.

If we assume that the observed variations in arrival time of the pulses are due to orbital motions of the source, then the amplitude of the sine curve directly measures the projected radius of the orbit which is 39.75 ± 0.04 light-seconds.

From the formula above, we can by differentiation derive an expression for the actual pulsation period dependence on time of the form

$$\tau = \tau_0 + A \cos(2\pi/T)(t - t_0), \tag{2}$$

where $\tau_0 = (1 + a)\tau^*$ is the average period, $A = \tau^* b(2\pi/T)$ is the amplitude of the period variations, and the other quantities have been previously defined. A plot of this function for the same data is shown in Figure 4b.

In Figure 4c we show the observed intensity values as well as a segment of the average light curve derived previously. The zeros of the period function are seen to coincide with the centers of the low and high states. We see the period of the $4\overset{s}{.}8$ pulsations

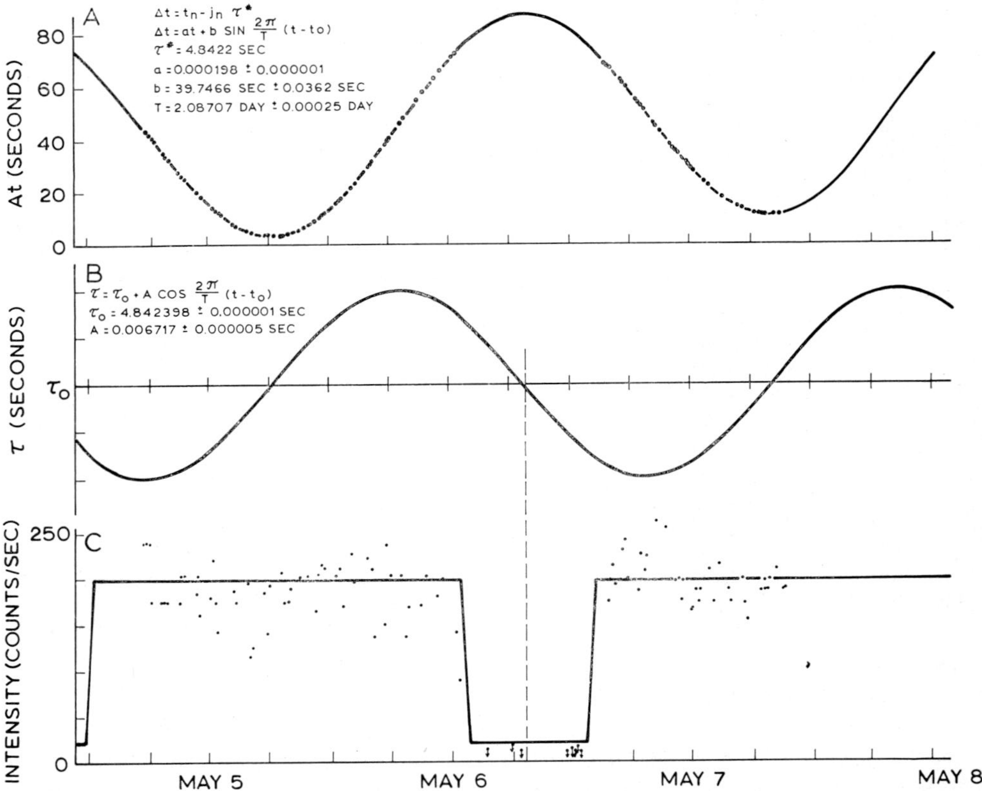

Fig. 4. (A) The difference Δt between the time of occurrence of a pulse and the time predicted for a constant period is plotted as a function of time. A best-fit function and the values of the parameters are given. (B) The dependence of the pulsation period τ on time as derived from the best-fit phase function above is shown and the values of the parameters given. (C) The intensity observed and the light curve predictions are shown for the same set of data. Note the coincidence of the null points of the period function with the centers of the high- and low-intensity states.

increasing as the object enters the low state and decreasing as it emerges. Thus, both the sign and the phase of the period variation are what would be predicted by Doppler effect. The values of the best-fit parameters derived from applying the phasing analysis to data in January, May, and June are shown in Table 1. It should be noted that a period of $2^{d}08712 \pm 0.00004$, independently derived from the phasing analysis by comparing phases of the Doppler curves, is consistent with the $2^{d}08710 \pm 0^{d}00015$ derived from the intensity variations.

There appears to be a decrease of 1 part in 4000 in the average period of pulsation from 1971 January to 1971 May which cannot be explained in terms of the Earth's motion about the Sun. This period change must be occurring at the source, although we cannot yet determine whether any secular trend exists.

4. Discussion

We may summarize the preceding observational data on Cen X-3 as follows: (1) There exist two distinct average intensity levels differing by a factor of 10 and repeating with a period of $2^{d}08710 \pm 0^{d}00015$; (2) there are times when the source remains at its low level for extended periods, but there are never high-level states when a low is predicted; and (3) the $4^{s}842$ pulsation period shows a sinusoidal variation occurring every $2^{d}08712 \pm 0^{d}00004$ and with null points of variation coinciding with the centers of the high and low states to within $0^{d}003 \pm 0^{d}006$.

We conclude that a single physical mechanism is reponsible for both changes in the period of the pulsations and the changes in intensity. The time of occurrence of the null point of period variation coinciding with the center of the low state is

$$\mathrm{JD}\ (2441132.081 \pm 0.001) \pm n(2.08712 \pm 0.00004).$$

On the basis of the above, we adopt the working hypothesis that we are observing an occulting binary system. The short-period ($4^{s}84$) X-ray pulsations originate from a pulsating or rotating compact object. This X-ray source orbits a large occulting object. The cyclic variations in the average intensity are due to occultation, and the cyclic changes in the pulsation period are due to Doppler effect.

The agreement between our trial sinusoidal fit and the measured phase shown in Figure 4 implies that the orbit is circular to within our statistical accuracy. We estimate an upper limit on the eccentricity of 0.05. We can then proceed to compute the various parameters of the system from the measured quantities. If i is the inclination of the orbit, r its radius in the center-of-mass system, R the radius of the central occulting object, M and m the masses of the objects, v the orbital velocity, c the speed of light, and G the gravitational constant, then

$$v \sin i \equiv \frac{Ac}{\tau_0} = 415.1 \pm 0.4 \ \mathrm{km\ s}^{-1},$$

$$r \sin i \equiv \frac{T}{2\pi} v \sin i = (1.191 \pm 0.001) \times 10^{12} \ \mathrm{cm},$$

$$\frac{M^3 \sin^3 i}{(M + m)^2} \equiv \frac{(2\pi)^2}{GT^2} (r \sin i)^3 = (3.074 \pm 0.008) \times 10^{34} \text{ g}.$$

Considering, in addition, that the occulted fraction of the orbit is one-quarter of the total orbital period, and assuming that the radius of the X-ray emitting object is negligible, we can write that

$$\sin i = \frac{r}{R} (1 + m/M) \sin 45° \, ;$$

thus the inclination must satisfy $45° < i \leq 90°$.

Although we cannot measure the inclination independently, we can use this range to place limits on the several parameters; thus,

$$415 \text{ km s}^{-1} < v < 588 \text{ km s}^{-1} \, ,$$

$$1.19 \times 10^{12} \text{ cm} < r < 1.69 \times 10^{12} \text{ cm} \, ,$$

$$3.07 \times 10^{34} \text{ g} < \frac{M^3}{(M + m)^2} < 8.81 \times 10^{34} \text{ g} \, ,$$

$$0.84 \times 10^{12} \text{ cm} < R < r(1 + m/M) \, .$$

From the existence of periodic pulsations of $4\overset{s}{.}8$ we had previously concluded that Cen X-3 is a collapsed object of radius less than 10^9 cm. Thus, the assumption that its size is negligible with respect to the radius of the orbit and that of the massive central star seems reasonable. If we assume that the mass of the compact source is $1 \, M_\odot$, the mass of the companion ranges from 17 to 46 $M_\odot$.

The radius of the occulting object cannot immediately be equated to the radius of a star. The occultation might be due to a tenuous atmosphere occupying a region much larger than the stellar dimensions. If we assume, however, that the radius of the star is of comparable dimensions to the occulting disk, then the range of mean density for this star is between 0.0015 and 0.01 g cm^{-3}. Thus, the values we find for the radius, mass, and mean density of the central object are reasonable for massive main-sequence stars and for B0-type supergiants. However, considerations of tidal instability would seem to argue against giant and supergiant stars.

We do have evidence which seems to indicate the existence of an atmosphere. This stems from the observation of the gradual nature of the occultation, which occurs over about 1 hour. The effect cannot be due to gradual occultation of the X-ray source which we believe to be compact. Thus, we explain the effect on the basis of progressively increasing absorption by an extended atmosphere. We would then expect that the spectrum of Cen X-3 should exhibit progressively more absorption at low energies as the occultation proceeds, and we have preliminary indications that this is indeed the case. The existence of the effect requires that the atmosphere exhibit a strong dependence of density on radius. In the assumption of an exponential atmosphere, we estimate a scale height of 5×10^{10} cm at the outer radius of the occulting region. The density in this region would be on the order of 10^{12} hydrogen atoms cm^{-3}. There also appears to be some indication that the duration of the occulted state varies. This might be due to expansion and contraction of the atmosphere.

It is clear from the persistence of X-ray emission even during occultation that an additional source of X-rays must be postulated to exist in the system. Since the level of intensity observed between peaks of the 4^{s}8 pulsations is also comparable, it is attractive to consider that the X-ray emission from the compact source may be close to 100 percent pulsed, and that the emission from a second quasi-constant source may explain the interpulse level, the emission observed during occultation, and the emission observed during those times in which the pulsating component seems to have disappeared. Further analysis of the energy spectrum of the radiation observed in these different states may establish if, indeed, the assumption of a unique source is valid, and help to differentiate between various physical processes which may give rise to this emission. A plausible process would appear to be the slow radiative cooling of the gas surrounding the system, which is heated by the pulsing source.

We believe that our data prove the binary nature of Cen X-3. We do not yet know in detail the origin of the 4^{s}8 pulsations. The absence of rapid changes in period allows as a clock mechanism either rotation or pulsation of a compact star. The energy source for the X-rays is also not known. Nuclear burning and accretion have been considered by various authors (Blumenthal *et al.*, 1972; Prendergast and Burbidge, 1968; Shklovsky, 1967; Zel'dovich and Shakura, 1969; Cameron and Mock, 1967). The presence of a close binary companion makes it very probable that accretion is taking place.

The position of Cen X-3 has been refined to R.A.: 11 19^{m}0$\pm$0^{m}4, δ: -60 19$'$2$\pm$3$'$ sufficiently precisely to attempt an intensive search for an optical candidate. We would expect this optical candidate to be quite luminous as is implied by the large mass of the system. It should be noted, however, that there is no reason in principle to expect that the visible light emission from the system, which is probably dominated by the central massive object, should exhibit either 4^{s}8 pulsations or a 2-day period photometrically. The object should appear as a spectroscopic binary with a velocity of a few tens of kilometers per second.

The recent discovery that the optical candidate for Cyg X-1 is a spectroscopic binary of 5^{d}6 period (Bolton, 1971; Webster and Murdin, 1971) in conjunction with our data on Cen X-3 greatly strengthens the view that a class of galactic X-ray sources may be associated with binary systems. The observation from *Uhuru* (Schreier *et al.*, 1971b; Tananbaum *et al.*, 1972) of a 1^{s}25 period pulsating X-ray source in Hercules (2U 1702+ +35), which also appears to show two distinct levels of intensity, may provide a third candidate for this class of objects.

We would like to acknowledge the assistance of Choon Shih in processing the data and of the Multi-Satellite Control Center and Attitude Control groups of the Goddard Space Flight Center for special maneuvering of the satellite. We would also like to thank Terry Matilsky, Wallace Tucker, and George Blumenthal for useful discussions.

This research was sponsored under NASA contract NAS5-11422.

References

Blumenthal, G., Cavaliere A., Rose, W., and Tucker, W.: 1972, *Astroph. J.* **173**, 213.

Bolton, C.: 1971, *Bull. AAS* **3**, No. 4 (Winter Meeting, San Juan).

Cameron, A. and Mock, M.: 1967, *Nature* **215**, 464.

Giacconi, R., Gursky, H., Kellogg, E., Schreier, E., and Tananbaum, H.: 1971, *Astrophys. J.* (*Letters*) **167**, L67.

Margon, B., Lampton, M., Bowyer, S., and Cruddace, R.: 1971, *Astrophys J.* (*Letters*) **169**, L23.

Prendergast, K. and Burbidge, G.: 1968, *Astrophys. J.* (*Letters*) **151**, L83.

Schreier, E., Gursky, H., Kellogg, E., Tananbaum, H., and Giacconi R.: 1971a, *Astrophys. J.* (*Letters*) **170**, L21.

Schreier, E., Tananbaum, H., Kellog, E., Gursky, H., and Giacconi, R.: 1971b, Winter Meeting, High Energy Astrophysics Division AAS, San Juan.

Shklovsky, I. 1967, *Astrophys. J.* (*Letters*) **150**, L48.

Tananbaum, H., Gursky, H., Kellogg, E., Levinson, R., Schreier, E., and Giacconi, R.: 1972, *Astrophys. J.* **174**, L143.

Webster, B. L. and Murdin, L.: 1972, *Nature* **235**, 37.

Zel'dovich, Ya. B. and Shakura, N.: 1969, *Soviet Astron.–AJ* **13**, 175.

MAXIMUM MASS OF A NEUTRON STAR*†

CLIFFORD E. RHOADES, JR.** and REMO RUFFINI

Abstract. On the basis of Einstein's theory of relativity, the principle of causality, and Le Chatelier's principle, it is here established that the maximum mass of the equilibrium configuration of a neutron star cannot be larger than 3.2 $M_\odot$. The extremal principle given here applies as well when the equation of state of matter is unknown in a limited range of densities. The absolute maximum mass of a neutron star provides a decisive method of observationally distinguishing neutron stars from black holes.

The estimate of the range of the critical mass for a neutron star varies from 0.32 to 1.5 $M_\odot$. The greatest uncertainty comes from the equation of state at nuclear densities and above. In fact the knowledge of physical properties of neutron-star material at densities smaller than 10^{13} g cm^{-3}, essential to describe the properties of the crust of neutron stars [1] and perhaps the change in period of the pulsars [2], is of no relevance for the determination of the maximum mass of a neutron star. The reason is that on increase of the central density the star becomes more and more compact and its crust becomes only a few tens of meters thick, or even less, depending on the models [3]. At nuclear densities and above, the equation of state is very poorly known because of the presence of strong interactions [4] between nucleons and threshold effects in the creation of resonances [5, 6] because of unavailability of phase space.

In recent times it has become clear that the most powerful tool in determining the difference between neutron stars and black holes relies on the possible difference in mass of the two objects [7]. No possibility exists of differentiating them on the basis of electrodynamic properties [8]. Moreover, the recent discovery of X-ray sources in binary systems gives the possibility of determining the mass of a collapsed object with great accuracy [9]. We therefore have the clear need of establishing on solid ground the maximum mass of a neutron star. Instead of trying to analyze the details of nuclear interactions we follow here a different approach. We take that most extreme equation of state that produces the maximum critical mass compatible solely with these three conditions: (1) standard general-relativity equation of hydrostatic equilibrium, (2) Le Chatelier's principle, and (3) the principle of causality. While no suggestion is made that the resultant equation of state accurately represents the actual physical behavior of matter, it does illustrate a point of principle by yielding a maximum mass for the critical mass. It is not altogether new to approach the equation of state from the side of hydrostatic theory rather than from the side of the structure of matter. Gerlach [10] has shown that from a set of measurements on a sequence of stars at the end point of thermonuclear evolution one can, in principle, work back to deduce the equation of state, without any call on nuclear or elementary particle theory. We present here, first,

* Reprinted from *Physical Review Letters* 32. Original article submitted 30 October, 1972.
† Work partially supported by National Science Foundation Grant No. GP-30799X to Princeton University.
** Present address: U.S. Air Force Base, Kirtland, N. Mex. 87117.

H. Gursky and R. Ruffini (eds.), Neutron Stars, Black Holes and Binary X-Ray Sources, 427–432. All Rights Reserved.

an extremization technique for the case in which we assume that the equation of state
is known everywhere apart from a finite range of pressure and densities

$$\varrho_0 \leqslant \varrho \leqslant \varrho_1 , \tag{1a}$$

$$p(\varrho_0) \leqslant p \leqslant p(\varrho_1) ; \tag{1b}$$

in order to avoid a 'supraluminous' equation of state (velocity of sound greater then
the speed of light, violation of causality; $c = 1$ in our units), we demand that

$$dp/d\varrho \leqslant 1 . \tag{2}$$

Also, we require that pressure be a monotonically nondecreasing function of the density (Le Chatelier's principle, which implies no spontaneous collapse of matter locally,
speed of sound real):

$$dp/d\varrho \geqslant 0 . \tag{3}$$

This last condition could, indeed, be relaxed; however, the proof is more straightforward assuming the validity of both conditions (2) and (3). We wish to choose the
equation of state between ϱ_0 and ϱ_1, so as to extremize the mass of the neutron star.
We integrate the standard general relativistic equations

$$dm(r)/dr = 4\pi\varrho r^2 = H(\varrho, r), \tag{4.1}$$

$$\frac{dp}{dr} = - \frac{(\varrho + p)}{r(r - 2m)} \left[4\pi r^3 p(r) + m(r)\right] = G(\varrho, p, m, r). \tag{4.2}$$

That is, the total mass of the star is given by

$$M = \int_0^R 4\pi\varrho(r) r^2 \, dr, \tag{5}$$

with R the radius of the star.

This problem may be formulated in terms of the standard calculus of variations
with inequality constraints. [11] We give here an alternative formulation in terms of
control theory, [12] which yields both necessary and sufficient conditions. We take ϱ
as the independent variable. Then Equation (5) can also be written

$$M = M_1(\varrho_c, \varrho_1) + \int_{\varrho_1}^{\varrho_0} 4\pi r^2 \varrho \, (dr/d\varrho) \, d\varrho + M_0(\varrho_0, r_0, m_0). \tag{6}$$

Here with M_1 we have indicated the mass contained in the range of densities $\varrho_1 \leqslant \varrho \leqslant \varrho_c$,
which is clearly a constant, and with M_0 the mass contained in the range of densities
$\varrho \leqslant \varrho_0$. Clearly M_0 is a function not only of ϱ_0 but also of r_0 and m_0, which in turn are
functions of the equation of state adopted in the range of densities $\varrho_0 \leqslant \varrho \leqslant \varrho_1$

Letting the primes denote derivatives with respect to ϱ, we have

$$u = p', \tag{7.1}$$

$$u/G = r', \tag{7.2}$$

$$Hu/G = m', \tag{7.3}$$

where $u(\varrho)$ is given by the known equation of state in the ranges $\varrho_0 \geqslant \varrho$ and $\varrho \leqslant \varrho_1$ and has to have

$$0 \leqslant u \leqslant 1 \text{ in the range } \varrho_0 \leqslant \varrho \leqslant \varrho_1. \tag{7.4}$$

Equations (7) replace Equations (1)–(5). Here u becomes the so-called control variable [12, 13]. Since we seek the maximum of M given by equation (6) with the constraints given by Equations (1)–(3), we can introduce the generalized Hamiltonian [13]

$$\mathfrak{H}(\varrho, p, m, r, y_1, y_2, y_3, u) = \{y_1 + y_2 H/G + y_3/G\} u +$$
$$+ [M_0(\varrho, r, m)]_{\varrho = \varrho_0} + \mu[(p - a_1)]_{\varrho = \varrho_0}, \tag{8}$$

where a_1 indicates the value of the pressure at $\varrho = \varrho_0$ and y_1, y_2, y_3 are a set of Lagrange multipliers which satisfy the Euler-Lagrange equations

$$y_1' = -\left(y_2 \frac{\partial}{\partial p} \frac{H}{G} + y_3 \frac{\partial}{\partial p} \frac{1}{G}\right) u, \tag{9.1}$$

$$y_2' = -\left(y_2 \frac{\partial}{\partial m} \frac{H}{G} + y_3 \frac{\partial}{\partial m} \frac{1}{G}\right) u, \tag{9.2}$$

$$y_3' = -\left(y_2 \frac{\partial}{\partial r} \frac{H}{G} + y_3 \frac{\partial}{\partial r} \frac{1}{G}\right) u. \tag{9.3}$$

Since u has to fulfill the inequality (7.4), we introduce the new variable

$$u = \sin^2 \omega. \tag{10}$$

The corresponding Euler-Lagrange equation then gives

$$2\{y_1 + y_2 H/G + y_3/G\} \sin\omega \cos\omega = 0, \tag{11}$$

which allows the solutions $\omega = 0$ and $\omega = \pi/2$ or $u = 0$ and $u = 1$. It is clear that in the extremization technique of Equation (6) we have to take into account not only the integral in the range of densities $\varrho_0 \leqslant \varrho \leqslant \varrho_1$ but also the mass M_0; this contribution is automatically taken into account by the transversality conditions [13]

$$(y_3 + \partial M_0/\partial r)_{\varrho = \varrho_0} = 0, \tag{12.1}$$

$$(y_2 + 1 + \partial M_0/\partial m)_{\varrho = \varrho_0} = 0, \tag{12.2}$$

$$(y_1 + \partial M_0/\partial p + \mu)_{\varrho = \varrho_0} = 0, \tag{12.3}$$

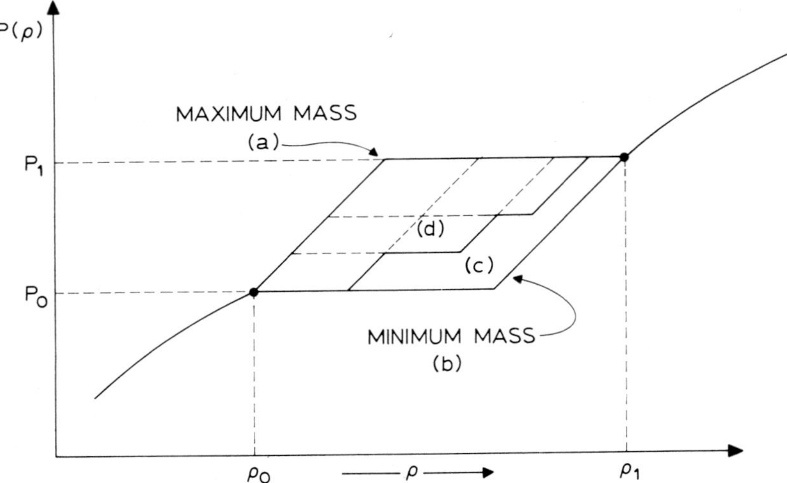

Fig. 1.   'Allowed rhomboid' in the $p\,\varrho$ plane. The equation of state is known for $\varrho < \varrho_0$ and $\varrho > \varrho_1$. In the range $\varrho_0 \leqslant \varrho \leqslant \varrho_1$ all equations of state compatible with the principle of causality and Le Chatelier's principle have to be contained inside the rhomboid. Path a (b) maximizes (minimizes) the mass of the neutron star; see Reference 14.

which are the boundary conditions to be fulfilled in the integration of the system of Equation (9). We can then conclude, in complete generality, that the desired extremum has to lie on the boundary of the allowed range of the control variable u, namely on path with $u=0$ and $u=1$ (see Figure 1). To see which one of the paths maximizes the mass and at which point the 'switch' of conditions from $u=0$ to $u=1$ has to be applied, it is necessary to proceed to a direct integration of Equations (7) and (9). We can then conclude that the path which maximizes the mass is given by path a in Figure 1 [14].

If we now turn from this general problem to the special case of establishing an absolute upper limit to the mass of a neutron star, our variational principle applies much more directly and the problem greatly simplifies. The extremization of the Hamiltonian (8) together with the constraints (7), the differential Equations (7), (9), and (11), and the transversality conditions simply tell us that the maximum mass is obtained for that equation of state which maximizes at $each$ density the velocity of sound of the material. We know from general arguments that at densities below $\varrho_0 = 4.6 \times 10^{14}$ g/cm^3 the equation of state of free degenerate neutrons, neglecting all nuclear interactions, maximizes the velocity of sound of neutron-star material [3], in the sense that any realistic equation of state has smaller values of the sound velocity [3]. At densities larger than $\varrho_0 = 4.6 \times 10^{14}$ g/cm^3 very little is known about the possible description of the interactions between nucleons at supranuclear densities. Therefore we assume the equation of state with the highest conceivable velocity of sound, namely the one with velocity of sound equal to the speed of light. By direct integration of the equations of equilibrium for selected values of the central density (see Figure 2) we can then conclude that no matter what the details of the equation of state at nuclear or supranuclear density, a neutron star can never have a mass larger than $3.2M_\odot$.

Finally, it is important to realize that although our arguments are presented here in the case of neutron stars with zero angular momentum they can indeed be applied to the case of pulsars and X-ray sources. The reason is that all the pulsars [16] and the pulsating binary X-ray sources [7] are rotating very slowly ($P \gtrsim 33$ ms), and in this region their masses can be affected by rotation only by a factor $\delta M/M < 0.2\omega^2/(M/R^3)$, where ω is the angular velocity, M the mass, and R the radius of the neutron star. Moreover, rapidly rotating neutron stars ($P \lesssim 0.1$ ms) would have a decay time $\tau = E_{\rm rot}/[-(dE/dt)_{\rm diss}]$ of approximately one week if dissipation is due to dipole magnetic radiation, or $\tau \lesssim 10$ min if dissipation is due to emission of gravitational radiation [17]. Even in this very extreme case ($P \lesssim 0.1$ ms) the critical mass of a neutron star would be changed by a factor smaller than 1.5 [18].

It is a pleasure to thank Professor D. Christodoulou, Professor A. Miele, and Professor J. A. Wheeler for discussions.

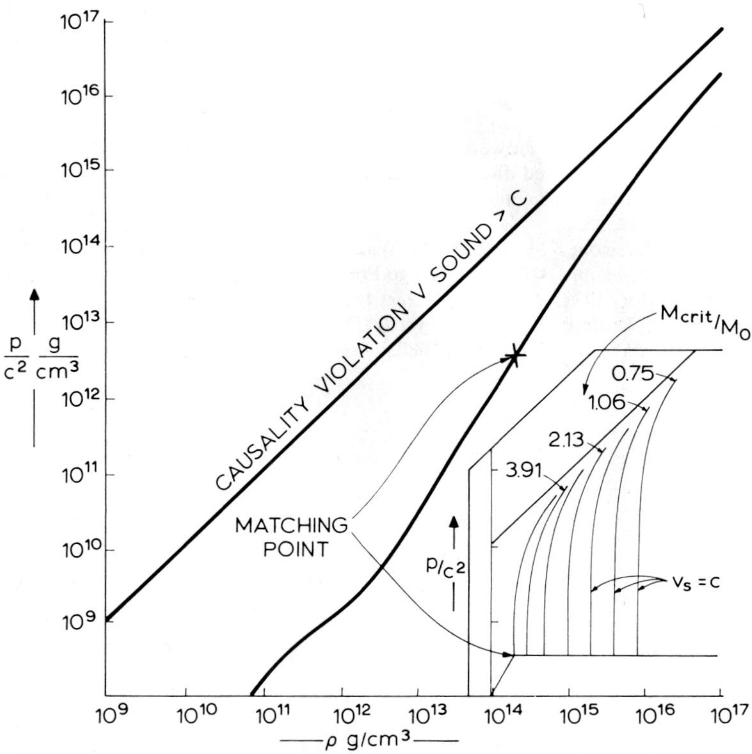

Fig. 2.   Matching of the 'allowed rhomboid' to the Harrison-Wheeler equation of state (Reference 15) for a value of the density $\varrho_0 = 2 \times 10^{14}$ g cm^{-3} (lower right-hand side of the figure). The different paths followed correspond to an equation of state with $u=1$ (path a in Figure 1) or to a combination of a path with $u=0$ and then $u=1$ (path b in Figure 1). The difference in inclination of the lines with $u=1$ between Figures 1 and 2 is due to the difference in scale: linear in Figure 1 and logarithmic here. In these computations the integrations are carried out up to the value of critical central density at which the value of the critical mass is reached.

References

[1] G. Baym, C. J. Pethick, and D. Pines: 1969, *Nature* (*London*) **224**, 674; C. E. Rhoades, Jr: 1971, Ph.D. thesis, Princeton University, (unpublished); M. Ruderman: 1969, *Nature* (*London*) **223**, 547, and **218**, 1128 (1968); R. Smoluchowski: 1970, *Phys. Rev. Letters* **24**, 923, 1191.

[2] P. E. Boynton *et al.*: 1969, *Astrophys. J.* **157**, L197.

[3] R. Ruffini: 1973, in *Black Holes* (ed. by B. DeWitt and C. deWitt), Gordon and Breach, New York.

[4] See, e.g., V. R. Pandharipande: 1971, *Nucl. Phys.* **A178**, 123; also Ruffini, Ref. 3.

[5] V. A. Ambartsumian and G. S. Saakian; 1960, *Astron. Zh.* **37**, 193 [*Sov. Astron.* **4**, 187 (1960)].

[6] R. Sawyer: 1972, *Phys. Rev. Letters* **29**, 382.

[7] R. Leach and R. Ruffini: 1973, *Astrophys. J.* **180**, L15.

[8] D. Christodoulou and R. Ruffini: 1973, in *Black Holes* (ed. by B. DeWitt and C. DeWitt), Gordon and Breach, New York.

[9] H. Gursky: 1973, in *Black Holes* (ed. B. DeWitt and C. deWitt), Gordon and Breach, New York.

[10] U. H. Gerlach: 1968, *Phys. Rev.* **172**, 1325.

[11] Rhoades, Ref. 1; F. A. Valentine: 1946, in *Contributions to the Calculus of Variation* (*1933–37*) (ed. by G. A. Bliss) Univ. of Chicago Press, Chicago, Ill.

[12] See, e.g., L. S. Pontryagin, V. G., Boltyanskii, R. V. Gamkrelidze, and E. F. Mischenko: 1962, *The Mathematical Theory of Optimal Processes* Interscience, New York; see also L. C. Young: 1969, *Calculus of Variations and Optimal Control Theory*, W. B. Saunders Co., Philadelphia, Penn.

[13] See, e.g., *Theory of Optimum Aerodynamic Shapes* (ed. by A. Miele), Academic, New York, 1965; and, more specific to our problem, A. Miele, R. E. Pritchard, and J. N. Damoulakis: 1970, *J. Opt. Theor. Appl.* **5**, 235.

[14] Situations can be conceived however in which this is not the case and the maximum occurs in path *b* of Figure 1. A detailed discussion of these pathological cases and a few explicit examples constructed *ad hoc* are presented and discussed by L. Pietronero and R. Ruffini: 1974, *Bull. Amer. Phys. Soc.* **19**, 95.

[15] See, e.g., B. K. Harrison, K. S. Thorne, M. Wakano, and J. A. Wheeler: 1965, *Gravitation Theory and Gravitational Collapse*, Univ. of Chicago Press, Chicago Press, Chicago, Ill.

[16] See, e.g., M. Taylor: 1972, *Astrophys. Letters* **10**, 167.

[17] A. Ferrari and R. Ruffini: 1969, *Astrophys. J.* **158**, L71.

[18] J. Wilson: 1973, *Phys. Rev. Letters* **30**, 1082.

DISCOVERY OF A PULSAR IN A BINARY SYSTEM*

R. A. HULSE and J. H. TAYLOR

Abstract. We have detected a pulsar with a pulsation period that varies systematically between $0\overset{s}{.}058967$ and $0\overset{s}{.}059045$ over a cycle of $0\overset{d}{.}3230$. Approximately 200 independent observations over 5-minute intervals have yielded a well-sampled velocity curve which implies a binary orbit with projected semimajor axis $a_1 \sin i = 1.0\ R_\odot$, eccentricity $e = 0.615$, and mass function $f(m) = 0.13\ M_\odot$. No eclipses are observed. We infer that the unseen companion is a compact object with mass comparable to that of the pulsar. In addition to the obvious potential for determining the masses of the pulsar and its companion, this discovery makes feasible a number of studies involving the physics of compact objects, the astrophysics of close binary systems, and special- and general-relativistic effects.

1. Introduction

We wish to report the detection of an unusual pulsar discovered during the course of a systematic survey for new pulsars being carried out (Hulse and Taylor, 1974) at the Arecibo Observatory in Puerto Rico. The object has a pulsation period of about 59 ms – shorter than that of any other known pulsar except the one in the Crab Nebula – and periodic changes in the observed pulsation rate indicate that the pulsar is a member of a binary system with an eccentric orbit of $0\overset{d}{.}3230$ period. Thus for the first time it is possible to observe the gravitational interactions of a pulsar and another massive object, and additional observations should make it possible to determine the masses of the two objects unambiguously.

2. Discovery of the Binary Pulsar

The equipment and searching method used in the pulsar survey have been described previously (Hulse and Taylor, 1974). Forty pulsars have now been detected in this work, of which 32 were not previously known; the parameters of the 21 most recently discovered will be given in another paper (Hulse and Taylor, 1975). The 59-ms pulsar, PSR 1913+16, was first detected in 1974 July. Attempts to measure its period to an accuracy of $\pm 1\ \mu s$ were frustrated by apparent changes in period of up to $\sim 80\ \mu s$ from day to day, and sometimes by as much as $8\ \mu s$ over 5 minutes. Such behavior is quite uncharacteristic of other pulsars: the largest known secular changes of period are of order $10\ \mu s$ per *year*, and irregular changes of period are many orders of magnitude smaller (Manchester and Taylor 1974). It soon became clear that Doppler shifts resulting from orbital motion of the pulsar could account for the observed period changes, and by the end of September an accurate velocity curve of this 'singleline spectroscopic binary' had been obtained (see Figure 1).

The parameters of the pulsar are given in Table I. In the table, celestial and galactic coordinates are followed by P_{cm}, the 'center of mass' pulsar period (corrected for the

* Reprinted from *Astrophysical Journal* **195**. Original article submitted 18 October, 1974.

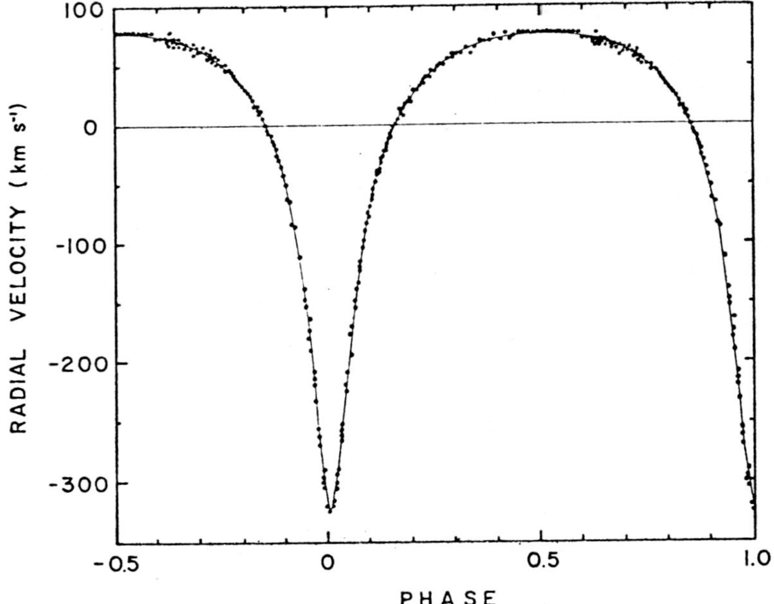

Fig. 1.   Velocity curve for the binary pulsar. Points represent measurements of the pulsar period distributed over parts of 10 different orbital periods. The curve corresponds to Equations (1) – (4), with parameters from Table II.

TABLE I

Parameters of the binary pulsar

$$\alpha(1950.0) = 19^\mathrm{h}13^\mathrm{m}13^\mathrm{s} \pm 4^\mathrm{s}$$
$$\delta(1950.0) = +16°00'24'' \pm 60''$$
$$l = 49°9$$
$$b = 2°1$$
$$P_\mathrm{cm} = 0^\mathrm{s}059030 \pm 0^\mathrm{s}000001$$
$$\mathrm{d}P_\mathrm{cm}/\mathrm{d}t < 1 \times 10^{-12}$$
$$\mathrm{DM} = 167 \pm 5 \ \mathrm{cm}^{-3} \ \mathrm{pc}$$
$$S_{430} = 0.006 \pm 0.003 \ \mathrm{Jy}$$
$$W_e < 10 \ \mathrm{ms}$$

orbital motion of the pulsar and for the motion of the observer in the solar system); an upper limit for $\mathrm{d}P_\mathrm{cm}/\mathrm{d}t$, the first derivative of the period; DM, the dispersion measure; S_{430}, the average flux density at 430 MHz; and an upper limit to W_e, the effective pulse width. (The pulses observed at 430 MHz are probably significantly broadened by multipath scattering in the interstellar medium.)

The elements of the binary orbit are given in Table II. K_1 is the semiamplitude of radial velocity variation of the pulsar with respect to the center of mass of the system; P_b is the period of the binary orbit, corrected for the motion of the observatory; e is the eccentricity of the orbit; ω is the longitude of periastron; T is the time of periastron

TABLE II

Elements of the orbit

$$K_1 = 199 \pm 5 \text{ km s}^{-1}$$
$$P_b = 27908 \pm 7 \text{ s}$$
$$e = 0.615 \pm 0.010$$
$$\omega = 179° \pm 1°$$
$$T = \text{JD } 2{,}442{,}321.433 \pm 0.002$$
$$a_1 \sin i = 1.00 \pm 0.02 \ R_\odot$$
$$f(m) = 0.13 \pm 0.01 \ M_\odot$$

passage; $a_1 \sin i$ is the projected semimajor axis of the pulsar orbit, i being the inclination between the orbit and the plane of the sky; and $f(m) = (M_2 \sin i)^3/(M_1 + M_2)^2$ is the mass function. These quantities were evaluated from the velocity measurements shown as filled circles in Figure 1. The velocity curve also shown in the figure was then computed from the elements using the equations (Aitken, 1964)

$$V_{r_1} = K_1 \left[\cos(\theta + \omega) + e \cos\omega \right], \tag{1}$$

$$\tan \tfrac{1}{2}\theta = \left[(1 + e)/(1 - e) \right]^{1/2} \tan \tfrac{1}{2}E, \tag{2}$$

$$M = E - e \sin E, \tag{3}$$

$$\phi = M/2\pi = (t - T)/P_b, \tag{4}$$

where V_{r_1} is the radial velocity of the pulsar (the 'visible' member of the binary pair); M, E, and θ are respectively the mean, eccentric, and true anomaly of the orbit of the pulsar about the center of mass; ϕ is the orbital phase; and t is the time.

The orbital elements given in Table II were obtained from direct measurements of the pulsar period over about 200 different 5-minute intervals distributed over 10-days. The 5-minute intervals are long enough that the period can be measured to an accuracy of about 1 μs, but short enough that the period does not change too drastically within the interval.

3. Physical Parameters of the Binary Pair

The mass of the pulsar is, of course, a quantity of great interest, as is the size and mass of the unseen companion. The observed mass function permits a wide range of values for M_1 and M_2. However, if we restrict attention to values of M_1 thought to be reasonable for neutron stars, the picture becomes clearer. Table III gives the required values for V_1, the maximum velocity of the pulsar, and M_2, the mass of the companion, for assumed inclinations $i = 90°$, $60°$, $30°$, $20°$, and $10°$, and pulsar masses $M_1 = 0.3$, 1.0, and 1.5 $M_\odot$. Evidently the mass ratio M_1/M_2 cannot be very different from unity unless the inclination i is rather small, which seems unlikely in view of the large observed radial velocity ($\sim 10^{-3}c$). Furthermore, the orbit is such that if the inclination were close to 90° and the size of the companion were large enough, eclipses of the pulsar would occur at orbital phase $\phi = 0.93$. No eclipses are observed, which requires

TABLE III

Possible parameters of binary pulsar system

i (degrees)	V_1 (max)	$M_1=0.3$		$M_1=1.0$		$M_1=1.5$	
		M_2	R_2	M_2	R_2	M_2	R_2
90	$0.0011c$	0.4	0	0.7	0	0.9	0
60	$0.0012c$	0.5	<0.6	0.9	<0.8	1.1	<0.8
30	$0.0021c$	1.5	<1.3	2.2	<1.6	2.6	<1.8
20	$0.0031c$	3.8	<1.9	4.8	<2.1	5.4	<2.3
10	$0.0061c$	26	<3.5	27	<3.7	28	<3.7

the radius of the companion to be less than

$$R_{2,\mathrm{max}} = (a_1 + a_2)(1 - e^2)\sin i/\tan i$$
$$= R_\odot(1 + M_1/M_2)(1 - e^2)/\tan i, \tag{5}$$

where a_2 is the semimajor axis of the orbit of the companion about the center of mass and M_1 and M_2 are the masses of the two objects. Comparison of these upper limits for R_2 with the corresponding values of M_2, together with the known dependence of radius on mass for main-sequence stars (Allen, 1973), virtually rules out the possibility that the companion is a main-sequence star. We conclude that the companion must be a compact object, probably a neutron star or a black hole. A white companion cannot be ruled out, but seems unlikely for evolutionary reasons.

4. Additional Observations

We cannot at present rule out the possibility that the unseen companion is also a radiofrequency pulsar. If pulsations from the companion can be found, the system will be in effect a 'double-line' spectroscopic binary and the mass ratio of the two bodies will be directly measurable. This is an exciting possibility, because then only the inclination would have to be determined in order to solve for the two masses.

Timing data much more accurate than that already available can in principle be obtained by recording the absolute time of arrival of the pulses. Observations of this sort done on other pulsars yield absolute arrival times accurate to $\sim 10^{-4}$ s. Measurements of comparable quality are now being acquired for PSR $1930+16$, and in due course the data will yield greatly improved accuracies for the celestial coordinates and for the orbital elements of the binary system. This in turn will allow a number of interesting gravitational and relativistic phenomena to be studied. The binary configuration provides a nearly ideal relativity laboratory including an accurate clock in a high-speed, eccentric orbit and a strong gravitational field. We note, for example, that the changes of both v^2/c^2 and GM/c^2r during the orbit are sufficient to cause changes in observed period of several parts in 10^6. Therefore, both the relativistic Doppler shift and the gravitational redshift will be easily measurable. Furthermore, the general-relativistic advance of periastron should amount to about $4°$ per year, which will be detectable in a short time. The measurements of these effects, not usually observable in spectroscopic binaries, would allow the orbit inclination and the individual masses to be obtained.

The star field in the direction of the pulsar is crowded, and the observed dispersion measure suggests that PSR 1913+16 is about 5 kpc distant. Probably there are some 5 to 10 mag of optical absorption along the line of sight, so we should expect the apparent visual magnitude of the pulsar (and its companion) to be some 18 to 23 mag fainter than the absolute magnitudes. Thus, the prospects for optical observations do not seem good unless a large fraction of the observed dispersion is the result of ionized material close to the pulsar. No changes in dispersion measure exceeding ± 20 cm^{-3} pc have been observed over the binary period, so it is clear that at most a small fraction of the dispersion can arise from electrons within the binary orbit.

We thank the staff of the Five College Radio Astronomy Observatory for assistance in construction of the pulsar search apparatus, and the staff of the Arecibo Observatory for assistance with the observations. This work is supported by the National Science Foundation under grants GP-37917 and GP-32414X. The Arecibo Observatory is part of the National Astronomy and Ionosphere Center and is operated by Cornell University under contract with the National Science Foundation. This paper is contribution number 196 of the Five College Observatories.

References

Aitken, R. G.: 1964, *The Binary Stars*, New York-Dover, pp. 79, 158.
Allen, C. W.: 1973, *Astrophysical Quantities* (3d ed.), Athlone Press, London, p. 209.
Hulse, R. A. and Taylor, J. H.: 1974, *Astrophys. J. (Letters)* **191**, L59.
Hulse, R. A.: 1975, in preparation.
Manchester, R. N. and Taylor, J. H.: 1974, *Astrophys. J. (Letters)* **191**, L63.

EDITORS' COMMENT ON
THE NEW BINARY RADIO PULSAR

R. A. Hulse and J. H. Taylor (*Astrophys. J.* (*Letters*) **195**, 51, 1975) have reported the discovery of a radio pulsar in a binary system. The pulse period is 0.059 s and the orbital period is 0.3230 days. The binary nature is revealed by the Doppler shift of the radio pulses. The orbit is highly elliptical and exhibits an eccentricity of 0.615. This remarkable discovery opens several new possibilities for future research. Classical predictions of general relativity such as the peristron precession and the spin orbit coupling of the central star are likely to be tested with unprecedented accuracy in this system. These and other effects are discussed in a set of recent papers (see e.g., T. Damour and R. Ruffini, *Compt. Rend.* **279**, 1971, 1974; A. R. Masters and D. H. Roberts, *Astrophys. J.* (*Letters*) **195**, 107, 1975; K. Brecher, *Astrophys. J.* (*Letters*) **195**, 113, 1975; L. V. Esposito and E. R. Harrison, *Astrophys. J.* (*Letters*) **196**, 1, 1975; C. M. Will, *Astrophys. J.* (*Letters*) **196**, 3, 1975; D. M. Eardley, *Astrophys. J.* (*Letters*) **196**, 59, 1975; R. V. Wagoner, *Astrophys. J.* (*Letters*) **196**, 63, 1975; G. C. Wheeler, *Astrophys. J.* (*Letters*) **196**, 67, 1975).

This system is also likely to reveal to be very important for the understanding of the formation of collapsed objects and of the evolution of binary systems. A possible evolutionary sequence has been presented by B. Flannery and E. van den Heuvel, *Astron. Astrophys.*, in press, 1975.

There is great interest in identifying the nature of the companion star of the pulsar which, as of today, can still be either a neutron star, a white dwarf, or a black hole.

ASTROPHYSICS AND SPACE SCIENCE LIBRARY

Edited by

J. E. Blamont, R. L. F. Boyd, L. Goldberg, C. de Jager, Z. Kopal, G. H. Ludwig, R. Lüst,
B. M. McCormac, H. E. Newell, L. I. Sedov, Z. Švestka, and W. de Graaff

1. C. de Jager (ed.), *The Solar Spectrum. Proceedings of the Symposium held at the University of Utrecht, 26–31 August, 1963.* 1965, XIV + 417 pp.
2. J. Ortner and H. Maseland (eds.), *Introduction to Solar Terrestrial Relations, Proceedings of the Summer School in Space Physics held in Alpbach, Austria, July 15–August 10, 1963 and Organized by the European Preparatory Commision for Space Research.* 1965, IX + 506 pp.
3. C. C. Chang and S. S. Huang (eds.), *Proceedings of the Plasma Space Science Symposium, held at the Catholic University of America, Washington, D.C., June 11–14, 1963.* 1965, IX + 377 pp.
4. Zdeněk Kopal, *An Introduction to the Study of the Moon.* 1966, XII + 464 pp.
5. B. M. McCormac (ed.), *Radiation Trapped in the Earth's Magnetic Field. Proceedings of the Advanced Study Institute, held at the Chr. Michelsen Institute, Bergen, Norway, August 16–September 3, 1965.* 1966, XII + 901 pp.
6. A. B. Underhill, *The Early Type Stars.* 1966, XII + 282 pp.
7. Jean Kovalevsky, *Introduction to Celestial Mechanics,* 1967, VIII + 427 pp.
8. Zdeněk Kopal and Constantine L. Goudas (eds.), *Measure of the Moon. Proceedings of the 2nd International Conference on Selenodesy and Lunar Topography, held in the University of Manchester, England, May 30–June 4, 1966.* 1967, XVIII + 479 pp.
9. J. G. Emming (ed.), *Electromagnetic Radiation in Space. Proceedings of the 3rd ESRO Summer School in Space Physics, held in Alpbach, Austria, from 19 July to 13 August, 1965.* 1968, VIII + 307 pp.
10. R. L. Carovillano, John F. McClay, and Henry R. Radoski (eds.), *Physics of the Magnetosphere, Based upon the Proceedings of the Conference held at Boston College, June 19–28, 1967.* 1968, X + 686 pp.
11. Syun-Ichi Akasofu, *Polar and Magnetospheric Substorms.* 1968, XVIII + 280 pp.
12. Peter M. Millman (ed.), *Meteorite Research. Proceedings of a Symposium on Meteorite Research, held in Vienna, Austria, 7–13 August, 1968.* 1969, XV + 941 pp.
13. Margherita Hack (ed.), *Mass Loss from Stars. Proceedings of the 2nd Trieste Colloquium on Astrophysics, 12–17 September, 1968.* 1969, XII + 345 pp.
14. N. D'Angelo (ed.), *Low-Frequency Waves and Irregularities in the Ionosphere. Proceedings of the 2nd ESRIN-ESLAB Symposium, held in Frascati, Italy, 23–27 September, 1968.* 1969, VII + 218 pp.
15. G. A. Partel (ed.), *Space Engineering. Proceedings of the 2nd International Conference on Space Engineering, held at the Fondazione Giorgio Cini, Isola di San Giorgio, Venice, Italy, May 7–10, 1969.* 1970, XI + 728 pp.
16. S. Fred Singer (ed.), *Manned Laboratories in Space. Second International Orbital Laboratory Symposium.* 1969, XIII + 133 pp.
17. B. M. McCormac (ed.), *Particles and Fields in the Magnetosphere. Symposium Organized by the Summer Advanced Study Institute, held at the University of California, Santa Barbara, Calif., August 4–15, 1969.* 1970, XI + 450 pp.
18. Jean-Claude Pecker, *Experimental Astronomy.* 1970, X + 105 pp.
19. V. Manno and D. E. Page (eds.), *Intercorrelated Satellite Observations related to Solar Events. Proceedings of the 3rd ESLAB/ESRIN Symposium held in Noordwijk, The Netherlands, September 16–19, 1969.* 1970, XVI + 627 pp.
20. L. Mansinha, D. E. Smylie, and A. E. Beck, *Earthquake Displacement Fields and the Rotation of the Earth. A NATO Advanced Study Institute Conference Organized by the Department of Geophysics, University of Western Ontario, London, Canada, June 22–28, 1969.* 1970, XI + 308 pp.
21. Jean-Claude Pecker, *Space Observatories.* 1970, XI + 120 pp.
22. L. N. Mavridis (ed.), *Structure and Evolution of the Galaxy, Proceedings of the NATO Advanced Study Institute, held in Athens, September 8–19, 1969.* 1971, VII + 312 pp.
23. A. Muller (ed.), *The Magellanic Clouds. A European Southern Observatory Presentation: Principal Prospects, Current Observational and Theoretical Approaches, and Prospects for Future Research.*

Based on the Symposium on the Magellanic Clouds, held in Santiago de Chile, March 1969, on the Occasion of the Dedication of the European Southern Observatory. 1971, XII+189 pp.

24. B. M. McCormac (ed.), *The Radiating Atmosphere. Proceedings of a Symposium Organized by the Summer Advanced Study Institute, held at Queen's University, Kingston, Ontario, August 3–14, 1970.* 1971, XI+455 pp.

25. G. Fiocco (ed.), *Mesopheric Models and Related Experiments. Proceedings of the 4th ESRIN-ESLAB Symposium held at Frascati, Italy, July 6–10, 1970.* 1971, VIII+298 pp.

26. I. Atanasijević, *Selected Exercises in Galactic Astronomy.* 1971, XII+144 pp.

27. C. J. Macris (ed.), *Physics of the Solar Corona. Proceedings of the NATO Advanced Study Institute on Physics of the Solar Corona, held at Cavouri-Vouliagmeni, Athens, Greece, 6–17 September 1970.* 1971, XII+345 pp.

28. F. Delobeau, *The Environment of the Earth.* 1971, IX+113 pp.

29. E. R. Dyer (general ed.), *Solar-Terrestrial Physics 1970. Proceedings of the International Symposium on Solar-Terrestrial Physics, held in Leningrad, U.S.S.R., 12–19 May 1970.* 1972, VIII+938 pp.

30. V. Manno and J. Ring (eds.), *Infrared Detection Techniques for Space Research, Proceedings of the Fifth ESLAB-ESRIN Symposium, held in Noordwijk, The Netherlands, June 8–11, 1971.* 1972, XII+344 pp.

31. M. Lecar (ed.), *Gravitational N-body Problem, Proceedings of IAU Colloquium No. 10, held in Cambridge, England, August 12–15, 1970.* 1972, XI+441 pp.

32. B. M. McCormac (ed.), *Earth's Magnetospheric Processes. Proceedings of a Symposium Organized by the Summer Advanced Study Institute and Ninth ESRO Summer School, held in Cortina, Italy, August 30–September 10, 1971.* 1972, VIII+417 pp.

33. Antonin Rükl, *Maps of Lunar Hemispheres.* 1972, V+24 pp.

34. V. Kourganoff, *Introduction to the Physics of Stellar Interiors.* 1973, XI+115 pp.

35. B. M. McCormac (ed.), *Physics and Chemistry of Upper Atmospheres. Proceedings of Symposium Organized by the Summer Advanced Study Institute, held at the University of Orléans, France, July 31–August 11, 1972.* 1973, VIII+389 pp.

36. J. D. Fernie (ed.), *Variable Stars in Globular Clusters and in Related Systems. Proceedings of the IAU Colloquium No. 21, held at the University of Toronto, Toronto, Canada, August 29–31, 1972.* 1973, IX+234 pp.

37. R. J. L. Grard (ed.), *Photon and Particle Interaction with Surfaces in Space. Proceedings of the 6th ESLAB Symposium, held at Noordwijk, the Netherlands, 26–29 September, 1972.* 1973, XV+577 pp.

38. Werner Israel (ed.), *Relativity, Astrophysics and Cosmology. Proceedings of the Summer School, held 14–26 August, 1972, at the BANFF Centre, BANFF, Alberta, Canada.* 1973, IX+323 pp.

39. B. D. Tapley and V. Szebehely (eds.), *Recent Advances in Dynamical Astronomy. Proceedings of the NATO Advanced Study Institute in Dynamical Astronomy, held in Cortina d'Ampezzo, Italy, August 9–12, 1972.* 1973, XIII+468 pp.

40. A. G. W. Cameron (ed.), *Cosmochemistry. Proceedings of the Symposium on Cosmochemistry, held at the Smithsonian Astrophysical Observatory, Cambridge, Mass., August 14–16, 1972.* 1973, X+173 pp.

41. M. Golay, *Introduction to Astronomical Photometry.* 1974, IX+364 pp.

42. D. E. Page (ed.), *Correlated Interplanetary and Magnetospheric Observations. Proceedings of the Seventh ESLAB Symposium, held at Saulgau, W. Germany, 22–25 May, 1973,* 1974, XIV+662 pp.

43. Riccardo Giacconi and Herbert Gursky (eds.), *X-Ray Astronomy.* 1974, X+450 pp.

44. B. M. McCormac (ed.), *Magnetospheric Physics. Proceedings of the Advanced Summer Institute, held in Sheffield, U.K., August 1973.* 1974, VII+399 pp.

45. C. B. Cosmovici (ed.), *Supernovae and Supernova Remnants. Proceedings of the International Conference on Supernovae, held in Lecce, Italy, May 7–11, 1973.* 1974, XVII+387 pp.

46. A. P. Mitra, *Ionospheric Effects of Solar Flares.* 1974, XI+294 pp.

49. Z. Švestka and P. Simon (eds.), *Catalog of Solar Particle Events 1955–1969.* 1975, IX+428 pp.

50. Z. Kopal and R. W. Carder, *Mapping of the Moon: Past and Present,* 1974, VIII+237 pp.

51. B. M. McCormac (ed.), *Atmosphere of Earth and the Planets. Proceeding of the Summer Advanced Study Institute, held at the University of Liège, Belgium, July 29–August 9, 1974.* 1975, VII+454 pp.

52. V. Formisano (ed.), *The Magnetospheres of the Earth and Jupiter. Proceedings of the Neil Brice Memorial Symposium, held in Frascati, May 28–June 1, 1974.* 1975, XI+485 pp.